ILLUSTRATED DICTIONARY *of*

ENVIRONMENTAL HEALTH & OCCUPATIONAL SAFETY

Herman Koren

NATIONAL
ENVIRONMENTAL
HEALTH ASSOCIATION

LEWIS PUBLISHERS

Boca Raton   New York   London   Tokyo

Library of Congress Cataloging-in-Publication Data

Koren, Herman.
   Illustrated dictionary of environmental health and occupational
safety / Herman Koren.
     p. cm.
   ISBN 0-87371-420-2 (alk. paper)
   1. Environmental health--Dictionaries.   2. Industrial safety-
-Dictionaries.   I. Title.
   RA566.K59   1995
   616.9'8'003--dc20
                                           95-17134
                                              CIP

© 1996 by CRC Press, Inc.
Lewis Publishers is an imprint of CRC Press

No claim to original U.S. Government works
International Standard Book Number 0-87371-420-2
Library of Congress Card Number 95-17134
Printed in the United States of America 1 2 3 4 5 6 7 8 9 0
Printed on acid-free paper

Foreword

People can do some strange things in the course of a professional career. The exercise of writing a foreword for Dr. Koren's *Illustrated Dictionary of Environmental Health and Occupational Safety* constitutes one of my top ten strangest things. I say that only because I doubt that there are that many people who can honestly say — as I now can — that I have read a dictionary! (I did hear once that if one was to read anything, read a dictionary. Why? All the other books are in there!)

When I approached this task, I thought that I would be functioning more in the mode of a reviewer. Does that definition make sense? Why wasn't such and such a term included, etc.? After (literally) reading the first several pages of this dictionary, however, my sense of what I was doing began to change. In fact, I enjoyed it and believe it or not, actually felt as if I were reading a book — complete with a story line! How in the world could I come to such a bizarre point of view? Actually, it was quite easy.

Once into Dr. Koren's dictionary and once into the minutia of all of the terms that he has included in this unusual reference book, it quickly became apparent to me just how incredibly broad this field actually is.

I've worked in the environmental field for my entire professional career, which now spans 23 years. Moreover, I've worked on or have become familiar with virtually every environmental issue imaginable. Yet, until I methodically proceeded through this dictionary, I confess that I really didn't appreciate the enormous breadth of this field of practice.

That Dr. Koren illuminated me as to the distant horizons of this field is something that I shall be thankful for. Moving from word to word, thus, read as a story, I was virtually spellbound over how encompassing and exciting this field of science actually is.

Now I wouldn't recommend that anyone other than a student or reviewer literally do what I did! I mention the impact that this book had on me only to emphasize to the user of the dictionary that they have in their hands a comprehensive and environmental health-focused publication that should just about cover any issue they might be interested in.

If I might turn from the big picture that caught my attention to the detail of the dictionary, there are several specific observations that I would like to share with you here as well. First, the cross-referencing is excellent. Often within a definition, other terms are used that also appear in the dictionary. I found it easy and instructive to a fuller comprehension of the term to move back and forth between all the terms that were appearing in any of the words that I was interested in.

I also liked the health-based orientation of the dictionary. Environmental health is not environmental conservation. Environmental health professionals are focused on the health implications of environmental issues. Dr. Koren does justice to this special orientation and the orientation that makes this field the unique one that it is.

Representing a profession as I do, I'm always interested in the application side of things. In other words, what is the useability index of this work? With this concern in mind, I asked questions as I read, such as "are the definitions expressed in such a way that the user can apply what is learned?" Consistently, I thought they were. In other words, this book WILL be useful to the practitioner.

I thought the book was particularly strong on chemical substances. That's good! So much of what this profession is doing these day requires an understanding in this area.

I also liked the fact that Dr. Koren did the research to make sure that organizations were included in his dictionary. How often have we all been frustrated by not knowing who the players are — or worse, knowing the initials of the players but not knowing what the initials stand for?! You'll find all the major players, initials and all, in this material.

Dr. Koren also includes many definitions pertaining to environmental law. That's great! It also serves to make this a reference book that can be consulted for just about any issue the professional is working on.

Though the draft that I saw did not have but a handful of sample illustrations, I have no doubt that illustrated definitions sprinkled liberally throughout the text will enhance even further the comprehension of the terms being defined. An illustrated dictionary is steps above one that is without this useful learning tool.

In the introduction, I recommend for the serious student that they read the third paragraph. That just about summarizes the comprehensive nature of this piece of work. My only other recommendation for "must reading" would be the definition of **environmental health professional**. Thanks, Hank, for so expansive a definition. That's who we are. And we need to never forget that.

About the only thing that I didn't care for in the dictionary was the absence of a pronunciation guide.

In short, I think it's great that professionals in this field now have such a comprehensive reference book and one that is so clearly health oriented. It is certainly going up on my reference shelf. I hope the cover is made of sturdy material because it is going to get a lot of use!

Nelson Fabian, ME
Executive Director
National Environmental
* Health Association*
Denver, Colorado

Introduction

Environmental health and occupational safety is not a single topic, but rather a colorful, complex, and diversified range of interrelated topics including basic science, law, government, disease, injury identification, prevention, and control. Physical, chemical, microbiological, psychological, and social stresses are always present and can create or enhance the potential for disease or injury. People are exposed to these stresses in home, community, occupational, and ambient environments. The sum total of exposures, past exposures, personal factors, including existing disease and/or injury, genetic factors, and general health may mean the difference between sickness and death or health and a fine quality of life called wellness.

The purpose of this reference is to fill the need created by the rapidly expanding area of the environmental sciences. It is meant to be a quick, user-friendly reference for professionals in environmental health, environmental control, medicine, nursing, law, planning, loss prevention, safety management, insurance, government, industry, students, and teachers. News agencies, civic associations, and the public at large will gain valuable information without having to go to several different references.

The over 7,500 terms selected come from thousands of references and represent as inclusive a list as possible. The definitions are in letter-by-letter alphabetical order. However, it is recognized that some terms may have been overlooked. The terms are drawn from varied specialized and technical fields, but are related in such a manner as to be accessible to both interested professionals and general readers. The terms provide general background information in anatomy, biology, chemistry, genetics, mathematics, microbiology, physics, and physiology to assist in the understanding of environmental issues and consequences. Epidemiology, toxicology, risk assessment, statistics, computer science, geographic information systems, mapping, and instrumentation terminology are needed to understand how much and why environmental problems cause disease and injury. Medical terminology explains the symptoms of a variety of exposures, and the discussion of various diseases explains the negative endpoint of exposure. Environmental subject areas include ecosystems, energy, environmental economics, food protection, housing, indoor air pollution, home injury control, hospitals, nursing homes, other institutions, recreational environment, occupational environment, noise, light, radiation, ergonomics, industrial accidents, occupational disease, air pollution, air toxics, solid waste, hazardous waste, drinking water, swimming areas, plumbing, on-site sewage, public sewage, soils, water pollution, environmental emergencies, and the laws related to these areas. Terminology related to equipment and environmental control are also used. Resources identified include governmental agencies, civic groups, professional organizations, and a variety of people working on environmental control, abstract services, and databases.

Detailed illustrations enhance these terms and provide needed visual assistance to make the terms more meaningful and understandable. Where applicable, definitions are supplemented with synonyms, acronyms, and abbreviations that are alphabetized and cross-referenced.

The diseases selected were those most commonly related to the environment and industry. The chemicals selected were those most frequently encountered by environmental health and safety management personnel or the public.

Further, all chemicals included are those considered to be hazardous by the Environmental Protection Agency, the National Institute of Occupational Safety and Health, or potentially toxic or carcinogenic by the Agency for Toxic Substances and Disease Registry.

Author

Dr. Herman Koren, R.E.H.S., M.P.H., H.S.D., is Professor and former Director of the Environmental Health Science Program, and Director of the Supervision and Management Program I & II at Indiana State University at Terre Haute. He has been an outstanding researcher, teacher, consultant, and practitioner in the environmental health field, and in the occupational health, hospital, medical care, and safety fields, as well as in management areas of the above and in nursing homes, water and wastewater treatment plants, and other environmental and safety industries for the past 41 years. In addition to numerous publications and presentations at national meetings, he is the author of six books, entitled *Environmental Health and Safety*, Pergamon Press, 1974; *Handbook of Environmental Health and Safety*, Volumes 1 and 2, Pergamon Press, 1980 (now published in updated and vastly expanded format by Lewis Publishers, CRC Press, as a 3rd edition); *Basic Supervision and Basic Management*, Parts I & II, Kendall Hunt Publishing, 1987; *Illustrated Dictionary of Environmental Health and Occupational Safety*, Lewis Publishers, CRC Press, 1995. He has served as a district environmental health practitioner and supervisor at the local and state level. He was an administrator at a 2000 bed hospital. Dr. Koren was on the editorial board of the *Journal of Environmental Health* and the former *Journal of Food Protection*. He is a founder diplomate of the Intersociety Academy for Certification of Sanitarians, a Fellow of the American Public Health Association, a 41-year member of the National Environmental Health Association, founder of the "National" Student National Environmental Health Association, and the founder and advisor of the Indiana State University Student National Environmental Health Association (Alpha chapter). Dr. Koren developed the modern internship concept in environmental health science. He has been a consultant to the U.S. Environmental Protection Agency, the National Institute of Environmental Health Science, and numerous health departments and hospitals, and has served as the keynote speaker and major lecturer for the Canadian Institute of Public Health Inspectors. He is the recipient of the Blue Key Honor Society Award for outstanding teaching and the Alumni and Student plaque and citations for outstanding teaching, research, and service. The National Environmental Health Association has twice honored Dr. Koren with presidential citations for "Distinguished Services, Leadership and Devotion to the Environmental Health Field" and "Excellent Research and Publications."

Acknowledgments

To Professor Alma Mary Anderson for her excellent art work and direction of the artwork product, which has thoroughly enhanced this *Illustrated Dictionary of Environmental Health and Occupational Safety*.

To Brian Flynn, Kayvan Koie, Lori Duda, and Brad Justus, Indiana State University graphic design students for their assistance in making the many illustrations possible.

To Kristin LaVigueur Clauss, a most notable young professional who proofread the manuscript and offered numerous excellent suggestions.

To the librarians and libraries of Indiana State University, Indiana University, Purdue University, University of Pennsylvania, and the many other libraries where I was permitted to conduct my research for this book.

To Brian Lewis for coming up with the idea for an illustrated dictionary and asking me to write it.

To Kathy Walters, former Associate Editor, Lewis Publishers, for her unique skills, her superb organizational work, and her dedication to a most difficult project. Without her help, innovative recommendations, guidance, and patience, this book would never have been completed. Kathy has asked that her efforts on this book be dedicated to the memory of her mother, Jennie Melnyk. "Thanks Mom for teaching me to be strong, compassionate, patient, and independent. You have instilled in me the need to maintain a passion for life, for our time here is priceless — and much too short."

Dedication

To my wife, Donna, a life-long friend, confidant, and sweetheart, whose love and support has motivated me to write six books, while continually sharing me with my students and colleagues.

A — *See* ampere.

Å — *See* angstrom unit.

AA — acronym for atomic absorption.

AADI — acronym for adjusted acceptable daily intake.

AAIH — *See* American Academy of Industrial Hygiene.

AAMI — acronym for Association for the Advancement of Medical Instrumentation.

AAOM — *See* American Academy of Occupational Medicine.

AAQS — *See* ambient air quality standard.

AAR/BOE — acronym for Association of American Railroads/Bureau of Explosives.

AASHTO — *See* American Association of State Highway and Transportation Officials.

abandoned vehicle — any unregistered motor vehicle deemed by applicable state laws to be an actual or potential safety hazard or which may be an actual or potential rodent harborage.

abate — (*law*) to put an end to; demolish; to do away with; to nullify or make void.

abatement — (*asbestos*) the removal or other treatment of asbestos-containing materials to prevent asbestos contamination of a building or environment; the reduction in degree of intensity of a substance, such as a pollutant.

abattoir — a slaughter-house for animals intended for human consumption such as cattle, pigs, and sheep.

aberration — deviation from the normal or usual.

ABIH — *See* American Board of Industrial Hygiene.

abiotic — referring to the non-living components of ecosystems.

abiotic factors — the physical and non-living chemical factors such as temperature, light, water, minerals, and climate which influence living organisms.

aboveground storage facility — a tank or other storage container, the bottom of which is on a plane not more than 6 inches below the surrounding surface.

ABPM — *See* American Board of Preventive Medicine.

abrasion — the wearing away of any surface material by the scouring action of moving solids, liquids, gases, or combination thereof; a

spot rubbed bare of skin or mucous membrane, or superficial epithelium by rubbing or scraping.

abrasion hazard — (*swimming pools*) a sharp or rough surface that would scrape the skin by chance impact or normal use of a facility.

abrasive — any of a wide variety of natural or manufactured substances of great hardness used to smooth, scour, rub away, polish, or scrub other materials; having qualities associated with abrasion.

abrasive blasting — the process of cleaning or finishing surfaces using sand, alumina, or steel in a blast of air.

ABS — *See* alkylbenzene sulfonate.

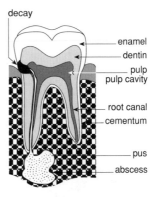

Abscess

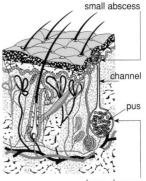

abscess — a localized collection of pus in a cavity formed by the disintegration of tissue. Abscesses are usually caused by specific microorganisms that invade the tissues through small wounds or breaks in the skin. An abscess is a natural defense mechanism by which the body attempts to localize an infection and isolate microorganisms to prevent spreading throughout the body.

absolute risk — an estimate of health risk by determining the rate of occurrence of the disease in the study population.

absolute temperature — the temperature in Celsius degrees relative to the absolute zero at –273.16°C or in Fahrenheit degrees relative to the absolute zero at –459.69°F.

absolute viscosity — a measure of a fluid's tendency to resist flow regardless of its density.

absolute zero — the temperature of –273.16°C or –459.69°F at which all gases theoretically contract to zero volume if Charles' law were obeyed exactly.

absorbed dose — (*biology*) the amount of a substance penetrating across an absorption barrier of an organism by physical or biological process; (*radiation*) the energy absorbed per unit mass in an irradiated medium. Also known as gray.

Countercurrent Packed Tower Absorber

absorbers — (*scrubbers*) a general classification of devices used in air pollution control to remove, treat, or modify one or more of the gaseous or vaporous constituents of a gas stream by bringing the stream into intimate contact with an absorbing liquid.

absorption — a route of entry into the body through the broken or unbroken skin; any chemical can enter through a cut or abrasion; organic lead compounds, nitro compounds, and organic phosphate insecticides can enter through hair follicles, whereas toluene and xylene can be absorbed in fats and oils in the skin; (*radiation*) the phenomenon by which radiation imparts

some or all of its energy to any material through which it passes; (*sound*) the ability of a material to absorb sound energy; (*physiology*) the process by which the end products of digestion, as well as other dissolved liquids and gases, enter the fluids and cells of an organism.

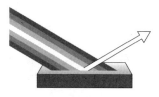

Absorption

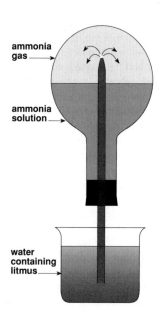

Absorption

absorption barrier — any of the body's barriers, such as the skin, lung tissues, or gastrointestinal tract, that allows differential diffusion of substances across a boundary.

absorption coefficient — (*radiation*) the fractional decrease in the intensity of a beam of x- or gamma radiation per unit thickness, per unit mass, or per atom of absorber due to deposition of energy in the absorber; (*acoustics*) the ratio of the sound energy absorbed by a surface of a medium or material to the sound energy incident on the surface; (*physics*) the constant for a material indicating the degree to which it absorbs radiation, atomic particles, rays of light, or sound; (*physiology*) the rate at which the human body absorbs a particular substance, usually expressed in milligrams per kilogram of body weight per hour.

absorption material — any material which absorbs small amounts of sound energy, changing it into small amounts of heat energy, such as with the use of acoustical ceiling tile.

Ac — *See* altocumulus clouds.

AC — *See* alternating current.

Acanthoscelides obtectus — *See* Bean Weevil.

acaricide — a pesticide used to destroy mites on domestic animals, crops, and humans. Also known as miticide.

accelerated erosion — the erosion of soil at a faster than natural rate.

acceleration — the rate of change of velocity with respect to time.

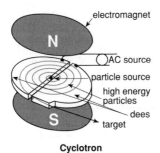

Cyclotron

Linear Accelerator

Accelerators

accelerator — (*chemistry*) a chemical additive which increases the speed of a chemical reaction; a device for providing a very high velocity to charged particles such as electrons or protons.

accelerometer — an instrument which measures the acceleration of a moving body; (*vibration*) an instrument which measures vibrations.

acceptable quality level (AQL) — the maximum percentage of regulated product that, for purposes of sampling inspection, can be considered as a process average.

acceptance test — (*geographic information system*) a test for evaluating a newly purchased computer system's performance and conformity to specifications.

acceptor molecule — a molecule that accepts a phosphate.

accessible — easily exposed for inspection and replacement of materials or parts; (*sewage*) a sewer located in a public right-of-way or easement contiguous to property; the service stub can be reached without excessive tunneling or boring under a roadway, building, or flowing stream, and the sewer or service stub is no farther than 200 feet from the residential building or 1,000 feet from the commercial/industrial building to be served.

access time — (*computer science*) a measure of the time interval between the instant that data are called from storage and the instant that delivery is complete.

accident — an unplanned and usually unexpected occurrence which has the potential to cause damage to persons and/or property.

accident site — (*hazardous materials*) the location of an unexpected occurrence, failure, or loss resulting in a release of hazardous materials either at a plant or along a transport route.

acclimation — the process of biochemical adaptations enabling a species or population to withstand environmental changes such as changes in temperature, humidity, and pressure. Also known as acclimatization.

acclimatization — *See* acclimation.

accommodation — (*optics*) the ability of the eye to adjust its focus for varying distances.

accordance agreement — the understood agreement by which a user of a pesticide agrees to follow labeled directions for proper application of the pesticide.

accreditation — the process by which an agency or organization evaluates a program of study or an institution as meeting certain predetermined standards.

accuracy — (*scientific technology*) the degree of agreement between a measured value and the true value of a calculation or reading; the extent to which a result is true when compared against a recognized standard.

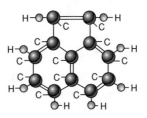

Acenaphtylene

acenaphtylene — a polycyclic aromatic hydrocarbon formed during the incomplete combustion of coal, oil, gas, garbage or other organic substances and found especially at hazardous waste sites. See polycyclic aromatic hydrocarbons.

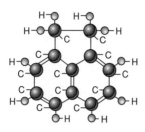

Acenapthene

acenapthene — a polycyclic aromatic hydrocarbon formed during the incomplete combustion of coal, oil, gas, garbage or other organic substances and found especially at hazardous waste sites. See polycyclic aromatic hydrocarbons.

acetaldehyde — CH_3CHO — a colorless liquid or gas (above 69°F) with a pungent, fruity odor; MW: 44.1, BP: 69°F, Sol: miscible, Fl.P: –36°F, sp gr: 0.79. It is used in the production of intermediates during the synthesis of acetic acid, acetic anhydride, and aldol compounds; during the manufacture of synthetic resins, in the synthesis of intermediates in the production of pesticides and pharmaceuticals, and in the synthesis of rubber processing chemicals; in coating operations in the manufacture of mirrors; as a hardening agent in photography and in the manufacture of gelatin, glue, and casein products; as a preservative in food products and leather. It is hazardous to the respiratory system, skin, and kidneys; and is toxic by inhalation and ingestion; eye, nose, and throat irritant. Symptoms of exposure include conjunctivitis, cough, central nervous system depression, eye and skin burns, dermatitis, delayed pulmonary edema; carcinogenic. OSHA exposure limit (TWA): 100 ppm [air].

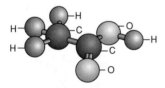

Acetic Acid

acetic acid — CH_3COOH — a colorless liquid or crystal with a sour, vinegar-like odor; the pure compound is a solid below 62°F; it is often used in aqueous solution; MW: 60.1, BP: 244°F, Sol: miscible, Fl.P: 102°F, sp gr: 1.05. It is used in the production of acetic anhydride, vinyl acetate, and acetic esters; as an esterifying, acetylating, acidifying, and neutralizing agent; as a

food additive or flavorant; in textile and dye industries as a solvent and intermediate, and in the production of nylon and acrylic fibers; in the manufacture of photographic chemicals; in the production of vitamins, antibiotics, and hormones; in the manufacture of rubber chemicals, accelerators, and as a coagulant of natural latex; in dry cleaning to remove rust; as a laboratory reagent in chemical and biochemical analysis; in field testing of lead fume, vinyl chloride determination, aniline vapors, and separation of gases; as a solvent for organic compounds. It is hazardous to the respiratory system, skin, eyes, and teeth; and is toxic by inhalation. Symptoms of exposure include conjunctivitis, lacrimation, nose and throat irritation, pharyngeal edema, chronic bronchitis, eye and skin burns, skin sensitivity, dental erosion, black skin, hyperkeratosis. OSHA exposure limit (TWA): 10 ppm [air].

acetic anhydride — $(CH_3CO)_2O$ — a colorless liquid with a strong, pungent, vinegar-like odor; MW: 102.1, BP: 282°F, Sol: 12%, Fl.P: 120°F, sp gr: 1.08. It is used in the manufacture of cellulose esters, fibers, plastics, lacquers, protective coating solutions, photographic films, cigarette filters, magnetic tape, and thermoplastic molding compositions; in the manufacture of pharmaceuticals and their intermediates; in organic synthesis as an acetylating, bleaching, and dehydrating agent; in synthesis of perfume chemicals, explosives, and weed killers. It is hazardous to the respiratory system, eyes, and skin; and is toxic by inhalation, ingestion, and contact. Symptoms of exposure include conjunctivitis, lacrimation, corneal edema and opacity, nasal and pharyngeal irritation, cough, dyspnea, bronchitis, skin burns, vesiculation, sensitization, dermatitis. OSHA exposure limit (TWA): ceiling concentration 5 ppm [air].

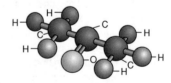

Acetone

acetone — CH_3COCH_3 — a colorless liquid with a fragrant, mint-like odor; MW: 58.1, BP: 133°F, Sol: miscible, Fl.P: 0°F, sp gr: 0.79. It is used during the application of lacquer, paints, and varnishes; during the use of solvents and cementing agents; during dip application of protective coatings; during fabric coating and dyeing processes. It is hazardous to the respiratory system and skin; and is toxic by inhalation, ingestion, and contact. Symptoms of exposure include eye, nose, and throat irritation, headache, dizziness, dermatitis. OSHA exposure limit (TWA): 750 ppm [air].

acetonitrile — CH_3CN — a colorless liquid with an aromatic odor; MW: 41.1, BP: 179°F, Sol: miscible, Fl.P (open cup): 42°F, sp gr: 0.78. It is used as a solvent; in organic synthesis in the preparation of vitamins, perfumes, water softeners, and plasticizers; as a catalyst; in low-temperature batteries, electrokinetic transducers, and angular accelerometers. It is hazardous to the kidneys, liver, cardiovascular system, central nervous system, lungs, skin, and eyes; and is toxic by inhalation absorption, ingestion, and contact. Symptoms of exposure include asphyxia, nausea, vomiting, chest pain, weakness, stupor, convulsions, eye irritation. OSHA exposure limit (TWA): 40 ppm [air].

2-acetylaminofluorene — $C_{15}H_{13}NO$ — a tan, crystalline powder; MW: 223.2, Sol: insoluble, BP: not known; Fl.P: not known. It is used in research and laboratory facilities. It is hazardous to the liver, bladder, kidneys, pancreas, skin, and lungs; and is toxic by inhalation, absorption, ingestion, and contact. Symptoms of exposure include reduced function of liver, kidneys, bladder, and pancreas; an occupational carcinogen. OSHA exposure limit (TWA): lowest feasible concentration regulated use.

acetylation — a process for introduction acetyl groups into an organic compound containing –OH, –NH_2, or –SH groups.

acetylcholine — $C_7H_{17}O_3N$ — the acetic acid ester of choline normally present in many parts of the body and having important physiological functions. It is a neurotransmitter at the cholinergic synapses in the central sympathetic and parasympathetic nervous systems of many vertebrates.

acetylene — *See* alkynes.

acetylene tetrabromide — $CHBr_2CHBr_2$ — a pale yellow liquid with a pungent odor; a solid below 32°F; MW: 345.7, BP: 474°F (decomposes), Sol: 0.07%, sp gr: 2.97. It is used as a catalyst or catalytic initiator in synthetic fibers; as a polymer additive in flame-proofing, flame-retardant polystyrenes, polyurethanes, and polyolefins; as a mercury substitute in the manufacture of gauges and balancing equipment; as a floatation agent in the processing and separation of mineral oils. It is hazardous to the eyes, upper respiratory system, and liver; and is toxic by inhalation, ingestion, and contact; symptoms of exposure include eye and nose irritation,

anorexia, nausea, severe headache, abdominal pain, jaundice, monocytosis. OSHA exposure limit (TWA): 1 ppm [air].

acetylene welding — *See* oxyacetylene welding.

ACGIH — *See* American Conference of Governmental Industrial Hygienists.

acid — a corrosive substance; any of a class of chemical compounds whose aqueous solutions turn blue litmus paper red, yields hydrogen ions, reacts with and dissolves certain metals to form salts, and reacts with bases to form water and salt.

acid aerosol — an acidic liquid or solid particle small enough to become airborne; high concentrations can be irritating to the lungs and are associated with respiratory diseases such as asthma.

acid-base balance — a state of equilibrium between acidity and alkalinity of the body fluids. Also known as hydrogen ion balance.

acid bottom and lining — a melting furnace's inner bottom and lining composed of refractory materials that produce an acid reaction in the melting process.

sulfur dioxide and nitrogen oxide interact with sunlight, chemical oxidants and water vapor

industrial wastes and vehicle exhaust

dry deposits, acid-rain or snow

Acid Deposition

acid deposition — a serious environmental and economic problem which occurs when an emission of sulfur dioxide and/or nitrogen oxide interacts with sunlight, chemical oxidants, and water vapor in the upper atmosphere to form acidic substances which then fall to earth as acid rain, acid snow, or dry acid deposition.

acid detergent — a material made up of synthetic detergents and hydrochloric acid used for the removal of mineral deposits from surfaces containing water deposits, milk, stone, and any other hard water films. It is also used for deliming dishwashing machines and similar equipment and for sanitizing toilet bowls and urinals.

acid extraction treatment system (AETS) — a process of soil washing that uses hydrochloric acid to extract contaminants from soils.

acid-fast bacteria — bacteria, especially *Mycobacterium tuberculosis*, that resist decolorizing by acids after being stained with basic dyes.

acid-forming bacteria — organisms fermenting in the lactose of milk which produce mainly lactic acid. The lactic acid combines with the protein calcium caseinate to liberate free protein casein, which is insoluble in water and is precipitated in the form of smooth gelatinous curds.

acidic anhydride — an oxide that reacts with water to form an acid.

acid mine drainage — water low in dissolved oxygen containing dissolved iron pyrites (ferrous sulfide) left behind from coal mining operations that becomes acidic and deposits ferric hydroxide on stream bottoms.

acidophilic — referring to any substance, tissue, or organism easily stained with acid dyes; an organism showing a preference for an acid environment.

acidosis — a pathological condition resulting from an accumulation of acid or depletion of the alkaline reserve in the blood and body tissues and characterized by increase in hydrogen ion concentration or a decrease in pH.

acid pickling — *See* pickling.

acid rain — rain that contains relatively high concentrations of acid-forming air pollutants, such as sulfur oxides and nitrogen oxides; acid rain may have a pH level as low as 2.8 as compared to normal rain pH of 6.

acid salt — a salt formed by replacing a part of the hydrogen ions of a polybasic acid with positive ions.

ACL — acronym for alternate concentration limit.

ACLF — acronym for adult congregate living facility.

ACM — acronym for asbestos-containing material.

acne — an inflammatory disease of sebaceous glands of the skin.

acneiform — resembling acne.

acoustic — related to or associated with sound.

acoustical calibrator — a device compatible with the microphone of a personal noise dosimeter and which emits one or more sound pressure levels to provide a calibration check on the general operation of a personal noise dosimeter set.

acoustic coupler — a device that enables a computer terminal to be linked to another over the telephone system via the handset of a conventional telephone.

acoustics — the science of sound, including its generation, transmission, and effects on the environment; the characteristics found within a structure that determine the quality of sound in it relevant to hearing.

acoustic trauma — a hearing loss caused by a sudden, loud noise in one ear, or a sudden blow to the head causing temporary loss or permanent damage.

acquired immunity — any type of immunity that is not inherited.

acquired immunodeficiency syndrome (AIDS) — (*disease*) a disease which is insidious and causes nonspecific symptoms such as anorexia, chronic diarrhea, weight loss, fever, and fatigue; incubation time is not known; it is caused by a retrovirus called T-lymphotropic virus type III (HTLV-III) and occurs throughout the world with the reservoir of infection being people; it is transmitted through sexual contact, through the use of unclean needles, and through blood transfusions; communicability, receptability, and resistance is not known; its spread is controlled by the practice of safe sex, the screening of blood donors, and the avoidance of contact with the body fluids of an infected individual.

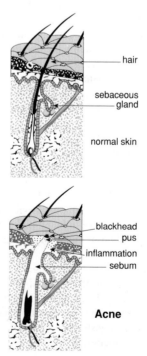

Acne

acre — a unit of measure equivalent to 43,560 square feet.

acrolein — $CH_2{=}CHCHO$ — a colorless or yellow liquid with a piercing, disagreeable odor; MW: 56.1, BP: 127°F, Sol: 40%, Fl.P: –15°F, sp gr: 0.84. It is used as a fungicide. It is hazardous to the heart, eyes, skin, and respiratory system; and is toxic by inhalation, ingestion, and contact. Symptoms of exposure include eye, skin, and mucous membrane irritation, abnormal pulmonary function, delayed pulmonary edema, chronic respiratory disease. OSHA exposure limit (TWA): 0.1 ppm [air].

acrylamide — $CH_2{=}CHCONH_2$ — a white crystalline, odorless solid; MW: 71.1, BP: 347–572°F (decomposes), Sol (86°F): soluble in water, Fl.P: 280°F, sp gr (86°F): 1.12. It is used in the manufacture of copolymers and polyacrylamides; as a grouting material in oil well drill holes, basements, tunnels, mine shafts, and dams. It is hazardous to the central nervous system, peripheral nervous system, skin, and eyes; and is toxic by inhalation, absorption, ingestion, and contact. Symptoms of exposure include ataxia, numbness of the limbs, paresthesia, muscle weakness, absent deep tendon reflex, hand sweating, fatigue, lethargy, eye and skin irritation; carcinogenic. OSHA exposure limit (TWA): 0.03 mg/m³ [skin].

acrylic — a family of synthetic resins made by polymerizing esters of acrylic acids.

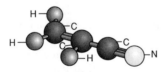

Acrylonitrile

acrylonitrile — $CH_2{=}CHCN$ — a colorless to pale yellow liquid with an unpleasant odor that can only be detected above the permissible exposure limit; MW 53.1, BP: 171°F, Sol: 7%, Fl.P: 30°F, sp gr: 0.81. It is used in the manufacture of ABS resin, SAN resin, plastic, or surface coating materials; in the manufacture and transfer of monomer to other reaction vessels or tank cars; as a chemical intermediate. It is hazardous to the cardiovascular system, liver, kidneys, central nervous system, skin, brain (tumor), lung and bowel (cancer); and is toxic by inhalation absorption, ingestion, and contact. Symptoms of exposure include asphyxia, eye irritation, headache, sneezing, nausea, vomiting, weakness, lightheadedness, skin vesication, scaling dermatitis; carcinogenic. OSHA exposure limit (TWA): 2 ppm [air].

ACS — *See* American Chemical Society.

act — (*law*) the general term describing the legislative enactment which provides the authority for a particular regulatory activity.

ACTH — *See* adrenocorticotropic hormone.

Actinomyces israelli
(causal agent of human actinomycosis)

Actinomycetes
(not proportionate)

Streptomyces

actinomycetes — a large group of mold-like microorganisms that give off an odor characteristic of rich earth and are the significant organisms involved in the stabilization of solid wastes by composting; any of any order of filamentous or rod-shaped bacteria including the actinomyces and streptomyces.

action level — the level or concentration of chemical residue in food or feed above which adverse health effects are possible, and above which corrective action is taken by the FDA; the level of toxicant which requires medical surveillance.

activated alumina — a granular adsorbent consisting mostly of highly porous aluminum oxide formed by heating aluminum hydroxide or most aluminum salts of oxy acids; it is especially useful for drying gases.

activated carbon — a specially treated and finely divided form of amorphous carbon which possesses a high degree of adsorption due to its large surface area per unit volume. Also known as activated charcoal.

activated charcoal — *See* activated carbon.

activated sludge — a semi-liquid mass removed from the liquid flow of sewage which was subjected to aeration and aerobic microbial action.

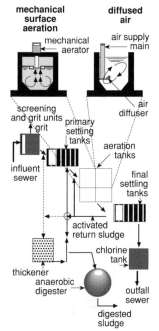

Activated Sludge Process

activated sludge process — a system of biological treatment that utilizes bacteria and other organisms to consume impurities in the sewage where activated sludge is mixed with sewage in aerators and impurities are absorbed by flocculent particles and consumed by the bacteria living in the floc.

activation — the process of causing a substance to become artificially radioactive by bombarding it with protons or neutrons in a reactor.

activator — a material added to a pesticide chemical formulation to increase its effect or toxicity.

active immunity — disease resistance in an individual due to antibody production after exposure to a microbial antigen following disease.

active immunization — stimulation with a specific antigen to promote antibody formation in the body.

active ingredient — (*pesticides*) the chemical or agent in a pesticide mixture that will destroy or prevent damage by pests.

active maintenance time — the time during which one or several persons work on a piece of equipment to carry out maintenance.

active material — (*pesticides*) *See* active ingredient.

active transport — from a lower to a higher concentration by means of energy.

activity — (*radiation*) the number of nuclear disintegrations occurring in a given quantity of material per unit time. *See also* curie.

act of God — an extraordinary, unpredicted, and unpreventable interruption, such as a flood or tornado, of normal activities.

Act to Require Aircraft Noise Abatement Regulation — a law passed in 1968 and amended several times until it became the Quiet Communities Act of 1978; it is used to abate aircraft noise and sonic booms.

actual dosage — the amount of a pure pesticide chemical in proper dilution used per unit area.

acuity — the sensitivity of receptors in hearing or vision.

acute care — care for individuals whose illnesses or health problems are of a serious nature where the care lasts until the patient is stabilized.

acute dermal LD_{50} — a single dose of a substance, expressed as milligrams per kilogram of body weight, that is lethal to 50% of the test population of animals under specific test conditions.

acute diarrhea — a clinical syndrome of diverse etiology associated with loose or watery stools and often vomiting and fever.

acute dose — a quantity which is administered for one day or in one dose.

acute exposure — exposure to a chemical for a duration of 14 days or less.

acute health effect — a single exposure to a toxin or microorganism that results in an adverse effect on a human or animal and produces severe symptoms which develop rapidly and become critical quickly.

acute LC_{50} — a concentration of a substance, expressed as parts per million parts of medium, that is lethal to 50% of the test population of animals under specified test conditions.

acutely toxic chemicals — chemicals which can cause both severe short- and long-term health effects after a single brief exposure; they can cause damage to living tissue, impairment of the central nervous system, severe illness, or, in extreme cases, death when injected, inhaled, or absorbed through the skin.

acute nonbacterial infectious gastroenteritis — a generic term for an infection caused by astrovirus, calicivirus, enteric adenovirus, parvovirus, or viral gastroenteritis.

acute toxicity — relating to the acute effects resulting from a single dose of or exposure to a poisonous substance; the amount of a pesticide that will seriously affect or destroy a test animal in a single dose.

adaptation — (*evolution*) the occurrence, by natural selection, of genetic or behavioral changes in a population or species in response to a new or altered environment.

ADC — *See* adult day care.

additive — a substance introduced into a basic medium to enhance, modify, suppress, or otherwise alter some property of that medium; a substance added to a food to enhance flavor, taste, or quality.

additive effect — an effect in which the combined effect of two chemicals may be greater than the sum of the agents acting alone.

addressability — (*geographic information system*) the number of positions in the X and Y axis on a visual display unit or graphics screen.

addressable point — (*geographic information system*) a position on a visual display unit that can be specified by absolute coordinates.

adenitis — inflammation of a gland or lymph node.

adenocarcinoma — a malignant tumor with glandular elements.

adenoma — an epithelial tumor, usually benign, with a gland-like structure.

adenosine diphosphate (ADP) — $C_{10}H_{15}N_5O_{10}P_2$ — a low energy compound found in cells which functions in energy storage and transfer; a nucleotide produced by the hydrolysis of adenosine triphosphate.

adenosine monophosphate (AMP) — a component of nucleic acid consisting of adenine, ribose, and phosphoric acid involved in energy metabolism and nucleotide synthesis. Also known as adenylic acid.

adenosine triphosphate (ATP) — $C_{10}H_{16}N_5O_{13}P_3$ — a nucleotide in all cells that stores energy from oxidation and glucose in the form of high-energy phosphate bonds. The release of these phosphate bonds supplies free energy to drive metabolic, especially involving muscular activity, reactions, or to transport molecules against concentration gradients (active transport), when ATP is hydrolyzed to ADP, an inorganic phosphate, or AMP, an inorganic pyrophosphate. ATP is also used to produce high-energy phosphorylated intermediary metabolites, such as glucose 6-phosphate.

adenosis — any disease of a gland, especially one involving the lymph nodes.

adenylic acid — *See* adenosine monophosphate.

adequately wetted — referring to a substance which has been sufficiently mixed or coated with water or an aqueous solution to prevent dust emissions.

ADH — *See* antidiuretic hormone.

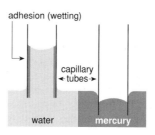

Adhesion

adhesion — the force of molecular attraction that holds the surfaces of two or more substances in contact, such as water and rock particles. The ability of one substance to stick to another.

adhesive — a substance that causes close adherence of adjoining surfaces.

ADI — acronym for acceptable daily intake of a substance, such as a contaminant in water or food.

adiabatic lapse rate — the –5.4°F theoretical change in air temperature for each thousand foot increase in altitude. Also known as the dry-adiabatic lapse rate.

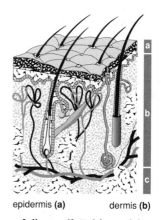

epidermis **(a)**   dermis **(b)**

Adipose (fatty) layer (c)

adipose — fatty; of or relating to fat.

adjacent — neighboring; having a common border.

adjoining — *See* adjacent.

adjournment — (*law*) the putting off or postponing of business until another time or place; the act of a court; to defer; closing formally.

adjudicate — to carry a lawsuit to a conclusion.

adjusted decibel (dBA) — sound level in decibels read on the A-scale of a sound-level meter.

adjusted rate — a rate calculated in such a way as to permit a more accurate comparison of the

A
B

state of health of two or more populations that differ substantially in some important respect, such as distribution by age or sex. *See also* standard rate.

adjuvant — a chemical or agent added to a pesticide mixture which enhances the action of the active ingredient (e.g., wetting agent, spreader, adhesive, emulsifying agent, penetrant).

ADME — acronym for the process called absorption, distribution, metabolism, and excretion.

administrative hearing — a hearing held by an authorized agency in the executive branch of government to correct a violation of the rules and regulations of that agency.

administrative law — rules and regulations in a given area established by an agency authorized by the legislature.

admittance (Y) — the ratio of current to voltage which includes information on the relative phase of these quantities; the reciprocal of impedance, it is expressed in mhos.

admonition — a reprimand from a judge to a person; warning of the consequences of conduct and punishment with greater severity if the same problem reoccurs.

adrenaline — *See* epinephrine.

adrenocorticotropic hormone (ACTH) — a pituitary hormone that stimulates the adrenal cortex to secrete its hormones. Also known as adrenotropic hormone.

adrenotropic hormone — *See* adrenocorticotropic hormone.

adsorbent — a substance that is used to collect other substances, such as gaseous pollutants.

adsorption — the process by which gases, liquids, and dissolved substances are attracted to the surface of a solid and retained there; the process in which the molecular or ionic species are accumulated on a surface and bound to the surface by forces of molecular attraction.

adsorption coefficient (Koc) — the ratio of the amount of a chemical adsorbed per unit weight of organic carbon in soil or sediment to the concentration of the chemical in solution at equilibrium.

adsorption ratio (Kd) — the amount of a chemical adsorbed by a sediment or soil in the solid phase divided by the amount of chemical in the solution phase which is in equilibrium with the solid phase at a fixed solid/solution ratio.

adult — (*pests*) a full-grown, sexually-mature insect, mite, nematode, or other animal.

adult day care (ADC) — a community-based, structured, comprehensive program for adults offered in a protective setting during daylight hours.

adulterants — chemicals or substances that by law do not belong in a food, plant, animal, or pesticide formulation.

adulterate — to debase or make impure by the addition of a foreign or inferior substance.

adulterated — food which is made impure by the addition of a foreign or inferior substance.

adulterated and misbranded milk and milk products — any milk or cream to which water has been added; milk or milk products which contain any unwholesome substance.

adulterated device — *See* adulterated drug.

adulterated drug — a drug or device consisting in whole or in part of any filthy, putrid, or decomposed substance. Also known as adulterated device.

adulterated food — food bearing or containing any poisonous or deleterious substance which may render injurious to health.

adulteration — the process of adulterating.

ad valorem tax — a tax on goods imposed at a rate percent of value.

advanced waste treatment — any biological, chemical, or physical treatment process used during any stage of treatment that employs unconventional, yet acceptable techniques. Also known as tertiary treatment.

advanced wastewater treatment (AWT) — a series of treatment techniques for the removal of additional contaminants from wastewater which has passed through a secondary treatment process to produce an effluent equivalent to potable water and which will meet drinking water standards. Also known as tertiary wastewater treatment.

advance notice of proposed rule-making (ANPRM) — a preliminary notice that an agency is considering a regulatory action.

advection — (*meteorology*) the transfer of heat, cold, or other atmospheric properties by the horizontal motion of an air mass.

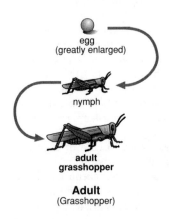

egg
(greatly enlarged)

nymph

**adult
grasshopper**

Adult
(Grasshopper)

adverse level — the measurement of the first discernible effects of air pollution likely to lead to symptoms of discomfort in humans.

AEC — *See* Atomic Energy Commission.

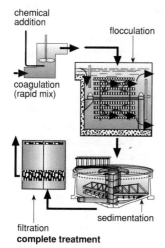

chemical addition

flocculation

coagulation (rapid mix)

filtration

sedimentation

complete treatment

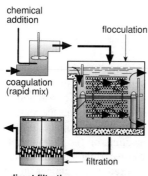

chemical addition

flocculation

coagulation (rapid mix)

filtration

direct filtration

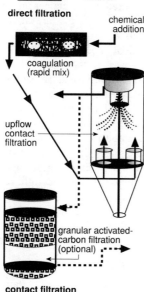

chemical addition

coagulation (rapid mix)

upflow contact filtration

granular activated-carbon filtration (optional)

contact filtration with optional granular activated carbon adsorption

Advanced Wastewater

AECD — *See* auxiliary emission control device.

Aedes — a genus of mosquitos of which one species transmits the virus that causes yellow fever in humans.

aerate — to bring in contact with air; to loosen or stir up soil mechanically for the purpose of introducing air; to expose to chemical action with oxygen; oxygenate by respiration.

aerated lagoon — *See* aerated pond.

aerated pond — a natural or artificial wastewater treatment pond or basin in which mechanical or diffused air aeration is used to supplement oxygen supply. Also known as aerated lagoon.

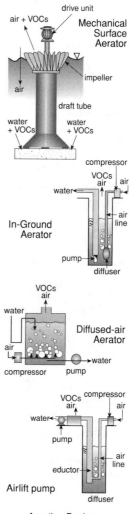

drive unit

air + VOCs

Mechanical Surface Aerator

impeller

air

draft tube

water + VOCs

water + VOCs

compressor

VOCs air

air

water

air line

In-Ground Aerator

pump

diffuser

VOCs air

water

Diffused-air Aerator

air

compressor   pump   water

VOCs air   compressor

air

water

pump

air line

eductor

Airlift pump

diffuser

Aeration Devices

Aeration

aeration — the purifying of water by spraying it into the air or by passing air through it; the process of exposing a bulk material, such as compost, to air or of charging a liquid with a gas or a mixture of gases.

aeration tank — a chamber for injecting air into water.

aerator — (*hospitals*) a device for the removal of ethylene oxide from sterilized materials by its exposure to the circulation of air.

aeroallergen pollutants — any airborne particulate matter that can induce allergic, infectious, and toxic responses in sensitive persons. Pollutants can include living organisms, such as animal dander, fungi, molds, infectious agents, bacteria, viruses, and pollen. Allergic reactions include watery eyes, runny nose and sneezing, nasal congestion, itching, coughing, wheezing, difficult breathing, and fatigue.

Aerobacter — in earlier classifications, a genus of the family *Enterobacteriaceae* consisting of gram-negative facultative anaerobic motile rods.

aerobe — a microorganism that requires the presence of free oxygen to live and grow.

aerobic digester

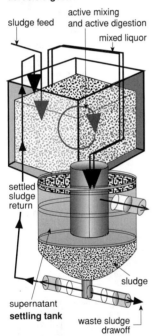

**Continuous-Flow
Aerobic Digestion System**

aerobic digestion — the biological stabilization of sludge through partial conversion of putrescible matter into liquid, dissolved solids, and gaseous by-products with some destruction of pathogens in an oxygenated environment.

aerobic sewage treatment tank — any unit incorporating as part of the treatment process the means of introducing air and oxygen into sewage held in a storage tank or tanks, so as to provide aerobic biochemical stabilization during a detention period prior to its discharge.

aerobic treatment unit — a sewage treatment unit which incorporates a means of introducing air into sewage so as to provide aerobic biochemical stabilization during a detention period.

aerodynamic diameter — the diameter of a unit density sphere having the same settling velocity as a particle no matter what its shape and density may be.

aerodynamic force — the force exerted on a particle in suspension by either the movement of air or gases around the particle or by the resistance of the gas or air to the movement of the particle through the medium.

aero filter — a filter bed for sewage treatment in which sewage is processed by maintaining a continuous rain-like application of sewage over the filter bed.

aeronomy — a study of the physical and chemical conditions of the upper atmosphere specifically concerning those regions upward of approximately 50 kilometers where dissociation and ionization are fundamental properties.

aerosol — a particle of solid or liquid matter under one micrometer in diameter that can remain suspended in the air because of its size; (*air pollution*) may cause irritation to the nose and throat, ulceration of the nasal passages, damage to the lungs, damage to other internal tissues depending on the chemical present, and may also cause dermatitis; (*pesticides*) pesticide chemicals stored in a container under pressure. A propellant forces the mixture out of the can as tiny droplets of liquid or solid particles light enough to float in air. *See also* particulate matter.

aerosol flammable — an aerosol that, when tested, yields a flame projection exceeding 18 inches at the valve opening, or a flashback at any degree of valve opening.

aerosol photometry — a system of measurement employing a direct-reading physical instrument that uses an electrical impulse generated by a photo cell to detect light scattered by a particle.

AETS — *See* acid extraction treatment system.

a–f — *See* audio-frequency.

AFDO — *See* Association of Food and Drug Officials.

affidavit — (*law*) a written sworn statement of persons having knowledge of certain facts.

affinity — the tendency of two substances to form strong or weak chemical bonds, forming molecules or complexes.

aflagellar — without flagella.

aflatoxin — any of a group of carcinogenic mycotoxins that are produced by some strains of the fungus *Aspergillus Flavus* in stored agricultural crops, such as peanuts.

AFSCME — *See* American Federation of State, County, and Municipal Employees.

afterburner — a device for augmenting the thrust of jet engines that includes an axillary fuel burner and combustion chamber to burn combustible gaseous substances; a device for burning or catalytically destroying unburned or partially burned carbon compounds in exhaust. *See also* secondary burner.

agar — a dried hydrophilic colloidal substance extracted from various species of red algae. It is used in cultures for bacteria and other microorganisms.

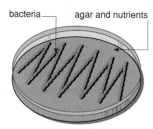

Agar Culture Plate

agar culture plate — a culture plate, usually a Petri dish, with an agar medium.

age adjusted rate — the number of events occurring to a given age group, divided by the population at risk of that age group, times one hundred.

Agency for Toxic Substances and Disease Registry (ATSDR) — an agency within the Public Health Service whose mission is to carry out health-related responsibilities of the Comprehensive Environmental Response Compensation and Liability Act of 1980 and its amendments by collecting, maintaining, analyzing, and disseminating information relating to serious diseases, mortality, and human exposure to toxic or hazardous substances.

age-specific death rate — the mortality rate for a specific age class.

age-standardization direct method — standardization of age-specific rates observed in a studied population are applied to an arbitrarily selected standard population to obtain the expected total number of events in the standard population. This expected total number is divided by the total number of persons in the standard population to obtain the age-standardized rates. Any difference between populations in age-adjusted rates based on the same standard population must be due to differences in age-specific rates and not due to differences in distribution by age.

age-standardization indirect method — age-specific rates obtained from a standard population are applied to age-appropriate groups of the studied population to obtain an expected number of events in the studied population. The observed number of events in the studied population is divided by the expected number to obtain a standardized mortality or morbidity ratio.

agglomerate — (*geology*) a rock composed of angular volcanic fragments fused by heat.

agglomeration — the consolidation of solid particles into larger forms by means of agitation or the application of heat.

agglutination — aggregation of separate particles into clumps or masses, especially the clumping together of bacteria by the action of a specific antibody directed against a surface antigen; the clumping together of blood cells in the presence of agglutinins.

agglutinin — any substance causing agglutination of cells, particularly a specific antibody in the blood combining with its homologous antigen.

agglutinogen — a protein substance on a corpuscle surface responsible for blood types; an antigen that stimulates production of a specific antibody (agglutinin) when introduced into an animal body.

aggregate — crushed rock or gravel screened to sizes for use in road surfaces, concretes, or bituminous mixes.

aggressive sampling — a sampling method using blowers and/or fans to keep particulates suspended during the sampling period.

agitate — to keep a solution from separating or settling by the use of mixing, stirring, or shaking; to keep a pesticide chemical mixed.

agitator — a device for keeping liquids or solids in liquids in motion to prevent settling; a device using a paddle, air, or other object to keep a pesticide chemical mixed in a spray tank.

agricultural solid waste — the solid waste that results from the rearing and slaughtering of animals and the processing of animal products and orchard and field crops.

AHA — *See* American Hospital Association.

AHERA — *See* Asbestos Hazard Emergency Response Act.

A
B

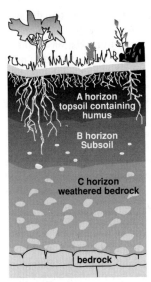

A Horizon
and other layers of soil

A horizon — (*soils*) properties that reflect the influence of accumulating organic matter or eluviation alone or in combination.

AIChE — *See* American Institute of Chemical Engineers.

AIDS — *See* acquired immunodeficiency syndrome.

AIHA — *See* American Industrial Hygiene Association.

air — a mixture of gases which contains approximately 78% nitrogen, 21% oxygen, less than 1% carbon dioxide and inert gases, and varying amounts of water vapor all forming the earth's atmosphere.

Air and Waste Management Association (AWMA) — an association of industrialists, researchers, equipment manufacturers, governmental personnel, educators, and meteorologists sponsoring continuing education and a library to seek economical answers to air pollution and waste management problems. Formerly known as Air Pollution Control Association (APCA).

air-bone gap — the difference in decibels between the hearing levels for a particular frequency determined by air conduction and bone conduction.

airborne particles — the liquid and solid aerosols as well as atmospheric particulates found in air.

airborne sound — sound that reaches the point of measurement through the air.

airborne toxics — toxic inorganic and organic chemicals released into the air and may cause short- or long-term health effects in humans.

air classifier — a system that uses a forced-air stream to separate mixed material according to size, density, and aerodynamic drag of the pieces.

air cleaner — any device used to remove atmospheric airborne impurities.

air cleaning — an indoor air quality control strategy to remove various airborne particulates and/or gases from the air; the three types of air cleaning most used are particulate filtration, electrostatic precipitation, and gas sorption.

air conditioning — a method of treating air to control temperature, humidity, cleanliness, and proper distribution within the space.

air conduction — the process by which sound is conducted to the inner ear through air in the outer ear canal.

air contaminant — any smoke, soot, fly ash, dust, dirt, fume, non-naturally occurring gas, odor, toxin, or radioactive substance occurring within an environment.

air-cooled wall — a refractory wall with a lane directly behind it through which cool air flows.

air curtain — a method of containing oil spills in which air bubbles pass through a perforated pipe that causes an upward flow slowing the spread of oil; (*mechanical engineering*) a stream of high velocity air directed downward to allow for the air-conditioning of a space with an open entrance.

air deficiency — a lack of air in an air-fuel mixture needed to supply the quantity of oxygen stoichiometrically required to completely oxidize fuel.

air emission — the release or discharge of a pollutant into the ambient air.

air exchange rate — the number of times that the outdoor air replaces the volume of air in a building per unit time, typically expressed as air changes per hour.

air filter — a common air-cleaning device used to reduce the concentration of light particulate matter from air to a level that can be tolerated.

air flush — part of the sterilizer cycle in which the vacuum pump operates continuously and a valve opens admitting filtered air into the sterilizer chamber.

air furnace — a reverberatory-type furnace in which metal is melted by the flame from fuel burning at one end of the hearth, passing over the bath, and exiting at the other end of the hearth.

air gap — the unobstructed vertical distance through the free atmosphere between the lowest opening from any pipe or faucet supplying water to a tank, plumbing fixture, or other device, and the flood level rim of the receptacle. In all cases, the air gaps should be two times the diameter of the pipe.

airglow — the continuous emission of upper atmosphere radiation during the day and at night over middle and low latitudes.

air hammer — a percussion-type pneumatic tool used for breaking through rock or concrete.

air heater — a heat exchanger through which air passes and is heated by a medium of a higher temperature, such as hot combustion gases or heat filaments.

air jets — streams of high velocity air from nozzles in an incinerator enclosure to provide turbulence, combustion air, or a cooling effect.

air line respirator — a respirator connected to a compressed breathing air source by a hose of small inside diameter.

air mass — a vast body of air often covering hundreds of thousands of square miles in which the conditions of temperature and moisture are about the same at all points in a horizontal direction. Also known as high-pressure cell.

air monitoring — the process of continuous or periodic ambient air sampling to measure the concentration and type of pollutants present.

air mover — any piece of equipment which is capable of moving air from one space to another.

air moving device (AMD) — a piece of equipment, such as a fan, that artificially moves air within a structure.

air pollution — a mixture of solid, liquid, and gaseous contaminants discharged into the atmosphere by nature and the activities of humans, and which may contribute to a variety of health or economic concerns and problems.

Air Pollution Control Act — a law passed in 1955 and updated through 1962; it states national policy to preserve and protect the primary responsibility and right of states and local governments in controlling air pollution; it initiated research by the United States Public Health Service and provided federal government technical assistance to states, established training programs in air pollution and inhouse and external air pollution research, and directed the Surgeon General to conduct studies of motor vehicle exhausts related to effects on human health.

Air Pollution Control Association (APCA) — *See* Air and Waste Management Association (AWMA).

air pollution episode — the occurrence of exceptionally high concentration levels of air pollution during a period of a few days or more.

air purifying respirator — a respirator that uses chemicals to remove specific gases and vapors

from the air or that uses a mechanical filter to remove particulate matter.

Air Quality Act — a law passed in 1967 establishing eight air quality control regions in the United States and issuing air quality criteria for specific pollutants. The law provided for the development and issuance of information on recommended air pollution control techniques and required state and local agencies to establish air quality standards and state implementation of plans; provided for federal action in cases of a state's non-compliance giving federal government emergency authority.

air quality assessment — the collection, handling, evaluation, analysis, and presentation of data needed to understand the air pollution problem of a given area and its causes.

air quality control region (AQCR) — an area designated by the federal government in which communities share a common air pollution problem, sometimes involving several states.

air quality criteria — the levels of pollution and lengths of exposure set by federal regulation at which specific adverse effects to health and welfare may occur.

air-quality specialist — an environmental health practitioner responsible for assuring air quality that protects the health, welfare, and comfort of people and the community.

air quality standard — the prescribed concentration that a pollutant may reach in the outside air without exceeding the limits set forth by various federal rules and regulations; the concentration represents the approximate level at which certain effects, clinical or sub-clinical, may be expected to begin to occur or the approximate level of concentration below which the effects defined should not ordinarily occur; in developing an air quality standard, it is necessary to identify the possible effects, to establish a criteria for the selection of the most sensitive group of receptors, to identify the existing burden of pollutants on the receptors, to specify a method for measurement of the concentration of the pollutant, and to specify an acceptable exposure limit to the pollutant.

air-regulating valve — an adjustable valve used to regulate air-flow to the face-piece, helmet, or hood of an air-line respirator.

air sampling — the process by which air samples or air pollution samples are collected.

air setting — the characteristic of some materials, such as refractory cements, core pastes, binders, and plastics, to take permanent set at normal air temperatures of 20–25°C.

airshed — a geographical area that shares the same air mass because of topography, meteorology, and climate.

air-supplied respirator — a respirator that uses a filter or sorbent to remove harmful substances from the air.

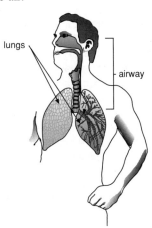

Airway

airway — the passage by which air enters and leaves the lungs.

airway resistance — the narrowing of the air passages of the respiratory system in response to the presence of irritating substances.

ALARA — acronym for as low as reasonably achievable.

albedo — a measure of the reflecting power of a surface being that fraction of the incident radiation which is reflected back in all directions.

albumin — a protein material found in animal and vegetable fluids that is soluble in pure water.

albuminuria — the presence of albumin, serum globulin, or other proteins in the urine; usually indicating renal disease.

alcohol — C_2H_5OH — a colorless, volatile liquid; BP: 78.3°C; Sol: soluble in water, chloroform, and methyl alcohol. It is used as a solvent and in the manufacture of many chemicals and medicines; an organic hydroxyl compound formed by replacing one or more hydrogen atoms of a hydrocarbon with an equal number of hydroxyl (OH) groups; primary alcohol contains a –CH_2OH group, secondary alcohol contains a –CHOH– group.

aldehyde — an organic compound formed by dehydrating oxidized alcohol; it contains the characteristic –CHO group.

aldrin — $C_{12}H_8Cl_6$ — a colorless to dark brown crystalline solid with a mild chemical odor; MW: 364.9, BP: decomposes, Sol: 0.003%, sp gr:

1.60. It is used as an insecticide. It is hazardous to the central nervous system, liver, kidneys, and skin; and is toxic by inhalation, absorption, ingestion, and contact. Symptoms of exposure include headache, dizziness, nausea, vomiting, malaise, myoclonic jerks of the limbs, clonic and tonic convulsions, coma, hematopoietic azotemia; carcinogenic. OSHA exposure limit (TWA): 0.25 mg/m³ [skin].

alert level — that concentration of pollutants at which first stage control action is to begin.

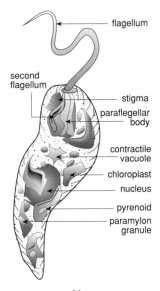

Algae
(*Euglena*)

algae — the general name for simple single-cell or multi-celled plants with chlorophyll which have no true roots, stems, or leaves and live in aquatic habitats and moist places on land.

algicide — a chemical used for the control and prevention of algae growth in swimming pools.

algorithm — a prescribed set of well-defined rules or processes for solving a problem in a given number of steps; frequently employed in electronic data processing.

alimentary canal — organs composing the food tubes in animals and humans; they are divided into zones for ingestion, digestion, and absorption of food, and for the elimination of indigestible material.

aliphatic — of or pertaining to any open chain carbon compound; three subgroups of such compounds are alkanes, alkenes, and alkynes.

aliquot — a sample that is representative of a larger quantity.

alkali — referring to any strong base, such as sodium hydroxide and potassium hydroxide.

alkalies — caustic substances that dissolve in water to form a solution with a pH higher than 7. This includes either solid or concentrated liquid forms of ammonia, ammonium hydroxide, calcium hydroxide, potassium hydroxide, and sodium hydroxide. Alkalies are more hazardous and destructive to tissue than most acids and cause irritation of the eyes and respiratory tract and lesions in the nose.

alkaline — referring to any substance having basic properties; having a pH greater than 7.

alkaline earths — usually considered to be the oxides of magnesium, barium, calcium, strontium, and radium.

alkane — a series of aliphatic saturated hydrocarbons that have no carbon-carbon multiple bonds with the general formula C_nH_{2n+2}.

alkapton — a class of substances with an affinity for alkali found in the urine and causing the metabolic disorder known as alkaptonuria.

alkene — one of a class of unsaturated aliphatic hydrocarbons in which two carbon atoms are connected by a double bond, with the general formula C_nH_{2n}. Also known as olefin.

alkyl — a monovalent radical C_nH_{2n+1} obtained from a saturated hydrocarbon by removing one hydrogen atom from an alkane, e.g., methyl CH_{3^-}, derived from methane.

alkylation — the combination of a saturated and unsaturated hydrocarbon.

alkylbenzene sulfonate (ABS) — a major class of alkylaryl sulfonate surfactants used in detergents.

alkynes — a group of unsaturated hydrocarbons in which two carbon atoms are connected by a triple bond, with the general formula $C{\equiv}C$. Also known as acetylene.

allegation — (*law*) an assertion undertaken to be proven with evidence by a party to a legal action.

allergen — a substance, such as pollen or a drug, that activates the immune system and causes an allergic response in sensitized humans or animals.

allergic reaction — an abnormal physiological response by a sensitive person to a chemical or physical stimuli that causes no response in nonsensitive individuals.

allergic rhinitis — an inflammation of the mucous membranes in the nose.

allergic sensitizer — any chemical acting as a sensitizer, especially epoxy monomers and their amine hardeners, potassium dichromate, nickel, and formaldehyde.

allergy — a condition of abnormal sensitivity in certain individuals to contact with substances such as proteins, pollens, bacteria, and certain foods. This contact may result in exaggerated physiologic responses such as hay fever, asthma, hives, and in severe enough situations, anaphylactic shock. Re-exposure increases the capacity to react.

allethrin — $C_{19}H_{26}O_3$ — a clear or amber, fairly viscid liquid with a mild pleasant odor; highly miscible in kerosene, xylene, and other organic solvents but insoluble in water; an insecticide. Also known as synthetic pyrethrum.

allochthonous — food material reaching an aquatic community originating elsewhere in the form of organic detritus.

allotrope — one of the two or more forms of an element differing in either or both physical and chemical properties. In the case of oxygen, Ozone (O_3) is an oxidant, molecular oxygen (O_2) is needed in respiratory processes, and elemental oxygen (O) changes into the other two forms.

allotropy — the existence of different forms of the same element in the same physical state with distinctly different properties due to differences in their energy content or arrangement of atoms.

alloy — any of a large number of materials composed of two or more metals, or a metal and a nonmetal mixed together and having metallic properties.

alloying elements — any of a number of chemical elements constituting an alloy.

alluvial — transported and deposited by running water.

alluvial cones — *See* alluvial fans.

alluvial fans — the conical deposit of sediment laid down by a swift-flowing stream as it enters a plain or an open valley. Also known as alluvial cones.

alluvial plane — a plane formed by the deposition of material in rivers and streams that periodically overflow.

alluvium — the surplus rock material consisting mainly of sand, silt, and gravel which a river or stream has eroded, carried in suspension, and deposited.

allyl alcohol — $CH_2{=}CHCH_2OH$ — a colorless liquid with a pungent, mustard-like odor; MW: 58.1, BP: 205°F, Sol: miscible, Fl.P: 70°F, sp gr: 0.85. It is used in the preparation of various allyl esters; in preparation of chemical derivatives used in perfumes, flavorings, and pharmaceuticals; as a fungicide, herbicide, and nematicide; in refining and dewaxing of mineral oil. It is hazardous to the eyes, skin, and respiratory system; and is toxic by inhalation, absorption, and ingestion. Symptoms of exposure include eye and skin

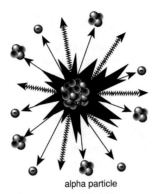

alpha particle

**Alpha Particle
Alpha Ray**

irritation, tissue damage, upper respiratory system irritation, and pulmonary edema. OSHA exposure limit (TWA): 2 ppm [air].

allyl chloride — $CH_2=CHCH_2Cl$ — a colorless, brown, yellow, or purple liquid with a pungent, unpleasant odor; MW: 76.5, BP: 113°F, Sol: 0.4%, Fl.P: –25°F, sp gr: 0.94. It is used in the manufacture of glycerin, epichlorohydrin, allyl alcohol, allylamines, allyl ethers of starch, and pharmaceuticals. It is hazardous to the respiratory system, skin, eyes, liver, and kidneys; and is toxic by inhalation, absorption, ingestion, and contact. Symptoms of exposure include eye, nose, and skin irritation, pulmonary edema, and liver and kidney damage in animals. OSHA exposure limit (TWA): 1 ppm [air].

allyl glycidyl ether — $C_6H_{10}O_2$ — a colorless liquid with a pleasant odor; MW: 114.2, BP: 309°F, Sol: 14%, Fl.P: 135°F, sp gr: 0.97. It is used as a reactive diluent in the formulation of epoxy resins; as a copolymer for vulcanization of rubber, surface coatings, and epoxy resins. It is hazardous to the respiratory system and the skin; and is toxic by inhalation, absorption, ingestion, and contact. Symptoms of exposure include dermatitis, eye and nose irritation, pulmonary irritation, edema, and narcosis. OSHA exposure limit (TWA): 5 ppm [air].

alopecia — a condition where there is an absence of hair from the skin areas where it is normally found; caused by trauma, cutaneous or systemic disease, drugs, chemical, or ionizing radiation.

alpha-destructive distillation — the distillation of organic soils or liquids where the substance decomposes during distillation leaving a solid or viscous liquid in the still.

alpha-emitter — a radioactive substance which gives off alpha particles.

alphanumeric code — (*computer science*) machine-processable letters, numbers, and special characters.

alpha particle — a positively charged particle emitted by several radioactive substances; having a massive charge identical to the nucleus of a helium atom and consisting of two protons and two neutrons.

alpha ray — a stream of fast moving helium nuclei or alpha particles that is strongly ionizing and penetrates weakly.

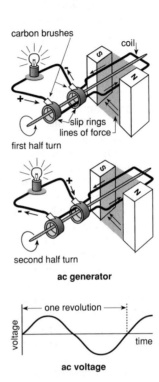

ac generator

ac voltage

Alternating Current

alternating current (AC) — current that reverses its flow with a constant frequency.

alternative system — any approved on-site sewage disposal systems used in place of, including modification to, a conventional subsurface system.

alternative technology — any proven wastewater treatment process and technique which provides for the reclaiming and reuse of water, recycling wastewater constituents, or otherwise eliminating the discharge of pollutants or recovering of energy.

alternative wastewater treatment works — a wastewater conveyance and/or treatment system other than a conventional system, including small diameter pressure and vacuum sewers and small diameter gravity sewers carrying partially- or fully-treated wastewater.

altimeter — an instrument used to measure altitude with respect to a fixed level.

Altocumulus Clouds

altocumulus clouds (Ac) — patches or layers of puffy, roll-like gray or white clouds made up of water droplets; bases average 10,000 feet above the earth.

Altostratus Clouds

altostratus clouds (As) — dense veils or sheets of gray or blue that very often totally cover the sky; bases average 10,000 feet above the earth.

alum — any group of double sulfates of a monovalent metal and a trivalent metal such as $K_2SO_4 \cdot Al_2(SO_4)_3$ $24H_2O$. Alum is a common name for commercial aluminum sulfate which is used in pools to form a gelatinous floc on sand filters or to coagulate and precipitate suspended particles in water. *See also* potassium alum.

aluminosis — a form of pneumoconiosis caused by respiratory exposure to aluminum-bearing dust.

aluminum — Al — a silver-like malleable and ductile metal found in soil; a flexible metal and a natural element in the earth; present in air, water, soil, and most foods; the principle source is bauxite; enters the environment through the weathering of rocks and minerals; releases are associated with industrial processes such as smelting; MW: 26.98, BP: 2307°C, insoluble in water, Fl.P: not known. It is used as a structural material in the construction, automotive, and aircraft industries; extensively used in the production of metal alloys; in the electrical industry in overhead distribution and power lines, electrical conductors and insulated cables and wiring; in auto parts, cooking utensils, decorations, highway signs, fences, cans, food packaging, foil, corrosion resistance chemical equipment, dental crowns, and denture materials; through inhalation exposure, the lungs receive Al^{+3} as particles of poorly soluble compounds; some of the particles stay within the lung tissue, others are transported up the respiratory system and eventually swallowed; may be distributed to the brain, bone, muscle, and kidneys and to some extent to the milk of lactating mothers and to cross the placenta and accumulate in the fetus; not thought to be toxic to humans when ingested from cooking utensils and as part of the daily intake in food; however sensitive subpopulations may include pregnant mothers and Alzheimer's patients; there appears to be a potential risk in the breathing in of the dust and the creation of respiratory symptoms; it may interact with neuronal DNA to alter gene expression and protein formation; the nervous system may possibly be a target for aluminum; exposure limits in air for 8 hour periods for the aluminum metal, total dust 15 mg/m³, respirable fraction 5 mg/m³, pyro powders 5 mg/m³, welding fumes 5 mg/m³, soluble salts 2 mg/m³, alkyls 2 mg/m³, for aluminum oxide the OSHA air regulation for total dust is 10 mg/m³, respirable fraction 5 mg/m³.

aluminum potassium sulfate — *See* potassium alum.

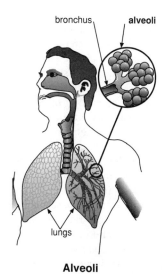

Alveoli

alveoli — the tiny sacks in the lungs where the actual exchange of oxygen and carbon dioxide takes place.

A
B

amalgam — an alloy of mercury with some other metal or metals.

amalgamation — the process of alloying metals with mercury to separate them from ore.

ambient air — the air surrounding a particular area in which climate, meteorology, and topography influence its ability to dilute and disperse pollutants present.

ambient air quality standard (AAQS) — the limit on the amount of a given pollutant which will be permitted into the ambient air.

ambient concentration — any air pollution concentration that occurs in the outdoor environment.

ambient noise level — the all-encompassing noise associated with a given environment, being a composite of sounds from all sources.

AMD — *See* air moving device.

ameba (amoeba) — the common name for a number of the simplest kinds of single-celled protozoans.

amebiasis — (*disease*) a parasitic infection of humans caused by the amoeba *Entamoeba histolytica*. Symptoms include acute dysentery, fever, chills, and blood or mucous diarrhea; it occurs especially in areas of poor sanitation; the reservoir of infection is humans; it is transmitted through the ingestion of water contaminated with feces containing amoebic cysts, and is communicable as long as the cyst is passing with feces; general susceptibility; it is prevented by the proper control of human feces, sewage protection of public and private water supplies, proper personal hygiene, and the control of flies; incubation time is 2–4 weeks but can vary from a few days to several months.

amended water — water containing wetting agents, penetrants, and/or other agents to enhance the wetting of asbestos-containing materials to reduce the generation of dust.

amenorrhea — absence of menstruation.

American Academy of Industrial Hygiene (AAIH) — a professional society of industrial hygienists.

American Academy of Occupational Medicine (AAOM) — a professional association of physicians working in occupational medicine full-time to promote maintenance and improvement of the health of industrial workers.

American Association of Poison Control Centers — an association which keeps up-to-date information on the ingredients and potential acute toxicity of substances that may cause accidental poisoning and the proper management of a person poisoned.

American Association of State Highway and Transportation Officials (AASHTO) — an association of public officials in state agencies responsible for highways and transportation.

American Board of Industrial Hygiene (ABIH) — a specialty board authorized to certify properly qualified industrial hygienists.

American Board of Preventive Medicine (ABPM) — the medical specialty board authorized to certify properly qualified specialists in preventive medicine.

American Chemical Society (ACS) — a scientific, educational, and professional society of chemists and chemical engineers.

American Conference of Governmental Industrial Hygienists (ACGIH) — a professional society of government and university employees engaged in full-time programs of industrial hygiene.

American Federation of State, County, and Municipal Employees (AFSCME) — a labor union of federal, state, county, and municipal employees.

American Hospital Association (AHA) — an association of hospitals, specialists, and individuals dedicated to the promotion of patient care through education, lobbying, publications, and provision of consultative services.

American Industrial Hygiene Association (AIHA) — an association promoting the study and control of environmental stresses arising in or from the work place or its products in relation to the health or well-being of workers and the public.

American Institute of Chemical Engineers (AIChE) — a member of the Accreditation Board for Engineering and Technology, the American National Standards Institute, and related organizations. It produces numerous publications and conducts annual educational meetings.

American National Standards Institute (ANSI) — a privately funded voluntary membership organization that identifies industrial and public needs for National Consensus Standards and coordinates development of such standards.

American Petroleum Institute (API) — a major trade association which is an umbrella organization for the major oil companies, independent oil producers, fuel distributors, service station owners, and other related concerns. They set standards and performance requirements for the industry and publish petroleum statistics.

American Public Health Association (APHA) — an association representing health professionals in over 40 disciplines in the development of health standards and policies.

American Public Health Association-Public Health Service housing code (APHA-PHS) — a code developed by the association and the federal government providing for local housing code requirements.

American Red Cross (ARC) — a voluntary, humanitarian organization that provides relief to victims of disasters and helps people prevent, prepare for, and respond to emergencies.

American Society for Testing and Materials (ASTM) — a nonprofit organization that develops standard testing methods by the consensus of volunteers from manufacturers, users, and others.

American Society of Clinical Pathologists (ASCP) — a professional society of clinical pathologists, clinical scientists, chemists, microbiologists, medical technologists, and medical technicians whose purpose is to promote a wider application of pathology and laboratory medicine to the diagnosis and treatment of disease, to conduct education programs and publish educational materials in the field of clinical and anatomic pathology, and to conduct an examination and certification program of medical laboratory personnel.

American Society of Heating, Refrigerating, and Air Conditioning Engineers (ASHRAE) — a society of professionals dedicated to the establishment of standards in refrigeration, heating, and air conditioning, and to the promotion of education through research and publications.

American Society of Mechanical Engineers (ASME) — a technical society with extensive programs in the development of safety codes, equipment standards, and educational guidance for students.

American Society of Safety Engineers (ASSE) — a society of safety related professionals dedicated to improving the occupational area through the promotion of standards, education, and professionalism.

American Standard Code for Information Interchange (ASCII) — a widely-used industry standard code for exchanging alphanumeric codes in terms of bit-signatures.

Ames assay — a test performed on bacteria to assess the capability of environmental chemicals to cause mutations; named for Bruce Ames.

amicus curiae — a friend of the court.

amide — one of a group of organic compounds formed by the reaction of any organic acid with ammonia or an amine, and containing the $CONH_2$ radical, $RCONH_2$ (primary); $(RCO)_2NH$ (secondary); $(RCO)_3N$ (tertiary), where R is a hydrocarbon group.

amine — one of a class of organic compounds, such as CH_3NH_2, derived from ammonia by substituting one or more hydrocarbon radicals for hydrogen atoms.

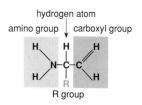

Amino Acid

amino acid — any of the organic compounds containing an amino group and a carboxyl group that are polymerized to form peptides and proteins; the nitrogen-containing molecules that plants and animals combine in great variety in building cell proteins.

4-aminodiphenyl — $C_6H_5C_6H_4NH_2$ — colorless crystals which turn purple in contact with air and have a floral odor; MW: 169.2, BP: 576°F, Sol: slight, Fl.P: not known, sp gr: 1.16. It is used in research and laboratory facilities. It is hazardous to the bladder and skin; and is toxic by inhalation, absorption, ingestion, and contact. Symptoms of exposure include headache, dizziness, lethargy, dyspnea, ataxia, weakness, methemoglobinemia, urinary bleeding, acute hemorrhagic cystitis; carcinogenic. OSHA exposure limit (TWA): regulated use through engineering controls, work practices, and personal protective equipment, including respirators.

aminoethane — *See* ethyl amine.

2-aminopyridine — $NH_2C_5H_4N$ — white powder or crystals with a characteristic odor; MW: 94.1, BP: 411°F, Sol: soluble Fl.P: 154°F. It is used in the production of chemical intermediates for pharmaceuticals; used in the manufacture of dyes, lubricant antioxidants, and herbicides. It is hazardous to the central nervous system, and the respiratory system; and is toxic by inhalation, absorption, ingestion, and contact. Symptoms of exposure include headache, dizziness, excitement, nausea, high blood pressure, respiratory distress, weakness, convulsions, stupor. OSHA exposure limit (TWA): 0.5 ppm [air].

A
B

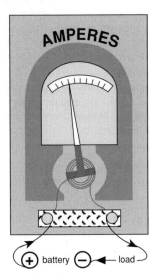

Ammeter
showing galvanometer

ammeter — an instrument used to measure the amount of the electric current in a circuit.

ammonia — NH_3 — a colorless gas with a very sharp odor; MW: 17.03, BP: –33.35°C, Sol: 42.8% at 0°C in water, 33.1% at 20°C in water, 34% at 25°C in water, flammability, limits in air 16–25%, density as a gas 0.7710 gram/L; a naturally occurring compound which is a key intermediate in the nitrogen cycle; it is essential for many biological processes and a background concentration is present in most environmental media; it is synthesized commercially and released into the atmosphere by volatization from decaying organic matter, animal livestock excreta, fertilization of soil, venting of gas, leaks or spills during commercial synthesis, production or transportation from sewage or wastewater effluent and the burning of coal, wood, other natural products, and through volcanic eruptions; largest and most significant use is the agricultural application of fertilizers; also used in production of fibers, plastics, explosives, urea, ammonium phosphates, nitric acid, ammonium nitrate, ammonium sulfate, as a refrigerant, corrosion inhibitor, in purification of water supplies, a component of household cleaners, in the pulp and paper, metallurgy, rubber, food and beverage, textile and leather industries, in pharmaceuticals, and as chemical intermediates. It may cause death at short-term exposure of 5,000–10,000 parts per million due to airway obstruction; infections and other secondary complications may occur; chemical burns and edema of exposed tissues including the respiratory tract, eyes, and exposed skin occur after exposure to lethal levels; an upper respiratory irritant to humans; exposures to levels exceeding 50 parts per million result in immediate irritation to nose and throat, although a tolerance level may develop at much higher levels, especially when exposed to anhydrous ammonia where nasopharyngeal and tracheal burns occur, airway obstruction and respiratory distress as well as bronchiolar and alveolar edema occur; acute exposure to highly concentrated aerosols of ammonia may cause elevated pulse and blood pressure and cardiac arrest in humans; bradycardia, hypertension and cardiac arrhythmias leading to cardiovascular collapse following acute exposure to concentrations exceeding 5,000 parts per million; highly concentrated aerosols of anhydrous ammonia can produce burns of the lips, oral cavity, and pharynx along with edema; cyanosis, elevated white blood cell count, and pulmonary artery thrombosis; spasms of the muscles of the extremities, hemorrhagic necrosis of the liver, dermal and ocular irritations, and secondary infections as well as blurred vision, diffuse nonspecific encephalopathy, loss of consciousness, and decreased deep tendon reflexes occur. OSHA exposure limit (TWA): 50 ppm [air].

ammonia stripping — the physical process in which ammonia is condensed and removed from alkaline aqueous waste solutions after contact with steam at atmospheric pressure.

ammonification — the release of ammonia, especially into the soil, from decaying protein by means of bacterial action.

ammonium sulfamate — $NH_4SO_3NH_2$ — a colorless to white crystalline, odorless solid; MW: 114.1, BP: 320°F (decomposes), Sol: soluble in water, Fl.P: 77°F, sp gr: 0.88. It is used as a herbicide; in the manufacture of fire-retardant compositions; in the generation of nitrous oxide gas; in the manufacture of electroplating solutions. It is hazardous to the upper respiratory system and eyes; and is toxic by inhalation, ingestion, and contact. Symptoms of exposure include eye, nose, and throat irritation, coughing, and difficulty breathing. OSHA exposure limit (TWA): 10 mg/m³ [total] or 5 mg/m³ [respiratory].

amniocentesis — a procedure during pregnancy by which a transabdominal perforation of the amniotic sac is made for the purpose of obtaining a sample of the amniotic fluid to determine maturation and viability of the fetus; also used to determine the presence of certain genetic disorders.

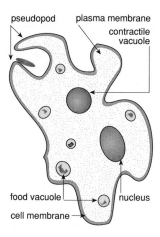

Amoeba

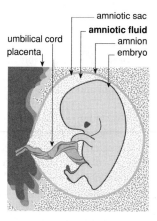

Amniotic Fluid

amniotic fluid — the watery fluid that surrounds the fetus or unborn child in the uterus to protect it from desiccation and shock.

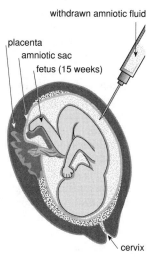

Amniocentesis

Amoeba (ameba) — a genus of single celled ameboid protozoa most of which are free-living; parasitic in humans.

amorphous — describing a noncrystalline solid having no orderly arrangement of molecules; having neither definite form nor structure.

amorphous silica — $SiO_2 + 2H_2O = Si\,(OH)_4$ — a form of silica which does not exhibit a well-defined, characteristic diffraction pattern when an X-ray beam is passed through it.

Amp — *See* ampere.

AMP — *See* adenosine monophosphate.

ampacity — the current-carrying capacity of electric conductors, expressed in amperes; used as a rating for power cables.

ampere (A) or (Amp) — the meter-kilogram-second unit of electrical current equal to 6.25×10^{18} electrons (or one coulomb) per second; used for measuring the rate of flow of electricity.

amperometry — a method of chemical analysis by techniques which involve measuring electric currents.

amphiboles — grayish-brown minerals present in many igneous and metamorphic rocks, and made up of 50% silica, 2% magnesium, and 40% iron. The formula typically is $5.5FeO$, $1.5MgO$, $8SiO_2$, H_2O.

amplification — (*physics*) an increase in the magnitude or strength of an electric current or other physical quantity or force.

amplifier — a device consisting of one or more vacuum tubes or transistors and associated circuits used to increase the strength of an electrical signal where the weak signal is changed into a stronger signal with larger waves.

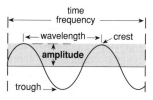

Amplitude

amplitude — (*physics*) the maximum vertical distance the particles in a wave are displaced from the normal position at rest; one half the extent of a vibration, oscillation, or wave where the amplitude depends upon its energy.

ampoule — a small glass vial containing ethylene oxide.

amu — *See* atomic mass unit.

A
B

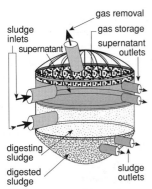

Anaerobic Digester

n-amyl acetate — $CH_3COO[CH_2]_4CH_3$ — a colorless liquid with a persistent banana-like odor; MW: 130.2, BP: 301°F, Sol: 0.2%, Fl.P: 77°F, sp gr: 0.88. It is used as a vehicle solvent in the manufacture of cellulose nitrate lacquers, lacquer thinners, adhesives, nail enamel, nail enamel removers, and paint; as an extractant of penicillin during production of antibiotics. It is hazardous to the eyes, skin, and respiratory system; and is toxic by inhalation, ingestion, and contact. Symptoms of exposure include eye and nose irritation, narcosis, and dermatitis. OSHA exposure limit (TWA): 100 ppm [air].

sec-amyl acetate — $CH_3COOCH(CH_3)C_3H_7$ — a colorless liquid with a mild odor; MW: 130.2, BP: 249°F, Sol: slight, Fl.P: 89°F, sp gr: 0.87. It is used during spray and dip application and manufacture of lacquers, varnishes, enamels, and metallic paints; in textile-sizing; as a solvent for chlorinated rubber; in the repair of motion picture film. It is hazardous to the respiratory system, eyes, and skin; and is toxic by inhalation, ingestion, and contact. Symptoms of exposure include eye and nose irritation, narcosis, and dermatitis. OSHA exposure limit (TWA): 125 ppm [air].

amylase — a class of enzymes that accelerates the hydrolysis of starches and other carbohydrates.

anabolic — pertaining to anabolism.

anabolism — the constructive process of the metabolism by which living cells convert simple substances into more complex compounds, especially into living matter; the method of synthesis of tissue structure.

anaerobic bacteria — any bacteria that live and grow in the absence of free oxygen; two types are facultative and obligate.

anaerobic contact process — an anaerobic waste treatment process in which the microorganisms responsible for waste stabilization are removed from the treated effluent stream by sedimentation or other means and held in or returned to the process to enhance the rate of treatment.

anaerobic digestion — the biological stabilization of sludge through partial conversion of putrescible matter into liquid, dissolved solids, and gaseous by-products with some destruction of pathogens in an oxygen-free environment.

anaerobic respiration — the process by which an organism is capable of releasing energy without the use of oxygen.

analgesic — any substance, such as salicylates, morphine, or opiates, used to relieve pain.

analog — a continuously varying electronic signal; (*biology*) an organ that is similar to another organ; (*chemistry*) a substance which resembles another substance structurally but differs from it in function; especially of concern when it interferes in a reaction because of its structural similarity. Also spelled analogue.

analogous — referring to structures that are similar in function and general appearance but not in origin.

analogue — *See* analog.

analysis — (*law*) an examination of evidence; (*chemistry*) a determination of the nature or amount of one or more components of a substance.

analytical air sampler — the sensor of a ventilating device which reacts directly with the contaminant or indirectly through a reagent; the change in chemical or physical properties of the sensor or the reagent is a measure of the concentration of the contamination in the air sample.

analytical epidemiological studies — research used to test a specific hypothesis regarding the etiology of the disease being studied.

analytical epidemiology — the study of causes, determining the relatively high or low frequency of disease in specific groups by formulating hypotheses based on actual observations of existing diseases, conducting studies of outbreaks of disease, and evaluating the data to determine the accuracy of the original hypothesis. Age, sex, ethnicity, race, social class, occupation, and marital status are the principal variables.

analytical methods — methods of chemically analyzing the quantities of concentrations of air pollutants in an air sample.

anaphase — the stage of mitosis or meiosis in which the chromosomes move from the equatorial plate toward the poles of the cell.

anaphylaxis — an unusual or exaggerated allergic reaction of an organism to foreign protein or other substances following a primary injection of the protein. It can produce reactions ranging from local swelling and redness to the systemic collapse of the vascular network by permitting the loss of fluid from the blood vessels into the interstitial fluid area.

anatomy — the scientific study of the structures of plants and animals.

Ancylostomiasis — (*disease*) *See* hookworm disease.

androgen — any of a class of steroid hormones that promotes male secondary sexual characteristics.

anemia — a condition marked by a low level of hemoglobin or circulating red blood cells; a symptom of a variety of disorders due to poor diet, blood loss, exposure to industrial poisons, diseases of the bone marrow, or any other upset in the balance between blood and blood production.

Cup Anemometer

anemometer — an instrument to measure and indicate wind velocity.

anemometry — the study of measuring and recording the direction and speed of wind.

aneroid barometer — a barometer which uses an aneroid capsule, that is a corrugated metal container from which the air has been removed; a spring inside of the container prevents air pressure from collapsing it completely; as air pressure increases, the top of the box bends in, as pressure decreases the top bows out; gears and levers transmit these changes to a pointer on a dial; the pull-down lever moves the pointer forward.

anesthesia — the entire or partial loss of sensation or consciousness following administration of an anesthetic.

anesthetic — a substance, such as ether, used to desensitize a reaction to pain or produce unconsciousness by inhibiting nerve activity.

aneurysm — a sac formed by the localized, abnormal dilatation of a blood vessel which increases pressure against adjacent tissue or organs and creates the possibility of a rupture.

angina pectoris — acute pain in the chest resulting from the increased blood supply to the heart muscle. It may be brought on by physical activity or emotional stress that places added burden on the heart and increases the need for additional blood supply to the myocardium; may be accompanied by pain radiating down the arms, up into the jaw, or to other sites.

angiostrongyliasis — (*disease*) a disease of the central nervous system involving the meninges due to a nematode, *Angiostrongylus contonensis*. Symptoms include severe headaches, stiffness of the neck and back, and occasionally temporary facial paralysis. Incubation time is usually 1–3 weeks; found in Hawaii, the Pacific Islands, the Philippines, Australia, and Asia. The reservoir of infection is the rat. It is transmitted through the ingestion of raw or improperly cooked snails, prawns, fish, and land crabs, but is not transmitted from person to person; those suffering from malnutrition or disease are especially susceptible. It is controlled through the destruction of rats and through the proper cooking of snails, fish, crab, and prawn.

Angiostrongylus — a genus of nematode parasites.

angle — the space formed by two diverging lines emanating from a common point and measured by the number of degrees.

angle of abduction — the angle between the longitudinal axis of a limb and a sagittal plane.

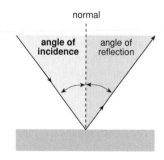

Angle of Incidence

angle of incidence — (*physics*) the angle formed between the direction of a ray or wave of sound emanating from a common point and striking a surface and the perpendicular to that surface at the point of arrival.

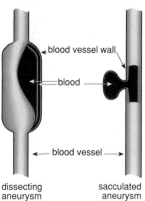

Aneurysm

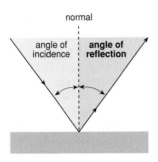

Angle of Reflection

angle of reflection — (*physics*) the angle formed between the direction of light or sound reflected by a surface and the line perpendicular to that surface at the point of reflection.

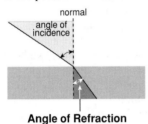

Angle of Refraction

angle of refraction — (*physics*) the angle that a ray of light makes with a line perpendicular to a surface separating two media.

Angoumois Grain Moth (*Sitotroga cerealella*) — a light grayish-brown or straw-colored moth with a satiny luster and wing expanse of 1/2"–2/3"; the hind wings are fringed with long, dark setae and have a point at the tip like a finger which distinguishes the insect from the Crows Moth; a stored food product insect.

angstrom unit (Å) — a unit of measure of wavelength of optical spectra equal to 10^{-10} meters or 0.1 nanometers.

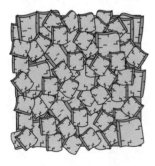

Angular Blocky Soil

angular blocky — cube-like soil structure where the aggregates have sides at nearly right angles and tend to overlap nearly block-like with six or more sides, all three dimensions about the same, usually found in subsoil or B horizon.

anhydride — a compound formed from an acid by the removal of water.

anhydrous — containing no water.

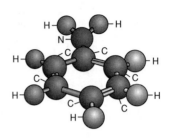

Aniline

aniline and homologs — $C_6H_5NH_2$ — a colorless to brown oily liquid with an aromatic amine-like odor; a solid below 21°F; MW: 93.1, BP: 363°F, Sol: 4%, Fl.P: 158°F, sp gr: 1.02; an aromatic amine compound that is a pale brown liquid at room temperature. One of the most important substances produced from coal tar, it is used in chemical synthesis and intermediates for rubber processing; in the synthesis of pharmaceuticals; in the manufacture of inks; in the synthesis of intermediates for artificial sweetening agents; in the synthesis of catalysts and stabilizers for hydrogen peroxide and cellulose. It is hazardous to the cardiovascular system, liver, kidneys, and blood; and is toxic by inhalation, absorption, ingestion, and contact. Symptoms of exposure include headache, weakness, dizziness, cyanosis, ataxia, dyspnea on effort, tachycardia, and eye irritation; carcinogenic. OSHA exposure limit (TWA): 2 ppm [air] or 8 mg/m³ [skin].

animal bedding — material, usually organic, that is placed on the floor of livestock quarters for animal comfort and to absorb excreta.

animal dander — *See* dander.

anion — a negatively charged ion; during electrolysis it moves to the anode.

anionic — characterized by an active, especially surface-active, anion.

anisidine (o-, p-isomers) — $NH_2C_6H_4OCH_3$ — o: a red or yellow oily liquid with an amine-like odor; a solid below 41°F. p: a yellow to brown crystalline solid with an amine-like odor; MW: 123.2, BP: 437/475°F, Sol: insoluble/moderate, Fl.P: 244°F (oc)/86°F, sp gr: 1.10/1.07. It is used in the synthesis of hair dyes; in the preparation of organic compounds; as a corrosion inhibitor for steel storage. It is hazardous to the cardiovascular system, blood, kidneys, and liver; and is toxic by inhalation, absorption, ingestion, and contact. Symptoms of exposure include headache, dizziness, cyanosis; carcinogenic. OSHA exposure limit (TWA): 0.5 mg/m^3 [skin].

anisotropy — the characteristic of a substance for which a physical property depends upon direction with respect to the structure of the material; doubly refracting or having a double polarizing power.

anneal — to treat a metal, alloy, or glass by heat with subsequent cooling to soften and render metals less brittle and make manipulation easier.

annealing — the process of cooling glass slowly to prevent brittleness.

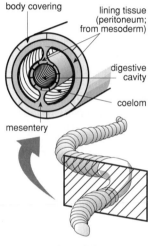

Annelid

Annelida — a diverse phylum of metazoan invertebrates comprising segmented worms.

annihilation — (*physics*) the destruction of a particle and its anti-particle as a result of their collision, their energy being shared by lighter particles that emerge from the collision.

annual operating expenses — the costs of operation and maintenance as well as capital related charges, except interest or return on capital.

annual plant — a plant that completes its growth cycles in one growing season.

annualized cost — the equivalent annual cost equal to the revenue requirement, including annual operating expenses, plus interest and return on capital.

annular space — (*wells*) the space between the side of an excavation and the casing or wall of a well.

anode — the positive terminal of an electrolytic cell at which current enters and to which negative ions are attracted.

anorexia — the lack or loss of appetite for food.

anoxia — absence of oxygen in the tissues; the condition is accompanied by deep respirations, cyanosis, increased pulse rate, and impairment of coordination.

ANPRM — *See* advance notice of proposed rule-making.

ANS — *See* autonomic nervous system.

ANSI — *See* American National Standards Institute.

answer — (*law*) a legal proceeding made after filing a claim of ownership in a seizure; claimants may file an answer in which they deny any or all of the allegations of the complaint for forfeiture; if this filing is made, a contest may ensue.

antagonism — the situation in which two chemicals, organisms, muscles, or physiologic actions, upon interaction, interfere in such a way that the action of one partially or completely inhibits the effects of the other.

antagonist — (*anatomy*) a muscle opposing the action of another muscle; necessary for control and stability of action.

antemortem — performed or occurring before death.

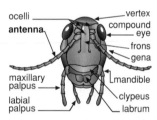

Grasshopper Antenna

antenna — one of the sensory appendages on the head of many arthropods; a device for radiating or receiving radio waves.

anterior — situated near the head or front end of an object.

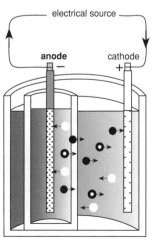

Voltaic Cell showing Anode

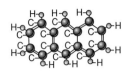

Anthracene

anthracene — a polycyclic aromatic hydrocarbon formed during the incomplete combustion of coal, oil, gas, garbage or other organic substances and found especially at hazardous waste sites. See polycyclic aromatic hydrocarbons.

anthracite — a hard, dense, metamorphic coal with a glassy texture formed from soft coal by heat and pressure within the earth; cannot be used for making coke; burns with a nonluminous flame and little smoke or odor. Also known as hard coal; kilkenny coal; stone coal.

anthrax — (*disease*) an acute bacterial disease of sheep and cattle usually affecting the skin; incubation time is within a 7 day period. Symptoms of exposure are mild and nonspecific initially and are like a common upper respiratory infection, followed by acute symptoms of respiratory distress, fever, and shock; death may occur; caused by *Bacillus anthracis*; found in most industrial countries in the occupational setting where people process hides, hair, bone and bone products, and wool. The reservoir of infection is contaminated soil with spores of *B. anthracis* which come from the skins and hides of infected animals. It is transmitted by contact with the tissues of the animals or the skins; it is not communicable from person to person; susceptibility is not known. It is prevented by immunization; dust control; thorough washing, disinfection, or sterilization of hair, wool, or hides; vaccination of animals; and control of waste.

anthrocosis — (*disease*) a disease of the lungs caused by prolonged inhalation of dust containing particles of carbon and coal.

anthropometry — (*ergonomics*) the measurement of the size and proportions of the human body to provide the dimensional data needed for the positioning of controls and the size of work spaces.

antibacterial — having the capacity to destroy or suppress the growth or reproduction of bacteria.

antibiotic — a chemical substance produced synthetically or by a bacterium mold or other fungus plant and used to kill bacteria and stop the growth of disease microorganisms in the body.

antibody — an immunoglobulin protein molecule, produced by certain white blood cells, that reacts with a specific antigen inducing its synthesis with similar molecules. It is classified according to its mode of action as an agglutinin, bacteriolysin, hemolysin, opsonin, or precipitin. Antibodies are synthesized by B lymphocytes that have been activated by the binding of an antigen to a cell surface receptor.

anticoagulant — any substance that suppresses, delays, or prevents coagulation of the blood; a chemical used in a bait to destroy rodents; it affects the walls of the small blood vessels and keeps the blood from clotting causing the animals to bleed to death.

anti-degradation clause — a regulatory concept that limits deterioration of existing air or water quality by restricting the addition of pollutants.

antidiuretic — any agent that prevents the excretion of urine.

antidiuretic hormone (ADH) — a polypeptide hormone from the posterior lobe of the pituitary gland that suppresses the secretion of urine. Also known as vasopressin.

antidote — an agent to relieve, prevent, or counteract the effects of a poison.

antigen — any substance capable, under appropriate conditions, of inducing a specific immune response and of reacting with the products of that response. Antigens may be soluble substances such as toxins, and foreign proteins or particulates, such as bacteria and tissue cells.

antihelminthic — *See* antihelmintic.

antihelmintic — a chemical agent used to destroy tapeworms in domestic animals. Also spelled antihelminthic.

antihistamine — a drug that counteracts the effects of histamine; a normal body chemical that is believed to cause the symptoms of persons who are hypersensitive to various allergens; used to relieve the symptoms of allergic reactions, especially hay fever and other allergic disorders of the nasal passages.

antimicrobial agent — a chemical compound that kills microorganisms or suppresses their multiplication or growth. Antimicrobial agents may interfere with the synthesis of the bacterial cell wall, which results in cell lysis and a weakening or rupturing of the cell wall, interferes with the synthesis of nucleic acids, changes the permeability of the cell membrane, and interferes with metabolic processes.

antimony and compounds — Sb — metal: a silver-white, lustrous, hard, brittle solid; scale-like crystals; or a dark gray, lustrous powder; MW: 121.8, BP: 2975°F, Sol: insoluble, Fl.P: N/A, sp gr: 6.69. It is found during the crushing and transferring of antimony ore; during production of lead/antimony alloys; during machining, grinding, buffing, and polishing of metal products containing antimony; during the manufacture of paints, pigments, enamels, glazes, ceramics, and glass. It is hazardous to the respiratory system, cardiovascular system, skin, and eyes; and is toxic by inhalation and contact. Symptoms of exposure include nose, throat, and mouth irritation, cough, dizziness, headache, nausea, vomiting, diarrhea, stomach cramps, insomnia, anorexia, skin irritation, inability to smell properly, cardiac abnormalities in antimony trichloride exposures. OSHA exposure limit (TWA): 0.5 mg/m^3 [air].

antineoplastic — inhibiting the maturation and proliferation of malignant cells.

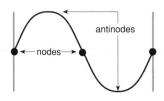

Antinode

antinode — the position of maximum amplitude in a standing wave. Also known as a loop.

antioxidant — a compound used to prevent deterioration by oxidation; used in food preservation as a retardant against rancidity of fats and to preserve the appearance and taste of fruits by preventing discoloration.

antiseptic — any substance that inhibits the growth of infectious bacteria.

antisiphone air gap — a device used to prevent the backflow of contaminated water into the potable water system.

antitoxin — a particular kind of antibody produced in the body in response to the presence of a toxin.

ANTU — *See* alpha-naphthyl thiourea.

anuria — complete suppression of urine output by the kidney.

AOAC — *See* Association of Official Analytical Chemists.

AOAC Use Dilution Confirmation Test — *See* Association of Official Analytical Chemists Use Dilution Confirmation Test.

aorta — the large artery arising from the left ventricle, being the main trunk from which the systemic arterial system proceeds.

APCA — *See* Air Pollution Control Association.

aperture — an opening; (*optics*) the diameter of the opening through which light passes in a camera, telescope, or other optical instrument.

APF — *See* assigned protection factor.

APHA — *See* American Public Health Association.

APHA-PHS housing code — *See* American Public Health Association-Public Health Service housing code.

aphasia — the partial or total loss of the ability to articulate ideas in any form as a result of brain damage to the cerebral cortex.

API — *See* American Petroleum Institute.

apical — related to or situated at an apex.

aplastic anemia — a condition in which the bone marrow fails to produce an adequate number of red blood corpuscles.

appeal — (*law*) a formal request for the review of a lawsuit and reversal of the verdict or decree.

application — the placing of a substance in a plant, animal, building, or soil, or the releasing of it into the air or water to destroy pests; (*computer science*) a task addressed by a computer system.

application program package — a set of computer programs designed for a specific task.

applicator — a person or piece of equipment that applies pesticides to destroy pests or prevent their damage.

Applied Science and Technology Index — a reference which indexes over 300 journals by subject from 1958 to the present.

apply uniformly — to spread or distribute a pesticide chemical evenly.

approved — acceptable to the regulatory authority based on the determination of conformity with widely accepted public health principles, practices, and generally recognized industry standards.

AQCR — *See* air quality control region.

AQL — *See* acceptable quality level.

AQM — acronym for air quality maintenance.

aquatic plants or leaves — plants or leaves that grow on, in, or near water.

aquatic toxicity — the adverse effects to marine life that result from being exposed to a toxic substance.

aqueduct — any artificial canal or passage for conveying water.

aqueous — relating to, containing, or prepared with water.

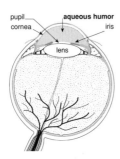

Aqueous Humor

aqueous humor — the transparent fluid filling the cavity between the cornea and lens of the eye which helps maintain the shape of the eye.

aqueous solution — a solution in which water is the solvent.

aquiclude — *See* aquitard.

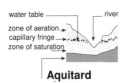

Aquifer

aquifer — the geological formation of a porous, water-bearing layer of sand, gravel, and rock below the earth's surface that is capable of transmitting and storing economically important quantities of water to wells and springs.

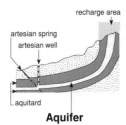

Aquitard

aquitard — a geological formation adjacent to an aquifer capable of transmitting and storing only negligible amounts of water. Also known as an aquiclude.

Arachnid
(Golden Orb Spider)

arachnid — an arthropod with four pairs of thoracic appendages.

Arachnida — the common name for members of the class of arthropods to which spiders, mites, and ticks belong; characterized by four pairs of thoracic appendages.

arbovirus — any of several viruses containing RNA spread by mosquitoes, ticks, and other arthropods; approximately 90 viruses classified as arboviruses produce disease in people including yellow fever, dengue fever, and equine encephalitis.

ARC — *See* American Red Cross.

arc — a continuous part of the circumference of a circle or regularly curved line; a discharge of electricity through a gas; a line connecting a set of points that form one side of a polygon.

arc cutting — a type of thermal cutting of metal using temperature generated by the heat of an arc between an electrode and the base metal.

arch drop — a form of construction that supports a vertical refractory furnace wall and serves to deflect downward.

arch furnace — a nearly horizontal structure that extends into a furnace and serves to deflect gases.

arch ignition — a refractory furnace, arch, or surface located over a fuel bed to radiate heat and to accelerate ignition.

arc welding — the process producing coalescence of metals by heating with an arc; the application of pressure and the use of gases or filler metal are not necessary in this process.

area — any particular extent of a surface; a geographical region or tract.

area monitoring — the routine monitoring of the level of radiation or radioactive contamination of any particular area, building, room, or equipment.

area of a circle — measurement of size found using the formula πr^2.

area of a rectangle — measurement of size found using the formula length × width.

area sanitary landfill method — a method in which waste is spread and compacted on the surface of the ground and a cover material is spread and compacted over the waste.

arene — *See* aromatic hydrocarbon.

arithmetic mean — the sum of a set of values or numbers divided by the quantity of those values or numbers. Also known as mean.

aroclor™ — *See* polychlorinated biphenyl.

aromatic hydrocarbon — a member of a class of organic compounds with a chemistry similar to that of benzene, characterized by large resonance energies. Also known as arene.

arraignment — (*law*) the appearance of a defendant in any criminal prosecution before the court to answer the allegations made against him or her and to enter a plea.

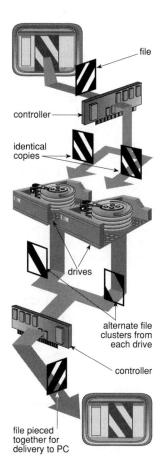

Array
(Mirrored Drive Array)

array — (*computer science*) a series of address-able data elements in the form of a grid or matrix.

arrhythmia — variation from the normal rhythm, especially of the heart beat or respiration.

arsenic — As — metal: a silver-gray or tin-white, brittle, odorless solid; MW: 74.9, BP: sublimes, Sol: insoluble, Fl.P: na, sp gr: 5.73. It is used during the manufacture of insecticides, weed killers, and fungicides; as a wood preservative; in the manufacture of electrical semiconductors, diodes, and solar batteries; as an addition to alloys to increase hardening and heat resistance; during the smelting of ores. It is hazardous to the liver, kidneys, skin, lungs, and lymphatic system; and is toxic by inhalation, absorption, ingestion, and contact. Symptoms of exposure include ulceration of the nasal septum, dermatitis, gastrointestinal disturbances, peripheral neuropathy, respiratory irritation, hyperpigmentation of the skin; carcinogenic. OSHA exposure limit (TWA): 0.010 mg/m³ [air].

arsenical — containing or pertaining to arsenic.

arsine — AsH_3 — a colorless gas with a mild garlic-like odor; MW: 78.0, BP: –81°F, Sol: 20%, Fl.P: N/A (gas). It is used during the manufacture of semiconductors and gallium arsenide; during the refining of metal ores containing arsenic; during the cleaning of metal equipment, electroplating of metals, metallic pickling, soldering and etching, photo-duplication. It is hazardous to the blood, kidneys, and liver; and is toxic by inhalation. Symptoms of exposure include headache, malaise, weakness, dizziness, dyspnea, back and abdominal pain, nausea, vomiting, bronze skin, jaundice, peripheral neuropathy; carcinogenic. OSHA exposure limit (TWA): 0.05 ppm [air] or 0.2 mg/m³.

arteriole — a tiny artery which branches to become capillaries.

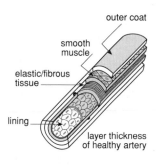

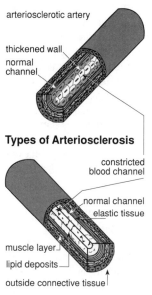

Types of Arteriosclerosis

arteriosclerosis — thickening of the lining of arterioles, usually due to hyalinization of fibromuscular hyperplasia.

A
B

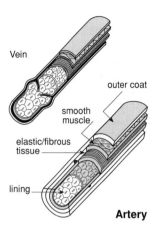

Artery

artery — a large, muscular vessel that carries blood away from the heart toward other body tissue.

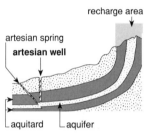

Artesian Well

artesian well — a pressurized aquifer located between two nonpermeable rock layers penetrated by a pipe or conduit causing the water to come to the surface under its own internal pressure.

arthritis — any inflammation affecting joints or their component tissue. Arthritis and other rheumatic diseases are the major cause of chronic disability in the United States.

Arthropoda — the largest phylum in the animal kingdom, comprising of invertebrates with jointed legs including insects, spiders, and crabs.

artifactual association — a statistically significant association that is spurious because of some error in sampling, measurement, or analysis of data.

artificial fiber — a filament made from materials such as glass, rayon, or nylon.

artificial radioactivity — radioactivity produced by particle bombardment or electromagnetic irradiation as opposed to natural radioactivity. Also known as induced radioactivity.

artificial recharge — an assortment of techniques used to increase the amount of water infiltrating an aquifer that has been depleted by abnormally large withdrawals.

artificial respiration — any method of forcing air into and out of the lungs to maintain breathing in the absence of normal respiration.

As — *See* altostratus clouds.

asbestos — a generic term for a group of six naturally occurring impure magnesium silicate fibrous materials which have high tensile strength and are heat and chemical resistant, inert, and do not evaporate or dissolve; amosite: $\{(Mg,Fe)_7Si_8O_{22}(OH)_2\}_n$, chrysotile: $Mg_3Si_2O_5(OH)_4$, tremolite: $\{Ca_2Mg_5Si_8O_{22}(OH)_2\}_n$, actinolite: $\{Ca_2(Mg,Fe)_5Si_8O_{22}(OH)_2\}_n$, anthophyllite: $\{(Mg,Fe)_7Si_8O_{22}(OH)_2\}_n$, crocidolite: $\{NaFe_3{}^{2+}Fe_2{}^{3+}Si_8O_{22}(OH)_2\}_n$. The basic unit of asbestos-class minerals is the silicate (SiO_4) group; this group can form a variety of polymeric structures through formation of Si–O–Si bonds; the amphibole class of asbestos has a characteristic polymeric structure consisting of a linear double chain which crystallize into long, thin, straight fibers. This class includes amosite, crocidolite, tremolite, and anthophyllite; the serpentines class is chrysotile, the polymeric form, an extended sheet forming a tubular fiber structure. Asbestos, mainly chrysotile, has been used in over 3,000 different products in the United States, including paper products, asbestos cement products, friction products, textiles, packaging and gaskets, coatings, and asbestos-reinforced plastics for automobile clutchs, brakes, and transmission components, asbestos cement pipe, roofing products, coatings, and sealants. Disposal may occur only in asbestos-containing waste special landfills approved and regulated by the federal government; currently the standard method for measuring asbestos concentrations in workplace air uses phase contrast microscopy (PCM); a particle visible under PCM is counted as a fiber if it is 5 or more micrometers long and has a length/thickness ratio of 3:1 or more; health hazards include respiratory effects where inhalation exposure to asbestos fibers can lead to a characteristic pneumoconiosis called asbestosis, which is an inflammatory response created in the lung by the deposition of asbestos fibers, with the symptoms of shortness of breath, râles or cough, and in severe cases, impairment of respiratory function which may ultimately lead to death; indirectly affects the cardiovascular system by reducing the blood flow through the pulmonary capillary bed because of increased resistance; increased risk of gastrointestinal cancer in humans since asbestos fibers deposited in the lung during inhalation exposure are transported by mucociliary action to the pharynx where they are swallowed and expose the gastrointestinal epithelium directly to the fibers; cell-mediated immunity is depressed in workers who have radiological evidence of asbestosis; concentrations of autoantibodies tend to be abnormally high which may lead to rheumatoid arthritis; mesothelioma is rare in the overall population but more frequent in individuals who have been exposed to asbestos; ingestion of asbestos in water causes little or no risk of non-carcinogenic injury, however, chronic oral exposure may lead to increased risk of gastrointestinal cancer; dermal exposure may lead to the formation of small warts or corns from the asbestos fibers. OSHA exposure limit PEL (TWA): at the action level 0.1 fiber/cc, at the excursion limit (30 minutes) 1 fiber/cc.

Asbestos Hazard Emergency Response Act (AHERA) — a law requiring the EPA to establish a comprehensive, regulatory system for controlling asbestos hazards in schools.

asbestosis — chronic lung inflammation caused by inhalation of asbestos fibers over an extended period of time. The onset of this disease is characterized by shortness of breath and some chest pain. Later, basal râles, clubbing of the fingers and occasionally the toes, and radiographic changes are found. Bronchitis occurs and eventually a variety of pneumoconiosis may occur along with related heart problems.

Asbestos School Hazard Act (ASHA) — a law authorizing the EPA to provide loans and grants to schools to help with a severe asbestos hazard.

asbestos tailing — any solid waste product of asbestos mining or milling operations which contain asbestos.

A-scale sound level — a measurement of sound approximating the sensitivity of the human ear, used to note the intensity or annoyance of sound below 55 decibels.

Ascaris lumbricoides
(in adult intestinal stage
of ascariasis)

ascariasis — (*disease*) a helminthic infection of the small intestine caused by *Ascaris lumbricoides*; generally, few symptoms occur; however, live worms are passed in the stool or occasionally from mouth to nose; pneumonitis can occur; may aggravate nutritional deficiencies, and may become fatal if bowel obstruction occurs. Incubation time is two months after ingestion of embryonated eggs; the disease is found worldwide. The reservoir of infection is humans and round worm eggs in soil. It is transmitted by ingestion of the infected eggs from the soil contaminated with human feces, or from uncooked food contaminated with infected soil; not

transmitted from person to person; communicable for 8–18 months; general susceptibility. It is controlled through the proper disposal of feces and through good personal hygiene.

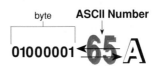

byte   ASCII Number

01000001 65 A

ASCII — *See* American Standard Code for Information Interchange.

Ascomycetes — a class of fungi including yeast and certain molds.

ASCP — *See* American Society of Clinical Pathologists.

asepsis — absence of septic matter; the state of being free from pathogenic microorganisms; freedom from bacteria or infectious materials.

aseptic processing and packaging — the filling of a commercially sterilized cooled product into presterilized containers followed by aseptic hermetical sealing with a presterilized closure in an atmosphere free of microorganisms.

aseptic technique — the performance of a procedure or operation in a manner that prevents infections from occurring.

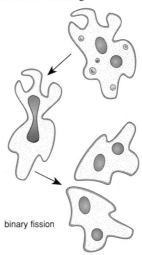

binary fission

Asexual Reproduction

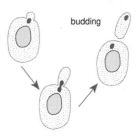

budding

asexual reproduction — reproduction from a single parent without eggs and sperm.

A
B

ash — inorganic residue less than 4 millimeters in diameter remaining after ignition of combustible substances, including cinders and fly ash.

ASHA — *See* Asbestos School Hazard Act.

ash pit — a pit or hopper located below a furnace or fireplace where residue is accumulated and from which it is removed.

ash sluice — a trench or channel in which water transports residue from an ash pit to a disposal or collection point.

ASHRAE — *See* American Society of Heating, Refrigerating, and Air Conditioning Engineers.

askarel — a generic term for a group of nonflammable, synthetic, chlorinated hydrocarbons used as electrical insulating material.

ASME — *See* American Society of Mechanical Engineers.

aspartame — $C_{14}H_{18}N_2O_5$ — a synthetic compound of two amino acids (L-aspartyl-L-phenylalanine methyl ester) used as a low calorie sweetener; it is 180 times as sweet as sucrose; it degrades into diketopiperazine at certain temperatures; the metabolism of aspartame, a peptide, proceeds in the same way as the metabolism of proteins. After metabolism, aspartame proteins become aspartic acid, phenylalanine, and methanol.

aspect ratio — the ratio of the longer dimension to the shorter in any rectangular configuration; length to width ratio.

aspergillosis — (*disease*) a rare fungus infection in humans and animals causing a variety of clinical symptoms including bronchial effects, allergic effects, lung abscesses, possible damage to the brain, kidney, and other organs, invasion of blood vessels; possibly fatal. Incubation time is a few days to weeks; caused by *Aspergillus fumigatus*, *A. niger,* and *A. flavus*; found worldwide; appears in compost piles, stored damp hay, decaying vegetation, cereal grains, and other food stocks starting to heat up undergoing fermentation and decay. It is transmitted by inhalation of airborne spores from grains, not people; controlled through prevention of growth of the fungus.

asphalt — black, solid, or semisolid bitumens which occur in nature or are obtained as residues during petroleum refining; used for paving and roofing and in paints and varnishes.

asphenosphere — the region of the lithosphere in which the rock is less rigid than that above and below but rigid enough to transmit transverse seismic waves.

asphyxia — suffocation due to a deficiency of oxygen in the blood and an increase in carbon dioxide in the blood and tissues.

asphyxiant — any chemical which may be a vapor or gas that excludes oxygen or actively interferes with oxygen uptake and distribution.

aspirator — an instrument that utilizes a vacuum to draw up substances.

assay — a qualitative or quantitative test for a particular chemical effect.

ASSE — *See* American Society of Safety Engineers.

assembler — (*computer science*) a computer program that converts program-written instructions into computer-executable (binary) instructions.

assessment — the critical analysis, evaluation, or judgment of the status or quality of a particular condition, situation, or other subject of appraisal.

assigned protection factor (APF) — the minimum anticipated protection provided by a properly functioning respirator or class of respirators to a given percentage of properly fitted and trained users.

assigned risk plan — a means for the insured who do not meet normal insurance standards to secure workers compensation insurance.

assimilation — conversion of nutritive material to living tissue; anabolism.

Association of Food and Drug Officials (AFDO) — an association of food and drug professionals promoting the passage and implementation of federal regulations to prevent the misbranding and adulteration of foods, drugs, cosmetics, and devices, and to promote uniformity among states in regard to regulations.

Association of Official Analytical Chemists (AOAC) — an independent international association devoted to the development, testing, validation, and publication of methods of analysis for foods, drugs, feeds, fertilizers, pesticides, water, forensic materials, and other substances.

Association of Official Analytical Chemists (AOAC) **Use Dilution Confirmation Test** — a test carried out by placing sterile stainless-steel ring carriers in broth culture of test organisms, drying the rings for 20–60 minutes, immersing in tub of use solution of germicide for 10 minutes at 20°C, immersing in nutrient media, incubating for 48 hours, and evaluating for growth of microorganisms.

asthenia — loss of strength; weakness.

asthma — a pulmonary disease characterized by shortness of breath, wheezing, thick mucus secretions, and constricted bronchioles resulting from abnormal responsiveness of the airways to certain substances or emotional stresses.

ASTM — *See* American Society for Testing and Materials.

astrovirus — (*disease*) an unclassified virus containing a single positive strand of RNA of about 7.5 kilobase surrounded by a protein capsid of 28–30 nanometer diameter which may cause sporadic viral gastroenteritis in children under 4 years of age, and about 4% of the hospitalized cases of diarrhea with additional symptoms of nausea, vomiting, malaise, abdominal pain, and fever. The mode of transmission is through the oral-fecal route, from person to person, or ingestion of contaminated food and water. The incubation period is 10–70 hours. The disease has been reported in England, Japan, and California. The reservoir of infection is people. Susceptibility found in very young children and the elderly in institutional settings. Also known as acute nonbacterial infectious gastroenteritis; viral gastroenteritis.

asymmetrical — the inability to be divided into like portions by hypothetical planes.

ataxia — failure of muscular coordination or irregularity of muscular action due to some disease of the nervous system.

atelectasis — the incomplete expansion of the lung at birth; the partial collapse of the lung later in life because of an occlusion of a bronchus or external compression due to a tumor or injury.

atheroma — a mass or plaque of degenerated, thickened arterial intima occurring in atherosclerosis.

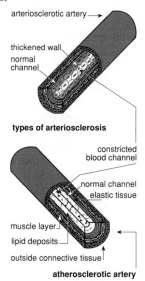

Atherosclerosis

atherosclerosis — a form of arteriosclerosis in which atheromas-containing cholesterol, lipoid material, and lipophages are formed within the intima and inner medial of large and medium-sized arterias.

Atm — *See* atmosphere.

atmometry — the science of measuring the evaporation of water from solid or liquid surfaces; evapotranspiration.

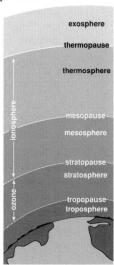

Atmosphere

atmosphere (Atm) — the gaseous envelope that surrounds a planet or celestial body and held in place by gravity; a unit of pressure equal to 760 millimeters of mercury (mm/Hg) at sea level.

atmosphere supplying respirator — a respirator that provides breathing air from a source independent of the surrounding atmosphere.

atmospheric pressure — the pressure exerted by the weight of the air above it at any point on the earth's surface; measurement equal to 14.7 pounds per square inch at sea level. Also known as barometric pressure.

atmospheric survey — an examination of the air of a given geographical area to determine the nature, sources, extent, and effects of air pollution.

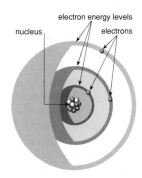

Structure of an Atom

atom — the smallest particle of a chemical that can take part in a chemical reaction without being permanently changed.

atomic absorption spectrophotometry — a technique used in the analysis of wastewater combining emission and absorption phenomena and resembling flame photometry.

atomic energy — the energy released in nuclear reactions. Also known as nuclear energy.

Atomic Energy Commission (AEC) — the agency now designated as the Nuclear Regulatory Commission in the Department of Energy.

atomic fluorescence spectrophotometry — a technique linked to atomic absorption spectrophotometry using atom fluorescence.

atomic mass — the mass of an atom expressed in terms of atomic mass units.

atomic mass unit (amu) — a unit of measure used to express relative atomic masses; it is equal to 1.660×10^{-27} kg which is 1/12 of the mass of carbon-12 atom, the most abundant form of carbon atoms. Also known as Dalton.

atomic number — the total number of protons in an atomic nucleus.

atomic oxygen — an atom of elemental oxygen.

atomic power — *See* nuclear power plant.

atomic waste — the radioactive ash produced by the splitting of nuclear fuel in a nuclear reactor.

atomic weight (at. wt.) — the relative mass of an atom expressed in atomic weight units as compared with another element (carbon-12), which is assigned a mass value of 12. Also known as relative atomic mass.

atom smasher — an accelerator for atomic and sub-atomic particles so they can be used as projectiles to blast apart the nuclei of atoms.

ATP — *See* adenosine triphosphate.

atrial fibrillation — cardiac arrhythmia marked by rapid, randomized contractions of the atrial myocardium causing an irregular ventricular rate. Also known as fibrillation. *See also* ventricular fibrillation.

at risk — the state of being subject to some uncertain loss or difficulty.

Atropa belladonna — *See* belladonna.

atrophy — a wasting or diminution in structure or function of a cell, tissue, or organ that was once of normal size.

atropine — $C_{17}H_{23}O_3N$ — an anticholinergic alkaloid occurring in belladonna, hyoscyamus, and stramonium that acts as a competitive antagonist of acetylcholine at muscarinic receptors, blocking stimulation of muscles and glands by parasympathetic and cholinergic sympathetic nerves; used as a smooth muscle relaxant and as an antidote to organophosphate poisoning.

ATSDR — *See* Agency for Toxic Substances and Disease Registry.

attack — an episode or onset of illness.

attack rate — the incidence rate of an acute disease.

attainment area — (*air pollution*) any area designated as having ambient air quality levels better than any national primary or secondary air quality standard for any pollutant.

attenuation — (*radiation*) the process by which a beam of radiation is reduced in intensity due to absorption and scattering when passing through some material; (*disease*) the change in the virulence of a pathogenic microorganism induced by passage through another host species decreasing its virulence; how live vaccines are produced.

attenuator — a resistive network that absorbs part of a signal and transmits the rest with a minimum of distortion or delay.

attractant — a chemical or agent that lures insects or other pests by stimulating their sense of smell.

attraction — the force or influence on one object which is drawn toward another one.

attributable risk — the arithmetic difference between the incidence of the disease in a group exposed to the factor and the incidence in an unexposed group, that is, the number of new cases of the disease among persons exposed to the factor that can be attributed to the factor; the amount of a disease that can be attributed to a particular factor and is commonly used to assess the public health importance of that factor as the cause of the disease.

attribute — (*geographic information system*) nongraphic information associated with a point, line, or area.

attrition — the wearing or grinding down of a substance by friction, which may lead to an increase in air pollutants.

at. wt. — *See* atomic weight.

audible range — (*sound*) the frequency range to which people with normal hearing respond; approximately 20–20,000 hertz.

audible sound — the sound which contains frequency components between 20–20,000 hertz.

audio-frequency (a–f) — the frequency of sound to which the ear can respond.

audiogram — a graph or table obtained from an audiometric examination showing hearing level as a function of frequency (usually 500–6,000 Hz).

audiologist — an individual who evaluates hearing functions to detect hearing impairment and/or hearing disorders.

audiology — the scientific study of hearing.

audiometer — an instrument for measuring in decibels the sensitivity of hearing; an apparatus used in audiometry.

audiometer setting — a setting on an audiometer corresponding to a specific combination of hearing level and sound frequency.

audiometric zero — the threshold of hearing; 0.0002 microbars of sound pressure.

audiometry — the measurement of the acuity of hearing at various frequencies of sound waves.

Audio Visuals On-Line (AVLINE) — a database maintained by the National Library of Medicine of educational aids in the health sciences.

audit — systematic review and evaluation of records and other data to determine the quality of services or products provided in a given situation.

audit gas — a primary standard bottle of calibration gas used to calibrate a secondary standard.

auditory — of or pertaining to the ear or sense of hearing.

Auger effect — the spontaneous ejection of an electron by an excited positive ion to form a doubly charged ion.

auricle — the part of the ear that projects from the head; one of the two upper, ear-shaped chambers of the heart.

aurora — a sporadic radiant emission from the upper atmosphere over middle and high altitudes seen most often along the outer realms of the Arctic and Antarctic.

autochthonous malaria — an indigenous strain of malaria acquired by mosquito transmission in an area where malaria regularly occurs.

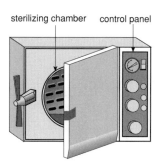

sterilizing chamber　　control panel

Autoclave

autoclave — a piece of equipment used for the sterilization of materials whereby steam under pressure flows around articles placed in a chamber, with the vapor penetrating cloth or papers used to package the articles being sterilized.

autogenous combustion — the burning of wet organic material where the moisture content is at such a level that the heat of combustion of the organic material is sufficient to vaporize the water and maintain combustion where no auxiliary fuel

is required except for start-up of a process. Also known as autothermic combustion.

autogenous incineration — a combustion characteristic of sludge having a composition, physical and chemical, such that no auxiliary fuel is required in incineration except at start-up and shut-down.

autoignition temperature — the lowest temperature at which a gas-air mixture or vapor-air mixture will ignite from its own heat source or contact heat source without a spark or flame.

automated cartography — the process of drawing maps with the aid of computer-driven display devices such as plotters and graphic screens.

automated filter tape — an air sampler which is a highly sensitive, portable instrument that operates with a high degree of repeatability.

automatic chlorinator — a device to feed a regulated amount of stabilized chlorine into pool water on a continuing basis.

automatic data processing — the use of any kind of automation in data processing.

automatic food vending machine — a machine designed and constructed so that with insertion of a coin, food or drink is delivered to a customer in a package, in bulk, or in a container.

automatic rain gauge — an automatic instrument used for measuring all types of precipitation.

automatic tape sampler — an air pollution sampler which collects ascended particulates by drawing air through a small area on a strip of filter paper at a rate of approximately 0.25ft³/min.

autonomic nervous system (ANS) — a division of the nervous system in vertebrates which regulates the vital internal organs in an involuntary manner.

autopsy — the examination of a body after death to determine the actual cause of death. Also known as postmortem; necropsy.

autoradiograph — a record of radiation from radioactive material in an object made by placing the object close to a photographic emulsion.

autosomal — pertaining to an autosome.

autosome — any of the 22 pairs of chromosomes found in an individual not concerned with determination of sex.

autothermic combustion — *See* autogenous combustion.

autotroph — an organism that synthesizes food molecules from inorganic molecules by using an external energy source; green plants which assimilate nutrient materials using energy and combine them into living organic substances.

auxiliary emission control device (AECD) — any element of design which senses temperature, vehicle speed, engine RPM, transmission

**A
B**

gear, manifold vacuum, or any other parameter for the purpose of modulating, delaying, or deactivating the operation of any part of the emission control system.

auxiliary fuel-firing equipment — equipment used in an incinerator to supply additional heat by burning an auxiliary fuel so that the resulting higher temperatures dry and ignite the waste material, maintain ignition, and combust the combustible solids, vapors, and gases.

auxochrome — (*chemistry*) any group of atoms capable of making a chromogen into a dye or pigment.

available chlorine — chlorine either free or combined which is active against bacteria in pool water.

available heat — the gross quantity of heat released within a combustion chamber minus the sensible heat carried away by the dry flue gases, the latent heat, and the sensible heat carried away in water vapor contained in the flue gases; the quantity of useful heat produced per unit of fuel if it is completely burned.

average head — the resistance of the flow of water in a pool recirculation system obtained by averaging the maximum and minimum resistance encountered in the course of a filter run.

average hot temperature — an average of the temperatures that occur during the hottest times of the year.

average-LET (linear energy transfer) — the average energy locally imparted to a medium by a charged particle of specified energy in traversing a short distance.

average lowest temperature — an average of the temperatures during the coldest times of the year.

average monthly discharge limitation — (*water*) the highest allowable average of "daily discharges" over a calendar month, calculated as the sum of all daily discharges measured during a calendar month divided by the number of daily discharges measured during that month.

average precipitation — an average of the amount of rain that falls in a given area at different times.

average temperature — an average of all the temperatures during the course of a year.

Aviation Safety and Noise Abatement Act — a law of 1979, enacted in 1980, and amended in 1982, providing assistance to airport operators to prepare and carry out noise compatibility programs; providing assistance to ensure continued safety in aviation; establishing a single reliable system for measuring noise; establishing a single system for determining individuals' exposure to noise at airports.

AVLINE — *See* Audio Visuals On-Line.

Avogadro's constant — 6.025×10^{23}; the number of atoms in a gram atomic weight of any element. It is also the number of molecules in the gram molecular weight of any substance; one mole of any gas occupies 22.414 liters of volume; formerly called Avogadro's number.

Avogadro's hypothesis — *See* Avogadro's law.

Avogadro's law — the law which states that equal volumes of ideal gases under the same conditions of temperature and pressure contain an equal number of molecules. Also known as Avogadro's hypothesis.

Avogadro's number — the number of molecules in a mole of any substance; equal to 6.02217×10^{23}; at standard temperatures and pressures.

avoirdupois weight — the system of units commonly used for measurement of the mass of any substance except medicines, precious metals, and precious stones; it is based on the pound (approximately 453.6 grams).

Aw — *See* water activity.

AWMA — *See* Air and Waste Management Association.

AWT — *See* advanced wastewater treatment.

axis of rotation — the true line about which angular motion takes place at any instance.

axis of thrust — the line along which thrust can be transmitted safely.

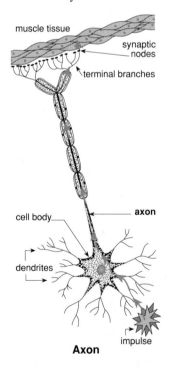

Axon

axon — a long, thread-like part of a nerve cell which carries an impulse away from the nerve body. Also known as a neuraxon; neurite.

axoneme — the central strand of the eukaryotic cilium or flagellum.

azinphos-methyl — $C_{10}H_{12}O_3PS_2N_3$ — colorless crystals or a brown, waxy solid; MW: 317.3, BP: decomposes, Sol: 0.003%, Fl.P: n/a, sp gr: 1.44. It is used as an insecticide and acaricide. It is hazardous to the respiratory system, central nervous system, cardiovascular system, and blood cholinesterase; and is toxic by inhalation, absorption, ingestion, and contact. Symptoms of exposure include miosis, aching eyes, blurred vision, lacrimation, rhinorrhea, headache, tight chest, wheezing, laryngeal spasm, salivation, cyanosis, anorexia, nausea, vomiting, diarrhea, sweating, twitching, paralysis, convulsions, low blood pressure, cardiac irregularities. OSHA exposure limit (TWA): 0.2 mg/m^3 [skin].

aziridine — *See* ethyleneimine.

azo dyes — a group of widely-used commercial dyes that carry the azo (–N=N–) group in the molecular structure and usually contain impurities of 20% or more; synthesized by diazotization or tetrazotization of aromatic monomine or aromatic diamine compounds with sodium nitrite in an HCl medium; some of the dyes are carcinogenic to animals.

azotemia — an excess of urea or other nitrogenous compounds in the blood.

azurophilic — the quality of being stained with azure or similar metachromatic dye.

A
B

B

B — *See* magnetic field.

babbit — an alloy of tin, antimony, copper, and lead; used for bearings.

bacillary dysentery — (*disease*) *See* shigellosis.

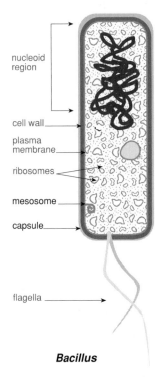

nucleoid region

cell wall

plasma membrane

ribosomes

mesosome

capsule

flagella

Bacillus

Bacillus — a genus of aerobic or facultatively anaerobic spore-forming rods, most of which are gram-positive and motile; includes some pathogenic species.

bacillus serous **food poisoning** — (*disease*) a gastrointestinal condition with symptoms including the sudden onset of nausea and vomiting in some people, and in others colic and diarrhea usually lasting no more than 24 hours. Incubation time is 1–6 hours where vomiting occurs, 6–16 hours where diarrhea is most prominent; caused by *Bacillus cerus*, an aerobic spore former which produces two enterotoxins, one heat stable causing vomiting, the other heat labile causing diarrhea; found regularly in Europe and occasionally in the United States. The reservoir of infection is an organism found in the soil and in raw, dried, and processed foods. It is transmitted by ingestion of food that has been kept at room temperature after cooking allowing the multiplication of the bacteria; especially prominent in vegetable and meat dishes; not communicable; susceptibility not known; controlled with proper cooking of foods and rapid cooling in refrigerators.

bacitracin — a group of antibacterial substances useful in a wide range of infections.

backend loader — a refuse truck that loads refuse from the rear and compacts it.

backfill — the material used to refill a ditch or other excavation, or the process of doing so.

backfire — *See* flashback.

backflow — the flow of water or other liquids, mixtures, or substances into the distribution pipes of a potable water supply from any source or sources other than the source of the potable water supply; back-siphonage is one form of backflow.

backflow connection — any arrangement of plumbing whereby backflow can occur.

backflow preventer — a device or means to prevent backflow.

backflow prevention device — any device, method, or type of construction to prevent backflow of water, liquids, mixtures, or substances into the distributing pipes of a potable supply of water from any source other than its intended source.

backflow valve — a mechanical device installed in a waste pipe to prevent the reversal of flow under conditions of back pressure.

background level — the amount of pollutant present in the ambient air due to natural sources such as marsh gases, pollen, volcanoes, and fire; may also refer to the amount of radiation or noise present, etc., so the term is used in a variety of situations, always as the constant or natural amount of whatever the pollutant is present.

background noise — all of the sources of interference in a system used for the production, detection, measurement, or recording of a signal, independent of the presence of a signal; noise coming from sources other than the noise source being monitored.

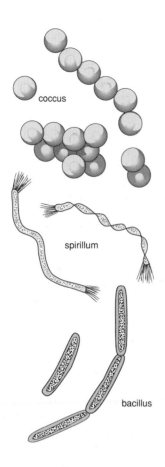

Bacteria

background radiation — low intensity ionizing radiation arising from radioactive material other than that being directly considered; may possibly be due to the presence of radioactive substances in other parts of the building; (*construction*) in the building materials, radon gas, and naturally occurring radiation; (*physics*) the low-level, natural radiation from cosmic rays and trace amounts of radioactive substances present in a given area.

backscattering — the scattering of radiation in a reverse direction from an irradiated substance.

back-siphonage — the backflow of used, contaminated, or polluted water from a plumbing fixture or vessel or other source in a potable water supply pipe as a result of negative pressure in the pipe.

bactericide — an agent that kills bacteria.

back-up — (*computer science*) duplication of a copy of a file or disk for safe keeping.

backwash — a reverse flow of water through a system of granular media to dislodge and remove solids that have accumulated on the bed.

backwashing — cleaning a filter or ion exchanger by reversing the flow of liquid and washing out the captured material.

backwash rate — the rate of flow in gallons per minute per square foot of filter surface area required for efficient filter cleaning.

bacteria — a group of microscopic, single-celled organisms without chlorophyll and lacking a distinct nuclear membrane.

bacterial food poisoning — an illness due to the consumption of food containing a toxin created by bacterial growth within the food.

bacterial growth curve — a four phase system, including a lagphase, lasting about two hours, a logarithmic or growth phase where growth is at a maximum rate; stationary or resting phase where bacteria die at the same rate as they are produced; (spores are produced in this phase) and a death phase, which may occur 18–24 hours after the resting phase.

bactericide — a substance that kills bacteria.

bacteriolysin — an antibacterial antibody that lysis bacterial cells.

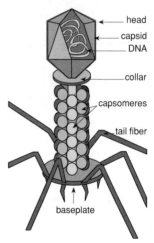

bacteriophage T4

Bacteriophage

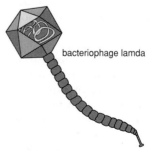

bacteriophage lamda

bacteriophage — one of several kinds of viruses that can destroy bacteria.

A
B

bacteriostat — a product that prevents bacteria from multiplying without actually killing the bacteria; ineffective on hard surfaces such as walls, floors, and countertops.

baffle — a deflector vane, guide, grid, grading, or similar device constructed or placed in air or gas flow systems, flowing water, or slurry systems to effect a more uniform distribution of velocities, to absorb energy, or to divert, guide, or agitate fluids, and check eddies.

baffle chamber — a settling chamber in which baffles regulate the velocity or direction of the combustion gases to promote the settling of fly ash or coarse particulate matter.

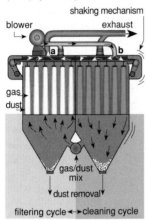

Bag Filter

bag filter — a device containing one or more fabric bags for recovering particles from dust-laden gas or air.

baghouse collector — a device in which dust-laden gas is filtered through the use of bags made of porous cloth.

bait — a food or other substance which will attract pests to a trap; a pesticide chemical that will destroy pests.

bake oven — a device that uses heat to dry or cure coatings.

baking soda — the common name for sodium bicarbonate; a mild alkali which may be helpful in removing acid soils.

balantidiasis — (*disease*) a protozoan infection of the colon producing diarrhea along with abdominal colic, nausea, and vomiting; stools may contain blood and mucous but little pus. Incubation time is not known but may be a few days; caused by *Balantidium coli*, a large ciliated protozoan; occurring worldwide but especially in waterborne outbreaks of disease where there are poor environmental water controls. The reservoir of infection is pigs and possibly other animals. It is transmitted by ingestion of cysts from feces of the infected host, from contaminated water, and occasionally from hands contaminated with feces or particularly contaminated food. It is communicable as long as the infection persists; debilitated people may have a serious or fatal case; controlled through proper disposal of hog feces, protection of public water, sewage, fly control, and good personal hygiene.

baler — a machine used to compress and bind solid waste or other materials into packages.

baling — a technique used to compact solid waste into blocks to reduce volume.

ballistic separator — a device that drops mixed materials having different physical characteristics onto a high-speed rotary impeller and hurls them off at different velocities because of their mass to separate collecting bins.

band application or treatment — the placing of a pesticide chemical on or in the soil in a narrow strip over or next to each row of plants in a field.

barium (soluble compounds) — $Ba(NO_3)_2$ — BaO — $BaCO_3$ — $BaCl_2$ — barium nitrate and barium chloride are white odorless solids; MW: 261.4/208.3, BP: decomposes/2840°F, Sol: 9/38%, Fl.P: N/A, sp gr: 3.24/3.86. They are used in the manufacture of pressed and blown glassware and flint and crown optical glass; in the manufacture of ceramic products; in the manufacture of magnets, vacuum tubes, cathodes, X-ray fluorescent screens, TV picture tubes, and dry cell vaporizers; in the manufacture of photographic papers, dyes, and chemicals; as pesticides, rodenticides, and disinfectants; in the manufacture of explosives, matches, and pyrotechnics; as catalysts, analytical reagents, and purifying agents; in the manufacture of pigments, paints, enamels, and printing inks; for treatment of textiles, leather, and rubber; as a smoke suppressant in diesel fuel. They are hazardous to the central nervous system, heart, respiratory system, skin, and eyes; and are toxic by inhalation, ingestion, and contact. Symptoms of exposure include upper respiratory irritation, gastroenteritis, muscle spasm, slow pulse, extrasystoles, hypokalemia, eye and skin irritation, skin burns. OSHA exposure limit (TWA): 0.5 mg/m³ [air].

barometer — an instrument used for measuring atmospheric pressure.

barometric pressure — the pressure at a given temperature and altitude due to the atmosphere; atmospheric pressure.

barometry — the science of pressure measurement.

barotrauma — an injury to the ear caused by a sudden alteration in barometric pressure.

barrel — a unit of liquid volume for petroleum equal to 42 United States gallons at 60°F.

barrier application or treatment — the use of a pesticide chemical or other agent to stop pests from entering a container, area, field, or building.

barrier substances — (*industrial safety*) creams and lotions applied to the skin in a thin film to act as a barrier to skin irritants; includes water-resistant, oil-resistant, film-forming, and specific-use substances.

bar screen — (*wastewater*) a filtering device that removes large solids at the beginning of the treatment process.

BART — *See* best available retrofit technology.

barye — *See* microbar.

basal application or treatment — the placing of an herbicide on stems or trunks of trees and brush just above the soil line.

basal metabolism — the activities required to maintain the body and to supply the energy necessary to support the basic life processes.

basalt aquifers — water stored within the layers or sheets of volcanic rock and unconsolidated sediments; found most often in areas where volcanic activity caused lava to flow across the surface.

**Basaltic Dome
(Shield Volcano)**

basaltic domes — mountain peaks consisting of a dark, dense, igneous rock of a lava flow or minor intrusion composed essentially of labradorite and pyroxene. Also known as shield volcano.

base — a substance which turns litmus blue and produces hydroxyl ions in water solution; it reacts with acids to form water and salts.

base exchange capacity — *See* cation-exchange capacity.

base temperature — an arbitrary reference temperature for determining liquid densities for adjusting the measured volume of a liquid quantity.

BASIC — *See* Beginner's All-Purpose Symbolic Instruction Code.

basic oxygen furnace — a furnace in which molten metal, scrap steel, and a flux are the charge.

basic premium — the premium arrived at by multiplying the rate by the number of $100 of payroll.

basin — a low-lying area in which water may accumulate as with rivers and their tributaries, streams, coastal waters, sounds, estuaries, bays, and lakes, as well as the lands drained by these waters.

Basle Nomina Anatomica (BNA) — a system of anatomic nomenclature adopted at the annual meeting of the German Anatomic Society in 1895.

basophil — a granulocytic white blood cell characterized by cytoplasmic granules that stain blue when exposed to a basic dye.

BAT — *See* best available technology.

batch fed incinerator — an incinerator that is periodically charged with solid waste; one charge is allowed to burn down or burn out before another is added.

batch processing — (*computer science*) the processing of a group of similar jobs on the computer without operator intervention.

batch sample — the collection of substances or products of the same category, configuration, or subgroup drawn from a batch and from which test samples are taken.

bather — any person using a pool, spa, or other water area and the adjoining deck for the purpose of therapy, relaxation, recreation, and other activities.

bathing place — a body of natural water, impounded or flowing, of such size in relation to the bathing load that the quality and quantity of water confined or flowing need be neither mechanically controlled for the purpose of purification nor contained in an impervious structure.

bathyal — of or relating to the water from 600–6,000 feet below the surface of the water.

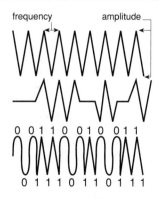

Baud Rate

baud rate — (*computer science*) a measure of the speed of data transmission between a computer and other devices; equivalent to bits per second.

Bedrock
and layers of soil

bauxite — the impure mixture of aluminum, oxides, and hydroxides.

BCF — *See* bioconcentration factor.

BCG vaccine (*bacillus Calmette*–Guerin vaccine) — a tuberculosis vaccine containing living avirulent bovine-strain tubercle bacilli. It offers some protection against tuberculosis but cannot be relied on for total control of the disease.

BCT — *See* best conventional pollutant control technology.

Be — *See* beryllium.

beam — unidirectional or approximately unidirectional flow of electromagnetic radiation or particles.

beam axis — a line from the source through the centers of the X-ray fields.

beam-limiting device — a piece of equipment used to restrict the dimensions of the X-ray field.

Bean Weevil (*Acanthoscelides obtectus*) — a short-snout beetle that feeds upon stored beans and peas; the adult is 1/8" long with reddish legs and a light olive-brown color mottled with darker brown and gray; a stored food product insect.

bearing capacity — the maximum load per unit area that a material can safely support before failing.

beats — periodic variations that result from the superposition of waves having different frequencies.

becquerel (Bq) — the International System of unit of activity equal to 1 transformation (disintegration) per second.

bed — elevated areas in fields where seedlings, cuttings, or crops have grown; (*institutions*) a unit of measurement used to determine size and/or capacity of health care institutions.

bedrock — the unweathered solid rock of the earth's crust overlaid by sand, gravel, or soil.

bedroom — (*sewage*) any room within a building that is used for sleeping purposes or could be converted into a room used for sleeping purposes such as a den or sewing room; is used as a means of calculating the amount of sewage coming from the home.

bed sore — *See* decubitus ulcer.

beef tapeworm disease — *See* taeniasis.

beetle — the common name give to the hard-shelled insect of the order *Coleoptera*.

Beginner's All-Purpose Symbolic Instruction Code — a simple, high-level computer programming language originally designed for inexperienced computer operators.

BEIs — *See* biological exposure indices.

bel — (*physics*) the amount of energy in the form of sound transmitted to one square inch of the ear.

belladonna (*Atropa belladonna*) — a perennial, poisonous herb that is the source of various alkaloids, atrophine, hyoscyamine; the leaf is used as an anticholinergic in the management of peptic ulcers and other gastrointestinal disorders. Also known as deadly nightshade.

belowground storage facility — a tank or other storage container whose base is located on a plant more than 6 inches below the surrounding surface.

benchmark test — (*computer science*) a test to evaluate the capabilities of a computer system in terms of the individual's needs.

bends — a condition resulting from a too-rapid decrease in atmospheric pressure, as when a deep-sea diver is brought too rapidly to the surface. Also known as caisson disease.

benefit — the sum of money provided in an insurance policy to be paid for certain types of loss under the terms of the policy.

benign — not recurrent or not tending to progress; of no danger to life or health.

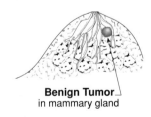

Benign Tumor
in mammary gland

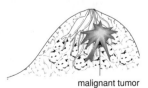

malignant tumor

benign tumor — a noncancerous or nonmalignant growth.

benthic — aquatic bottom-dwelling organisms including sponges, barnacles, muscles, oysters, insects, snails, certain clams, and worms.

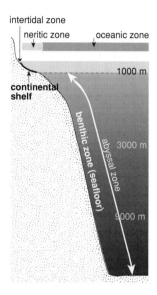

Benthic Zone

benthic zone — the area of the aggregate of organisms living on or at the bottom of a body of water.

bentonite — a soft, absorbent, swelling, and colloidal clay formed from the alteration of volcanic ash; used widely in industry as a bonding agent, sealant, thickener, filler, and as a liner in landfills; added to ordinary natural clay-bonded sands where extra strength is required.

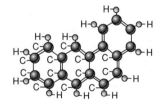

Benz (a) anthracene

benz {a} anthracene — $C_{18}H_{12}$ — one of the polycyclic aromatic hydrocarbon compounds formed when gasoline, garbage, or any animal or plant material burns, usually found in smoke and soot; MW: 228.29; BP: 435°C; sublimes, virtually insoluble in water, Fl.P: not known; density 1.274 (20°C); it has some

use as a research chemical; coal tar pitch is removed from the tar still as a residue; used as a binder for electrodes in the aluminum reproduction process, as an adhesive in membrane groups, for wood preservation, used in the clinical treatment of skin disorders such as eczema, dermatitis, and psoriasis. It is a possible cause of cancer through inhalation when the chemical is part of a mixture of PAHs. OSHA exposure limit standard in air 0.2 mg/m³ for an 8 hour time-weighted average.

benzene — C_6H_6 — a colorless, volatile, flammable, toxic, aromatic liquid hydrocarbon; MW: 78.11; BP: 80.1°C; Sol. in water 1,780 mg/L; Fl.P: –11°C. Benzene has an acute narcotic effect on the individual and also acts as a local irritant to the skin and mucous membranes. It may injure the blood-forming tissues, even at low concentrations. Benzene is considered to be a hazardous air pollutant. Benzene recovered from petroleum and coal is primarily used as an intermediate in the manufacture of other chemicals and in products. It is used in the production of ethylbenzene, cumene, and cyclohexane; ethylbenzene is an intermediate and a synthesis of styrene which is used in plastics and elastomers; cumene is used to produce phenol and acetone; phenol is used in the manufacture of phenolic resins and nylon intermediates; acetone is used as a solvent and in the manufacture of pharmaceuticals; cyclohexane is used to make nylon resins; benzene is also used to manufacture nitrobenzene which is used in the production of aniline, urethanes, linear alkylbenzene sulphonates, chlorobenzene, and maleic anhydride, a solvent and a reactant; benzene is a component of gasoline since it naturally occurs in crude oil and is a by-product of the oil refinery process; it is important for unleaded gasoline because of its anti-knock characteristics; 2% of the benzene produced is used as a solvent in industrial plants, rubber cements, adhesives, paint removers, artificial leather, and rubber goods; it has also been used in pesticides, shoe manufacturing, and the rotogravure printing industries; it may be detected in carpet glue, textured carpet liquid detergent, and furniture polish; inhalation and dermal exposure can occur through occupational or environmental exposure; however, the main route of entry to the body is through inhalation; in acute exposures, benzene can cause death in 5–10 minutes at 20,000 ppm; death is caused by asphyxiation, respiratory arrest, central nervous system depression, or cardiac collapse; non-fatal symptoms may include headaches, nausea, staggering gait, paralysis, convulsions and unconsciousness; where death occurs, cyanosis, hemolysis, and hemorrhage of the organs occur; it may cause adverse immunological effects in humans; it alters humoral immunity, that is, the ability to produce changes in blood levels of antibodies; it affects cellular immunity, that is changes in circulating leukocytes and lymphocytes; it can cross the human placenta and is present in the cord blood in amounts equal to or greater than those in maternal blood; it may impair fertility in women if exposed to high levels; high concentrations of benzene may produce chromosome abnormalities; a cause-effect relationship exists between benzene and leukemia; it can cause intense toxic gastritis and later pyloric stenosis; skin contact may cause swelling, edema, and central nervous system toxicity. Also known as benzol. OSHA exposure limits PEL TWA 1 ppm, action level (8-hour average) 0.5 ppm, STEL (15-minute average) 5 ppm.

benzene hexachloride (BHC) — $C_6H_6Cl_6$ — a crystalline solid not soluble in water but soluble in varying degrees in a wide variety of common solvents; used as an insecticide. *See also* hexachlorocyclohexane.

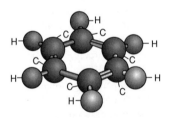

Benzene Ring

benzene ring — an arrangement of atoms in the shape of a hexagon where each of the points are made up of carbon atoms, and each carbon atom may be linked to other structural arrangements of atoms or to a single chemical.

benzidine — $NH_2C_6H_4C_6H_4NH_2$ — a grayish-yellow, reddish-gray, or white crystalline powder; MW: 184.3, BP: 752°F, Sol: (at 54°F): 0.04%, sp gr: 1.25. It is used in the manufacture of azo dyes; as a hardener in the rubber industry; in research and analytical laboratories for detection of blood and inorganics. It is hazardous to the bladder, kidneys, liver, skin, and blood; and is toxic by inhalation, absorption, ingestion, and contact. Symptoms of ex-

posure include secondary anemia from hemolysis, acute cystitis, acute liver disorders, dermatitis, painful and irregular urination; carcinogenic. OSHA exposure limit (TWA): regulated use.

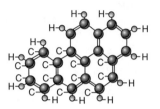

Benzo (a) pyrene

benzo {a} pyrene — $C_{20}H_{12}$ — one of the polycyclic aromatic hydrocarbons formed when any gasoline, garbage, or any animal or plant material burns; usually in smoke and soot; pale yellow fluoresces yellow-green in ultraviolet light; MW: 252.3; BP: 310–312°C at 10 millimeters of mercury, 495°C at 760 millimeters of mercury; soluble in water 3.8×10^{-6} g/L; FL.P: not known. Some use as a research chemical; coal tar pitch is removed from the tar still as a residue; primarily used as a binder for electrodes in the aluminum reduction process; used to bind the carbon electrodes in reduction pots and as pitch which is an adhesive in membrane roots; creosote is produced and is sold for wood preservation. People may be exposed through environmental sources such as air, water, soil, and from cigarette smoke and cooked food; studies in experimental animals have shown the ability of B{a}P to induce respiratory tract tumors following long-term inhalation exposure; short-term or intermediate exposure to very high levels results in the death of experimental animals fed B{a}P in the diet; there appears to be an induction of cancer to animals and possibly humans of long-term dermal exposure; OSHA exposure limit (TWA) 0.2 mg/m³ [air].

benzo {b} fluoranthene — $C_{20}H_{12}$ — a colorless substance; odor: not known; MW: 252.32; BP: not known; Sol: in water 14 micrograms/L; Fl.P: not known; one of the polycyclic aromatic compounds formed when gasoline, garbage, or any animal or plant material burns; usually found in smoke and soot; found in coal tar pitch used industrially to join electrical parts together and is also used to preserve wood; no information is available on problems in the various systems of the body; however the chemical, when part of a mixture of PAHs, has some relationship to cancer. OSHA exposure limits 0.2 mg/m³ [air].

2,3-benzofuran — C_8H_6O — a colorless, sweet-smelling oily liquid which does not mix with water; formed when coal is processed to make coal oil; MW: 118.14; BP: 175°C; Sol: insoluble; Fl.P: not known. No studies available on health effects due to human exposure, but some problems related in laboratory mice and rats such as mineralization of the pulmonary artery, chronic inflammation of the fore-stomach, liver damage, kidney damage, and cancer in rats and mice. OSHA exposure limits in food 200 ppm.

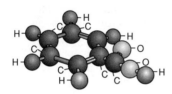

Benzoic Acid

benzoic acid — $C_7H_6O_2$ — white powder or crystals with a faint, pleasant, slightly aromatic odor; MW: 122.12, BP: 480°F, Sol: 0.290 g/100 ml, Fl.P: 250°F, sp gr: 1.27. It is used as a preservative in foods, juices, fats, and oils; in the manufacture of plasticizers, dyes, pharmaceuticals, and perfumes; as a laboratory reagent; and as an ingredient in antiseptic ointments to treat fungal infections. It is hazardous to the respiratory system, skin, and digestive system; and is toxic by inhalation, skin contact, or ingestion. Symptoms of exposure include irritation of the nose, throat, and upper respiratory system, reddening, swelling, and burning of skin, asthmatic sneezing, anaphylactic shock, violent cough, chest constriction, convulsions, collapse, and death. OSHA exposure limit: not established.

benzol — *See* benzene.

benzoyl peroxide — $(C_6H_5CO)_2O_2$ — colorless to white crystals or a granular powder with a faint benzaldehyde-like odor; MW: 242.2, Sol: <1%, Sp.Gr (77°F): 1.33. It is used in vulcanization of natural and synthetic rubber; as a bleaching agent for flour, cheese, fats, oils, and waxes; as an ingredient in skin creams for burns, dermatitis, poisoning, and external wounds; in the manufacture of fast-drying printing inks. It is hazardous to the skin, respiratory system, and

eyes; and is toxic by inhalation, ingestion, and contact. Symptoms of exposure include skin, eye, and mucous membrane irritation, sensitization dermatitis. OSHA exposure limit (TWA): 5 mg/m³ [air].

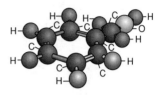

Benzyl Alcohol

benzyl alcohol — C_7H_8O — a colorless liquid with a pleasant, fruity odor; MW: 108.13, BP: 402.8°F, Sol: 3.5 g/100 ml, Fl.P: 200°F (closed cup), sp gr: 1.04535. It is used as a solvent for lacquers, cosmetics, and inks; used in the manufacture of soaps, perfumes, pharmaceuticals, and dyes. It is hazardous to the respiratory and digestive systems. It is toxic by inhalation or ingestion. Symptoms of exposure include mild depression of the central nervous system, headache, nausea, dizziness, drowsiness, incoordination, and confusion; may be aspirated into the lungs. OSHA exposure limit: not established.

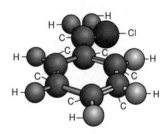

Benzyl Chloride

benzyl chloride — $C_6H_5CH_2Cl$ — a colorless to slightly yellow liquid with a pungent, aromatic odor; MW: 126.6, BP: 354°F, Sol: 0.05%, Fl.P: 153°F, sp gr: 1.10. It is used in rubber adhesives and TV tubes; as a raw material for pickling inhibitors, gasoline gum inhibitors, and synthetic tanning agents; in the processing of starch and the preparation of textile fibers. It is hazardous to the eyes, respiratory system, and skin; and is toxic by inhalation, ingestion, and contact. Symptoms of exposure include eye and nose irritation, weakness, irritability, headache, skin eruption, pulmonary edema. OSHA exposure limit (TWA): 1 ppm [air] or 5 mg/m³.

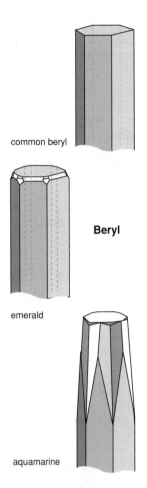

common beryl

Beryl

emerald

aquamarine

beryl — a silicate of beryllium and aluminum.

berylliosis — (*disease*) beryllium poisoning usually involving the lungs and less often the skin, subcutaneous tissues, lymph nodes, liver, and other structures causing a toxic or allergic reaction.

beryllium — Be — a rare, light-weight, gray metal. It is used in the manufacture of beryllium-copper alloys used in industry and science. It is hazardous to the respiratory system, lungs, liver, spleen, kidneys, and lymphatic system; and is toxic by inhalation. Symptoms of exposure include shortness of breath, general weakness, weight loss, acute pneumonitis, immunological disorders, and death. OSHA exposure limit (TWA): 0.002mg/m^3 [air].

best available retrofit technology (BART) — emission limitations imposed on existing sources of air pollutants to protect visibility in certain areas where visibility is a particularly valuable resource, such as in national parks.

best available technology (BAT) — level of control is described as "best of the best" technology in use and is to include controls on toxic pollutants.

best conventional pollutant control technology (BCT) — (*wastewater*) technology-based effluent limitations for conventional pollutants which direct dischargers had to meet as of July 1, 1984, pursuant to the Federal Water Pollution Control Act.

best practical control technology currently available (BPT) — technology-based effluent limitations direct dischargers had to meet as of July 1, 1977, pursuant to the Federal Water Pollution Control Act.

beta-chloroprene — $CH_2=CClCH=CH_2$ — a colorless liquid with a pungent, ether-like odor; MW: 88.5, BP: 139°F, Sol: slight, Fl.P: –4°F, sp gr: 0.96. It is used in the manufacture of a variety of neoprene elastomers; in the manufacture of neoprene. It is hazardous to the respiratory system, skin, and eyes; and is toxic by inhalation, absorption, ingestion, and contact. Symptoms of exposure include eye and respiratory irritation, nervousness, irritability, dermatitis, alopecia; carcinogenic. OSHA exposure limit (TWA): 10 ppm [skin] or 35 mg/m^3.

beta-naphthylamine — $C_{10}H_7NH_2$ — odorless, white to red crystals with a faint aromatic odor; darkens in air to a reddish-purple color; MW: 143.2, BP: 583°F, Sol: miscible in hot water, Fl.P: 315°F, sp.gr (208°F): 1.06. It is used in the manufacture of dyes, acids, and rubber. It is hazardous to the bladder and skin; and is toxic by inhalation, absorption, ingestion, and contact. Symptoms of exposure include dermatitis, hemorrhagic cystitis, dyspnea, ataxia, methemoglobinemia, dysuria; carcinogenic. OSHA exposure limit (TWA): regulated use.

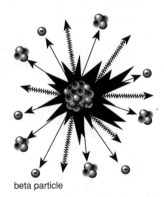

beta particle

Beta Particle
Beta Ray

beta particle — a charged particle emitted from the nucleus of an atom during radioactive decay having the mass and charge equal in magnitude to that of the electron.

beta-propiolactone — $C_3H_4O_2$ — a colorless liquid with a slightly sweet odor; MW: 72.1,

BP: 323°F (decomposes), Sol: 37%, Fl.P: 165°F, sp gr: 1.15. It is used in the manufacture of acrylic acids and esters; in the sterilization of blood plasma, tissue grafts, and surgical instruments; in research and laboratory facilities. It is hazardous to the kidneys, lungs, skin, and eyes; and is toxic by inhalation, absorption, ingestion, and contact. Symptoms of exposure include corneal opacity, frequent urination, dysuria, skin irritation, blistering, and burns; carcinogenic. OSHA exposure limit (TWA): regulated use.

beta ray — a stream of high-speed electrons or positrons of nuclear origin more penetrating and less ionizing than an alpha ray.

BeV — notation for one billion electronvolts (10^9).

beverage — any liquid intended for human consumption.

BHC — *See* benzene hexachloride.

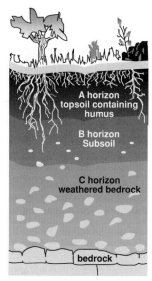

B Horizon
and other layers of soil

B horizon — (*soils*) the horizon immediately beneath the A horizon characterized by a higher colloid clay or humus content, or by a darker or brighter color than the soil immediately above or below, the color usually being associated with the colloidal materials; the colloids may be of alluvial origin as clay or humus, they may have been formed in place clays including sesquioxide, or they may have been derived from a texturally-layered parent material.

bias — (*statistics*) a systematic error that causes a population value to be under-estimated or over-estimated; may be due to improper sampling or to improper measurement of the variable under study; a preference or inclination that inhibits impartial judgment.

biennial — a plant that completes its growth cycle in two growing seasons.

bile — a bitter-tasting, greenish-yellow alkaline liquid substance secreted by the liver, stored in the gall bladder, and which flows into the small intestine where it aids in the digestion and absorption of fat.

bill of lading (B/L) — the written order from a shipper to a carrier to move goods from one place to another; when available, this is the best source of shipping dates, origin, and name of shipper.

binary — composed of or characterized by two elements or parts.

binary arithmetic — the mathematics of calculating in powers of two.

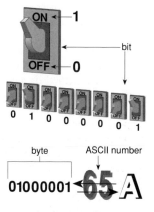

Binary Coded Decimal

binary coded decimal (BCD) — (*computer science*) a system of number representation in which each digit of a decimal number is represented by a binary number.

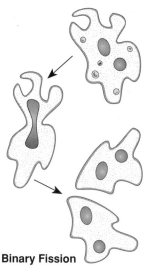

Binary Fission

binary fission — the division of cells into two approximately equal parts.

binary system — a number system with a base of two.

binder — a resin or other cement-like material, other than clay, that is added to foundry sand to provide mechanical strength or ensure uniform consistency.

binding energy — the energy represented by the differences in mass between the sum of component parts and the actual mass of the nucleus.

bioaccumulation — the uptake, retention, and concentration of environmental substances by an organism from its environment rather than uptake from food; (*ecology*) the accumulation of toxic chemicals in living things through the consumption of food or water.

bioassay — the determination of the concentration or dose of a given material necessary to affect the growth of a test organism under stated conditions; determination of the active power of a chemical by comparing its effects on a live animal or an isolated organ to a reference standard; (*pesticide*) the method of measuring the amount of pesticide chemical present in a test material by the action of a test plant or an animal's exposure to it.

bioaugmentation — the process of using specifically cultured microorganisms to decontaminate polluted systems speeding up the naturally occurring biodegradation processes by controlling the nature of the biomass.

bioavailability — the degree to which a chemical or other substance becomes available to a target tissue after being taken into the body.

biochemical oxygen demand (BOD) — the amount of oxygen reported in milligrams per liter that would be consumed by all the organics oxidized by bacteria and protozoa. The BOD is useful in predicting how low the levels of oxygen will go in a stream or river when organic wastes are oxidized in one liter of polluted water.

biochemistry — the study of the chemistry of living organisms and other chemical constituents and vital processes.

biocidal — destructive to living organisms.

biocide — a substance that is destructive to many different organisms. Also known as pesticide.

bioconcentration factor (BCF) — the quotient of the concentration of a chemical in aquatic organisms at a specific time or during a discrete time period of exposure, divided by the concentration in the surrounding water at the same time or during the same period.

bioconversion — a resource recovery method that uses the biological processes of living organisms to transform organic waste into useable material such as humus and compost.

biodegradability — the characteristic of organic matter to be broken down by microorganisms.

biodegradable — susceptible to being degraded by biological processes.

biodegrade — to decompose, especially by bacterial action.

bioengineering — the application of engineering principles to biological or medical science.

bioethics — a discipline dealing with the ethical applications of biological research and applications, especially in medicine.

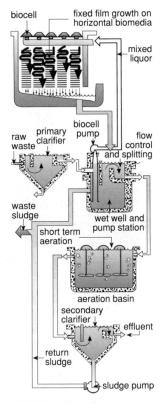

Activated Biofilter Process

biofiltration — a process that uses recirculation and a high rate of application to a shallow trickling filter to remove pollutants from effluent in a sewage treatment process.

biohazard area — any area in which work has been or is being performed with biohazardous agents or materials.

biohazard control — equipment and procedures utilized to prevent or minimize human and environmental exposure to biohazardous agents or materials.

biohazards — biological substances that constitute a human health or environmental threat. These hazards include but are not limited to infectious and parasitic agents; noninfectious microorganisms such as some fungi, yeast, algae, plants, and plant products; and animals and animal products which are either infectious microorganisms, toxic biological substances, biological allergens, or any combination of the three.

biolarvicide — a delta endotoxin produced by the bacterium *Bacillus thuningiensis* subspecies *israelensis* which, when ingested by mosquito larvae, reacts with stomach secretions and causes gut paralysis and disruption of the ionic regulation capacity of the gut epithelium leading to death of the larvae.

biological — a medicinal preparation made from living organisms and their products including serums, antitoxins, and vaccines.

Biological Abstracts — a bibliography of available journals of biology from 1926 to the present.

Biological and Agricultural Index — an index of available journal articles of biology and agriculture from 1916 to the present.

biological contaminants — living organisms or agents derived by viruses, bacteria, fungi, and mammal and bird antigens that can be inhaled to cause many health effects including allergic reactions, respiratory disorders, hypersensitivity diseases, and infectious diseases.

biological control — natural or applied methods using natural predators, parasites, or diseases to control populations of pest organisms.

biological diversity — genetic and ecological differences in species.

biological exposure indices (BEIs) — reference values intended as guidelines for the evaluation of potential health hazards in the practice of industrial hygiene.

biological half-life — the time required for the body to eliminate one half of an administered dose of any substance during normal processes of elimination. This time is about the same for both stable substances and radioactive isotopes. Also known as effective half-life.

biological hazardous waste — waste containing biohazardous materials.

biological indicator — a vial containing bacterial spores used for determination of sterilization.

biological magnification — *See* biomagnification.

biological monitoring — a technique for measuring the presence of a chemical or its metabolites in tissues or excreta, or for measuring pathological effects of the toxin on the person.

biological oxidation — the process by which microorganisms decompose complex organic materials as it occurs in activated sludge processes and self-purification of streams.

biological oxygen demand (BOD) — a standard test to determine the amount of dissolved oxygen utilized by organic material over a five-day period.

biological residue — any substance, including metabolites, remaining in livestock at the time of slaughter or in any of its tissues after slaughter as the result of treatment or exposure to a pesticide, organic or inorganic compound, hormone, hormone-like substance, growth promoter, antibiotic, anthelmintic, tranquilizer, or other therapeutic or prophylactic agent.

biological waste — waste derived from living organisms.

biologic dose — a measure of the concentration of a substance and its biologic effect.

biology — a branch of the natural sciences dealing with the study of living organisms and their vital processes.

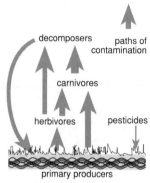

Biomagnification

biomagnification — the process by which impurities found in water are concentrated in lower forms of life and reconcentrated substantially during their movement through the food chain (e.g., whereas 0.001 micrograms of mercury might be found in the surrounding water, 30–50 micrograms of mercury may be found in the fish which had bioconcentrated the mercury during their feeding process). Also known as biological magnification.

biomarker — the measure of an individual's internal exposure or dose from a chemical. Also known as biomonitoring.

biomass — plant materials and animal waste used as a direct source of fuel. These wastes include forest products and residues, animal manure, sewage, municipal waste, agricultural crops, residues, and aquatic plants; the total mass or weight of living organisms expressed in terms of a given area of volume of the habitat. Also known as biomass fuel.

biomass fuel — *See* biomass.

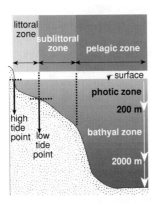

Marine Biome Strata

biome — a major ecological community containing distinctive plant and animal life in a given climate usually extending over a large region of the earth; examples are tundras and tropical rain forests.

biomechanics — the discipline dedicated to the study of the living body as a structure which can function properly only within the confines of both the laws of Newtonian Mechanics as well as the biological laws of life, including muscular activity such as those used in moving or exercise; (*ergonomics*) the application of mechanical laws to human structures.

biomedical — relating to or involving biological, medical, and physical science.

biomedicine — a branch of medical science concerned especially with the capacity of human beings to survive and function in abnormally stressing environments and with the protective modification of these environments.

biometeorology — the study of the relationship between living beings and atmospheric conditions.

biometrics — *See* biometry.

biometry — the statistical analysis of biological observations and phenomena. Also known as biometrics.

biomonitoring — a means of surveillance of water quality by observing the biota from the field and laboratory standpoints; (*medicine*) *See* biomarker.

bionics — the science concerned with the application of data about the functioning of biological systems to the solution of engineering problems.

biophysics — the hybrid branch of science concerned with the application of physical principles and methods to biological problems.

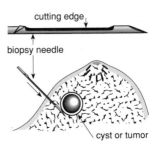

Breast Biopsy
in a closed operation

biopsy — the removal and examination of tissues, cells, or fluids from the living body for the purpose of diagnosis.

biorational control agents — *See* insect growth regulator.

biorhythm — an inherent rhythm that appears to control or initiate various biological processes.

bioscience — the study of biology wherein all of the applicable sciences (physics, chemistry, and math) are applied.

bioscrubber technology — a system used to digest hazardous organic emissions from soil, water, and air decontamination processes.

Biosis Previews — a database covering 1969 to the present and including biological abstracts, biological abstract reports, reviews, and meetings, as well as coverage of environmental biology.

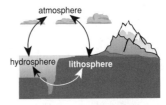

Structure of the Biosphere

biosphere — the zone of air, land, and water at the surface of the earth that is occupied by actively metabolizing matter and receives its supply of energy from an external source, which is ultimately the sun. However, spores may be found in areas outside the biosphere that are too dry, too cold, or too hot to support organisms that metabolize.

biostatistics — statistical processes and methods applied to the analysis of biological data.

biosynthesis — the production of a chemical compound and the organization of organic molecules by living organisms.

biota — animal and plant life characterizing a given area.

biotechnology — the use of genetic engineering to improve the quantity or quality of medical products, agricultural products, chemicals, and other products; applied biological science.

biotelemetry — the remote detection and measurement of a condition, activity, or function relating to the behavior and physiology of living organisms.

biotic — of or pertaining to the living environment of a community, made up of autotrophs, or producer organisms that construct organic substances, and heterotrophs, or consumer or reducer organisms that destroy organic substances.

biotic community — all living organisms comprising an ecosystem.

biotic factors — those factors, such as weather and population, which affect the living organisms in an environment.

biotic index — a numerical index using various aquatic organisms to determine their degree of tolerance to differing water conditions.

biotransformation — the transformation of chemical compounds within a living system.

biotransfusion — a collective term for the various reactions such as in the soil when sewage is broken down by bacteria that occur as a result of metabolism by organisms.

biotype — a group of organisms sharing a specified genotype.

biphasic — possessing both a sporophyte and a gametophyte generation in the life cycle; changing from positive to negative.

bipolar — having or involving the use of two poles.

birth rate — *See* natality.

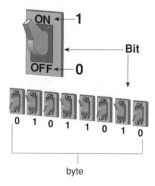

byte

bit — the smallest unit of information that can be stored and processed in a computer; equivalent to the result of a choice between two alternatives, that is yes/no, true/false, or on/off.

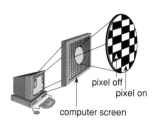

Bit Map

bit map — (*geographic information system*) a pattern of bits on the grid stored in memory and used to generate an image on a raster scan display.

bit plane — (*geographic information system*) a grided memory in a graphics device used for storing information for display.

bits per inch (BPI) — (*computer science*) the density of bits recorded on a magnetic tape.

bitumen — any of various mixtures of hydrocarbons (as tar) often together with their nonmetallic derivatives occurring naturally or obtained as residue after heat-refining naturally-occurring substances such as petroleum.

bituminous coal — a dark brown to black soft coal with high volatile gas and ash content.

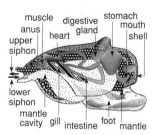

Bivalve (Clam)

bivalve — the common name for a number of bilaterally symmetrical animals, including mollusks; crustaceans, possessing a two-part valve or shell.

B/L — *See* bill of lading.

black death — *See* plague.

black lead — *See* graphite.

black light — ultraviolet light radiation, between 3,000 and 4,000 angstrom units or 0.3–0.4 micrometers, which falls on fluorescent materials causing them to emit visible light.

black lung disease — an inflammation and fibrosis caused by accumulations and subsequent calcification of coal dust in the lungs or airways resulting in permanent reduced lung capacity. Also known as pneumoconiosis.

blackwater — liquid and solid human body waste and the carriage waters generated through toilet usage.

blanching — an operation in which raw food material is immersed in hot water or exposed to live steam; the product shrinks and the respiratory gases contained in the plant cells are expelled; release of gas prevents strain on the container during heat-processing and favors development of a higher vacuum in the finished product; also serves as an added cleaning measure and may remove raw flavors from foods.

blanket application or treatment — *See* broadcast application or treatment.

coke, iron ore, limestone (charge)

exhaust gases (CO, CO_2, N_2)

exhaust gases

(oxygen-enriched air) blast nozzle

slag outlet

slag — iron — pig iron outlet

Blast Furnace

blast furnace — a furnace in which the cooked fuel is ignited by a blast of hot, compressed air to increase the combustion rate; the heat is used to reduce iron oxide ores to cast iron or pig iron; the raw materials are iron ore, coke, and limestone.

blast gate — a sliding metal damper in a duct usually used to regulate the flow of forced air.

bleach — a chlorine product that cleans, whitens, brightens, and remove stains from fabrics and other surfaces.

bleeders — (*food*) openings used to remove air that enters retorts and steam chambers; promotes circulation of steam in such retorts and steam chambers; may serve as a means of removing condensation.

blende — *See* sphalerite.

blepharism — spasm of the eyelids causing continuous, involuntary blinking.

blight — a general term for the sudden and wilting death of a plant or a plant part caused by fungi, bacteria, and viruses.

blood — the fluid that circulates through the heart, arteries, capillaries, and veins. It is the body's chief means of transport of oxygen from the lungs to the body tissues, and of carbon dioxide from the tissues to the lungs. It transports nutritive substances and metabolites to the tissues and removes waste products from the kidneys and other organs of excretion. Blood is composed of fluid plasma, suspended formed elements, blood cells, and platelets.

blood bank — a place where whole blood or plasma is typed, processed, and stored for future use.

blood count — a count of the number of corpuscles per cubic millimeter of blood.

blood fluke — *See* schistosome.

blood poisoning — a common, non-medical term referring to the presence of infecting agents, such as bacteria or toxins, in the blood stream. *See also* septicemia.

blood pressure — the pressure of the blood against the walls of the blood vessels.

blood urea nitrogen — the urea concentration of serum or plasma; an important indicator of renal function.

bloom — (*phytoplankton*) a large and sometimes rapid growth of phytoplankton often associated with hypoxic conditions giving a distinct color to the water.

blowdown — the discharge of water from the boiler or cooling tower by pressure to dispose of accumulated salts.

blower — a fan used to force air or gas under pressure.

blowout — a sudden, violent escape of gas and oil from an oil well when high pressure gas is encountered.

BNA — *See* Basle Nomina Anatomica.

Board of Certified Safety Professionals — a professional board establishing minimum academic and experience attainments needed to qualify as a safety professional; issues certificates to qualified individuals.

BOD — *See* biological oxygen demand.

bodyburden — the existing level of pollutant or radioactivity an individual has been subjected to and the potential level of severity and seriousness of the effect of this pollutant; the amount of radioactive material in the body at a given time.

body coat — (*swimming pools*) diatomaceous earth which builds up on a filter element during the course of water running through a filter to help maintain filter porosity.

body feed — (*swimming pools*) diatomaceous earth fed constantly or intermittently during the time water is running through a filter to produce a body coat.

body fluids — fluids produced in the body such as semen, blood, vaginal secretions, and breast milk.

body scanner — a diagnostic, radiological instrument technique used for diagnosis of a patient.

BOF — *See* basic oxygen furnace.

bog — a permanently wet, spongy land, which is usually poorly drained, highly acidic, and rich in plant residue. Also known as moor; quagmire.

boil — a painful nodule containing a central core and formed on the skin by inflammation of the dermis and subcutaneous tissue. It may develop wherever friction, irritation, or a scratch or break in the skin occurs and allows bacteria to penetrate the outer layer of the skin. Most boils are caused by *Staphylococcus aureus*. Also known as a furuncle.

boiler feedwater — water provided to a boiler for conversion to steam and the steam generation process.

boiling — movement which occurs in any liquid when pressure produced by the vapor within the liquid equals the pressure on its surface. The lower the pressure, the lower the temperature needed to bring a liquid to a boil. Each pure liquid has its own definite boiling point.

boiling point (BP) — the temperature at which the vapor pressure of a liquid equals the atmospheric pressure.

boiling water reactor (BWR) — (*energy*) a reactor in which water, used as both a coolant and a moderator, is allowed to boil in the core, resulting in steam which can be used to drive a turbine.

BOM — acronym for the Bureau of Mines, Department of Interior.

bond — (*law-food*) a sum of money or other thing of value held by the court pending proper execution of a consent decree of condemnation in a seizure in which the goods are being reconditioned by a claimant or during import reconditioning, or while goods are being detained pending decision as to admission. (*chemistry*) *See* valence.

bonding — the interconnecting of two objects by means of a clamp and bare wire to equalize the electrical potential between the objects to prevent a static discharge when transferring a flammable liquid from one container to another.

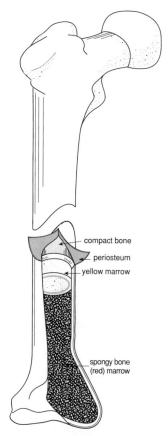

Bone Marrow

bone marrow — the soft material which fills the cavities in most bones of vertebrates and manufactures most of the formed elements of the blood.

bone seeker — any compound or ion which tends to migrate preferentially into bone when introduced in the body.

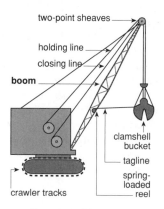

Boom on Clamshell

boom — any heavy mechanical support beam that is hinged at one end and carries a weight-hooking device at the other.

boot stage — the appearance of a grass or a cereal seed head emerging from the sheaf covering it.

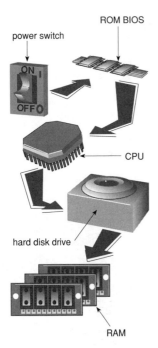

Boot Up

boot-up — (*computer science*) to start up a computer system.

borax — $Na_2B_4O_7 \cdot 10H_2O$ — a white crystalline, mildly alkaline water soluble salt (sodium borate) used as a laundry additive; provides moderate alkalinity buffering.

boron (B) — a brown, amorphous non-metallic powder that can be fused to a brittle mass; atomic number: 5; atomic weight: 10.811. It is used as borox, a cleaning material.

boron oxide — B_2O_3 — colorless, semi-transparent lumps or hard, white, odorless crystals; MW: 69.6, BP: 3380°F, Sol: 3%, sp gr: 2.46. It is used in the manufacture of metal borates and in the preparation of fluxes; in the manufacture of glass for heat-resistance; as a herbicide; in the production of surface coatings as fire resistant additives in enamel and paints. It is hazardous to the skin and eyes; and is toxic by inhalation. Symptoms of exposure include nasal irritation, conjunctivitis, erythema. OSHA exposure limit (TWA): 10 mg/m³ [air].

boron trifluoride — BF_3 — a colorless gas with a pungent, suffocating odor; forms dense white fumes in moist air; MW: 67.8, BP: –148°F, Sol: soluble in water. It is used as an acid catalyst for alkylation of aromatic compounds; in polymer technology; in the synthesis of other boron-containing organic and inorganic compounds; in the purification of hydrocarbons; in nuclear technology for separation of boron isotopes; as a filling gas for neutron counters; in metallurgy as a flux and antioxidant; as a flame-coloring agent for liquefied petroleum gas. It is hazardous to the respiratory system, kidneys, eyes, and skin; and is toxic by inhalation and contact. Symptoms of exposure include nasal irritation, epistaxis, eye and skin burns; pneumonia, kidney damage, and epistaxis in animals. OSHA exposure limit (TWA): ceiling 1 ppm [air] or 3 mg/m³.

botanical pesticide — a pesticide chemical produced by a plant in nature.

bottled drinking water — water, including mineral water, sealed in bottles, packages, or other containers and offered for sale for human consumption.

bottom ash — the ash that drops out of the furnace gas stream in the furnace and economizer sections.

botulism — (*disease*) a severe intoxication resulting from ingestion of toxin contained in contaminated food. Symptoms of exposure usually relate to the nervous system, including blurred or double vision, also dry mouth, sore throat, paralysis, vomiting, diarrhea, and possible respiratory failure and death. Incubation time is 12–36 hours for neurological symptoms to appear; caused by toxins produced by *Clostridium botulinum* types A, B, E, and occasionally F and G; found worldwide. Reservoirs of infection include soil, sediment in the water, and the intestinal tract of animals, including fish. It is transmitted by ingestion of food containing the toxin; not communicable; general susceptibility. It is controlled by the careful processing of all foods and disposal of foods which appear to have off-color, off-odors, or where gas has been produced.

botulism antitoxin — an equine antitoxin administered intravenously and used against the toxins produced by the type A, B, E *Clostridium botulinum*.

Boyle's law — the law stating that the volume of a confined gas at a constant temperature is inversely proportional to the pressure to which it is subjected; true only for ideal gases. Also known as Mariotte's law.

boxboard — paperboard used to manufacture boxes and cartons.

BP — *See* boiling point.

BPI — *See* bits per inch.

BPT — *See* best practical control technology currently available.

Bq — *See* becquerel.

brackish water — a mixture of fresh and salt water; undrinkable.

bradycardia — (*physiology*) abnormal slowness of the heart rate.

branching — (*physics*) the occurrence of competing decay processes in the disintegration of a particulate radionuclide; (*botany*) a shoot or secondary stem on the trunk or on the limb of a tree.

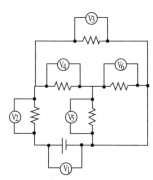

Branching Circuit
V_x =Voltmeter

branching circuit — a portion of a wiring system in the interior of a structure that extends from a final overload protection device to a plug receptacle, motor, or heater.

brass — an alloy of copper and zinc which may contain a small portion of lead.

braze — to solder with a relatively infusible alloy such as brass.

breakpoint chlorination — chlorine added to water containing ammonia that reacts with the ammonia to form chloramines; as additional chlorine is added, the total chlorine residual continues to rise until the concentration reaches

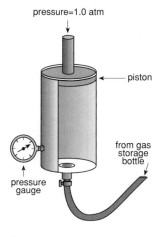

pressure=1.0 atm

piston

from gas storage bottle

pressure gauge

Boyle's Law
apparatus for investigation

a point that forces the reaction with ammonia to go rapidly to completion; the compounds of nitrogen and chlorine are released from the water and the apparent residual chlorine decreases; the breakpoint is that point at which the residual suddenly drops off; passed the breakpoint, all combined chlorine compounds disappear, eye-irritation potential and chlorine odors disappear and the chlorine that remains in the water is all in the free state.

breakthrough — the localized penetration of unfiltered materials through a gas or vapor air-purifying element.

breakthrough time — the elapsed time between the initial contact of a chemical with the outside of the material and the time at which the chemical is first detected on the inside surface of the material by means of a chosen analytical instrument.

breakwater — an offshore structure used to protect a harbor or beach from the force of waves.

breast — either of the two protuberant milk-producing glands located on the chest of human females and some other mammals.

breathing — the inhaling and exhaling movements of the chest and lungs to ventilate the alveoli.

breathing tube — a tube through which air or oxygen flows to the face-piece, helmet, or hood.

breathing zone — the zone of the ambient environment in which persons perform the normal respiratory functions; the area of a room in which occupants breath as they stand, sit, or lie down.

breathing zone sample — (*industrial*) an air sample collected in the breathing area of a worker to assess exposure to airborne contaminants.

breeching — a passage that conducts the products of combustion from a furnace to a stack or chimney.

breeder — a nuclear reactor that produces more fuel than it consumes.

bremsstrahlung — electromagnetic radiation (photons) emitted by the negative acceleration of a charged particle, usually an electron, after collision with the nucleus of an atom.

BRI — *See* building-related illness.

bridge crane — (*mechanical engineering*) a lifting unit in which the hoisting apparatus is carried by a bridge-like structure spanning the area of question.

brightness ratio — the comparison of two different brightnesses of light in an area.

brine — water saturated with saline.

briquet — coal or ore dust pressed into an oval or brick-shaped block. Also spelled briquette.

briquette — *See* briquet.

briquetter — a machine that compresses a material, such as metal turnings or coal dust, into small pellets.

A
B

British thermal unit (Btu) — originally the amount of heat needed to raise the temperature of one pound of water 1°F; Btu is now defined as 1055.06 joules (British Standards Institution).

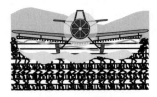

Broadcast Application
(as in crop dusting)

broadcast application or treatment — the placing of a pesticide chemical or agent over an entire field, lawn, or other vast area.

broadleafed plants or weeds — any plants or weeds with wide, flat leaves having net veins, not the parallel veins of grasses or conifers.

broad-spectrum antibiotic — an antibiotic effective against a wide variety of pathogenic microbes.

brodifacoum — 3-[3-(4¹-bromo-[1,1¹-biphenyl]-4-yl)-1,2,3,4-tetrahydro-1-naphthalenyl]-4-hydroxy-2H-1-benzopyran-2-one — a potent anticoagulant poison used to exterminate warfarin-resistant Norway rats, roof rats, and house mice. It is hazardous to children, pets, domestic animals, and nontarget animals by reducing the blood clotting ability and causing hemorrhaging. It can be used around homes, industrial, commercial, agricultural, and public buildings, modes of transportation, but not in sewers.

bromethalin — N-methyl-2,4-dinitro-N-(2,4,6-tribromophenyl)-6-(trifluoromethyl)benzenamine — (*rodenticides*) a rodenticide used for control of Norway rats, roof rats, and mice which works through the uncoupling of oxidative phosphorylation in the mitochondria and leads to fluid build up between the myelin sheaths of the nerves, increased spinal fluid pressure, and pressure on the nerve axons. Acute exposure in humans shows symptoms of headache, confusion, personality changes, seizures, coma, and death.

bromide — a chemical compound containing bromine.

brominator — a device which allows fresh or recirculated water to be introduced into a bromine jar at atmospheric pressure to form a solution which is then conveyed to water.

bromine — Br_2 — a dark, reddish-brown, fuming liquid with suffocating, irritating fumes; MW: 159.8, BP: 139°F, Sol: 4%, sp gr: 3.12. It is used during the synthesis of ethylene

dibromide; in the manufacture of pesticides; as a laboratory reagent; as a disinfecting, sanitizing, and bleaching agent; in the preparation of flame retardants in plastics and fibers; in the manufacture of organic and inorganic compounds for use in photography, pharmaceuticals, fungicides, intermediates, dyes, and bleaching agents. It is hazardous to the respiratory system, eyes, and central nervous system; and is toxic by inhalation, ingestion, and contact. Symptoms of exposure include dizziness, headache, lacrimation, epistaxis, cough, feeling of oppression, pulmonary edema, pneumonia, abdominal pain, diarrhea, measle-like eruptions, severe skin and eye burns. OSHA exposure limit (TWA): 0.1 ppm [air] or 0.7 mg/m³.

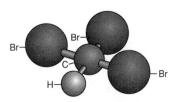

Bromoform

bromoform — $CHBr_3$ — a colorless to yellow liquid with a chloroform-like odor; a solid below 24°F; MW: 252.8, BP: 301°F, Sol: 0.1%, sp gr: 2.89. It is used as a heavy liquid flotation agent in mineral separation-sedimentary petrographical surveys; in chemical and pharmaceutical synthesis; as an industrial solvent in liquid-solvent extractions in nuclear magnetic resonance studies; as a flame retardant; as a catalyst, initiator, or sensitizer in polymer production, irradiation reactions, and vulcanization of rubber. It is hazardous to the skin, liver, kidneys, respiratory system, central nervous system; and is toxic by inhalation, absorption, and ingestion. Symptoms of exposure include respiratory system and eye irritation, central nervous system depression, liver damage. OSHA exposure limit (TWA): 0.5 ppm [skin] or 5 mg/m³.

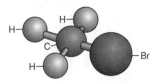

Bromomethane

bromomethane — CH_3Br — a colorless gas with little odor; MW: 94.95; BP: 3.6F°C: Sol. in water at 20°C, 0.9 g/L; Fl.P: not flammable; density at 20°C 3.97. It is primarily used as a soil or space fumigant for the control of insects, fungi, and rodents; can cause accidental death through acute inhalation exposure during manufacturing, packaging operations, in the use of fire extinguishers containing bromomethane or fumigation activities; death is not immediate, usually occurring within 1–2 days of exposure; cause of death is not certain but is probably due to neurological and lung injury. Symptoms of exposure include edema accompanied by focal hemorrhagic lesions which severely impair respiratory function and lead to hypoxia, cyanosis, and complete respiratory failure; the liver may become swollen and tender in some cases; renal effects include congestion, anuria or oliguria, and proteinuria; irritating to the skin and eyes and may cause conjunctivitis, arrhythmia, rashes, or blisters; it may also cause neurological problems with initial symptoms including headache, nausea, confusion, weakness, numbness, and visual disturbances regressing to ataxia, tremor, seizures, paralysis, and coma; OSHA exposure limits: 5 ppm (20 mg/m^3) [air].

bromthymol blue — a chemical dye sensitive to changes in pH which can test the pH over a range of 6.0–7.6; the color changes from yellow to blue as the pH increases.

bronchial asthma — abnormal responsiveness of the air passages to certain inhaled or ingested allergens. An attack of bronchial asthma consists of a wide-spread narrowing of the bronchioles by muscle spasm, swelling of the mucous membranes, or thickening and increasing mucous secretion accompanied by wheezing, gasping, and sometimes coughing.

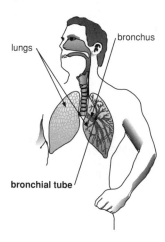

Bronchial Tube

bronchial tube — a subdivision of a bronchus within a lung.

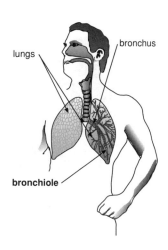

Bronchiole

bronchiole — a small, thin-walled branch of the bronchial tubes within a lung.

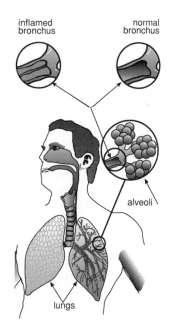

Bronchitis

bronchitis — an inflammation of the bronchial linings causing a persistent cough, copious amounts of sputum, and involuntary muscle spasms that constrict the airways.

bronchoconstrictor — an agent that causes a reduction in the diameter of a bronchus or bronchial tube.

bronchopneumonia — (*disease*) an inflammation of the lungs which usually begins in the terminal bronchioles. Also known as locular pneumonia.

A
B

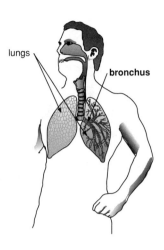

Bronchus

bronchus — one of the larger passages conveying air to and within the lungs.

Brownian movement — the incessant and random movements of particles less than one micrometer in diameter in a fluid caused by collision of the particles with surrounding molecules.

Brucella — (*microbiology*) a genus of gram-negative, aerobic, nonmotile, cocci, or rod-shaped bacteria, all of which are parasites and pathogens of mammals.

brucellosis — (*disease*) a systemic bacterial disease with acute or insidious onset of irregular fever, headache, weakness, profuse sweating, chills, depression, weight loss, and aching; may last for several months. Incubation time is highly variable from 5–30 days, or occasionally several months; caused by *Brucella abortus* biotypes 1–7 and 9, *B. melitensis* biotypes 1–3, *B. suis* biotypes 1–4, and *B. canis*; found worldwide. The reservoir of infection is cows; transmitted by contact with tissues, blood, urine, vaginal discharges, aborted fetuses, and especially placentas through breaks in the skin, ingestion of raw milk or dairy products from infected animals; may also be an airborne infection; not communicable from person to person; wide variety of susceptibility. It is controlled through educating farmers, serological testing of animals, and pasteurization of milk.

Btu — *See* British thermal unit.

bubble chamber — a chamber containing a liquified gas, such as liquid hydrogen, and a charged particle which passes through the liquid and creates a visible path of bubbles.

bubble concept — a means of judging the amounts of air pollutants emitted from smoke stacks within a given area.

bubble tube — a glass tube filled with liquid and bubble used to calibrate air-sampling pumps.

bubo — (*medicine*) an inflammatory swelling of a lymph gland.

bubonic plague — (*disease*) *See* plague.

bucket — an open container affixed to the moveable arms of a wheeled or track vehicle to spread solid waste and cover material, and to excavate soil.

budding — asexual reproduction in which a portion of the cell body is thrust out and then becomes separated forming a new cell.

buffer — a substance that stabilizes pH; any substance in a fluid which tends to resist the change in pH when an acid or alkali is added.

buffer capacity — (*chemistry*) the ability of a solution to maintain its pH when stressed chemically.

buffer solution — (*chemistry*) a solution to which small amounts of an acid or base can be added without appreciatively changing its hydrogen ion concentration.

bug — (*computer science*) an error in a computer program or in a piece of electronics that causes it to function improperly.

building envelope — elements of the building including all external building materials, windows, and walls that enclose the internal space.

building-related illness — a diagnosable illness with identifiable symptoms and causes that can be directly attributed to airborne building pollutants.

bulk density — the mass of powdered or granulated solid material per unit of volume.

bulk food — unpackaged or unwrapped processed or unprocessed food in aggregate containers from which quantities desired by the consumer are withdrawn.

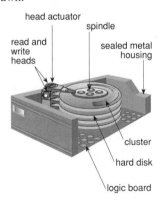

Bulk Memory

bulk memory — (*computer science*) an electronic device, such as a disk or tape, that allows the storage of large amounts of data.

bulky waste — items whose large size complicates its handling by normal collection, processing, or disposal methods.

bulla — a blister; a circumscribed, fluid-containing, elevated lesion of the skin usually more than 5 mm in diameter.

bull clam — a tracked vehicle with a hinged, curved bowl on the top of the front of the blade.

bulldozer — (*mechanical engineering*) a track vehicle equipped with an earth blade.

bullous — of or pertaining to bulla.

BUN — *See* blood urea nitrogen.

bundle — (*EPA definition*) a structure composed of three of more fibers in a parallel arrangement with each fiber closer than one fiber diameter; (*NIOSH definition*) a compact arrangement of parallel fibers in which separate fibers or fibrils may only be visible at the ends of the bundle; asbestos bundles having aspect ratios of 3:1 or greater, and less than 3 millimicrons in diameter are counted as fibers.

Bunyavirus — an arbovirus spread by mosquitos and ticks causing fever, rashes, encephalitis, hemorrhage, meningitis, and hemorrhagic fever.

burden of proof — the necessity or responsibility of proving a fact or facts in dispute on an issue.

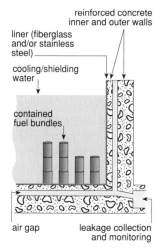

Burial Ground

burial ground — (*radiation*) a disposal site for unwanted radioactive materials using earth or water for a shield.

burn — injury to tissues caused by contact with heat, steam, chemicals, electricity, lightning, or radiation. A first degree burn involves a reddening of the skin area; a second degree burn involves blistering of the skin; a third degree burn, the most serious type, involves damage to the deeper layers of the skin with necrosis through the entire skin.

burning area — a horizontal projection of a grate, a hearth, or both.

burning hearth — a solid surface without air openings to support the solid fuel or solid waste in a furnace during drying, ignition, or combustion.

burning rate — the quantity of solid waste incinerated or the amount of heat released during incineration; usually expressed in pounds of solid waste per square foot of burning area per hour or in BTUs per square foot of burning area per hour.

burr — a thin, rough edge of a machined piece of metal.

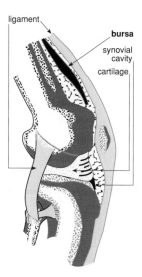

Bursa

bursa — a sac filled with fluid and situated at a place in the tissues at which friction of movement has been reduced.

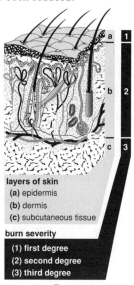

Burns

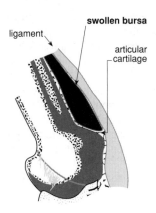

Bursitis in Knee

bursitis — inflammation of the bursa resulting in severe pain and limitation of motion of the affected joint.

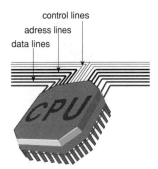

Bus

bus — (*computer science*) a circuit or group of circuits that provide a communication path between the computer and peripherals.

Business Periodicals Index — indexes of business literature from 1958 to the present.

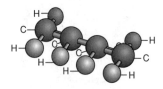

1,3–Butadiene
(Butadiene)

1,3-butadiene — $CH_2{=}CHCH{=}CH_2$ — a colorless gas with a mild aromatic or gasoline-like odor; a liquid below 24°F; MW: 54.1, BP: 24°F, Sol: insoluble, sp gr: 0.65. It is used in molding and vulcanizing operations in the processing of rubber products; in the manufacture of high-impact polystyrene and elastomers. It is hazardous to the eyes, respiratory system, and central nervous system; and is toxic by inhalation and contact. Symptoms of exposure include eye, nose, and throat irritation, drowsiness, light-headedness, frostbite; carcinogenic. OSHA exposure limit (TWA): 1,000 ppm [air] or 2,200 mg/m³.

2 Butanone
(Methyl Ethyl Ketone)

2-butanone — $CH_3COCH_2CH_3$ — a colorless liquid with a moderately sharp, fragrant, mint- or acetone-like odor; MW: 72.1, BP: 175°F, Sol: 28%, Fl.P: 16°F, sp gr: 0.81. It is used in spray application of vinyl and acrylic coatings; during mixing of dye solutions; in laboratories; in the application of adhesives for artificial leather, during sponge and brush application of solvent for cleaning operations; during mixing of waterproofing compounds. It is hazardous to the central nervous system and lungs; and is toxic by inhalation, ingestion, and contact. Symptoms of exposure include eye and nose irritation, headache, dizziness, and vomiting. OSHA exposure limit (TWA): 200 ppm [air] or 590 mg/m³.

2-butoxyethanol — $C_4H_9OCH_2CH_2OH$ — a colorless mobile liquid with a mild ether-like odor; MW: 118.2, BP: 339°F, Sol: miscible, Fl.P: 143°F, sp gr: 0.90. It is used in spray and brush application of varnishes, lacquers, and enamels; as an industrial solvent; in the production of plasticizers; as a stabilizing agent in metal cleaners and liquid household cleaners; in hydraulic fluids, insecticides, herbicides, and rust removers. It is hazardous to the liver, kidneys, lymphoid system, skin, blood, eyes, and respiratory system; and is toxic by inhalation, absorption, ingestion, and contact. Symptoms of exposure include eye, nose, and throat irritation, hemolysis, hemoglobinuria. OSHA exposure limit (TWA): 25 ppm [skin] or 120 mg/m³.

butter — a fatty food product made from fresh or ripened, partially sour milk or cream and a small portion of other constituents natural to milk, with or without common salt, with or without additional harmless food coloring.

butter fat — *See* milk-fat.

buttermilk — a fermented fluid product resulting from the churning of milk or cream containing not less than 8-1/4% milk solids (not fat).

n-butyl acetate — $CH_3COO(CH_2)_3CH_3$ — a colorless liquid with a fruity odor; MW: 116.2, BP: 258°F, Sol: 1%, Fl.P: 72°F, sp gr: 0.88. It is used during application of nitrocellulose by spraying, brushing, or dipping; in the production of lacquer thinner; as a solvent for oils, pitch, camphor, ethyl cellulose acetate, vinyl, polystyrene, methacrylate plastics, and chlorinated rubber; as a solvent in the production of artificial leather; in the manufacture of safety glass; in the production of flavorings, perfumes, cosmetics, adhesives, shoe polishes, and stain removers. It is hazardous to the eyes, skin, and respiratory system; and is toxic by inhalation, ingestion, and contact. Symptoms of exposure include headache, drowsiness, dryness and irritation of the eyes, upper respiratory system, and skin. OSHA exposure limit (TWA): 150 ppm [air] or 710 mg/m³.

sec-butyl acetate — $CH_3COOCH(CH_3)CH_2CH_3$ — a colorless liquid with a pleasant, fruity odor; MW: 116.2, BP: 234°F, Sol: 0.8%, Fl.P: 62°F, sp gr: 0.86. It is used as a solvent for coating paper, leather, and artificial leather material; as a lacquer and adhesive; in the manufacture of photographic film. It is hazardous to the eyes, skin, and respiratory system; and is toxic by inhalation, ingestion, and contact. Symptoms of exposure include eye irritation, headache, drowsiness, dryness of the upper respiratory system and skin. OSHA exposure limit (TWA): 200 ppm [air] or 950 mg/m³.

tert-butyl acetate — $CH_3COOC(CH_3)_3$ — a colorless liquid with a fruity odor; MW: 116.2, BP: 208°F, Sol: insoluble, Fl.P: 62–72°F, sp gr: 0.87. It is used as a dry cleaning agent; in the formulation of paint systems; as an additive to improve antiknock properties of leaded aliphatic gasoline; as an activator in alkaline polymerization of caprolactam. It is hazardous to the respiratory system, eyes, and skin; and is toxic by inhalation, ingestion, and contact. Symptoms of exposure include itching, inflamed eyes, irritation to the upper respiratory tract, headache, narcosis, dermatitis. OSHA exposure limit (TWA): 200 ppm [air] or 950 mg/m³.

n-butyl alcohol — $CH_3CH_2CH_2CH_2OH$ — a colorless liquid with a strong, characteristic, mildly alcoholic odor; MW: 74.1, BP: 243°F, Sol: 9%, Fl.P: 99°F, sp gr: 0.81. It is used in the leather industry; in the manufacture of safety glass; as a solvent or diluent in the manufacture of brake fluids, perfumes, detergents, adhesives, denatured alcohol, and surface coatings; in photographic processing; in laboratory analysis. It is hazardous to the skin, eyes, and respiratory system; and is toxic by inhalation, absorption, ingestion, and contact. Symptoms of exposure include eye, nose, and throat irritation, headache, vertigo, drowsiness, corneal inflammation, blurred vision, lacrimation, photophobia, dry and cracked skin. OSHA exposure limit (TWA): ceiling 50 ppm [skin] or 150 mg/m³.

sec-butyl alcohol — $CH_3CHOHCH_2CH_3$ — a colorless liquid with a strong, pleasant odor; MW: 74.1, BP: 211°F, Sol: 16%, Fl.P: 75°F, sp gr: 0.81. It is used in industrial cleaning compounds; in ethyl cellulose-based surface coatings; in paint removers, perfumes, and hydraulic fluids. It is hazardous to the eyes and central nervous system; and is toxic by inhalation, ingestion, and contact. Symptoms of exposure include narcosis and eye irritation. OSHA exposure limit (TWA): 100 ppm [air] or 305 mg/m³.

tert-butyl alcohol — $(CH_3)_3COH$ — a colorless liquid with a camphor-like odor; pure compound is a liquid above 77°F; MW: 74.1, BP: 180°F, Sol: miscible, Fl.P: 52°F, sp gr: 0.79. It is used in industrial cleaning compounds; in lacquer surface coatings; as a chemical intermediate; as a solvent for drug extraction, water removal, wax solvent; in laboratory procedures. It is hazardous to the eyes and skin; and is toxic by inhalation, ingestion, and contact. Symptoms of exposure include drowsiness and eye and skin irritation. OSHA exposure limit (TWA): 100 ppm [air] or 300 mg/m³.

butylamine — $CH_3CH_2CH_2CH_2NH_2$ — a colorless liquid with a fishy, ammonia-like odor; MW: 73.2, BP: 172°F, Sol: miscible, Fl.P: 10°F, sp gr: 0.74. It is used as a vulcanizing accelerator and reaction initiator in rubber and polymer industries; as a chemical intermediate in the production of emulsifying agents; in chemical synthesis of developers for photography, pharmaceuticals, antioxidants for gasoline, synthetic tanning materials, curing agents, gum inhibitors; as a stabilizer in aviation fuels; on stored

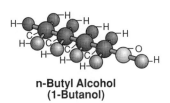

**n-Butyl Alcohol
(1-Butanol)**

agricultural crops as fungicide and to prevent decay. It is hazardous to the respiratory system, skin, and eyes; and is toxic by inhalation, absorption, ingestion, and contact. Symptoms of exposure include eye, nose, and throat irritation, headache, flushed and burned skin. OSHA exposure limit (TWA): ceiling 5 ppm [skin] or 15 mg/m³.

tert-butyl chromate — $((CH_3)_3CO)_2CrO_2$ — a liquid which solidifies at 32–23°F; MW: 230.3, BP: not known, Sol: not known, Fl.P: not known. It is used in oxidation of steroids as a means of identification on paper chromatography; in the manufacture of catalysts used for polymerization of olefins. It is hazardous to the respiratory system, skin, eyes, and central nervous system; and is toxic by inhalation, absorption, ingestion, and contact. Symptoms of exposure include lung, sinus cancer; carcinogenic. OSHA exposure limit (TWA): ceiling 0.1 mg/m³ [skin].

n-butyl glycidyl ether — $C_7H_{14}O_2$ — a colorless liquid with an irritating odor; MW: 130.2, BP: 327°F, Sol: 2%, Fl.P: 130°F, sp gr (77°F): 0.91. It is used as a reactive diluent of epoxy resins; as a chemical intermediate for preparation of ethers, surfactants, polymers, and resins; as a stabilizing agent for organic chemicals. It is hazardous to the eyes, skin, respiratory system, and central nervous system; and is toxic by inhalation, ingestion, and contact. Symptoms of exposure include eye, nose, and skin irritation, sensitization, and narcosis. OSHA exposure limit (TWA): 25 ppm [air] 135 mg/m³.

butyl mercaptan — $CH_3CH_2CH_2CH_2SH$ — a colorless liquid with a strong garlic-, cabbage-, or skunk-like odor; MW: 90.2, BP: 209°F, Sol: 0.06%, Fl.P: 35°F, sp gr: 0.83. It is used as an odorant for natural gas; as a chemical intermediate in the manufacture of agricultural chemicals, herbicides, and defoliants; in the manufacture of polymerization catalysts, stabilizers, modifiers, and chain transfer agents; as a solvent. It is hazardous to the respiratory system (in animals: the central nervous system, liver, and kidneys); and is toxic by inhalation, ingestion, and contact. Symptoms of exposure in animals include narcosis, incoordination, weakness, cyanosis, pulmonary irritation, liver and kidney damage. OSHA exposure limit (TWA): 0.5 ppm [air] or 1.5 mg/m³.

p-tert-butyltoluene — $(CH_3)_3CC_6H_4CH_3$ — a colorless liquid with a distinct aromatic odor, somewhat like gasoline; MW: 148.3, BP: 379°F, Sol: insoluble, Fl.P: 155°F, sp gr: 0.86. It is used as a primary intermediate in chemical and pharmaceutical industries. It is hazardous to the cardiovascular system, central nervous system, skin, bone marrow, eyes, upper respiratory system; and is toxic by inhalation, ingestion, and contact. Symptoms of exposure include eye and skin irritation, dry nose and throat, headache, low blood pressure, tachycardia, abnormal cardiovascular stress, central nervous system depression, hematopoietic depression. OSHA exposure limit (TWA): 10 ppm [air] or 60 mg/m³.

BWR — *See* boiling water reactor.

by-product — a material produced or generated in an industrial process in addition to the principle product.

byssinosis — (*disease*) a pneumoconiosis disease occurring in workers from prolonged exposure to heavy air concentrations of cotton or flax dust.

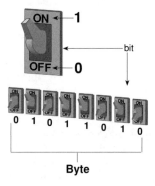

Byte

byte — (*computer science*) a group of contiguous bits, usually eight, that represent a character and which are operated on as a unit.

C

c — *See* capacitance; ceiling level; centi-.

C — (*chemistry*) *See* carbon; carbon black; (*computer science*) a high-level programming language used in graphics; (*electricity*) *See* coulomb.

cachexia — a profound and marked state of wasting of the body resulting from general ill health, malnutrition, or terminal illnesses.

CAD/CAM — acronym for computer-aided design/computer-aided manufacturing.

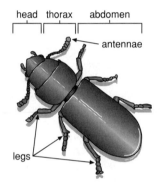

Cadelle
(Tenebroides mauritanicus)

Cadelle — a small black beetle destructive to stored grain.

cadmium — CD — a soft, silver-white metal in its pure state (rarely found this way in the environment). MW: 112.40, BP: 767°C, Sol: insoluble, density: 8.462. It is used in the production of nickel-cadmium batteries; in metal plating, plastics, and synthetics; and for alloys and other miscellaneous uses. It is hazardous to the respiratory system, kidneys, and bones. Symptoms of exposure include severe tracheobronchitis, pneumonitis, pulmonary edema, calcium deficiency, osteoporosis, proteinuria, renal damage, and death. OSHA exposure limit (TWA): 0.1 mg/m^3 [air].

caecum — *See* cecum.

CAER — *See* Community Awareness and Emergency Response Program.

caisson disease — *See* bends.

cake — (*wastewater*) the solids discharged from a dewatering apparatus.

cal — *See* calorie.

calcareous soil — soil containing sufficient calcium carbonate, often with magnesium carbonate, to effervesce visibly when treated with cold 0.1N hydrochloric acid.

calcination — the heat treatment of solid materials to bring about thermal decomposition or a phase transition other than melting.

calcine — to heat to a high temperature without fusing in order to decompose, oxidize, and so on.

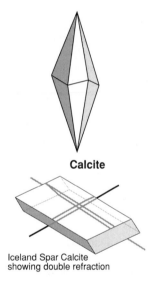

Calcite

Iceland Spar Calcite
showing double refraction

calcite — $CaCO_3$ — one of the most common minerals, it is usually colorless or white and is a major constituent in limestones and marbles; a calcium carbonate ($CaCO_3$); the major mineral in limestone; is quite soluble which accounts for its usual presence in water; one of the three crystalline forms of calcium carbonate, the others being aragonite and vaterite. Also known as calcspar.

calcium arsenate — $Ca_3(AsO_4)_2$ — a colorless to white, odorless solid; MW: 398.1, BP: decomposes, Sol: (77°F): 0.01%, sp gr: 3.62. It is used as an insecticide and herbicide. It is hazardous to the eyes, respiratory system, liver, skin, lymphatics, and central nervous system; and is toxic by inhalation, absorption, ingestion, and contact. Symptoms of exposure include weakness, gastrointestinal disturbance, skin hyperpigmentation, palmar planter hyperkeratoses; dermatitis; carcinogenic; liver damage in animals. OSHA exposure limit (TWA): 0.010 mg/m³ [air].

calcium carbonate — $CaCO_3$ — a white powder occurring naturally as calcite. It is used in paint manufacture, as a dentrifice, as an anticaking medium for table salt, and in the manufacture of rubber tires.

calcium hypochlorite — $Ca(OCl)_2 \cdot 4H_2O$ — (*inorganic chemistry*) a compound of chlorine and calcium used in white granular or tablet form as a bactericide in pools; in water solution it releases 70% of its weight as available chlorine.

calcium oxide — CaO — (*inorganic chemistry*) caustic white or gray, odorless lumps or granular powder; MW: 56.1, BP: 5162°F, Sol: reacts, sp gr: 3.34. It is used in the preparation of lime; in the manufacture of Portland cement, mortar, stucco, and plaster; in the manufacture of iron and steel in wire-drawing; as a flux in metal refining and smelting; as a softening, purifying, coagulating, and suspending compound in water treatment and purification; during food processing; in the manufacture of silicate and nonsilicate glass; in the manufacture of pesticides and fungicides. It is hazardous to the respiratory system, skin, and eyes; and is toxic by inhalation, ingestion, and contact. Symptoms of exposure include eye and upper respiratory tract irritation, ulcer, perforated nasal septum, pneumonia, dermatitis. OSHA exposure limit (TWA): 5 mg/m³ [air].

calcspar — *See* calcite.

calendar call — (*law*) a public calling of the docket or list of cases at the commencement of term of court for setting a time for trial or entering orders of continuance, default, nonsuit, etc. Also known as calling the docket.

calibrate — (*science and technology*) to adjust mathematically the pollution concentration predicted by a model so as to match the results obtained from model validation field studies; to adjust instruments to a standard; to control the amount of a pesticide chemical applied by each nozzle or opening of a sprayer, duster, or granular applicator to a given area, plant, or animal; the determination of variation from standards for accuracy of a measuring instrument to ascertain necessary correction.

calibrating solution — a solution containing a known amount of a substance having an effect equivalent to a pollutant concentration that has passed through the detection component during the static calibration of an analyzer.

calibration error — the difference between the pollution concentration indicated by the measurement system and the known concentration of the test gas mixture.

calibration gas — a gas used to calibrate an air monitoring instrument.

caliche — *See* hardpan.

calicivirus — (*disease*) a virus classified in the family *Caliciviridae* containing a single strand of RNA surrounded by a protein capsid of 31–40 nanometers in diameter, which may cause viral gastroenteritis in children 6–24 months of age, and accounts for approximately 3% of the hospitalized cases of diarrhea with additional symptoms of nausea, vomiting, malaise, abdominal pain, and fever. The mode of transmission is the oral-fecal route, from person to person, or ingestion of contaminated food and water. The incubation period is 10–70 hours. The disease has been reported in England and Japan. The reservoir of infection is humans. Susceptibility found in the very young. Also known as acute nonbacterial infectious gastroenteritis; viral gastroenteritis.

California waste — a group of liquid hazardous wastes, including ones with polychlorinated biphenyls, heavy metals, and halogenated organic compounds that the EPA evaluated as of July 8, 1987 to determine if they should be banned from land disposal, or if restrictions should be placed on the land disposal of these wastes.

calling the docket — *See* calendar call.

calorie (cal) — (*thermodynamics*) the amount of heat necessary to raise the temperature of one gram of water 1°C; a measurement now largely replaced by the joule; 1 calorie = 4.1868 joules. *See also* gram-calorie.

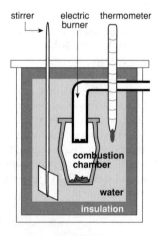

Calorimeter

calorimeter — an instrument used for the measurement of the amount of heat liberated or absorbed during a change in temperature caused by a chemical reaction, change of state, or formation of solution.

calyx — the outermost part of a flower which covers the bud.

CAM — *See* continuous air monitoring.

camphor (synthetic) — $C_{10}H_{16}O$ — colorless or white crystals with a penetrating, aromatic odor; MW: 152.3, BP: 399°F, Sol: insoluble, Fl.P: 150°F, sp gr: 0.99. It is used in the preparation and loading of explosives; during embalming procedures; in mixing and blending of lacquers and varnishes; as a catalyst or chemical intermediate; in the packaging of moth and insect repellents. It is hazardous to the central nervous system, eyes, skin, and respiratory system; and is toxic by inhalation, ingestion, and contact. Symptoms of exposure include eye, skin, and mucous membrane irritation, nausea, vomiting, diarrhea, headache, dizziness, excitement, irrational epileptiform convulsions. OSHA exposure limit (TWA): 2 mg/m³ [air].

Campylobacter — a genus of gram-negative microaerophilic to anaerobic, motile, curved, or spiral rod-shaped bacteria found in the oral cavity, intestinal tract, and reproductive organs of humans and animals.

campylobacter enteritis — an acute enteric disease of varying severity characterized by diarrhea, abdominal pain, malaise, fever, nausea, and vomiting. Incubation time is 3–5 days that ranges from 1–10 days; caused by *Campylobacter jejuni* and *C. coli*. It is found in all parts of the world; the reservoir of infection is animals. It is transmitted by ingestion of organisms in food or in unpasteurized milk or water, contact with infected pets, or from uncooked or poorly refrigerated food. It is communicable from several days to several weeks; universal susceptibility. It is controlled through isolation of patients and other individuals, as well as disinfection of articles soiled with feces, and control of sewage disposal.

canal caps — devices that rest on the ear canal opening and are supported by a headband.

cancellation — (*pesticides*) an act of disallowing the use of a pesticide by the U.S. Environmental Protection Agency when the chemical adversely affects the environment or does not comply with the requirements of the Federal Insecticide, Fungicide, and Rodenticide Act.

cancer — any malignant cellular tumor including carcinoma and sarcoma. It encompasses a group of neoplastic diseases in which there is a transformation of normal body cells into malignant ones, probably involving some change in the genetic material of the cells (DNA), possibly as a result of faulty repair of damage to the cell

caused by carcinogenic agents or ionizing radiation. The altered cells pass on inappropriate genetic information to their offspring and begin to proliferate in an abnormal and destructive manner. The mass of abnormal tissue enlarges, ulcerates, and begins to shed cells that spread the disease locally or to other sites called metastasis. Environmental, hereditary, and biological factors are important in the development of cancer.

cancer effect level (CEL) — the lowest dose of chemical in a study or group of studies that produces significant increases in the incidence of cancer or tumors between the exposed population and its appropriate control.

CANCERLIT — *See* Cancer Literature.

Cancer Literature (CANCERLIT) — a database containing more than 475,000 citations with abstracts to worldwide literature on oncology epidemiology, pathology, treatment, and research.

candela — the international reference standard for the unit of candle power defined as 1/60 of the luminous intensity per square centimeter of a black body radiator operating at the temperature of freezing platinum; formerly known as standard candle.

Candida — a genus of yeast-like fungi that are commonly part of the normal flora of the mouth, skin, intestinal tract, and vagina and that can cause a variety of infections.

candidiasis — a mycosis usually confined to the superficial layers of skin or mucous membrane; ulcers may be formed in the esophagus, gastrointestinal tract, or bladder; lesions may occur in the kidneys, spleen, lungs, liver, endocardium, eyes, meninges, or brain. Incubation time is 2–5 days; caused by *Candida albicans* and *C. tropicalis*; found worldwide as it is a part of normal human flora. The reservoir of infection is people. It is transmitted by excretions of the mouth, skin, vagina, and feces; patients are carriers, and it can also be carried from mother to infant during childbirth; communicable during the duration of the lesions; a wide-spread, low-level of pathogenicity occurs. It is controlled by disinfection of secretions and contaminated articles, and specific treatment for the individuals.

candlepower (cp) — the unit of intensity or brightness of a standard candle at its source.

cane — the woody stem growing directly from the box of a plant as in raspberries, blackberries, and some vines.

canister — the container which holds the filter used in a gas mask.

canned meats — meat products which are preserved and hermetically sealed in metal or glass containers.

C
D

capacitance (c) — the property of a system of conductors and dielectrics which permits the storage of electric charge and electric energy when potential differences exist between the conductors.

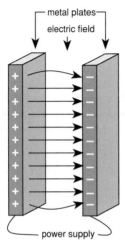

Capacitor

capacitor — a device for holding and storing charges of electricity.

capillaries — the smallest blood vessels of both the circulation and lymphatic systems.

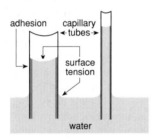

Capillarity

capillarity — (*physics*) the tendency of the surface of a liquid to rise or fall in contact with a solid. Also known as capillary action.

capillary action — *See* capillarity.

capillary attraction — the force of adhesion between a solid and a liquid in capillarity; a liquid's movement over or retention by a solid surface due to the interaction of adhesive or cohesive forces.

capillary fringe — a zone in the rocks just above the water table in which water is drawn upward into the spaces by capillary tension.

capillary tension — the tendency for water to be drawn upward into very narrow spaces.

capillary water — the water held in the small pores of a soil by capillarity usually with a tension greater than 60 cm. of water.

capital investment — investment for long-term use of over one year to treat as capital rather than as an expense.

capsid — the shell of protein that protects the nucleic acid of a virus.

capture system — the equipment, including hoods, ducts, fans, and dampers, used to capture or transport particulate matter generated by a pollutant source.

capture velocity — the air velocity at a point within or in front of an exhaust hood necessary to overcome opposing air currents particle inertia, causing contaminated air to flow into the hood; the air velocity needed to draw contaminants into the hood.

carbamate — a salt or ester of carbamic acid; a synthetic organic insecticide.

carbamic acid — an acid occurring only in the form of salts H_2NCOOH.

carbaryl (Sevin®) — $C_{10}H_7OOCNHCH_3$ — a white or gray odorless solid; MW: 201.2, BP: decomposes, Sol: 0.01%, sp gr: 1.23. It is used as a pesticide and acaricide for field crops, fruit, vegetables, ornamentals, livestock, poultry, pets, domestic dwellings, medical facilities, schools, commercial and industrial areas, urban and rural outdoor areas, and sewage treatment plants. It is hazardous to the respiratory system, central nervous system, cardiovascular system, and skin; and is toxic by inhalation, absorption, ingestion, and contact. Symptoms of exposure include miosis, blurred vision, tearing, nasal discharge, salivation, sweating, abdominal cramps, nausea, vomiting, diarrhea, tremors, cyanosis, convulsions, and skin irritation. OSHA exposure limit (TWA): 5 mg/m³ [air].

carbohydrate — any group of organic compounds consisting of a ring or chain of carbon atoms with hydrogen and oxygen attached usually in a ratio of 2:1; sugars, starches, cellulose, and glycogen are carbohydrates.

carbon (C) — a nonmetallic element found either in its elemental state or as a constituent of coal, petroleum, limestone, and other organic or inorganic compounds; atomic number: 6; atomic weight: 12.01115. Its three main allotropic forms are diamond, graphite, and coal.

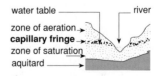

Capillary Fringe

carbonaceous matter — pure carbon or carbon compounds present in the fuel or residue of a combustion process.

carbon arc cutting — an arc cutting process in which metals are severed by melting them with the heat of an arc between a carbon electrode and the base metal.

carbon arc welding — an arc welding process that produces fusion of metals by heating them with an arc between a carbon electrode and the work.

carbonate aquifer — a type of aquifer made from limestone and dolomite rocks consisting of calcium and magnesium carbonate.

carbonate hardness — temporary hardness due to soluble bicarbonates of calcium, magnesium, and iron which can be removed by boiling.

carbon black — C — a black, odorless solid; MW: 12.0, BP: sublimes, Sol: insoluble, Fl.P: not known, sp gr: 1.8–2.1. It is used in the manufacture of natural and synthetic rubber, dry cells, explosives, plastics, and paper; in the manufacture of coating and printing inks; as a color pigment and source of carbon. It is hazardous to the respiratory system and eyes; and is toxic by inhalation and contact. A cough and irritated eyes occur; in the presence of polycyclic aromatic hydrocarbons it is carcinogenic. OSHA exposure limit (TWA): 3.5 mg/m^3 [air].

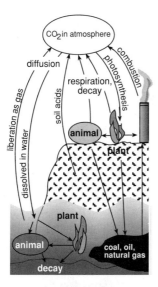

The Carbon Cycle

carbon cycle — (*geochemistry*) the natural sequence through which carbon circulates in the biosphere beginning with the conversion of atmospheric carbon dioxide to carbohydrates by plants; the carbon stored in plants is returned to the atmosphere in the form of carbon dioxide by several routes, including respiration by animals that subsist on plants, the burning of fossil fuels, primarily coal and oil, or the eventual decomposition of plants, animals, and animal waste products.

carbon 14 dating — *See* carbon dating.

carbon dating — the use of radioactive carbon-14 (C_{14}) isotope to estimate the approximate age and archeology of ancient organic materials. Also known as radioactive carbon dating; radiocarbon dating; carbon 14 dating.

carbon dioxide — CO_2 — colorless, odorless gas; normal constituent of air (about 3000 ppm); solid form is utilized as dry ice; MW: 44.0, BP: sublimes, Sol:(77°F) 0.2%. It is used in cooling and refrigeration for storage, preparation, and transfer of foods; as an inert gas in fire extinguishers; in shielded arc welding; in canned food production; as an inert pressure medium for aerosol packaging; pressure spraying, spray painting, to inflate life rafts; in the manufacture of carbonated beverages; as a neutralizing agent in textile processing and treatment of leather hides; in the manufacture of urea, aspirin, carbonates and bicarbonates, beer, and sugar. It is hazardous to the lungs, skin, and cardiovascular system; and is toxic by inhalation and contact. Symptoms of exposure include headache, dizziness, restlessness, paresthesia, dyspnea, sweating, malaise, increased heat rate and pulse, elevated blood pressure, coma, asphyxia convulsions at high concentrations, and frostbite (from dry ice); associated with the greenhouse effect. OSHA exposure limit (TWA): 10,000 ppm [air] or 18,000 mg/m^3.

carbon dioxide cycle — a series of processes by which the balance of carbon dioxide and oxygen in the atmosphere is maintained.

carbon disulfide — CS_2 — pure carbon disulfide is a colorless liquid with a pleasant chloroform-like odor; impure carbon disulfide, in most industry processes, is a yellowish liquid with an unpleasant odor like that of rotting radishes; MW: 76.14, BP: 46.5°C, Sol. in water 0.294% at 20°C, Fl.P: 125–135°C. It is used in viscose rayon industry, production of carbon tetrachloride which is an intermediate in the production of fluorocarbon propellants and refrigerants; increased levels of death occur to workers in viscose rayon plants where carbon disulfide as well as other chemicals are used; accidental inhalation may result in subtle and transient changes in pulmonary function with reduced vital capacity and decreased partial pressure of arterial oxygen; vascular atherosclerotic changes are a primary effect following long-term exposure; gastrointestinal effects may occur; inhibition of microsomal hepatic enzymes occur; severe ocular effects characterized by dot hemorrhages or microaneurysms of the retina have been observed; the primary target appears to be the nervous system with neuro, physiological, and behavioral effects as well as pathomorphology of peripheral nervous system structures. Symptoms of exposure include distal sensory shading, resting tremors, nerve conduction abnormalities, brain organic changes, and brain atrophy; systemic toxicity upon oral exposure usually occurs in the liver because of enzymatic disruptions. OSHA exposure limit: 4 ppm (12 mg/m³) [air].

carbonization — a process by which the remains of living organisms are partially decomposed leaving a residue of carbon.

carbon monoxide — CO — a colorless, odorless, very toxic gas produced by the incomplete combustion of carbon-containing substances such as coal, oil, gasoline, and natural gas; MW: 28.0; BP: –313°F; Sol: 2%. It is produced on an industrial scale for use as a reducing agent in metallurgy. Carbon monoxide has an affinity for hemoglobin at a rate of 200–300 times higher than oxygen, causing the body to become oxygen-starved. Cigarette smokers' blood contains 2–10% carboxy hemoglobin, while nonsmokers have an average of 1% carboxy hemoglobin. Carbon monoxide can cause severe health symptoms or death when individuals are exposed at levels of 100–10,000 ppm. The low end of the scale results in slight headaches, and the upper end of the scale results in death. It is hazardous to the cardiovascular system, lungs, blood, and central nervous system; and is toxic by inhalation and (liquid) contact. Symptoms of exposure include headache, tachypnea, nausea, weakness, dizziness, confusion, hallucinations, cyanosis, depressed S-T segment of electrocardiogram, angina, and syncope. OSHA exposure limit (TWA): ceiling 200 ppm [air] or 229 mg/m³.

carbonmonoxyhemoglobin — *See* carboxyhemoglobin.

carbon nitrogen ratio (C/N) — the ratio of the weight of carbon to the weight of nitrogen present in a compost or materials that are being composted.

carbon particulate matter — a product of incomplete combustion. Also known as soot.

carbon tetrachloride — CCl₄ — a colorless liquid with a characteristic ether-like odor; MW: 153.8, BP: 170°F, Sol: 0.05%, sp gr: 1.59. It is used in the manufacture of fluorocarbons for aerosols, refrigerants, and fire extinguishants; as an agricultural grain fumigant and pesticide; in polymer technology as a reaction medium, catalyst, chain transfer agent, and solvent for resins; in organic synthesis; as an industrial solvent for rubber cements, cable and semiconductor manufacture; as a laboratory solvent. It is hazardous to the central nervous system, eyes, lungs, liver, kidneys, and skin; and is toxic by inhalation, absorption, ingestion, and contact. Symptoms of exposure include central nervous system depression, nausea, vomiting, liver and kidney damage, skin irritation; carcinogenic. OSHA exposure limit (TWA): 2 ppm [air] or 12.6 mg/m³.

carborundum — a trademark used for various manufactured abrasives.

carboxyhemoglobin — hemoglobin combined with carbon monoxide when it is inhaled occupying the sites on the hemoglobin molecule that normally bind with oxygen. Also known as carbonmonoxyhemoglobin.

carboxylic acid — any of a family of organic acids which contain the carboxyl group –COOH.

carbuncle — (*bacteriology*) a necrotizing infection of skin and subcutaneous tissue composed of a cluster of boils usually due to *Staphylococcus aureus*.

carcass — all parts of a slaughtered animal that are capable of being used for human consumption, including viscera.

carcinogen — any substance capable of increasing the incidence of neoplasms, or decreasing the time it takes for them to develop.

carcinoma — a malignant new growth made up of epithelial cells tending to infiltrate surrounding tissues and give rise to metastases.

cardiopulmonary — of or pertaining to the heart and lungs.

cardiopulmonary resuscitation (CPR) — the re-establishment of heart and lung often applied in cases of cardiac arrest; the method of artificially oxygenating and circulating the blood of a person in cardiac arrest.

cardiovascular disease — diseases of or involving the heart or blood vessels.

carditis — inflammation of the heart tissues.

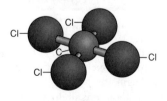

Carbon Tetrachloride

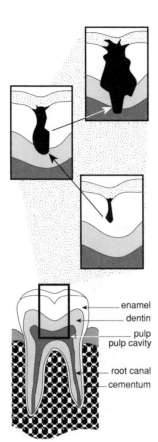

Caries

caries — a process of progressive destruction of the hard tissues of the teeth initiated by bacterially produced acids of the tooth surface.

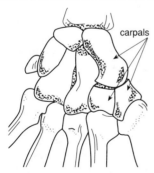

Carpals

carpal — of or pertaining to the carpus or eight bones of the human wrist.

carpal tunnel — the osteofibrous passage for the median nerve and the flexor tendons formed by the flexor retinaculum and the carpal bones.

carpal tunnel syndrome — a painful syndrome resulting from compression of the median nerve in the carpal tunnel with pain and burning or tingling paresthesia in the fingers and hand, sometimes extending to the elbow. The disorder is found most often in middle-aged women due to excessive wrist movements, arthritis, and swelling of the wrist.

carrier — (*medicine*) an individual harboring specific infectious agents in the absence of discernible clinical symptoms who serves as a potential source or reservoir of infection for other humans; (*pesticides*) a liquid, solid, or gas used to transport a pesticide chemical to the pest.

carrier gas — the gas or mixture of gases which contains and moves the contaminant material but does not react with it.

carrying capacity — the maximum biomass that a system is capable of supporting continuously without deterioration.

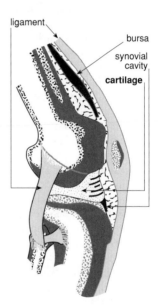

Cartilage

cartilage — a specialized, elastic connective tissue attached to articular bone surfaces.

cartography — the production of maps including construction of projections, design, compilation, drafting, and reproduction. Also known as map.

cartridge disk — a type of magnetic-memory disk enclosed in a plastic cartridge and used in computer technology.

CAS — *See* Chemical Abstract Services.

cascade impactor — a sampling device for particulates in air in which high velocity, dust-laden air striking a flat surface at a 90° angle is forced to suddenly change direction and momentum, and cause the particulates to settle out.

case-control study — type of analytical epidemiological study in which the past histories of affected persons (cases) and of non-affected persons (controls) are investigated to identify factors in the etiology or control of a disease.

case fatality rate — a percentage of the number of persons diagnosed as having a specified disease and who die as a result of that disease.

case-hardening — a process of hardening the surface of metals by raising the carbon or nitrogen content of the outer surface.

case history — the collected data concerning an individual, family, or environment including medical history, environmental conditions, and other information useful in analyzing and diagnosing for teaching or research purposes.

case-history study — an association between two variables such that change in one is followed by a predictable change of the other; in the absence of experimental evidence, the following criteria are used in judging whether or not an association is causal: strength of association, consistency of association, correctness of temporal sequence, specificity of association, and coherence with existing information. Also known as causal association.

casein — solid or semi-solid protein of milk obtained by precipitation of the milk solids by the additions of acids or whey.

cash flow — annual cash receipts in the form of net profits plus the depreciation charge.

casing — (*water*) a pipe or tube of varying diameter and weight lowered into a borehole, during or after drilling, in order to support the sides of the hole or prevent contamination from water, gas, or other fluids.

CAT — *See* computerized axial tomography.

catabolic — referring to any destructive process by which complex substances are converted by living cells into more simple compounds with release of energy.

catabolism — the decomposition process of metabolism in which relatively large food molecules are broken down to yield smaller molecules and energy. Catabolism releases energy as heat or chemical energy. The chemical energy released by catabolism is first converted into ATP molecules, and then the ATP molecules are easily broken down and release the energy which can be utilized in anabolism.

catalysis — the phenomenon in which a relatively small amount of a substance speeds up or slows down a chemical reaction but is regenerated at the end of the reaction in its initial form.

catalyst — a substance that facilitates or alters a chemical reaction without being changed itself.

catalytic — of or pertaining to a process in which a catalyst is present in the incineration system and the chemicals are destroyed by a direct-flame oxidation process which proceeds at a lower temperature and in the absence of flame.

catalytic afterburner — an afterburner in which a catalyst induces incineration at lower temperatures than are possible in a direct fire-afterburner; a platinum alloy system which adsorbs the gases to be burned along with oxygen that has been injected in sizeable amounts in order that the gases may react with each other on the catalyst and oxidation products; carbon dioxide and water are the end products.

catalytic combustion system — a process in which a substance is introduced into an exhaust gas stream to burn or oxidize vaporized hydrocarbons or odorous contaminants while the substance itself remains intact.

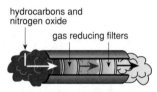

hydrocarbons and nitrogen oxide

gas reducing filters

Catalytic Converter

catalytic converter — an air pollution abatement device that removes organic contaminants by oxidizing them into carbon dioxide and water through chemical reaction.

catalytic oxidation — a sulfur dioxide eliminating process; flue gases are sent through a high-efficiency electrostatic precipitator to remove particulates; the gases then flow over a catalyst vanadium pentoxide that changes the sulfur dioxide to sulfur trioxide; the sulfur trioxide combines with water vapor in the flue gases to form sulfuric acid, which is condensed by cooling and can be used.

cataract — a clouding of the crystalline lens of the eye, obstructing the passage of light.

caterer — any food service establishment where food or drink is prepared for service elsewhere.

caterpillar — the worm-like stage of moths and butterflies that usually feeds on plants.

catheter — a tubular, flexible instrument passed through body channels for withdrawal or injection of fluids.

cathode — the negative terminal of an electrolytic cell to which positive ions are attracted.

cathode ray oscilloscope — *See* oscilloscope.

cathode rays — (*electricity*) streams of electrons given off by the cathode of a vacuum tube.

cathode-ray tube (CRT) — (*electricity*) a device attached to a computer that can visually display

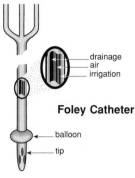

Foley Catheter

balloon

tip

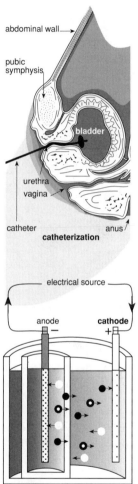

abdominal wall

pubic
symphysis

bladder

urethra
vagina

catheter anus

catheterization

electrical source

anode − cathode +

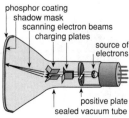

Voltaic Cell showing Cathode

phosphor coating
shadow mask
scanning electron beams
charging plates
source of
electrons

positive plate
sealed vacuum tube

Cathode Ray Tube

information contained in magnetic disk files; an electron tube in which a beam of electrons can be focused on a small area on a surface. Also known as an electron-ray tube; visual display terminal (VDT).

cation — (*chemistry*) the ion in the electrolyzed solution that migrates to the cathode, a positively charged ion.

cation-adsorption capacity — *See* cation-exchange capacity.

cation exchange — (*chemistry*) the chemical reaction interchange between a cation and solution and another cation and the surface of any surface-active material such as clay or organic colloids.

cation-exchange capacity — the sum total of exchangeable cations that a soil can adsorb; sometimes called total-exchange, base-exchange capacity, or cation-adsorption capacity; expressed in milliequivalents per hundred grams or per gram of soil. Also known as base exchange capacity; cation-adsorption capacity.

CAT scan — *See* computed tomography.

causal association — *See* case-history study.

causal organism — (*medicine*) a bacterium, fungus, nematode, or virus that can cause a given disease.

caustic — an alkali that strongly irritates, burns, corrodes, or destroys living tissue.

caustic soda — *See* sodium hydroxide.

caution — (*pesticides*) a warning to the user of pesticide chemicals; a message placed on labels of pesticide containers alerting the user to special precautions.

caveat emptor — a principle of commerce where a warranty does not exist and the buyer takes the risk of receiving inadequate quality.

cavitation — (*fluid mechanics*) the formation of cavities in a fluid downstream from an object moving in it as behind the moving blades of a propeller; (*pathology*) the formation of cavities in any body structure, especially in tubercular lungs.

cavitation corrosion — (*metallurgy*) the deterioration which occurs by the production, formation, and collapse of vapor bubbles on a metal surface.

CBA — *See* cost benefit analysis.

CBC — *See* complete blood count.

cc — *See* cubic centimeter.

CD — *See* criteria documents.

CDBG — *See* Community Development Block Grant.

CDC — *See* Centers for Disease Control and Prevention.

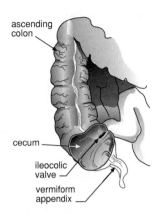

Cecum

cecum — (*anatomy*) the blind pouch-end of a cavity, duct, or tube. Also spelled caecum.

ceiling level (c) — the maximum allowable concentration of an airborne chemical in the workplace. Also known as ceiling limit.

ceiling limit — *See* ceiling level.

ceiling plenum — the space below the flooring and above the suspended ceiling that accommodates the mechanical and electrical equipment that is used as part of the air distribution system; the space is kept under negative pressure.

ceiling value (DL) — the concentration of a substance that should not be exceeded even instantaneously.

CEL — *See* cancer effect level.

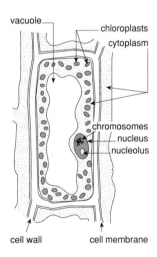

Plant Cell

cell — (*biology*) the fundamental structural and functional unit of living organisms. It is any of the minute protoplasmic masses making up organized tissue consisting of a nucleus, surrounded by cytoplasm, and enclosed in a plasma

membrane. The cell consists of the cytoplasmic membrane, endoplasmic reticulum (ER), Golgi apparatus, mitochondria, lysosomes, ribosomes, nucleus, and nucleoli; (*solid waste*) compacted solid wastes that are enclosed by natural soil or cover material in a sanitary landfill; (*geographic information system*) the basic element of spatial information on the roster (grid) description of spatial entities.

cell height — (*solid waste*) the vertical distance between the top or bottom of the compacted solid waste enclosed by natural soil or cover material in a sanitary landfill.

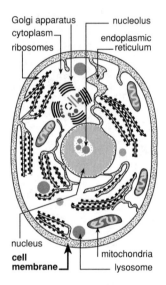

**Animal Cenl
Cell Membrane**

cell membrane — the thin, outer layer of cytoplasm of the cell consisting mainly of lipids and proteins.

cellosolve — *See* 2-ethoxyethanol.

cell thickness — (*solid waste*) the perpendicular distance between the cover materials placed over

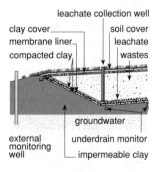

Landfill Cell

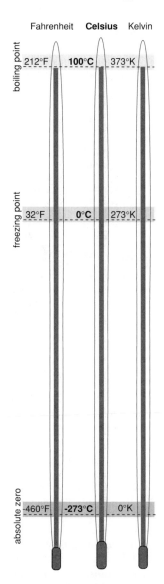

Fahrenheit **Celsius** Kelvin

boiling point   212°F   **100°C**   373°K

freezing point   32°F   **0°C**   273°K

absolute zero   -460°F   **-273°C**   0°K

Celsius Temperature Scale
compared to K and F scales
(not proportionate)

Celsius temperature scale — the freezing point of water at standard atmospheric pressure is nearly 0°C and the corresponding boiling point is nearly 100°C; formerly the Centigrade temperature scale.

cement — a substance made from limestone and clay which, when mixed with water, sets to a hard mass; used as an ingredient in concrete.

cementation — consolidation of loose sediments or sand by injection of a chemical agent or binder.

census days — the total number of days of patient care rendered by a health care facility in a given year or other specified period of time.

Centers for Disease Control and Prevention (CDC) — a federal health agency and branch of the Department of Health and Human Services located in Atlanta, Georgia; it provides national health and safety guidelines, statistical data, and specialized training in control of infectious diseases.

centi- (c) — prefix meaning 1/100 or .01 in the metric system.

Centigrade scale — *See* Celsius temperature scale.

centimeter (cm) — metric measurement of length equal to 0.3937 inches.

central heating — a single system supplying heat to one or more dwelling units or more than one rooming unit via pipes and ducts.

central nervous system (CNS) — the division of the vertebrate nervous system comprised of the brain and spinal cord and which coordinates all neural functions.

central nervous system depressants — poisons which depress the central nervous system resulting in headaches, vertigo, coma, and death; they include ethyl alcohol, wood alcohol, ether, chloroform, trichloroethylene, and benzene among others.

central neurons — those neurons in the brain and spinal cord that connect motor nerves to sensory nerves.

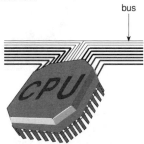

bus

**Central Processing Unit
CPU**

the last working faces of two successive cells in a sanitary landfill.

cell-type incinerator — an incinerator with grate areas divided into cells, each of which having its own ash drop, under fire, air control, and ash grate.

cellulose — $(C_6H_{10}O_5)_n$ — a polymer of glucose with a very high molar mass and containing a type of linkage not easily split by hydrolysis; a carbohydrate and the chief substance contained in the cell walls of plants.

cell wall — the tough, non-living outer layer of a plant cell consisting mainly of a cellulose secreted by the cytoplasm of the cell.

central processing unit (CPU) — (*computer science*) the part of the computer that operates the whole system.

central service (CS) — a department of a hospital responsible for sterilizing equipment and materials.

centrifugal — moving away from a center or axis of rotation.

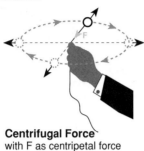

Centrifugal Force
with F as centripetal force

centrifugal force — force exerted by a revolving body in which movement extends away from the center of revolution.

Simple Centrifuge

centrifuge — a device used to separate solid or liquid particles of different densities by spinning them at high speeds.

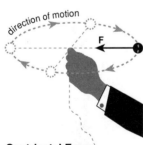

Centripetal Force

centripetal force — moving into the center toward the axis of rotation.

centromere — (*genetics*) a specialized area of a chromosome involved in its attachment to the fibers of the spindle during cell division. Also known as kinetochore.

CEPP — *See* Chemical Emergency Preparedness Program.

CEQ — *See* Council on Environmental Quality.

ceramic — calcined pottery, brick, and tile products molded from clay.

CERCLA — *See* Comprehensive Environmental Response Compensation and Liability Act.

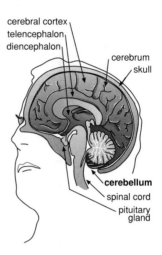

Cerebellum

cerebellum — the part of the vertebrate brain region between the cerebrum and medulla concerned with equilibrium and muscular coordination and consisting of three lobes.

cerebrospinal — of or pertaining to the brain and spinal cord.

cerebrospinal fluid (CSF) — a clear fluid in the brain ventricles and surrounding the spinal cord to protect the central nervous system from mechanical injury.

cerebrovascular accident (CVA) — a disorder of the blood vessels serving the cerebrum resulting from an impaired blood supply to, and ischemia in, parts of the brain. Also known as a stroke.

cerebrovascular disease — a disease of the blood vessels supplying the cerebrum.

cerebrum — the largest region of the vertebrate brain, considered to be the seat of emotions, intelligence, and other nervous system activities and consisting of two lateral hemispheres.

certification — the process by which a nongovernment agency or association grants recognition to a person meeting predetermined qualifications specified by that agency or association.

certified safety professional (CSP) — a person who meets specific educational and experience criteria and passes the appropriate examinations to do the work of an occupational safety specialist.

certiorari — (*law*) to be certified; a writ commanding a court to certify and return records to a superior court.

cerumen — earwax.

cervical — of or pertaining to the neck.

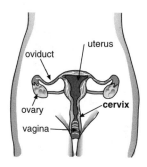

Cervix

cervix — the lower end of the uterus extending into the vagina.

Cestoda — a subclass of *Cestoidea* comprising the true tapeworm which has a head (scolex) and segments (proglottides); the adults are endoparasitic in the alimentary tract and associated ducts of various vertebrate hosts; their larvae may be found in various organs and tissues.

cestode — *See* tapeworm.

CFCs — *See* chlorofluorocarbons.

cfh — *See* cubic feet per hour.

cfm — *See* cubic feet per minute.

CFR — *See* Code of Federal Regulations.

cfs — *See* cubic feet per second.

CGI — *See* combustible gas indicator.

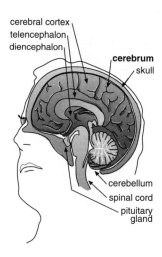

Cerebrum

chain — (*geographic information system*) a sequence of coordinates defining a complex line or boundary.

chain of infection — an organism leaving the reservoir of disease by means of a vector or vehicle and entering a host to cause disease.

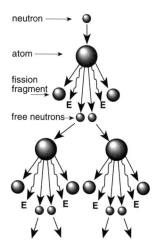

Chain Reaction

chain reaction — a self-sustaining reaction in which the fission of nuclei produces subatomic particles that cause the fission of other nuclei.

chamber — an enclosed space inside an incinerator; (*mining engineering*) the workplace of a miner.

char — a solid product of destructive distillation such as charcoal or coke; carbon.

character — (*geographic information system*) an alphabetical, numerical, or special graphic symbol that is treated as a single unit of data.

characteristic radiation — radiation originating from an atom after removal of an electron.

characteristics of hazardous waste — a present or potential hazard to human health and environment when improperly managed; measured by quick, available, standardized test methods, reasonably detected by generators of solid waste through knowledge of waste including ignitability, corrosivity, reactivity, and extraction procedure toxicity.

characters per second (CPS) — (*geographic information system*) a measure of the speed with which a device, usually a printer or visual display unit, can process data in the form of characters.

charcoal filter — a filtering device made of activated carbon capable of eliminating certain materials, especially those organic in nature.

charge — the basic property of elementary particles of matter; positive, negative, or zero; the quantity of solid waste introduced into a furnace at one time.

charged density — the charge per unit area on a surface or per unit volume in space.

charged particle — a particle which possesses at least one unit electrical charge and will not disintegrate upon loss of the charge.

charge sheet — specifications for the violation attached to a notice of hearing; the proposed statement charges that prosecution should ensue and becomes the government's allegation in the criminal case.

charging chute — an overhead passage through which waste materials drop into an incinerator.

charging gate — a horizontal moveable cover that closes the opening on the top charge of the furnace.

Charles' law — the law stating that when volume is held constant, the absolute pressure of a given mass of a perfect gas of a given composition varies directly as the absolute temperature. Also known as Gay-Lussac law.

check analysis — a second examination of a sample which has been examined and found to violate a law; usually conducted by an analyst other than the one performing the original analysis.

cheese — a food product made from the separated curd obtained by the casein of milk, skimmilk, or milk enriched with cream, and is then cut, shaped, pressed, and salted or brined.

chelate — an inert complex compound or ion in which a metallic atom or ion is bound at two or more points to a molecule or ion so as to form a ring.

chelating agent — any organic compound which will inactivate a metallic ion with the formation of an inner-ring structure in the molecule, where the metal ion becomes a member of the ring; a substance that removes organic material by forming a complex ion that does not have the chemical reactions of the ion that is removed.

chemical — a substance characterized by a specific molecular composition.

Chemical Abstracts — an abstract of chemical-related articles available in journals from 1907 to the present.

Chemical Abstract Services (CAS) — an organization of the American Chemical Society.

chemical asphyxiant — a substance in which the rough, direct chemical action can prevent the uptake of oxygen by blood, such as carbon monoxide, or interferes with the transportation of oxygen from the lungs to the tissues, or prevents normal oxygenation of tissues. Examples are hydrogen cyanide which inhibits enzyme systems affecting utilization of molecular oxygen, and hydrogen sulfide which paralyzes the respirator center of the brain and the olfactory nerve.

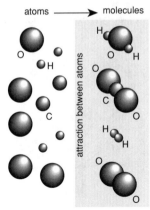

Chemical Bond

chemical bond — the force that holds atoms together in a molecule or crystal.

chemical burns — burns similar to those caused by heat but are instead caused by chemical contact and require special treatment.

chemical-cartridge respirator — a respirator, similar to a gas mask, that uses various chemical substances to purify inhaled air of certain gases and vapors; it is effective for concentrations up to ten times the threshold limit value of the contaminant.

chemical change — a transformation in which atoms are either rearranged to form new compounds or elements are separated from compounds.

chemical coagulation — the destabilization and initial aggregation of colloidal and finely divided suspended matter by the addition of a floc-forming chemical.

chemical decomposition indicators — indicators based on the fact that ionization produced by radiation in a chemical system is an indication of the amount of radiation received.

Chemical Dictionary On-Line (CHEMLINE) — the National Library of Medicine's on-line, interactive chemical dictionary file created by the Specialized Information Services in conjunction with the Chemical Abstracts Service.

chemical dose-response curve — the relationship between the potential toxicity inherent in a given chemical and kind of symptoms exhibited when the chemical interacts with the living, biological system.

chemical elements — a substance composed of atoms that are chemically alive and which cannot be separated into similar parts by chemical means.

Chemical Emergency Preparedness Program (CEPP) — a program developed by the EPA to address accidental releases of acutely toxic substances.

chemical energy — the energy contained in the chemical bond between atoms which can be released into the environment by a chemical reaction, such as combustion.

chemical engineering — the branch of engineering concerned with the development and application of manufacturing processes in which chemical or certain physical changes of materials are involved.

Chemical Exposure — a database from 1974 to the present providing comprehensive coverage of chemicals identified in both human tissues and body fluids and in feral and food animals; the service receives information on chemical properties, analytical methods, and toxic effects as well as references to selected journal articles, conferences, and reports.

chemical family — a group of simple elements or compounds with a common general name.

chemical feeder — (*swimming pools*) a mechanism for the automatic addition of chemicals, such as pH control agents and disinfectants to swimming pool water.

chemical food poisoning — an illness which develops rapidly and is due to chemicals accidentally introduced into the food or leached into food from a variety of containers; incubation period is less than 30 minutes.

Chemical Hazardous Information Profile (CHIP) — a profile of hazardous components found at an industrial site or disposal site including all information listed on Material Safety Data Sheets. *See also* Material Safety Data Sheet.

Chemical Hazards Response Information System/Hazard Assessment Computer System (CHRICS/HACS) — a program developed by the United States Coast Guard to provide chemical-specific data and determine how to handle chemical spills.

Chemical Information Database — a database available covering journal articles with information on pure chemicals, natural substances, and mixtures which result from industrial processes.

Chemical Manufacturer's Association (CMA) — an organization of chemical companies conducting specific research on chemicals germane to industrial activity.

chemical name — the name given to a chemical based on its structure in the nomenclature system developed by the International Union of Pure and Applied Chemistry (IUPAC) or the Chemical Abstract Service (CAS).

chemical oxygen demand (COD) — a measure of the amount of dissolved oxygen needed in the receiving water to dispose of the substances that can be oxidized chemically, but cannot be removed biologically.

chemical process — the particular method of manufacturing or making a chemical usually involving a number of steps or operations.

chemical protective clothing (CPC) — items of clothing used to isolate parts of the body from direct contact with potentially hazardous chemicals.

chemical reaction — a change in the arrangement of atoms or molecules to yield substances of different composition and properties.

chemical sensitivity — a sensitivity in some individuals to certain chemicals resulting in health problems with symptoms of dizziness, eye and throat irritation, chest tightness, and nasal congestion.

chemical sterilization — *See* cold sterilization.

chemical toilet — a toilet in which sewage is retained and treated with chemicals.

Chemical Transportation Emergency Center (CHEMTREC) — a center provided by the Chemical Manufacturers' Association to give information and/or assistance to emergency responders.

chemical waste — the waste generated by chemical, petrochemical, plastic, pharmaceutical, biochemical, or microbiological manufacturing processes.

chemical weathering — the decomposition of rocks and minerals by chemical reactions, usually through oxidation, carbonation, hydration, and solution in water above or below the surface.

chemiluminescence — emission of light as a result of a chemical reaction without an apparent change in temperature.

chemistry — the science that deals with the composition, structure, and properties of substances and their transformations.

CHEMLINE — *See* Chemical Dictionary On-Line.

CHEMNET — a mutual-aid network of chemical shippers and contractors.

chemoprophylaxis — the use of a chemical, including antibiotics, to prevent the development or progression of an infection.

chemosterilant — a pesticide chemical that controls populations by destroying their ability to reproduce.

chemosynthesis — the organization of carbohydrates of organisms by means of energy from inorganic chemical reactions instead of energy from light.

chemosynthetic bacteria — non-green plants that derive their energy for living from oxidizing inorganic compounds.

chemotaxis — the orientation or movement of a living organism in response to a chemical gradient.

chemotherapy — the use of chemicals of particular molecular structures in the treatment of specific disorders, where the structures exhibit an affinity for certain parts of the affected cell tissues or of the invaded bacterium, or in the case of cancer, on cancerous cells.

CHEMTREC — *See* Chemical Transportation Emergency Center.

chert — a microcrystalline form of silica.

Cheyne-Stokes respiration — breathing characterized by rhythmic waxing and waning of the depth of respiration; often observed in sleeping or unconscious people who seem to stop breathing for 5–40 seconds, start again, increase intensity of breathing, stop breathing, and repeat the process; commonly found in healthy infants.

chigger — the six-legged mite larva of the family Trombiculidae that attaches to the host's skin, and whose bite produces a wheal, usually with intense itching and severe dermatitis.

childhood lead poisoning — poisoning due to a variety of lead exposures, but primarily through the consumption and digestion of lead paint or other inedible materials containing lead; prenatal exposure may cause decreased IQ and other forms of impaired learning in the fetus; also causes increased blood pressure, brain and kidney damage, anemia, encephalopathy, and death.

CHIP — *See* Chemical Hazardous Information Profile.

chi-square — (*statistics*) a test of statistical significance which compares the observed number of cases falling into each category with the expected number of cases under the null hypothesis to determine whether the actual distribution differs significantly from what was expected.

chitin — a material present in the exoskeleton of insects and other arthropods.

chlamydia — any of a genus of coccoid to spherical gram-negative intracellular rickettsial parasites, including one that causes or is associated with various diseases of the eye, including trachoma and lymphogranuloma; also causes some forms of urethritis.

chloracne — an acneiform eruption caused by chlorinated napthalenes and polyphenyls acting on sebaceous glands.

chloramine-T — (sodium paratoluene sulfonchloramide) — $CH_3C_6H_4SO_2NClNa\cdot 3H_2O$ — a chlorine compound that decomposes slowly in air, used for the sanitization of utensils and equipment.

4-chloraniline — C_6–H_6–Cl–N — a white to pale yellow to tan colored crystalline solid with a slightly sweetish characteristic amine odor; MW: 127.57, BP: 158–162.5°F, Sol: practically insoluble in cold water, soluble in hot water, Fl.P: 220°F, sp gr: 1.427. Other information is not available. OSHA exposure limit (PEL-TWA): not established.

chlorasis — discoloration of normally green plant parts that can be caused by disease, lack of nutrients, or various air pollutants.

chlordane — $C_{10}H_6Cl_8$ — an amber-colored, viscous liquid with a pungent, chlorine-like odor; MW: 409.8, BP: decomposes, Sol: insoluble, sp gr (77°F): 1.56. It is used as an insecticide on pre-planting soil, fire ants, and harvester ants (banned by the EPA in 1976). It is hazardous to the central nervous system, eyes, lungs, liver, kidneys, and skin; and is toxic by inhalation, absorption, ingestion, and contact. Symptoms of exposure include blurred vision, confusion, ataxia, delirium, cough, abdominal pain, nausea, vomiting, diarrhea, irritability, tremors, convulsions, anuria; lung, liver, and kidney damage in animals. OSHA exposure limit (TWA): 0.5 mg/m^3 [skin].

CHLOREP — *See* Chlorine Emergency Plan.

chlorinate — to treat or cause to combine with chlorine or a chlorine compound.

chlorinated camphene — $C_{10}H_{10}Cl_8$ — an amber, waxy solid with a mild chlorine- and camphor-like odor; MW: 413.8, BP: decomposes, Sol: 0.0003%, Fl.P: 275°F, sp gr (77°F): 1.65. It is used in liquid, dust, powder, and granular insecticides. It is hazardous to the central nervous system and skin; and is toxic by inhalation, absorption, ingestion, and contact. Symptoms of exposure include nausea, confusion, agitation, tremors, convulsions, unconsciousness, and dry, red skin; carcinogenic. OSHA exposure limit (TWA): 0.5 mg/m^3 [skin].

chlorinated diphenyl oxide — $(C_6H_2CL_3)_2O$ — a light yellow, very viscous, waxy liquid; MW: 376.9, BP: 446–500°F, Sol: 0.1%, sp gr: 1.60. It is used in organic synthesis; in corrosion inhibitors, dry cleaning detergents, thermal lubricants, additives for soaps and lotions; in the manufacture of hydraulic fluids, pesticides, wood preservatives, and electric insulators. It is hazardous to the skin and liver; and is toxic by inhalation, ingestion, and contact. Symptoms of exposure include acne-form dermatitis and liver damage. OSHA exposure limit (TWA): 0.5 mg/m^3 [air].

chlorinated hydrocarbons — a group of organic chemicals containing chlorine and used as pesticides.

chlorinated polyethylene (CPE) — a very flexible thermoplastic produced by a chemical reaction between chlorine and polyethylene.

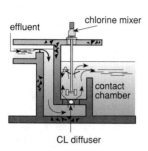

Mechanical Mixing Chamber

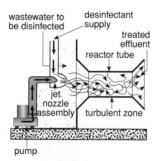

Chlorine Injector Mixer (Pump Type)

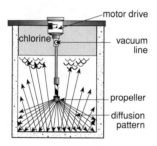

Chlorine Injector Mixer (Aspirating Type)

Chlorinator

chlorinator — a chemical feeder for the automatic addition of chlorine to water.

chlorine — Cl_2 — a heavy, greenish-yellow gas with a pungent, irritating odor; shipped in steel cylinders; MW: 70.9, BP: –29°F, Sol: 0.7%. It is used as a chlorinating and oxidizing agent in organic synthesis; in the manufacture of solvents, automotive antifreeze and antiknock compounds, plastics, resins, elastomers, pesticides, refrigerants, bleaching, and inorganic chemicals; as a fluxing, purification, and extraction agent in metallurgy; as a bacteriostat, disinfectant, odor control, and demulsifier in the treatment of sewage; as a bleaching and cleaning agent and disinfectant in laundries, dishwashers, and cleaning powders. It is hazardous to the respiratory system; and is toxic by inhalation and contact. Symptoms of exposure include lacrimation, rhinorrhea, cough and choking, substernal pain, nausea, vomiting, headache, dizziness, syncope, pulmonary edema, pneumonia, hypoxemia, dermatitis, burning of the eyes, skin, nose, and mouth. OSHA exposure limit (TWA): 0.5 ppm [air] or 1.5 mg/m³.

chlorine combined available residual — that portion of the total residual chlorine remaining in water, sewage, or industrial waste at the end of a specified contact period that will react chemically and biologically as chloramines or organic chloramines.

chlorine demand — the amount of chlorine necessary to oxidize all organic matter present in water at any given moment or over a period of time.

chlorine dioxide — ClO_2 — a yellow to red gas or a red-brown liquid (below 52°F) with an unpleasant odor similar to chlorine and nitric acid; MW: 67.5, BP: 52°F, Sol (77°F): 0.3%, sp gr: 1.6. It is used as a bleaching agent for wood pulp, paper, and textiles; in water treatment to remove tastes and odors by oxidation; as a bleaching and maturing agent in flour for baking; as a bleaching, cleaning, and dehairing agent in leather manufacture; in chemical synthesis as an oxidizing agent; as a bacteria and algae control agent in aquariums; and in packaged pharmaceuticals. It is hazardous to the respiratory system and eyes; and is toxic by inhalation, ingestion, and contact. Symptoms of exposure include eye, nose, and throat irritation, cough, wheezing, bronchitis and chronic bronchitis, and pulmonary edema. OSHA exposure limit (TWA): 0.1 ppm [air] or 0.3 mg/m³.

Chlorine Emergency Plan (CHLOREP) — operated by the Chlorine Institute as a 24-hour mutual aid program.

chlorine-free available residual — that portion of the total chlorine remaining in water, sewage, or industrial waste at the end of a specified contact period that would react chemically and biologically as hypochlorite ions.

chlorine oxidation — a means of control of odor and reduction of pathogens without significantly reducing sludge solids through the use of chlorine.

chlorine residual — the total amount of chlorine, combined and free available chlorine, remaining in water, sewage, or industrial waste at the end of a specified contact period following chlorination.

chlorine room — a separate room utilized for housing gas chlorine; it is adjacent to the filter room in swimming pool operations.

chlorine trifluoride — ClF_3 — a colorless gas or a greenish-yellow liquid (below 53°F) with a somewhat sweet, suffocating odor; MW: 92.5, BP: 53°F, Sol: reacts, sp gr: 1.77. It is used as a fluorinating agent in organic and inorganic chemical synthesis; as a rocket fuel oxidizer; as an incendiary; as a cutting agent for well castings in oil well drilling. It is hazardous to the skin and eyes; and is toxic by inhalation, ingestion, and contact. Symptoms of exposure include eye and skin burns; in animals: lacrimation, corneal ulcer, and pulmonary edema. OSHA exposure limit (TWA): ceiling 0.1 ppm [air] or 0.4 mg/m³.

chloroacetaldehyde — $ClCH_2CHO$ — a colorless liquid with an acrid, penetrating odor; MW: 78.5, BP: 186°F, Sol: miscible, Fl.P: 190°F, sp gr: 1.19. It is used during the manufacture of 2-aminothiazole; in the control of algae, bacteria, and fungi in water; during debarking operations; in acid media during chemical synthesis. It is hazardous to the eyes, skin, and respiratory system; and is toxic by inhalation, ingestion, and contact. Symptoms of exposure include eye, skin, and mucous membrane irritation, skin burns, eye damage, pulmonary edema, skin and respiratory system sensitization, narcosis, coma. OSHA exposure limit (TWA): ceiling 1 ppm [air] or 3 mg/m³.

alpha-chloroacetopheone — $C_6H_5COCH_2Cl$ — a colorless to gray crystalline solid with a sharp, irritating odor; MW: 154.6, BP: 472°F, Sol: insoluble, Fl.P: 244°F, sp gr (59°F): 1.32. It is used during denaturing of industrial alcohol; in solutions for aerosols for law enforcement and civilian protective devices. It is hazardous to the eyes, skin, and respiratory system; and is toxic by inhalation, ingestion, and contact. Symptoms of exposure include eye, skin, and respiratory system irritation, and pulmonary edema. OSHA exposure limit (TWA): 0.3 mg/m³ [air].

chloroacne — a disease caused by chlorinated naphthalenes and polyphenols acting on sebaceous glands.

chlorobenzene — C_6H_5Cl — a colorless liquid with an almond-like odor; MW: 112.6, BP: 270°F, Sol: 0.05%, Fl.P: 85°F, sp gr: 1.11. It is used as an intermediate in the manufacture of dye ingredients, DDT, aniline, picric acid, and other chemicals; in the manufacture of rubber adhesives; as a fiber swelling agent and dye carrier in textile processing; as a tar and grease remover in cleaning and degreasing operations; as an extractant in the manufacture of rubber, perfumes, and pharmaceuticals, and diisocyanates. It is hazardous to the respiratory system, eyes, skin, central nervous system, and liver; and is toxic by inhalation, ingestion, and contact. Symptoms of exposure include drowsiness, incoordination, and eye, skin, and nose irritation; in animals: liver, lung, and kidney damage. OSHA exposure limit (TWA): 75 ppm [air] or 350 mg/m³.

o-chlorobenzylidene malononitrile — $ClC_6H_4CH=C(CN)_2$ — a white crystalline solid with a pepper-like odor; MW: 188.6, BP: 590–599°F, Sol: insoluble. It is used as a harassing agent in civil disturbances and wars. It is hazardous to the respiratory system, skin, and eyes; and is toxic by inhalation, absorption, ingestion, and contact. Symptoms of exposure include painful, burning eyes, lacrimation, conjunctivitis, erythema eyelids, blepharism, throat irritation, cough, chest constriction, headache, vesiculation of the skin. OSHA exposure limit (TWA): ceiling 0.05 ppm [skin] or 0.4 mg/m³.

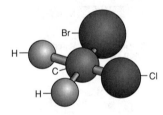

Chlorobromomethane
(Bromochloromethane)

chlorobromomethane — CH_2BrCl — a colorless to pale yellow liquid with a chloroform-like odor; MW: 129.4, BP: 155°F, Sol: insoluble, sp gr: 1.93. It is used as a fire extinguishing fluid in vaporizing fire extinguishers; in mineral and salt separations for flotation; as a grain fumigant. It is hazardous to the skin, liver, kidneys, respiratory system, and central nervous system;

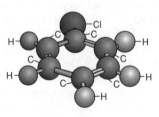

Chlorobenzene

and is toxic by inhalation, ingestion, and contact. Symptoms of exposure include disorientation, dizziness, pulmonary edema, and eye, skin, and throat irritation. OSHA exposure limit (TWA): 200 ppm [air] or 1,050 mg/m^3.

chlorodibenzofurans — $C_{12}H_4ClO$, $C_{12}H_3Cl_5O$, $C_{12}H_2Cl_6O$, $C_{12}HCl_7O$, $C_{12}Cl_8O$ — a family containing 135 individual, colorless compounds known as congeners with varying harmful health and environmental effects. They are not commercially manufactured or used in the United States or abroad, except on a laboratory scale, but are produced as unwanted compounds during the manufacture of several chemicals and consumer products such as wood treatment chemicals, some metals, and paper products; also produced from the burning of municipal and industrial waste in incinerators, from exhaust of leaded gasoline, heat, or production of electricity; and from fires or breakdowns involving capacitors, transformers, and other electrical equipment containing polychlorinated biphenyls (PCBs). They are hazardous to the respiratory system, gastrointestinal system, liver, musculoskeletal system, skin, and nervous system; and are toxic by inhalation, ingestion, and contact. Symptoms of exposure include frequent coughing, severe respiratory infections, chronic bronchitis, abdominal pain, muscle pain, acne rashes, skin color changes, unexpected weight loss, nonmalignant or malignant liver disease. Listed as hazardous waste pollutants by the Environmental Protection Agency; no exposure limits exist at this time.

chlorodiphenyl (42% chlorine) — $C_6H_4ClC_6H_3Cl_2$ (approximately) — a colorless to light colored, viscous, liquid with a mild hydrocarbon odor; MW: 258 (approx), BP: 617–691°F, Sol: insoluble, Fl.P: not known, sp gr (77°F): 1.39. It is used as a high-temperature transfer medium in chemical and food processing vessels and drying ovens; as a dielectric in the manufacture of transformers, capacitors, resistors, and other electrical apparatus; in the manufacture of impregnants for cloth, paper, fiberboard, and wood; in the manufacture of natural and synthetic waxes and polishes; in the manufacture of hot-melt and other adhesives; in the manufacture of pesticides and fungicides; as a sealer for gaskets of natural rubber and synthetics; as a pigment carrier; as a pressure adhesive for sign backings; in compounding mastics, and sealing and caulking materials; in compounding printing inks. It is hazardous to the skin, eyes, and liver; and is toxic by inhalation, absorption, ingestion, and contact. Symptoms of exposure include eye irritation, chloracne, and liver

damage; carcinogenic. OSHA exposure limit (TWA): 1 mg/m^3 [skin].

chlorodiphenyl (54% chlorine) — $C_6H_3Cl_2C_6H_2Cl_3$ (approximately) — a colorless to pale yellow, viscous (liquid or solid below 50°F) with a mild, hydrocarbon odor; MW: 326 (approx), BP: 689–734°F, Sol: insoluble, Fl.P: not known, sp gr (77°F): 1.38. It is used as a high-temperature transfer medium in chemical and food processing vessels and drying ovens; as a dielectric in the manufacture of transformers, capacitors, resistors, and other electrical apparatus; in the manufacture of impregnants for cloth, paper, fiberboard, and wood; in the manufacture of natural and synthetic waxes and polishes; in the manufacture of hot-melt and other adhesives; as a non-flammable working fluid in vacuum pumps, hydraulic systems, and expansion systems; in compounding and processing of plastics for flame retardancy; in the manufacture of pesticides and fungicides; as a sealer for gaskets of natural rubber and synthetics; as a pigment carrier; as a pressure adhesive; in compounding mastics, and sealing and caulking materials; in compounding of printing inks. It is hazardous to the skin, eyes, and liver; and is toxic by inhalation, absorption, ingestion, and contact. Symptoms of exposure include acneform dermatitis, eye and skin irritation; in animals: liver damage; carcinogenic. OSHA exposure limit (TWA): 0.5 mg/m^3 [air].

chloroethane — *See* ethyl chloride.

bis (2-chloroethyl) sulfide — *See* mustard gas.

chlorofluorocarbons (CFCs) — relatively nontoxic, nonflammable, stable combinations of chlorine, fluorine, and carbon used in air conditioners, refrigerators, and freezers; in insulation and foam products; in cleaning solvents and medical sterilizing agents; and as aerosol propellants. CFCs can exist for hundreds of years, eventually wafting up to the stratosphere, where bombardment by ultraviolet radiation releases chlorine atoms that consume ozone molecules and hinder formation of new ozone molecules; this process results in the depletion of the ozonosphere.

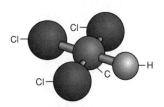

Chloroform

chloroform — CHCl$_3$ — a colorless liquid with a pleasant nonirritating odor and a slight sweet taste; MW: 119.38, BP: 61.3°C, Sol. in water at 25°C, 7.22×10^3 mg/L, Fl.P: none. It is used in producing fluorocarbon-22 (chlorodifluoromethane), fluoropolymers, as a solvent or an extraction solvent for fats, oils, greases, resins, lacquers, rubber, alkaloids, gums, waxes, penicillin, vitamins, flavors, floor polishes, adhesives, in artificial soot manufacturing, dry cleaning spot remover, fire extinguishers, as an intermediate, manufacture of dyes and pesticides, as a fumigant, and in chlorination of wastewater. It may be transported for long distances before being degraded by reacting with photochemically generated hydroxyl radicals. Acute exposure can result in death; changes in respiratory rates occur in individuals exposed to chloroform; they become depressed during sleep and prolong anesthesia; causes cardiac effects in patients under anesthesia with bradycardia, cardiac arrhythmia, hypotension, first degree atreo-ventricular block, or complete heart block, nausea and vomiting occurs as well as a fullness in the stomach; one of the major toxic effects is in the liver since chloroform induces hepatotoxicity with impaired liver function, jaundice, liver enlargement, tenderness, delirium, coma, and death; fatty degeneration of kidneys may also occur; the central nervous system is a target for chloroform toxicity, creating headaches and light intoxication, exhaustion, hallucinations, delusions, and convulsions. Also known as trichloromethane. OSHA exposure limits (TWA): 2 ppm (9.79 mg/m³) [air].

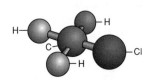

Chloromethane
(Methyl Chloride)

chloromethane — CH_3Cl — a clear colorless gas or vapor that is difficult to smell; has a faintly sweet nonirritating odor at high levels in the air; MW: 50.49, BP: –23.73°C, Sol. in water at 25°C; 0.5%, Fl.P: less than 0°C. Cardiovascular effects include electrocardiogram abnormalities, tachycardia, increased pulse rate and decreased blood pressure with death occurring in extreme exposures; gastrointestinal effects include nausea and vomiting accompanied by central nervous system toxicity as well as clinical jaundice and cirrhosis of the liver; renal effects include renal toxicity; blurred, double vision, fatigue, drowsiness, staggering, head-

aches, mental confusion, tremor, vertigo, muscular cramping and rigidity, sleep disturbances, and ataxia may occur. *See also* methyl chloride. OSHA exposure limit (TWA): 50 ppm (8 hours); (STL): 100 ppm.

bis-chloromethyl ether — $(CH_2Cl)_2O$ — a colorless liquid with a suffocating odor; MW: 115.0, BP: 223°F, Sol: reacts, Fl.P: <66°F, sp gr: 1.32. It is used in the manufacture of ion exchange resins and polymers; as a solvent for polymerization reactions; as a chloromethylation agent in chemical synthesis; in the treatment of textiles. It is hazardous to the lungs, eyes, and skin; and is toxic by inhalation, ingestion, and contact. Symptoms of exposure include pulmonary congestion and edema, corneal damage, necrosis, reduced pulmonary function, cough, dyspnea, wheezing, blood-stained sputum, bronchial secretions, irritation of the eyes, skin, and mucous membrane of respiratory system; carcinogenic. OSHA exposure limit (TWA): regulated use.

chloromethyl methyl ether — CH_3OCH_2Cl — a colorless liquid with an irritating odor; MW: 80.5, BP: 138°F, Sol: reacts, Fl.P (oc): 32°F, sp gr: 1.06. It is used in the manufacture of ion exchange resins and polymers; as a solvent for polymerization; as a chloromethylation agent in chemical synthesis; in the treatment of textiles. It is hazardous to the respiratory system, skin, eyes, and mucous membrane; and is toxic by inhalation, absorption, and contact. Symptoms of exposure include pulmonary edema and congestion, pneumonia, burns, necrosis, cough, wheezing, blood-stained sputum, weight loss, bronchial secretions, irritation of the eyes, skin, and mucous membrane; carcinogenic. OSHA exposure limit (TWA): regulated use.

1-chloro-1-nitropropane — $CH_3CH_2CHClNO_2$ — a colorless liquid with an unpleasant odor; MW: 123.6, BP: 289°F, Sol: 0.5%, Fl.P (oc): 144°F, sp gr: 1.21. It is used as a solvent and anti-gelling agent for rubber cements; as a fungicide. It is hazardous to the respiratory system, liver, kidneys, and cardiovascular system of animals; and is toxic by inhalation, ingestion, and contact. Symptoms of exposure in animals include eye irritation, pulmonary edema, and liver, kidney, and heart damage. OSHA exposure limit (TWA): 2 ppm [air] or 10 mg/m³.

2-chlorophenol — C_6H_5-Cl-O — a colorless to yellow-brown liquid with an unpleasant penetrating odor; MW: 128.56, BP: 347–348.8°F, Sol: 2.85 g/100 ml, Fl.P: 147°F, sp gr: 1.241. Other information is not available. Short exposure could cause serious temporary or residual injury. OSHA exposure limit (PEL-TWA): not established.

2,2-bis(para-chlorophenyl)-1,1-dichloroethane
(DDD) — $C_{14}H_{10}Cl_4$ — a colorless, crystalline
compound; MP: 109–111°C, Sol: insoluble. It
is used as an insecticide on fruits and vegetables.
An analogue of DDT, it is equally as toxic as a
mosquito larvacide and of lower toxicity to fish
and shellfish.

chlorophyll — the generic name for several green
pigmented substances in plants which capture
and transform light energy into chemical energy
that can be used by living organisms.

chloropicrin — CCl_3NO_2 — a colorless to faint
yellow, oily liquid with an intensely irritating
odor; MW: 164.4, BP: 234°F, Sol: 0.2%, sp gr:
1.66. It is used as a soil fumigant, disinfectant,
and sterilizer for control of fungi, nematodes,
and other injurious organisms; as a fumigant for
stored cereals, grains, and fruits; as a rodenticide and insecticide; as a chemical intermediate
in organic synthesis; as a warning agent in illu-
minating gas; as a nauseant in chemical warfare; as a chemical sterilant without high temperature. It is hazardous to the respiratory system,
skin, and eyes; and is toxic by inhalation, ingestion, and contact. Symptoms of exposure include eye irritation, lacrimation, cough, pulmonary edema, nausea, vomiting, skin irritation.
OSHA exposure limit (TWA): 0.1 ppm [air] or
0.7 mg/m³.

chloroplast — a type of cell plastid containing
chlorophyll and functioning in photosynthesis.

chlorosis — yellowing or whitening of normally
green plant parts because of disease organisms
including viruses; or due to lack of oxygen in
water-logged soil; or a lack of boron, iron, magnesium, manganese, nitrogen, or zinc.

cholera — (*disease*) an acute, bacterial, enteric
disease with sudden onset, profusely watery
stools, occasional vomiting, rapid dehydration,
acidosis, and circulatory collapse. Incubation
time is a few hours to five days, but usually 2–
3 days; caused by *Vibrio cholerae*, found worldwide. The reservoir of infection is people. It is
transmitted through the ingestion of water contaminated with feces or vomitus; or the ingestion of food contaminated with water or hands;
or the spread of disease by flies; communicable
during the stool-positive stage or as a carrier
state for many months; susceptibility is variable. It is controlled through isolation using
enteric precautions, hand washing, proper disposal of feces and vomitus as well as articles
contaminated, and the proper treatment and
chlorination of water.

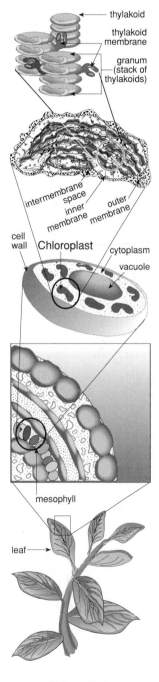

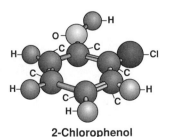

2-Chlorophenol

Chloroplast

C
D

choline — a quaternary amine that occurs in the neurotransmitter acetylcholine and is an important methyl donor in intermediary metabolism.

cholinesterase — an enzyme found in blood and other tissues that splits acetylcholine into acetic acid and choline; helps regulate the activity of nerve impulses and is destroyed or damaged when organic phosphates or carbamates enter the body.

cholinesterase inhibitor — any organo-phosphate, carbamate, or other pesticide chemical that can interrupt the action of enzymes, which inactivate the acetylcholine used in the nervous system.

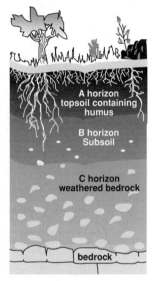

C Horizon
and other layers of soil

C horizon — (*soils*) the soil horizon that normally lies beneath the B horizon but may lie beneath the A horizon; it is unaltered or slightly altered parent material.

CHRICS/HACS — *See* Chemical Hazards Response Information System/Hazard Assessment Computer System.

chromatid — (*genetics*) the daughter strand of a chromosome that remains joined by the centromere to another chromatid following chromosomal duplication.

chromatograph — the apparatus used in chromatography to separate substances.

chromatography — a technique for the analysis of chemical substances. The substances to be analyzed are poured into a vertical glass tube containing an adsorbent where the various components of the substance move through the adsorbent at different rates of speed according to their degree of attraction to it, thereby producing bands of color at different levels of the adsorption column.

chromatolysis — the solution and disintegration of the chromatin of cell nuclei; disintegration of the Nissl bodies of a neuron as a result of injury, fatigue, or exhaustion.

chromic acid and chromates — H_2CrO_4 — appearance and odor vary depending on specific compound; properties vary depending on specific compound. They are used in metal and leather finishing; as corrosion inhibitors in radiator coolants, internal combustion and gas turbine engines, refrigerator and air conditioning systems, and water-cooled nuclear reactors; in photoreproduction processes; as coloring pigments; in the dyeing of fur, leather, fabrics, wool, and nylon; in textile and paper printing; in the manufacture of some glues; as a fungicide; to increase the shelf-life of batteries; in the manufacture of safety matches and explosives; as a chemical reagent, oxidizing agent, catalyst, indicator; in chemical synthesis and analytical chemistry; in the manufacture of cement. They are hazardous to the blood, respiratory system, liver, kidneys, eyes, and skin; and are toxic by inhalation, ingestion, and contact. Symptoms of exposure include respiratory system irritation, nasal septum perforation, liver and kidney damage, leukocytosis, leukopenia, monocytosis, eosinophilia, eye injury, conjunctivitis, skin ulcers, sensitization dermatitis; carcinogenic. OSHA exposure limit (TWA): ceiling 0.1 mg CrO_3/m^3 [air], supplementary exposure limits for chromium compounds.

chromium — Cr — a metallic element with oxidation states ranging from chromium II to chromium VI; the important balance states of chromium are 2, 3, and 4; elemental chromium (O) does not occur naturally; the divalent state II is relatively unstable and is readily oxidized to the trivalent state III, or which is the chromic state; the hexavalent state VI is the chromate state; a naturally occurring element found in rocks, animals, plants, soil, and in volcanic dust and gases; no known taste or odor; the metal is a steel-gray solid with a high melting point; MW: 51.996, BP: 2,672°C, Sol: insoluble in water, Fl.P: not known. It is used in the metallurgical, refractory, and chemical industries to produce stainless steels, alloy cast irons, nonferrous alloys, and other miscellaneous materials; also used in drilling, muds, water treatment, as rust and corrosion inhibitors, chemical manufacturing, textiles, toners for copying machines, magnetic tapes, and as catalysts; various and increased risk of death from noncancer respiratory disease by workers in chrome plating plants; the respiratory tract is a major target of inhalation exposure of chromium compounds; chromate-sensitive workers acutely exposed to chromium VI compounds may develop asthma and other signs of respiratory distress accompanied by erythema of the face, nasopharyngeal pruritus, nasal blocking, cough, wheezing, and dyspnea; exposure to chromium trioxide causes coughing, expectoration, nasal irritation, sneezing, rhinorrhea, nose bleed, and ulceration and perforation of the nasal septum; gastrointestinal effects have been associated with occupational exposure; hepatic effects of chromium VI cause severe liver defects in workers with derangement of the cells in the liver, necrosis, lymphocytic and histiocytic infiltration; acute systemic and dermal allergic reactions have been observed in chromium-sensitive individuals exposed to chromium via inhalation; it causes immunological effects with frequent skin eruptions, dyspnea, chest tightness, anaphylactic reactions, bronchospasms, and a tripling of the plasma histamine levels; neurological effects include dizziness, headache, anoxia, and weakness; occupational exposure to chromium compounds has been associated with increased risk of respiratory system cancers, primarily bronchogenic and nasal. OSHA exposure limits PEL (TWA) for chromium 0 and insoluble salts: 1.0 mg/m³; chromium II: 0.5 mg/m³; chromium III: 0.5 mg/m³; ceiling concentration chromium IV: 0.1 mg/m³.

chromium metal — Cr — a blue-white to steel-gray, lustrous, brittle, hard solid; MW: 52.0, BP: 4788°F, Sol: insoluble, sp gr: 7.14, atomic weight: 24. It is used in the fabrication of alloys; in the fabrication of plated products for decoration or increased wear-resistance; as chemical intermediates; in dyeing, silk treating, printing, and moth-proofing wool; in the leather industry; in photographic fixing baths; as catalysts for halogenation, alkylation, and catalytic cracking of hydrocarbons; as fuel additives and propellant additives. It is hazardous to the respiratory system; and is toxic by inhalation and ingestion. Symptoms of exposure include histologic fibrosis of the lungs, systemic and dermal allergic reactions, and neurological effects. OSHA exposure limit (TWA): 1 mg/m³ [air], supplementary exposure limits.

chromogen — (*chemistry*) an organic substance that forms a colored compound or becomes a pigment when exposed to air; a compound, not a dye, having color-forming groups and thus capable of being converted into a dye.

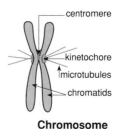

Chromosome

chromosomes — thread-like structures visible during cell division that carry the hereditary traits and that are contained in the nucleus of a cell; they are composed of chromatin.

chronic condition — a condition which has one or more of the following characteristics: (1) permanent, (2) leaving residual disability, (3) being caused by nonreversible pathological alteration, (4) requiring special training for rehabilitation. Also known as chronic disease.

chronic disease — *See* chronic condition.

chronic effect — a permanent adverse effect on a human or animal body with symptoms that develop slowly over a long period of time or which reoccur frequently as a result of exposure to a toxin or development of a disease.

chronic exposure — exposure to a chemical or other substance for a period of 365 days or more.

chronic obstructive pulmonary disease (COPD) — a functional category designating a chronic condition of persistent obstruction of the bronchial air flow. It is the most significant chronic pulmonary disorder in the United States in regard to the morbidity rate and is the second most common cause of hospital admissions in the country.

chronic toxicity — adverse chronic effects resulting from repeated doses of or exposures to a substance over a relatively long period of time.

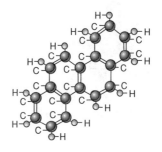

Chrysene

chrysene — $C_{18}H_{12}$ — a polycyclic aromatic hydrocarbon formed when gasoline, garbage, or any animal or plant material burns, usually found in smoke and soot; MW: 228.3, BP: 448°C (sublimes in vacuo), Sol: 1.5–2.2 mg/L, Fl.P: not known. It is used as a research chemical, as coal tar pitch, as a binder for electrodes in the aluminum reduction process, to bind the carbon electrodes in the reduction pots; also used as an adhesive in membrane roots; found in creosote wood preservative; used in paving and sound and electrical insulation. It is a weak carcinogen by dermal exposure and a weak mutagen in experimental systems. OSHA exposure limit (TWA): 0.2 $\mu g/m^3$ [air].

chrysotiles — $Mg_3Si_2O_5(OH)_4$ — a white-greenish hydrated magnesium silicate and a fibrous form of asbestos.

chute-fed incinerator — an incinerator that is charged through a chute that extends two or more floors above it. Also spelled shute-fed incinerator.

chyme — partially digested food as it leaves the stomach.

Ci — *See* curie.

CIB — *See* Current Intelligence Bulletins.

cigarette beetle (*Lasioderma serricorne*) — primarily a pest of tobacco but also a stored food product pest; a small, oval, light brown beetle, 1/10" long with smooth wing covers; the head is retracted to meet the thorax.

cilia — minute, short, hair-like cells capable of lashey movement to produce locomotion in unicellular organisms or a current in higher organisms. Singular is cilium.

CIP — *See* cleaning-in-place.

circuit *(electricity)* — the complete path, or a part of it, over which an electrical current flows, usually including the generating apparatus.

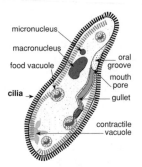

Cilia on *Paramecium*

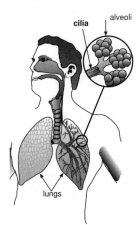

Cilia in Bronchiole

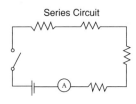

Series Circuit

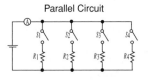

Parallel Circuit

Branching Circuit

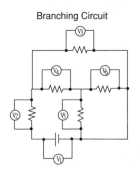

Circuit

S=Switch   A=Ammeter
V^x=Voltmeter R_x=Resistor

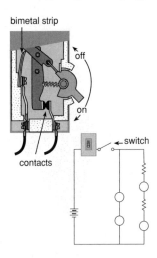

Circuit Breaker

circuit breaker — an electromagnetic switch that breaks an electric circuit when overloaded.

circulatory system — the series of organs composing the lymphatic and cardiovascular systems in vertebrates that enable the blood to be moved throughout the body.

circumference of a circle — distance around a circle equaling the diameter × 3.1416 (π).

cirrhosis — a progressive inflammatory disease of a tissue or organ, especially the liver, by excessive growth of hepatic connective tissue.

cirrocumulus clouds — a principal cloud type formed at 20,000–25,000 feet appearing as thin, patchy, clouds (often wave-like).

cirrostratus clouds — a principal cloud type formed at 20,000–25,000 feet appearing as thin sheets that look like fine veils or torn, wind-blown patches of gauze and made up of ice crystals.

cirrus clouds — a principal high-altitude cloud type appearing as white patches or bands usually formed at 25,000 feet above the earth and higher.

cis — a descriptive term meaning "on the same side," and also used to identify geometric isomers.

cistern — an underground box used for collecting water from the roof of the house for usage within the home.

CIS-trans isomers — *See* geometric isomers.

citation — *See* hearing.

citee — one who is cited. *See also* hearing.

Civilian-Military Contingency Hospital System (CMCHS) — a voluntary system of hospitals in the United States and other support services who reserve beds to handle large numbers of military casualties or civilian injuries caused by a disaster.

civil suit — generally any lawsuit other than a criminal prosecution.

claim — a demand by an individual or corporation to recover losses under a policy of insurance; a statement of ownership under oath entered in response to seizure of a lot, which gives the right to contest.

clamshell bucket — a two-sided vessel used to hoist and convey materials; when excavating it has two jaws that clamp together when it is lifted by specially attached cables.

clarifier — a settling tank used to remove settleable solids by gravity or colloidal solids by coagulation following chemical flocculation; also used to remove floating oil and scum through skimming.

clarity — (*swimming pools*) the degree of transparency of pool water.

class A explosive — a material capable of detonation by means of a spark or flame, or even with a small shock; includes nitroglycerine, lead azide, and black powder.

class A fire — a fire resulting from the combustion of ordinary cellulose materials which leaves embers or coals; generally involving wood, paper, and certain textiles; water is an effective extinguishant.

class A poison — a Department of Transportation term for extremely dangerous gas or liquid poisons that in very small amounts when mixed with air are dangerous to life; examples are phosgene, cyanagene, hydrocyanic acid, and nitrogen peroxide.

class B explosive — rapidly combustible materials, including photographic flash powder, that explode under extreme conditions of temperature.

Clamshell Bucket

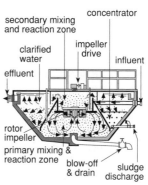

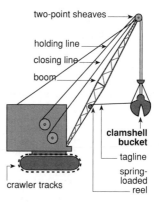

Suspended Solids Contact Clarifier

class B fire — a fire resulting from the combustion of flammable liquids such as gasoline, kerosene, greases, oils, and flammable gases such as methane and hydrogen; carbon dioxide and foam are effective extinguishants.

class B poison — a Department of Transportation classification for liquid, solid, paste, or semi-solid substances, other than class A poisons or irritating materials, that are known or presumed on a basis of viable tests to be toxic to humans and considered a hazard to health during transportation.

class C explosive — materials, such as flares, that do not ordinarily detonate in restricted quantities and are a minimum explosion hazard.

class C fire — a fire resulting from the combustion of materials occurring in or originating from live electrical circuits; carbon dioxide is the effective extinguishant.

class D fire — a fire resulting from the combustion of certain metals which possess unique chemical properties such as being reactive with water and carbon dioxide, includes titanium, magnesium, zirconium, and sodium; special extinguishants such as graphite are effective.

classification code number — the identifying number for a classification listed in the *Workers' Compensation Manual*.

clay — small mineral soil grains less than 0.0002 millimeter in diameter; a fine-textured soil material which breaks into very hard clods or lumps when dry and which is plastic and usually sticky when wet.

clay loam — a soil containing 27–40% clay, 20–45% sand, and the remaining portion silt.

Clean Air Act (CAA) — a law passed in 1963; the basis of current air pollution laws, it provides matching grants for establishing and expanding air quality management programs, provides for developing air quality criteria, initiates efforts to control air pollution from federal facilities, encourages automotive companies and fuel industries to prevent air pollution, establishes federal authorities to abate interstate air pollution, provides process for reviewing status of motor vehicle pollution. The initial Air Pollution Control Act was passed in 1955, and was amended in 1960 and 1962. The Motor Vehicle Air Pollution Control Act was passed in 1965, the Air Quality Act was passed in 1967, the Clean Air Act Amendments were passed in 1970, and updated in 1977, 1983, and 1990.

Clean Air Act Amendments of 1970 — a law passed to stimulate additional air pollution research authorizing state and regional grants, setting national ambient air quality standards, designating air quality control regions, setting fixed time-tables for state implementation plans, setting new source performance standards, setting national emission standards for hazardous air pollutants, requiring industry to monitor and maintain emission records.

Clean Air Act Amendments of 1977 — a law passed in 1977 requiring states with non-attainment areas to submit state implementation plan revisions; established emission offset policy; established banking policy; set modifications to emission standards for vehicles.

Clean Air Act Amendments of 1990 — a law passed in 1990 which recognizes existing severe ozone and carbon monoxide problems in certain metropolitan areas, creates tighter controls on tail pipe exhaust, reduction of acid rain, reduction of nitrogen oxides, reduction of air toxics, and protection of ozone layer by phasing out chlorofluorocarbons, carbon tetrachloride, methylchloroform, and hydrochlorofluorocarbons.

clean air standards — any enforceable rule, regulation, guideline, standard, limitation order, control, or prohibition pursuant to the Clean Air Act.

cleaning — the act or process of the physical removal of residues of foods, ingredients, and other soiling materials.

cleaning-in-place (CIP) — a technique for cleaning pipelines in a milk plant dependent upon circulating appropriate cleaning solutions as relatively high velocities of not less than 5 feet per second; the milk-pipelines go up from the pump to the outlet without any dips to provide self drainage of the solutions and for maintenance of full lines during circulation to ensure contact of solutions with all milk-contact surfaces.

clean room — the area within a hospital where washed materials (such as instruments) are dried, inspected, and packaged; an area where elaborate precautions are taken to reduce contaminants in the air.

Clean Water Act (CWA) — a law passed in 1977 and amended numerous times through 1988; emphasizes control of toxic pollutants in water, enhances National Pollution Discharge Elimination System, establishes criteria for permits including lists of pollutants regulated by federal or state law with permissible amounts of pollutants discharged per unit of time, monitoring requirements and schedules for implementing pollution concentration requirements, special conditions imposed on polluters by agencies, and establishes best available technology (BAT) requirements for toxic substances.

clean water standards — any enforceable rule, regulation, guideline, standard, limitation order, control, or prohibition pursuant to the Clean Water Act.

Cleveland open-cup test (COC) — a laboratory flashpoint test method.

climate — the weather at a given place over a period of time; involves averages, totals, and extremes to set a picture of the weather pattern.

climatic — having to do with weather conditions.

climax community — the stage of ecological development at which a community becomes mature, stable, self-perpetuating, and at equilibrium with the environment.

clinical trial — (*disease control*) an experiment to determine whether one procedure is more or less effective than another in preventing occurrence of a disease (prophylactic trial) or in treating an established disease process (therapeutic trial).

clinker — a hard, sintered, or fused piece of residue ejected by a volcano or formed in a furnace or by the agglomeration of ash, metals, glass, and ceramics.

clod — a small, hard mass of dirt formed by plowing or cultivating excessively wet or dry soil.

clonic — alternate involuntary muscular contraction and relaxation in rapid succession.

close-coupled valve — (*milk*) a valve whose seat is either flush with the inner wall of the holder or so closely coupled that no fluid in the valve pocket is more than 1°F colder than the fluid at the center of the holder at any given time during the holding period.

closed-loop system — a system of heat exchange where the ground acts as a condenser for the pipes from the house; the fluid within the closed loop consists of 20% glycol (antifreeze) and 80% water.

closed-position — (*milk*) any position of the valve seat which stops the flow of a liquid into or out of the holder.

Clostridium — a genus of gram-positive obligate anaerobic or microaerophilic spore-forming, rod-shaped, motile bacteria.

Clostridium botulinum — *See* botulism.

***Clostridium perfringens* food poisoning** — (*disease*) *C. welchii* or *C. perfringens* food poisoning; an intestinal disorder with sudden onset of colic followed by diarrhea, nausea of short duration (1 day or less). Incubation time is 6–24 hours, usually 10–12 hours; caused by *C. welchii* or type A strains of *C. perfringens* or *C. welchii*; produced by the toxins from the microorganisms; found worldwide. Reservoirs of infection

include the soil, and gastrointestinal tracts of healthy humans and animals. It is transmitted through ingestion of food in which a toxin has been produced by the microorganisms found in the soil or in feces; not communicable; general susceptibility. It is controlled by proper cooking and rapid cooling of food, especially large cuts of meat.

cloth filter collector — a dry collector working on the same principle as a vacuum cleaner bag; the gas stream carrying the particles to be filtered passes through the woven fabric bags of 30 feet or more in length, trapping the particles; depending on the type of cloth filter, up to 100% of the particles down to 0.4 microns in size may be trapped.

cloud — a visible mass of particles of water or ice in the form of fog, mist, or haze suspended in the air.

cloud chamber — a device which detects nuclear particles through the formation of cloud tracks. Also known as an expansion chamber; fog chamber.

cluster — (*EPA definition*) a structure with fibers in a random arrangement such that all fibers are intermixed and no single fiber is isolated from the group; groupings must have more than two intersections; (*NIOSH definition*) a network of randomly-oriented interlocking fibers arranged so that no fiber is isolated from the group; dimensions of clusters can only be roughly estimated and clusters are defined arbitrarily to consist of more than 4 individual fibers.

cm — *See* centimeter.

CMA — *See* Chemical Manufacturer's Association.

CMCHS — *See* Civilian-Military Contingency Hospital System.

CMV — *See* cytomegalovirus.

C/N — *See* carbon nitrogen ratio.

CNS — *See* central nervous system.

coagulation — the combination of small particles into fewer, larger particles; a water treatment process in which chemicals are added to combine with or trap suspended and colloidal particles to form rapidly-settling aggregates.

coal — natural, rock-like, brown to black consolidated peat formed by the decomposition of woody plant debris.

coalescence — the process by which small water droplets suspended in the air combine and grow into larger droplets.

coal gasification — a process for converting coal, char, or coke to gas by reaction with hydrogen in the presence of steam, high temperature, and pressure; a process where coal is exposed to molecular oxygen and steam at temperatures of 900°C or higher. Carcinogenic compounds are reduced drastically in this process, but there is a potential for producing carbon monoxide, hydrogen sulphide, and other carbon-nitrogen products.

coal liquefaction — any of a number of processes for converting coal to partially liquid form by heating with additives other than oxygen occurring in a temperature range of 400–450°C. Some of the compounds found in this process are carcinogenic such as benz (a) anthracene, chrysene, and benzopyrene.

coal-tar — a by-product obtained from destructive distillation of bituminous coal. *See also* naphtha.

coal-tar color — articles added or applied to a food, drug, cosmetic, the human body, or any part thereof and capable of imparting color and containing any substance derived from coal-tar, or any substance so related in its chemical structure to a constituent of coal-tar.

coal-tar pitch volatiles — black or dark-brown amorphous residue from the redistillation of coal-tar. It is used as a binding agent in the manufacture of coal briquettes used for fuel; as a dielectric in the manufacture of battery electrodes; in the manufacture of roofing felts and papers; for protective coatings for pipes for underground conduits and drainage; in road paving and sealing; as a coating on concrete as waterproofing; in the manufacture of refractory brick and carbon ceramic items. It is hazardous to the respiratory system, bladder, kidneys, and skin; and is toxic by inhalation and contact. Symptoms of exposure include dermatitis and bronchitis; carcinogenic. OSHA exposure limit (TWA): 0.2 mg/m³ [skin].

coal washing — the process of crushing coal and washing out soluble sulfur compounds with water or other solvents.

coarse texture — (*soils*) the texture exhibited by sands, loam sands, and sandy loams; being visibly crystalline.

Coastal Zone Management Act — a law passed in 1972 and updated through 1986; establishes effective management, beneficial use, protection, and development of the coastal zone including fish, shellfish, other living marine resources, and wildlife in an ecologically fragile area.

coating — any organic material that is applied to a surface producing a continuous film.

coaxial cable — a cable that consists of electrically-conducting material surrounding a central conductor held in place by insulators; used to transmit television and telephone signals of high frequency.

cobalt metal, dust, and fume — Co — an odorless, silver-gray to black metal; MW: 58.9, BP: 5,612°F, Sol: insoluble, sp gr (metal): 8.92, atomic weight: 27. It is used as a binder in the manufacture of cemented carbide items; during refining and concentration of ores; during manufacture of metal items from magnetic and super-temperature alloys; in the manufacture of dental prosthetics and osteosynthetic items; as pigments. It is hazardous to the respiratory system and skin; and is toxic by inhalation, ingestion, and contact. Symptoms of exposure include cough, dyspnea, decreased pulmonary function, weight loss, dermatitis, diffuse nodular fibrosis, respiratory hypersensitivity. OSHA exposure limit (TWA): 0.05 mg/m³ [air].

COC — *See* Cleveland open-cup test.

cocarcinogen — a noncarcinogenic agent that increases the effect of a carcinogen by direct concurrent local effect on the tissue.

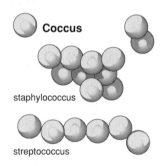

coccus — a sphere-shaped bacterium.

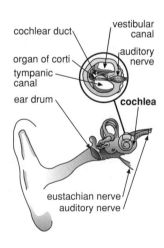

Cochlea

cochlea — the spiral-shaped structure that forms part of the inner ear of mammals and contains the essential organs of hearing.

cockroach — *See* roach.

COD — *See* chemical oxygen demand.

Code of Federal Regulations (CFR) — the compilation of the general and permanent rules published in the Federal Register by the executive departments and agencies in the federal government; published annually by the United States Government Printing Office.

coefficient — any factor in a product; a number placed in front of formulas to balance a chemical equation.

coefficient of entry — the ratio of the actual rate of air blown into an exhaust opening to the theoretical rate calculated by assuming that the negative static pressure in the exhaust opening is completely converted to velocity pressure.

coefficient of expansion — the ratio of increase in volume per degree of increase in temperature at constant pressure.

coefficient of friction — (*physics*) the ratio between the weight of an object being moved and the force pressing the surface together.

coefficient of haze (COH) — a measurement of visibility interference in the atmosphere.

coefficient of linear expansion — the fractional increase in length of a solid per degree rise in temperature at constant pressure.

coefficient of variation (CV) — the ratio of the standard deviation of a distribution to its arithmetic mean.

coefficient of volume expansion — the fractional increase in volume of a substance per degree rise in temperature at constant pressure.

coenocyte — a multinucleate mass of protoplasm resulting from repeated nuclear division unaccompanied by cell fission.

COH — *See* coefficient of haze.

Coherent Waves

coherent waves — waves whose crests and troughs are synchronized.

cohesion — the tendency of like molecules of a substance to hold together as a result of intermolecular attractive force; the act or state of sticking together tightly.

cohesive — exhibiting or production of cohesion or coherence.

cohort — (*statistics*) the entire group of people who shared a similar experience or acquired characteristics in a defined period of time.

cohort effect — systematic differences among two cohorts with respect to the distribution of some variable (e.g., height, weight at any specified attained age).

cohort study — a type of analytical epidemiologic study in which a cohort of unaffected persons is followed over time in order to investigate the incidence of a disease or health-related event in relation to a characteristic measured when the cohort was defined.

coke — bituminous coal from which the volatile components have been driven by heat, leaving fixed carbon and ash fused together.

coke-oven — an oven in which coal is converted to coke by destructive distillation.

coke-oven chemicals — organic compounds derived from bituminous coal during its conversion to metallurgical coke; it is a major source of raw materials for a number of chemicals.

coke-oven emissions — hazardous air pollutants emitted by coke-ovens including substantial quantities of carbon monoxide and all the by-products of the cleaning, grading, and combustion of coal. Polycyclic aromatic hydrocarbons may occur from the smoke escaping when the ovens are filled, or from leaks or from blow off. Respiratory cancer and skin cancer have been found in individuals who have been exposed to coke oven emissions.

cold — a condition of low temperature; the absence of heat.

cold air mass — a large mass of unstable air turbulent with gusty winds creating good surface visibility and the possibility of thunder showers.

Cold Front

cold front — an advancing edge of a cold air mass.

cold sterilization — a process conducted in a closed vessel using ethylene oxide gas for two hours to kill all bacteria and spores. Also known as chemical sterilization.

cold welding — a solid-state welding process in which pressure is used at room temperature to produce coalescence of metals with substantial deformation at the weld.

colic — acute abdominal pain usually caused by smooth muscle spasm, obstruction, or twisting.

coliform — pertaining to fermentative gram-negative enteric bacilli sometimes restricted to those fermenting lactose, that is *Escherichia, Klebsiella, Enterobacter,* and *Citrobacter.*

coliform bacteria — a group of bacteria inhabiting the intestines of animals, including people, also found elsewhere; includes all the aerobic nonspore forming rod-shaped bacteria that produce from lactose fermentation gas within 48 hours at 37°C; used to determine levels of pollution in water, soil, or milk.

coliform-group bacteria — a group of bacteria predominantly inhabiting the intestines of people or animals, or also occasionally found elsewhere which may contaminate potable water.

coliform index — a rating of the purity of water based on a count of fecal bacteria.

coliform test — a bacteriological test involving the taking of a sample, inoculating it into a tube, incubating it, and measuring the number of samples which contain gas production; when run on raw milk, the test may indicate problems in the initial production of the raw milk; when run on pasteurized milk, the test may indicate problems in the pasteurization plant.

colitis — inflammation of the colon.

collection efficiency — the amount of substance absorbed or detected divided by the amount sampled.

collection frequency — the number of times collection is performed in a given period of time.

collection stop — a stop made by a vehicle and crew to collect solid waste at one or more service sites.

collet — *See* cullet.

collimation — the process of restricting the useful beam of radiation to a predetermined cross-sectional area by attaching beam-limiting devices to the source of radiation.

collimator — a device which reduces the scatter of x-radiation while X-ray pictures are being taken of the body structures.

colloid — a chemical system composed of a continuous medium, such as protoplasm or albumen, throughout which are distributed small particles 1–1,000 nm in size, which do not settle out under the influence of gravity.

colloidal state — a condition obtained by dispersing submicroscopic particles of gases, liquids, or solids through a second substance in a medium which may be gaseous, liquid, or solid; a gas dispersed in a gas is not colloidal.

colluvial — (*soils*) of or pertaining to material that has moved down hill by the force of gravity or action of frost and local wash and accumulated on lower slopes or at the bottom of the hill.

colluvium — (*soils*) material deposited in footslope positions by the action of gravity, soil creep, or local wash; frequently silty and loamy material.

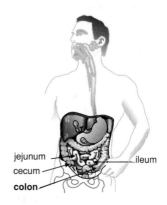

jejunum

cecum

colon

ileum

Colon

colon — the portion of the human intestine exclusive of the cecum and the rectum; it is divided into four sections: ascending, transverse, descending, and sigmoid.

colonization — (*infection control*) the presence of methicillin resistant *Staphylococcus aureus* on body tissue without symptoms or clinical manifestations of illness or infection.

color — a general term that refers to the property of light that is determined by its wavelength with particular reference to its visual appearance.

color additive — a dye pigment or other substance made by a process of synthesis or similar artifice or extracted, isolated, or otherwise derived with or without intermediate or final change of identity from a vegetable, animal, mineral, or other source, and when added or applied to a food, drug, cosmetic, or to the human body, is capable of imparting color.

Colorado tick fever — (*disease*) an acute, febrile, two-phase disease where the initial fever subsides and a second fever occurs later, lasting 2–3 days. Incubation time is 4–5 days; caused by viruses found in western Canada, Washington, Oregon, Idaho, California, and other western states. The reservoir of infection is small animals and ticks, primarily *D. andersoni*; transmitted by the bite of the tick and not person to person; general susceptibility. It is controlled by use of tick repellents and proper clothing in areas where ticks are prevalent.

colorimeter (*chemistry*) — a device used in chemical analysis for comparing the color of a liquid with a standard color.

colorimetric analysis — a method of determining the concentration of a chemical substance by measuring the intensity of its color or of a color produced by it.

colostrum — reddish-yellow, strongly odored, bitter-tasting milk with an increased number of proteins and decreased lactose; it is essential for calves' health, since it contains many antibodies; it should not be present in market milk.

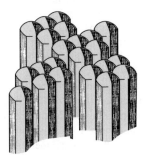

Columnar Soil

columnar soil — (*soils*) a soil structure with rounded caps, prism-like with the vertical axis greater than the horizontal, usually found in subsoil or B horizon.

coma — a state of unconsciousness from which a patient cannot be aroused even by powerful stimuli.

combined chlorine — chlorine which is available as a bactericide in water, but which is combined with another substance, usually ammonia; it is less effective against bacteria.

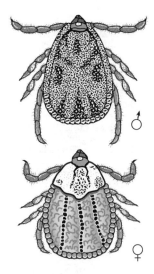

Dermacentor andersoni
(Rocky Mountain wood tick)
a vector for
Colorado Tick Fever

combined sewer — a sewer receiving both surface water run-off and sewage from homes, businesses, and/or industry.

combustible — possessing the ability to catch fire and burn; material that can be ignited at a specific temperature in the presence of air to release heat energy; a term used by the National Fire Protection Association, Department of Transportation, and others to classify certain liquids that will burn on the basis of flashpoints.

combustible dust — dust capable of undergoing combustion or of burning when subjected to a source of ignition.

combustible gas indicator (CGI) — an instrument used to determine the level of combustible gases present in a given environment before they reach their lowest explosive level.

combustible liquid — any liquid having a flashpoint at or above 100°F but below 200°F, except any mixture having components with flashpoints of 200°F or higher.

combustion — the production of heat and light energy through a chemical process, usually oxidation of the gas, liquid, or solid fuel; (*air analysis*) a technique of analysis of combustible gases and vapors where they are passed over a filament which is part of a Wheatstone bridge circuit; the combustion of the chemicals alter the resistance of the filament causing an imbalance in the circuit, which is then used as a measure of concentration.

combustion air — the air used for burning the fuel.

combustion gases — the mixture of gases and vapors produced by combustion.

combustion product — material produced or generated during the burning or oxidation of a material.

come-down-time — (*milk*) the time which elapses between the processing or pasteurization temperature and the temperature required to lower the product.

come-up-time — (*milk*) the time which elapses between the initial temperature of a product and the temperature required to process or pasteurize the product.

comfort ventilation — air flow intended to remove heat, odor, and cigarette smoke from an enclosure.

command — (*computer science*) an instruction sent from the keyboard or other control device to execute a computer operation.

command language — (*computer science*) an English-like language for sending commands for complicated program sequences to the computer.

command post — (*emergencies*) a facility at a safe distance upwind from an accident site where the on-scene coordinator, responders, and technical representatives can make response decisions, deploy people and equipment, maintain liaison with the media, and handle communications.

commensal — relating to or living in a state of commensalism.

commensalism — a relationship between two kinds of organisms in which one obtains food or other benefits and the other is neither damaged or benefited.

commercial processor — any person engaged in commercial or custom processing of food.

commercial solid waste — the solid waste generated by stores, offices, or activities that do not actually produce a product.

commercial sterility — (*food*) the condition achieved by application of heat, rendering food free of viable forms of microorganisms of public health significance; spores can exist during the commercial sterilization process.

commercial waste — a combination of garbage, refuse, ashes, demolition waste, urban renewal waste, construction waste, and remodeling waste.

commingled shipment — two or more separate shipments of the same goods at one common location and bearing no distinguishing marks.

comminuted — (*food*) reduced in size by chopping, flaking, grinding, or mincing; the food may be restructured or reformed as in sausage or gyros.

comminution — *See* pulverization.

comminutor — a device for catching and shredding heavy solid matter.

commissary — a place where food, containers, or supplies are stored, prepared, or packaged for transit to, and sale or service at, other locations.

common dining area — a central location in a group residence where people gather to eat.

common exposure route — a likely way by which a pesticide may reach and/or enter an organism; usually oral, contact, or respiratory.

common law — a body of unwritten principles based on custom and the precedent of previous legal decisions.

common mode rejection ratio — the difference in signal gain divided by the common mode signal gain.

common name — any identification such as code name, code number, trade name, brand name or generic name, other than the chemical name, used to identify a chemical.

communicable disease — an illness due to an infectious agent or its toxic products being transmitted directly or indirectly to a well individual from an infected individual or animal, or transmitted through an intermediate animal host, vector, or inanimate environment.

communicable period — the time or times during which the etiologic agent may be transferred from an infected individual or animal to an infected individual.

communications superhighway — an electronic information system composed of a merger of video, telephones, and computers where all audio and video communications can be translated into digital information, and stored and compressed to travel through existing phone and cable lines; fiber-optic wiring provides a large capacity for transmission of information over long distances, and coaxial cable wire carries this information effectively for about 1/4 mile into the home, thus eliminating rewiring of the home. Also known as the electronic superhighway and the information superhighway.

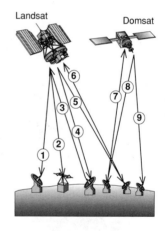

1 S-band (MSS/RBV data)
2 UHF (DCS data)
3 S-band (commands)
4 S-band/MSS/DCS/RBV
5 S-band/MSS/RBV/DCS
6 S-band VHF (commands)
7 Process MSS/RBV data
8 Video MSS/RBV data
9 Video MSS/RBV data

(Lansat) Communication and Processing System

communication system — any combination of devices permitting the passage or exchange of messages.

Community Awareness and Emergency Response Program (CAER) — a program designed by the Chemical Manufacturer's Association to help plant managers in taking the initiative to operate with local communities to develop integrated, hazardous material response plans.

community based residential facility — any building or buildings, sections of a building, or distinct part of a building or other place operated for profit or not, which undertakes through its ownership or management to provide for a period exceeding 24-hour housing, food service, and one or more personal services for persons not related to the owner or administrator by blood or marriage.

Community Development Block Grant (CDBG) — a federal program under the Department of Housing and Urban Development providing grants to carry out a wide range of community development activities directed toward neighborhood revitalization, economic development, and improved community facilities and services.

Community-Right-to-Know-Act — *See* Emergency Planning and Community-Right-to-Know Act.

community water system — a public water supply that serves at least 15 connections of 25 individuals on a regular basis.

compaction — the consolidation and reduction in size of solid particles by rolling, tamping, or other means of applying mechanical pressure.

compactor — a vehicle with an enclosed body containing mechanical devices that convey solid waste into the main compartment of the body and compress it; a machine designed to consolidate earth and paving materials.

compatible pesticides — two or more pesticide chemicals that can be safely mixed together without reducing their effectiveness or damaging the product, plant, or animal treated.

compensable injury — accidental and not self-inflicted injury arising both out of and during the course of employment.

compiler — (*computer science*) a computer program that translates a high-level programming language into machine-readable code.

complainant — the individual registering a complaint, as a consumer complaining about an article of merchandise or environmental situation.

complaint for forfeiture — the document furnished to the United States Attorney for filing with the clerk of the court to initiate the seizure of a lot. Formerly known as libel of information.

complete blood count (CBC) — a determination of the number of blood cells in a given blood sample expressed as the number of cells (red, white, platelet) in a cubic millimeter of blood.

complete metamorphosis — the four stages of development of certain insects: the egg, larva, pupa, and adult.

complex conductivity — a property of material which gives the ratio between current density and electric field and includes the phase difference between the two field quantities.

compliance — a legally enforceable action taken to correct pollutant problems and meet the requirements of the law.

composite map — a single map created by joining together several digitized maps.

compost — a mixture of decaying organic matter used for fertilizing and conditioning soil.

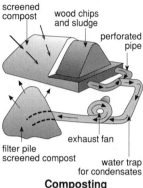

Composting
with Force Aeration

composting — a controlled process of degrading organic matter by microorganisms.

compound — a substance made up of two or more elements in union. The elements are united chemically, which means that each of the original elements loses its individual characteristics once it has combined with the other element or elements and cannot be separated by physical means. They combine in definite proportions by weight.

comprehensive care facility — a licensed health facility providing room, board, laundry, and

Complete Metamorphosis
(*Danaus plexippus*)

administration of medication under supervision and responsibility of an attending medical staff.

Comprehensive Environmental Response Compensation and Liability Act (CERCLA) — a federal act passed in 1980 requiring the Environmental Protection Agency to designate hazardous substances that can present substantial danger to the public, authorizes the clean-up of sites contaminated with these substances, and gives the federal government the power to respond to release or threatened release of any hazardous substance against the environment, as well as the release of any contaminant that may present imminent and substantial danger to public health or welfare. Also known as the Superfund Act.

compressed air nebulizer — a device used to produce aerosols as from liquids.

compressed gas — a gas or mixture of gases having in a container an absolute pressure exceeding 40 psi at 70°F of gas; a mixture of gases having in a container an absolute pressure exceeding 104 psi at 130°F regardless of the pressure; a liquid having a vapor pressure exceeding 40 psi at 100°F.

compressible flow — a flow of high-pressure gas or air which undergoes a pressure drop causing a significant reduction in density.

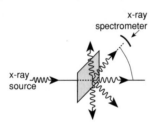

The Compton Effect

Compton effect — an attenuation process observed for x- or gamma-radiation in which an incident photon interacts with an orbital electron of an atom to produce a recoil electron and a scattered photon of energy less than the incident photon.

computed tomography (CT) — a diagnostic, radiological technique which permits physicians to view cross-sectional images of 1 cm thick slices of internal body structures.

computer graphics — a general term including any computer activity that results in a graphic image.

computerized axial tomography (CAT) — commonly called CAT scan. *See* computed tomography.

conc — *See* concentration.

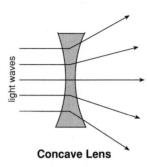

Concave Lens

concave lens — a lens with an inward curve which diverges parallel light rays.

concentrate — (*pesticides*) a pesticide chemical before dilution.

concentrated milk — a fluid product unsterilized and unsweetened resulting from the removal of a large portion of water from milk.

concentration (conc) — the mass or volume of a substance present in a unit volume of a gas, solid, or liquid.

concentration threshold — the pollutant concentration below which no receptors experience an ill effect.

condemned — (*food*) food or any food product which has been determined by inspection or analysis to be unsound, unhealthful, unwholesome, or otherwise unfit for human consumption; (*housing*) housing which has been determined by inspection and analysis to be unfit for human habitation.

condensate — the liquid obtained by condensation of a gas or vapor, such as steam.

condensation — the formation of water upon aerosol particles acting as nuclei by causing the water temperature to pass through its dew point; a reduction to a denser form as from steam to water.

condensation nucleus — a tiny particle suspended in air and upon which the condensation process begins.

condenser — a device in which the temperature of a gas is lowered to its condensation point, thereby converting it from its vapor to its liquid stage.

condiment — any food, such as ketchup, mayonnaise, mustard, and relish, that is used to enhance the flavor of other foods.

conditional maintenance — preventive maintenance which depends upon a predetermined schedule when a state of reduced function of the equipment may occur.

C
D

conditioned air — air that has been heated, cooled, humidified, or dehumidified to maintain an interior space within the comfort zone. Also known as tempered air.

conditioning — pretreatment of a sludge to facilitate the removal of water in a thickening or dewatering process.

conditions to avoid — (*chemicals*) conditions encountered during the handling or storage which could cause the substance to become unstable.

conductance — the ratio of current to voltage in a circuit which absorbs but does not store electrical energy; a measure of the ability of a circuit to conduct electricity; it is reciprocal of resistance.

conduction — the transfer of heat or electrical charge by physical contact between the molecules.

conductive flooring — floor material that has the ability to allow electric energy to flow through it with minimum resistance in ranges between 25,000 and 1,000,000 ohms.

conductive hearing loss — a type of hearing loss due to any disorder in the middle or external ear that prevents sound from reaching the inner ear; not caused by noise.

conductivity — the capacity for conduction.

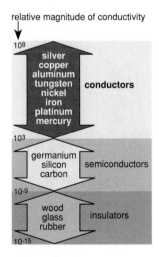

Conductor

conductor — an object or substance which allows a current of electricity to pass continuously, such as a wire or combination of wires suitable for carrying current.

confidence interval — (*epidemiology*) the interval which has the probability of including a given population being studied.

confidence limit — (*epidemiology*) the upper and lower limits of a confidence interval.

configuration — the three-dimensional, spatial arrangement of electrons in atomic orbitals in an atom; (*computer science*) a combination of computer hardware and software used for a specific task.

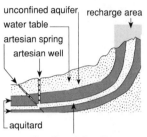

Confined Aquifer

confined aquifer — a geological formation in which the water-bearing material is located between two aquitards and therefore is well protected against contamination percolating from the surface.

confined areas — rooms, buildings, and greenhouses with limited or inadequate ventilation.

confined space — any area not intended or designated for continuous human occupancy which has limited openings for entry, exit, or ventilation, or would make escape difficult in an emergency.

confounder — a factor, such as age, known to be associated with a health event; for example, age not being taken into account when determining excess weight may be a cause of heart attack.

Confused Flower Beetle (*Tribolium confusum*) — an elongated reddish-brown, flightless beetle about 1/7" long with a distinct joint between the thorax and abdomen; has antennae which gradually enlarge toward the tip; a stored food product insect.

congenital — existing at or before birth.

congenital defect — an abnormality present at birth; it may be genetically determined or may result from some environmental insult during pregnancy.

conical burner — a hollow, cone-shaped combustion chamber that has an exhaust vent at its point and a space through which waste materials are charged; air is delivered to the burning solid waste inside the cone. Also known as a teepee burner.

needles

♀

female
cones

♂

♀ male
flowers

Eastern White Pine

Conifer

conifers — trees and shrubs with needle-like leaves such as pine, cedar, spruce, and hemlock.

conjugation — a form of sexual reproduction in unicellular organisms in which two individuals join in temporary union to transfer genetic material; (*biochemistry*) the joining of a toxic substance with a natural substance of the body to form a detoxified product which can be eliminated.

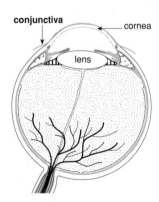

conjunctiva   cornea

lens

Conjunctiva

conjunctiva — the delicate membrane lining the eyelids and covering the eyeball.

conjunctivitis — an inflammation of the mucous membrane that lines the inner surface of the eye lid and the exposed surface of the eyeball; may

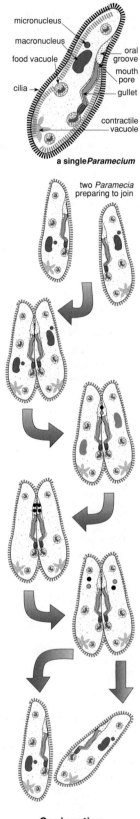

micronucleus

macronucleus

food vacuole

oral groove

mouth pore

cilia

gullet

contractile vacuole

a single *Paramecium*

two *Paramecia* preparing to join

Conjugation

be caused by bacteria, virus, allergic responses, or chemical or physical factors.

connate water — highly mineralized water which has been trapped in igneous rock formations at the time the rocks were formed.

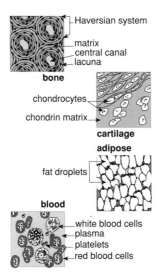

Connective Tissue

connective tissue — tissue that lies between groups of nerve, gland, and muscle cells and beneath epithelial cells; also includes bone, cartilage, blood, and lymph.

consent decree of condemnation — a court order agreed to by a claimant in a seizure in which the individual admits that the article seized is in violation as alleged in the complaint for forfeiture, and agrees to condemnation and either destruction or reconditioning as well as payment of costs; it commits the claimant to provide a bond and recondition the article under supervision of the Food and Drug Administration.

conservation of energy — the principle that the energy of the universe is constant and cannot be created or destroyed; there is no known violation of this principle. Also known as law of conservation of energy; first law of thermodynamics.

conservation of mass-energy — the theory that energy and mass are interchangeable in accordance with Einstein's equation ($E = mc^2$) where "E" is energy, "m" is mass, and "c^2" is the velocity of light.

conservation of matter — the notion that matter can neither be created nor destroyed or that weight remains constant in an ordinary chemical change; violated by microscopic phenomenon. Also known as law of conservation of matter.

consistence — (*soils*) the resistance of a material to deformation or rupture; the degree of cohesion or adhesion of the soil mass.

console — the main operating unit in which indicators and general controls of a radar or electronic group are installed; (*computer science*) a device that allows the operator to communicate with the computer.

consortism — *See* symbiosis.

constant air volume system — an air handling system that provides a constant air flow of varying temperature to meet the heating and cooling needs of a specific area.

constant-flow sampler — a pump which overcomes variations in the rate of flow of air by use of sophisticated flow-rate sensors with feedback mechanisms permitting the maintenance of the preset flow-rate during sampling.

constipation — infrequent and difficult evacuation of the feces resulting in discomfort.

construction and demolition waste — building materials and rubble resulting from construction, remodeling, repair, and demolition operations.

consumer — (*ecology*) a plant or animal that cannot derive its energy from inorganic matter but must depend upon other plants or animals for its energy for living. Also known as heterotroph.

Consumer Product Safety Commission (CPSC) — an agency created in 1970 to evaluate and regulate chemicals which may be carcinogenic and also to evaluate a number of products which may be hazardous to individuals.

Consumer Product Safety Commission Act (CPSCA) — a law passed in 1970, updated in 1984, establishing the Consumer Product Safety Commission as an independent regulatory agency, giving the commission the power to regulate consumer products and oppose unreasonable risk of injury or illness; to regulate consumer products except for foods, drugs, pesticides, tobacco, tobacco products, motor vehicles, aircraft, aircraft equipment, boats, and boat accessories regulated by other federal agencies; to publish consumer product safety standards to reduce the level of unreasonable risk.

contact aeration process — a secondary treatment process for sewage which depends on aerobic biological organisms breaking down the putrescible organic materials in sewage to simpler and more stable forms.

contact condensers — a device in which a vapor is condensed by being forced to give up its latent heat to a cooling liquid brought into contact with the vapor.

C
D

contact dermatitis — a skin rash marked by itching, swelling, blistering, oozing, or scaling; caused by direct contact between the skin and a substance to which the individual is allergic or sensitive.

contact pesticide — a pesticide chemical that controls or destroys a pest that touches a treated surface or is touched by the pesticide.

contagious — capable of being transmitted from one individual to another.

container — (*chemicals*) any bag, barrel, box, bottle, can, cylinder, drum, reaction vessel, or storage tank that contains a hazardous chemical; any portable device in which material is stored, transported, treated, disposed of, or otherwise handled.

containment — the confinement of radioactive material in such a way that it is prevented from being dispersed into the environment or is released only at a specified rate.

contaminant — the agent of contamination.

contamination — (*radiation*) the deposition of radioactive material in any place where it is not desired, especially where it may be harmful or constitute a hazard; (*microbiology*) the introduction of pathogenic organisms in or on a surface of an inanimate object.

contempt of court — any act which is calculated to embarrass, hinder, or obstruct a court in the administration of justice.

contest — (*law*) to make defense to an adverse claim in a court of law; to impose, resist, or dispute the case made by a plaintiff.

continental drift — the slow movement of land masses caused by the movement of molten rock underneath the earth's crust.

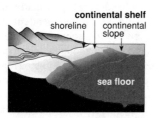

Continental Shelf

continental shelf — the relatively shallow ocean floor bordering a continental land mass.

Continental Shelf Lands Act — a law passed in 1983 and updated through 1988; it states that the outer continental shelf, including all submerged lands lying seaward and outside of areas of lands beneath navigable waters, subsoil, and seabed attached to the United States, is of vital importance to the country since it is a national resource reserve of a variety of potential

minerals, oils, etc.; exploration, development, and production of minerals is controlled by the federal government under this act.

contingency plan — a document setting out an organized plan and coordinated course of action to be followed in case of an unexpected event such as a fire, explosion, or release of hazardous waste that could threaten human life or the environment.

continuous air monitoring (CAM) — a process in which air samples are collected continuously at a known rate.

continuous discharge — discharge occurring without interruption throughout the operating hours of the facility.

continuous feed incinerator — an incinerator into which solid waste is charged almost continuously to maintain a steady rate of burning.

continuous monitoring — the taking and recording of measurements at regular and frequent intervals during operation of a facility.

continuous sampling — a process in which samples are collected continuously at a known rate.

continuous wave — a time-dependent function of constant amplitude.

continuous-wave laser (CW laser) — a laser in which the beam is generated continuously as required for communication or other application.

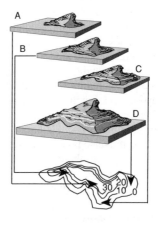

Contour Lines

contour — a line connecting points of equal elevation.

contraband — goods exported or imported to a country against the laws of that land.

control efficiency — the ratio of the amount of pollutant removed from effluent gases by a control device to the total amount of pollutant without control.

C
D

controlled-air incinerator — an incinerator with two or more combustion chambers in which the amount of distribution of air is controlled; partial combustion takes place in the first zone and gases are burned in a subsequent zone or zones.

controlled experiment — an experiment involving two or more similar groups; the control group is held as a standard for comparison while the other group is subjected to some procedure, treatment, measure, or disease whose effect is being tested.

control of infectious diseases — techniques directed against the reservoir of infection, such as isolation and quarantine; techniques used to interrupt the transmission of the agent such as separation of food and water from fecal material, eradication of vectors; techniques used to reduce susceptibility of the host, such as immunization.

control regulation — (*air pollution*) a rule of the Environmental Protection Agency intended to limit the discharge of pollutants into the atmosphere and thereby achieve a desired degree of ambient air quality.

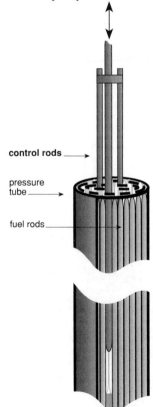

Control Rod

control rod — a rod, usually made of boron, used to control the power of a nuclear reactor.

control velocity — the air velocity required at the face of an enclosing hood to retain the contaminant within the hood.

convection — the transfer of heat by the movement of molecules of liquids and gases.

convectional precipitation — precipitation due to the uneven heating of the ground causing the air to rise, expand, and the vapor to condense and precipitate.

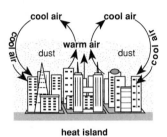

heat island

Convection Current

in earth's mantle

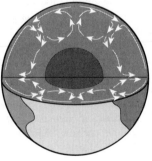

convection currents — rising or sinking air currents that mix air in an atmosphere and transport heat from area to area.

conventional septic system — *See* septic system.

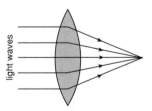

Convex Lens

convex lens — a lens with an outward curve which converges parallel light rays.

convulsant poisons — poisons that act directly on the central nervous system; they include strychnine and camphor; they produce convulsions, increased respiration, cardiac edema, vomiting, and death.

convulsion — a series of involuntary contractions of the voluntary muscles.

coolant — a liquid or gas used to reduce the heat generated by power production in nuclear reactors or electric generators; a transfer agent used in a flow system to convey heat from its source.

cooling air — ambient air that is added to hot combustion gases to cool them.

cooling sprays — water sprays directed into flue gases to cool them and to remove some fly ash.

COPD — *See* chronic obstructive pulmonary disease.

copolymer — a mixed polymer or heteropolymer which is a product of the polymerization of two or more substances.

copper dusts and mists — Cu — a reddish, lustrous, malleable, odorless metal; MW: 63.5, BP: 4,703°F, Sol: insoluble, sp gr (metal): 8.94. It is used in the manufacture of copper rod, wire, piping, and tubing for use in electrical, plumbing, and building; in the manufacture of domestic utensils; in fungicides, insecticides, and germicides; in paint pigments and coloring agents; in wood preservation, automotive emission controls, textile treatment, and organic synthesis. It is hazardous to the respiratory system, skin, liver, kidneys; and is toxic by inhalation, ingestion, and contact. Symptoms of exposure include nasal, mucous membrane, and pharynx irritation, nasal perforation, eye irritation, metallic taste, dermatitis, and increased risk of Wilson's disease; in animals: lung, liver, and kidney damage, anemia. OSHA exposure limit (TWA): 1 mg/m^3 [air].

copper fume — CuO/Cu — finely divided black particulate dispersed in air; MW: 79.5, BP: decomposes, Sol: insoluble, sp gr: 6.4. It is liberated during the construction and installation of material fabricated from copper metal or copper alloys. It is hazardous to the respiratory system, skin, eyes; and is toxic by inhalation and contact. Symptoms of exposure include metal fume fever, chills, muscle ache, nausea, fever, dry throat, cough, weakness, lassitude, eye and upper respiratory tract irritation, metallic or sweet taste, discoloration of the skin and hair, and increased risk of Wilson's disease. OSHA exposure limit (TWA): 0.1 mg/m^3 [air].

copper sulfate — CuSO$_4$ — an effective algicide used infrequently in swimming pools because of its toxicity and incompatibility with some other compounds found in water. Also known as cupric sulfate.

core — the uranium-containing center of a nuclear reactor where energy is released; (*geology*) the central part of the earth having a radius of approximately 2,100 miles and physical properties different from those of the surrounding area.

corium — the deeper skin layer containing the fine endings of the nerves and the finest divisions of the blood vessels. Also known as derma.

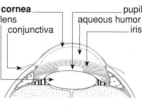

Cornea

cornea — a transparent bulge of the vertebrate eye's sclerotic layer which covers the iris and pupil and through which light rays pass.

corona — a luminous discharge due to ionization of the air surrounding a conductor occurring when the local electric field exceeds the dielectric strength of air; a series of colored rings surrounding luminaries when veiled; due to diffraction by water droplets.

corpus — a body of a person or animal when dead.

corpuscle — a red or white blood cell.

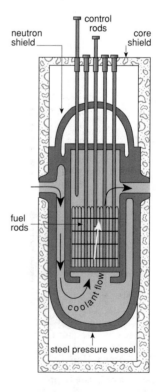

Reactor Core

corrective maintenance — unscheduled maintenance carried out following an equipment failure.

corrode — to weaken or destroy gradually; to deteriorate or erode by degrees.

corrosion — a chemical reaction at the surface of a substance, usually metal, causing an alteration or deterioration of the surface with the reactions approximately doubling with each 18°F increase in temperature.

corrosion resistant — (*food*) having the capacity to maintain original surface characteristics under the prolonged influence of the use environment, including the expected food contact and the normal use of cleaning compounds and sanitizing solutions; easily cleanable, readily accessible, and of such material and finish and so fabricated, cleaning can be accomplished by normal methods.

corrosive — (*food*) a substance that causes corrosion; (*biochemistry*) a chemical that causes visible destruction or irreversible alterations in living tissues by chemical action at the site of contact.

cortex — the outer layer of an organ or structure.

cortisone — $C_{21}H_{28}O_5$ — a hormone secreted by the cortex of the adrenal gland of vertebrates.

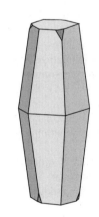

Corundum

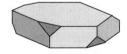

corundum — Al_2O_3 — an impure mineral form of aluminum oxide; gem varieties are ruby and sapphire.

Corynebacterium — a genus of gram-positive, nonmotile, straight to slightly curved, rod-shaped bacteria, including both pathogenic and nonpathogenic organisms widely distributed in nature.

coryza — profuse discharge from the mucous membrane of the nose.

cosmic rays — high-energy particulate and electromagnetic radiations which originate outside the earth's atmosphere and impinge upon the earth from all directions of space with nearly the speed of light.

cost benefit analysis (CBA) — a procedure in which a program's monetary cost is compared to the program's benefits expressed in dollars to determine the best investment; an evaluation of the costs and benefits of a proposed action; used as a tool in trying to make decisions concerning specific kinds of problems or programs.

cost center — an accounting device whereby all related costs attributable to some "center" within an institution, such as an activity, department, or program, are segregated for accounting or reimbursement purposes.

cost containment — the control of the overall cost.

cost-effective — a plan or combination of plans which produce the best results toward achieving stated objectives, along with the most prudent use of resources.

cost per ton per minute — a unit often used in cross comparisons between transfer and direct-haul operations.

cottage cheese — the soft, uncured cheese prepared from the curd obtained by adding harmless lactic acid-producing bacteria with or without enzymatic action to pasteurized skim-milk or pasteurized reconstituted skim-milk.

cotton dust — a colorless, odorless solid; MW: not known, BP: decomposes, Sol: insoluble, Fl.P: not known. It is used in the manufacture of cotton yarn. It is hazardous to the respiratory and cardiovascular systems; and is toxic by inhalation. Symptoms of exposure include tight chest, cough, wheezing, dyspnea, decreased forced expiratory volume, bronchitis, malaise, fever, chills, upper respiratory system symptoms after initial exposure. OSHA exposure limit (TWA): 1 mg/m³ [air] plus supplementary exposure limits.

coulomb (C) — (*electricity*) a unit of electrical charge equal to an ampere-second or 6.3×10^{18} electronic charges; the amount of electrical charge that crosses a surface in 1 second when a steady current of 1 absolute ampere is flowing across the surface.

Coulomb's law — the law that the force between two charged bodies is proportional to the product of their charges and inversely proportional to the square of the distance between them. Also known as the law of electrostatic attraction.

coulometry — a determination of the amount of an electrolyte released during electrolysis by measuring the number of coulombs used.

Council on Environmental Quality (CEQ) — a staff office of the Executive Office of the President responsible for monitoring overall environmental quality of the nation.

count — (*radiation measurements*) the external indication of a radiation-measuring device designed to enumerate ionizing events; (*law*) a statement of a violation of the act in a criminal information or indictment; the number of counts depends on the number of interstate shipments involved in the case.

counter — an instrument or apparatus by which numerical value is computed; (*radiology*) a device for enumerating ionizing events.

counterclaim — a claim presented by a defendant in opposition to or deduction from the claim of the plaintiff; it is in effect a new suit in which the party named as the defendant under the bill is plaintiff and the party named as plaintiff under the bill is the defendant.

counter-ion — the charge on the surface of a colloidal particle compensated for by an equal and opposite ionic charge in the liquid immediately in contact with the particle.

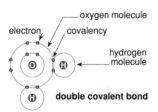

Covalent Bond

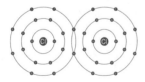

chlorine molecules

covalent bond — a kind of chemical bond in which one or more pairs of electrons are shared by different atoms, with each electron in a pair coming from a different atom.

cover material — soil used to cover compacted solid waste in a sanitary landfill.

cow pox — *See* vaccinia.

coxsackie virus — one of a heterogeneous group of enteroviruses producing in humans a disease resembling poliomyelitis, but without paralysis.

cp — *See* candlepower.

CPC — *See* chemical protective clothing.

CPE — *See* chlorinated polyethylene.

CPR — *See* cardiopulmonary resuscitation.

cps — *See* cycles per second.

CPS — *See* characters per second.

CPSA — *See* Consumer Product Safety Act.

CPSC — *See* Consumer Product Safety Commission.

CPU — *See* central processing unit.

cracking — the process of reducing large, complex hydrocarbon molecules into short-chain hydrocarbons, as in oil refining processes.

cradle-to-grave — a management technique for hazardous waste disposal required by the Resource Conservation and Recovery Act including five basic elements: identification of generators and type of waste; tracking of waste by a "uniform manifest"; requiring permits for all hazardous waste treatment, storage, and disposal facilities; restrictions and control on hazardous waste facilities; and enforcement of regulations and compliance by generators, transporters, and disposal facilities.

Crag™ herbicide — $C_6H_3Cl_2OCH_2CH_2OSO_3Na$ — a colorless to white crystalline, odorless solid; MW: 309.1, BP: decomposes, Sol (77°F): 26%, sp gr: 1.70. It is used in the manufacture of weedicides. It is not known which organs are harmed by Crag™ herbicide, but is toxic by inhalation, ingestion, and contact. No known symptoms of exposure. OSHA exposure limit (TWA): 10 mg/m³ [total] or 5 mg/m³ [resp].

cramps — painful muscular contractions that may affect almost any voluntary or involuntary muscle.

cream — a portion of milk which contains not less than 18% milk-fat.

credentialing — the recognition of professional or technical competence.

creosote — a translucent amber, brown, or blackish liquid; MW: not known, BP: >180°, Sol: practically insoluble in water, Fl.P: 75°C, sp gr: not known. Type 1 results from high-temperature treatment of coal (coal-tar creosote); Type 2 results from high temperature treatment of beech and other woods (beechwood creosote); Type 3 comes from the resin of the creosote bush (creosote-bush resin). Type 1 is the most widely used wood preservative in the United States; also used as a component of roofing pitch; fuel oil and lamp black; and as a restricted-use pesticide. Types 2 and 3 are used in industry. It is toxic by inhalation, absorption, ingestion, and contact; contains mutagenic polycyclic aromatic hydrocarbons creating a possible relationship between creosote exposure and the development of multiple myeloma. Symptoms of exposure include cancers of the nasal cavity, larynx, lungs, skin, and scrotum; ulceration of the oropharynx; petechial hemorrhages over the gastrointestinal serosal surfaces; degeneration and necrosis of hepatocytes; possible kidney failure; dermal and ocular irritation; burns and warts. OSHA exposure limit (TWA): 0.2 mg/m³ [air].

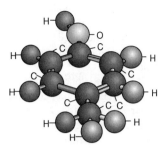

**p–Cresol
(4–Methylphenol)**

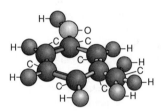

m-Cresol

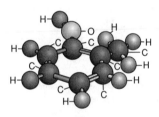

**o-Cresol
(2 Methylphenol)**

cresol (all isomers) — CH₃C₆H₄OH — a colorless, yellow, brown, or pinkish, oily liquid or solid with a sweet, tarry odor; MW: 108.2, BP: 376–397°F, Sol: 2%, Fl.P: 178–187°F, sp gr: (77°F): 1.03. It is used in the manufacture of antiseptics and disinfectants, phosphate ester, antioxidants, resins, herbicides, perfumes, explosives, and photographic developers; as a solvent and engine and metal cleaner. It is hazardous to the central nervous system, respiratory system, liver, kidneys, skin, and eyes; and is toxic by inhalation, absorption, ingestion, and contact. Symptoms of exposure include central nervous system effects, confusion, depression, respiratory failure, dyspnea, irregular and rapid respiration, weak pulse, skin and eye burns, dermatitis, and lung, liver, and kidney damage. OSHA exposure limit (TWA): 5 ppm [skin] or 22 mg/m³.

p-cresol — C₇H₈0 — colorless crystals with a phenolic odor; MW: 108.13, BP: 395.6°F, Sol: 2.4 g/100 ml, Fl.P: 187°F, sp gr: 1.018. Other information not available. Skin exposure may create health problems. OSHA exposure limit (PEL-TWA): 5 ppm (22 mg/m³). Also known as 4-methylphenol.

crest factor — the ratio of a peak voltage to the root-mean-square (RMS) voltage of a wave form where both values are measured and referenced to the arithmetic mean value of the wave form.

criminal law — the law of crimes and their punishments.

cristobalite — a crystalline, extremely hard and chemically inert form of free silica created when refractory bricks and amorphous silica in diatomaceous earth are altered by exposure to high temperatures.

criteria — plural of criterion.

criteria documents (CD) — recommended occupational safety and health standards for the Department of Labor, usually part of the recommended standard is a recommended exposure limit.

criteria pollutants — those pollutants which are measured by ambient air quality standards and include total suspended particulates, sulfur dioxide, carbon monoxide, nitrogen oxide, photochemical oxidants (primarily ozone), gaseous hydrocarbons, and lead.

criterion — a standard on which a judgment or decision is made.

criterion standards — measurable characteristics and the quantitative and/or qualitative value assigned the particular characteristic.

critical control point — (*food*) any point or procedure in a specific food system at which a loss of control would result in an unacceptable health risk.

critical defect — a defect that may result in hazardous or unsafe conditions for individuals using and depending upon the product.

C D

critical item — (*food*) a food, which if in non-compliance with the Food Code of 1993, is more likely than other violations to contribute to food contamination, illness, or environmental degradation.

critical limit — (*food*) one or more prescribed tolerances that must be met to ensure that a critical control point effectively eliminates or controls a microbiological hazard.

critical mass — the smallest amount of fissionable material which can sustain a chain reaction.

critical organ — an organ or tissue which when affected by physical, chemical, or microbiological factors will create the greatest hazard to an individual or an individual's descendants.

critical orifice — a flow-rate meter pump which uses a critical or limiting orifice to regulate the rate of airflow.

critical soils — (*soils*) those soil materials that have been disturbed and/or have natural limitations enough to require alternative systems engineering or are perhaps so limited as to preclude the practicality of on-site wastewater treatment.

critical temperature — the highest temperature at which a substance can exist as a liquid regardless of the pressure.

cross connection — any physical connection or arrangement between two otherwise separate piping systems, one of which contains potable water and the other any fluid or suspension other than potable water, whereby there may be a flow from one system to another, the direction depending on the pressure differential between the two systems.

cross-hatching — (*geographic information system*) the technique of shading areas on a map with a given pattern of lines or symbols.

cross infection — an infection acquired by a patient during hospitalization; symptoms may appear while still hospitalized or may not appear until after discharge.

cross sectional epidemiological study — an examination of a point in time of various health indices of a population.

cross-sectional study — (*epidemiology*) a one-time survey of a population generally intended to describe the current existing prevalence of a disease; this type of study can be markedly influenced by cohort effects.

crotonaldehyde — $CH_3CH=CHCHO$ — a water-white liquid with a suffocating odor; turns pale yellow on contact with air; MW: 70.1, BP: 219°F, Sol: 18%, Fl.P: 45°F, sp gr: 0.87. It is used as an intermediate in the manufacture of butyl alcohol; in polymer technology; in organic synthesis in the manufacture of dyestuffs, sedatives, pesticides, and flavoring agents; as a solvent for the purification of mineral and lubricating oils; in the manufacture of surface active agents, as bactericides, for petroleum well fluids, metal brighteners, and leather/paper sizing; in the manufacture of rubber; in the manufacture of chemical warfare agents; in photographic emulsions. It is hazardous to the respiratory system, eyes, and skin; and is toxic by inhalation, ingestion, and contact. Symptoms of exposure include eye and respiratory irritation; in animals: dyspnea, pulmonary edema, and skin irritation. OSHA exposure limit (TWA): 2 ppm [air] or 6 mg/m³.

CRT — *See* cathode-ray tube.

crucible — a heat-resistant, barrel-shaped pot used to hold metal in a furnace.

crude petroleum — a mixture of hydrocarbons that have a flash point below 150°F and have not been processed in a refinery.

crude rates — (*epidemiology*) overall rates calculated from the actual number of events, such as deaths or births, in a total population during a specified period of time.

crumb soil — (*soils*) a soil structure in which the aggregates are 1–5 millimeters in diameter, soft, porous, and weakly held together (nearly spherical) with many irregular surfaces; usually found in the surface soil or A horizon.

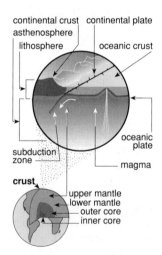

Earth's Crust

crust — (*earth*) the outermost solid layer of the earth containing soils ranging in thickness from a few millimeters to perhaps as much as an inch, that is much more compact, hard, and brittle when dry than the material immediately beneath it; (*medicine*) an outer layer of solid matter formed by drying of a body exudate or secretion.

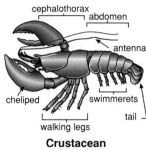

Crustacean
(lobster)

crustacean — any edible, commercially distributed shrimp, crab, lobster, or other member of the *Crustacea* and having jointed feet and mandibles, two pairs of antennae, and segmented bodies.

cryogenics — the science concerned with the production of ultra-low temperatures and the study of properties and materials at these temperatures.

Cryptolestes pusillius — *See* Flat Grain Beetle.

Cryptosporidiosis — (*disease*) a protozoan infection of the gastrointestinal tract associated with intestinal symptoms such as copious water diarrhea, abdominal cramps, malabsorption, and weight loss. Incubation time is 2–14 days with an average of 7 days, and the duration of infection is 10–14 days of watery diarrhea, followed by continuing shedding of oocyst for an additional 14–21 days. It is caused by *Cryptosporidium*, an aflagellar intestinal protozoan with motile sporozoites that are released in the host intestinal tract upon dissolution of the oocyst's outer wall. Sporozoites implant on the host epithelium and undergo sexual and asexual development. The parasite host relationship is unique in that it is intracellular, that is, enveloped by the host cell membrane but extracytoplasmic to the cytoplasm. Sporozoites develop into trophozoites and subsequently go through asexual multiplication, formation of macrogametes and microgametes, fertilization, and oocyst formation. Newly-formed oocysts that are expelled in the feces are immediately infective. It is found worldwide in day care centers, among immunocompromised individuals, as a nosocomial infection, and in the general population through outbreaks of waterborne disease with a reported level of 4–7% of the population, up to 65% of the population in areas of outbreaks of disease, and 10–15% asymptomatic individuals carrying the parasite; causes outbreaks of waterborne disease where the source may be surface water, water treated by rapid sand filtration and chlorination, contaminated by heavy rainfall and runoff, or animals, especially cattle. The reservoir of infection includes humans and domestic animals. It is transmitted by ingestion of oocysts in water contaminated by feces, person to person, and self recontamination. It is communicable for the entire period of the infection and at least 2–3 additional weeks. Susceptibility is high, especially in target populations, such as day care centers, and among the immunosuppresed. It is controlled through proper disposal of feces, vigorous personal hygiene, closing of contaminated water supplies, and superchlorination.

Cryptosporidium — an intestinal aflagellar protozoan, 4–5 nanometers in size, environmentally resistant, and of the class of sporozoa, suborder *Eimeriorina,* parasitic in the intestines of humans and animals which may cause severe watery diarrhea; can develop wholly in one host, and therefore creates a tremendous potential for reinfection, especially for immunosuppressed individuals.

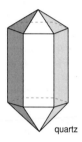

quartz

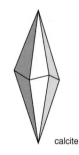

calcite

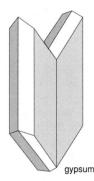

gypsum

Crystal

crystal — a mineral resulting from the arrangement of atoms, ions, or molecules in definite geometric patterns.

**C
D**

crystalline — having a definite molecular or ionic structure characteristic of crystals; crystalline materials need not exist as crystals.

crystalline aquifers — geological formations of granites and various types of igneous and metamorphic rocks which store water in tiny fractures and cracks within the solid masses; this type of aquifer produces very low volumes of water.

crystallization — the process of forming definitive shaped crystals when water has evaporated from a solution of the substance.

crystallography — the process of photographing the X-ray defraction pattern of a crystal.

CS — *See* central service.

CSF — *See* cerebrospinal fluid.

CSP — *See* certified safety professional.

CT — *See* computed tomography.

CT scan — *See* computed tomography.

CT scanner — an instrument designed for diagnostic radiological study of the head.

cubic centimeter (cc) — a volumetric measurement that is equal to one milliliter.

cubic feet per hour (cfh) — the amount of air flowing past a specific point in cubic feet per hour.

cubic feet per minute (cfm) — the number of cubic feet of air drawn through a device per minute of time.

cubic feet per second (cfs) — the number of cubic feet of air drawn through a device per second of time.

cubic foot (cu. ft.) — measurement equal to 7.48 gallons, 28.3162 liters, or 1,728 cubic inches.

cubic foot of water — measurement equal to approximately 62.5 pounds.

cu. ft. — *See* cubic foot.

cullet — clean, color-sorted crushed glass that is used in glassmaking to speed up the melting of silica sand. Also spelled collet.

culm — waste material resulting from the screening or removal of coal from the surrounding materials.

culture — the propagation of microorganisms or of living tissue cells in special media conducive to their growth.

culture medium — any substance or preparation of material used for the growth of cultures and cultivation of microorganisms.

cumene — $C_6H_5CH(CH_3)_2$ — a colorless liquid with a sharp, penetrating, aromatic odor; MW: 120.2, BP: 306°F, Sol: insoluble, Fl.P: 96°F, sp gr: 0.86. It is used as a constituent of solvents; in petroleum distillates; in the mixing and blending of aviation fuels. It is hazardous to the eye, upper respiratory system, skin, and central nervous system; and is toxic by inhalation, absorption, ingestion, and contact. Symptoms of exposure include eye and mucous membrane irritation, headache, dermatitis, narcosis, coma. OSHA exposure limit (TWA): 50 ppm [skin] or 245 mg/m^3.

cumulative dose — (*radiation*) the total dose resulting from repeated or continuous exposures to radiation.

cumulative effect — (*pesticides*) the build-up of pesticides and their storage in the body over a period of time which can sicken or kill an animal or human.

cumulative incident rate — (*epidemiology*) the number of new cases of disease during a specified period divided by the population initially free of the disease.

Cumulative Index to Nursing and Allied Health Literature — a reference index of approximately 300 nursing, allied health, and health-related journals from 1956 to the present.

cumulonimbus clouds — a principal cloud type, the bases of which almost touch the ground; violent up-drafts carry the tops to 75,000 feet; the most violent of these produce tornadoes; thunderheads.

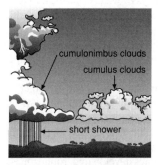

**Cumulus and
Cumulonimbus Clouds**

cumulus clouds — a principal cloud type formed by rising air currents; puffy and resembling cauliflower at the head; usually a sign of good weather.

cup — a measurement equal to 16 tablespoons.

cupric sulfate — *See* copper sulfate.

cured meat — meat preserved with chemicals or by drying including salted, smoked, pickled, and dried meats.

curettage — the cleansing of a diseased surface.

curie (Ci) — the unit of radioactivity in which the number of disintegrations is 3.700×10^{10} per second.

curing — the addition of sodium chloride to meat mixed with sugar and spices, developed to preserve meat without use of refrigeration; however, most processed meats necessitate refrigeration.

current density — (*electricity*) the current per unit area of a conductor.

current dollars — dollars at the most recent time.

Current Intelligence Bulletins (CIB) — important public health information and recommended protective measures to industry, labor, public interest groups, and academia concerning occupational health problems.

current meter — a device used to measure the velocity of flowing water.

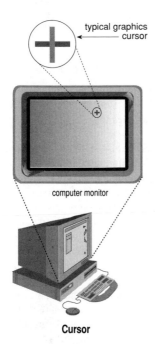

Cursor

cursor — (*computer science*) a visible symbol guided by the keyboard indicating a position on a visual display device.

curtain wall — a refractory, construction, or baffle that deflects combustion gases downward; an external wall that is not load bearing.

cut — (*land*) a portion of a land surface or area from which earth or rock has been or will be excavated.

cutaneous — pertaining to or affecting the skin.

cutaneous hazards — chemicals which affect the dermal layers of the body with defatting of the skin, rashes, and irritation.

cutaneous larva migrans — (*disease*) a disease caused by an infective larva of dog and cat hookworm *Ancylostoma braziliense* and *Ancylostoma caninum*, causing a dermatitis in individual in contact with damp, sandy soil contaminated with cat and dog feces; the larvae enter the skin, migrate intracutaneously for long

periods, and then penetrate to deeper tissues; a spontaneous cure occurs after several weeks or months. Also known as swimmer's itch.

cutaneous route — a route of entry into the body through the skin.

cutaneous toxicity — adverse effects resulting from skin exposure to a substance.

cutaneous toxins — chemicals that affect the dermal layer of the body; symptoms include defatting of the skin, rashes, and irritation; chemicals include ketones and chlorinated compounds.

cutie pie — a portable instrument equipped with a direct-reading meter used to determine the level of ionizing radiation in an area; a radiation dose-rate meter mounted on a pistol-type grip.

cutting fluid — an oil or oil-water emulsion used to cool and lubricate a cutting tool.

CV — *See* coefficient of variation.

CVA — *See* cerebrovascular accident.

CWA — *See* Clean Water Act.

CW laser — *See* continuous-wave laser.

cyanide — KCN and NaCN — potassium cyanide and sodium cyanide are white granular or crystalline solids with a faint almond-like odor; MW: 65.1/49.0, BP: not known/ 2725°F, Sol (77°F): 72/58%, sp gr: 1.55/1.60. It is used as fumigants and pesticides in greenhouses, ships, mills, and warehouses; in metal treatment; in the manufacture of intermediates in pharmaceuticals, dyes, vitamins, plastics, and sequestering agents; in cellulose technology, paper manufacture, and dyeing; as a photographic fixative. It is hazardous to the cardiovascular and central nervous systems, the liver, skin, and kidneys; and is toxic by inhalation, absorption, ingestion, and contact. Symptoms of exposure include weakness, headache, confusion, nausea, vomiting, increased rate of respiration, slow gasping respiration, eye and skin irritation; asphyxia and death. OSHA exposure limit (TWA): 5 mg/m^3 [air].

cyanosis — a bluish discoloration of the skin and mucous membranes due to excessive concentration of carbon monoxide and reduced hemoglobin in the blood.

cyclamate — a non-nutritive sweetener banned in the United States when animal testing showed it may be carcinogenic.

cycles per second (cps) — *See* hertz.

cyclohexane — C_6H_{12} — a colorless liquid with a sweet chloroform-like odor; a solid below 44°F; MW: 84.2, BP: 177°F, Sol: insoluble, Fl.P: 0°F, sp gr: 0.78. It is used as a solvent to dissolve cellulose, ethers, resins, fats, waxes, oils, bitumen, and crude rubber; in the manufacture of perfumes; in the synthesis of adipic acid for production of Nylon. It is hazardous to the eyes, respiratory system, skin, central nervous system; and is toxic by inhalation, ingestion, and contact. Symptoms of exposure include eye and respiratory system irritation, drowsiness, dermatitis, narcosis, coma. OSHA exposure limit (TWA): 300 ppm [air] or 1,050 mg/m^3.

cyclohexanol — $C_6H_{11}OH$ — a sticky solid or colorless to light-yellow liquid (above 77°F) with a camphor-like odor; MW: 100.2, BP: 322°F, Sol: 4%, Fl.P: 154°F, sp gr: 0.96. It is used in the synthesis of adipic acid and caprolactam; in surface coatings of natural and synthetic textile dyes, cotton mercerizing, paints, varnishes, lacquers, and shellacs; in the production of soaps, synthetic detergents, rubber cement, dicyclohexyl phthalate, and other cyclohexyl esters. It is hazardous to the eyes, respiratory system, and skin; and is toxic by inhalation, absorption, ingestion, and contact. Symptoms of exposure include eye, nose, and throat irritation, narcosis. OSHA exposure limit (TWA): 50 ppm [skin] or 200 mg/m^3.

cyclohexanone — $C_6H_{10}O$ — a water-white to pale yellow liquid with a peppermint- or acetone-like odor; MW: 98.2, BP: 312°F, Sol (50°F): 15%, Fl.P: 146°F, sp gr: 0.95. It is used in the surface coating of fabrics and plastics; in the cleaning of leather and textiles; as a solvent in crude rubber, insecticides, epoxy resins, as a sludge solvent in lubricating oils. It is hazardous to the respiratory system, eyes, skin, and central nervous system; and is toxic by inhalation, absorption, ingestion, and contact. Symptoms of exposure include eye and mucous membrane irritation, headache, narcosis, coma, dermatitis. OSHA exposure limit (TWA): 25 ppm [skin] or 100 mg/m^3.

cyclohexene — C_6H_{10} — a colorless liquid with a sweet odor; MW: 82.2, BP: 181°F, Sol: insoluble, Fl.P: 11°F, sp gr: 0.81. It is used in organic synthesis as a starting material or chemical intermediate; in the synthesis of polymers as polymer modifiers to control molecular weight; as a stabilizing agent. It is hazardous to the skin, eyes, and respiratory system; and is toxic by inhalation, ingestion, and contact. Symptoms of exposure include eye, skin, and respiratory system irritation, drowsiness. OSHA exposure limit (TWA): 300 ppm [air] or 1,015 mg/m^3.

cyclone — a cylindrical shaped air-cleaning apparatus through which a gas enters, swirls, and throws particles to the outside where they slide down into a hopper.

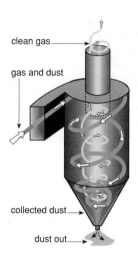

Cyclone Collector

cyclone collector — a device in which the direction of the gas flow is changed suddenly, causing the particles to flow straight ahead because of inertia.

cyclopentadiene — C_5H_6 — a colorless liquid with an irritating, terpene-like odor; MW: 66.1, BP: 107°F, Sol: insoluble, Fl.P (oc): 77°F, sp gr: 0.80. It is used in thermal cracking and chemical synthesis; in the production of modified oil; in the preparation of epoxy resins, polymers, varnishes; as a drying oil for varnishes; in the production of fire-resistant polyurethane foams; foundry-core bindings, and epoxidized olefins. It is hazardous to the eyes and respiratory system; and is toxic by inhalation, ingestion, and contact. Symptoms of exposure include eye and nose irritation. OSHA exposure limit (TWA): 75 ppm [air] or 200 mg/m^3.

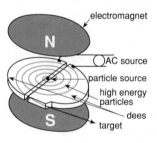

Cyclotron

cyclotron — a device for accelerating charged particles to high energy by means of an alternating electric field between electrodes placed in a constant spiraling magnetic field; used for bombarding the nuclei of atoms.

cyclotron D — *See* dee.

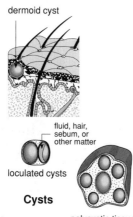

loculated cysts

fluid, hair,
sebum, or
other matter

Cysts

polycystic tissue

cyst — a closed epithelium-lined sac or capsule containing a liquid or semi-solid substance; a stage in the life cycle of certain parasites during which they are contained in a protective wall; (*physiology*) a small abnormal saclike growth in animals or plants usually containing liquid and diseased matter produced by inflammation.

cystitis — inflammation of the urinary bladder which may result from an infection, descending from the kidney, or coming from the exterior of the body by way of the urethra.

cytogenetics — the study of the structure and function of chromosomes as units of heredity and evolution.

cytology — the science that deals with cell function, growth, reproduction, and structure.

cytomegalovirus (CMV) — a viral infection that may occur without symptoms or result in mild flu-like symptoms; severe infections can result in hepatitis, mononucleosis, or pneumonia; the virus is shed in body fluids and is of particular concern for individuals suffering from some form of immune deficiency.

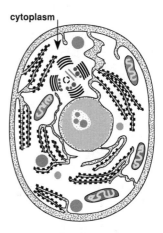

cytoplasm

Cytoplasm in Animal Cell

cytoplasm — the jelly-like protoplasm lying outside the nucleus of an animal or plant cell.

cytoplasmic membrane — the cell boundary which maintains the integrity of the cell.

cytotoxin — a substance having a specific toxic effect on cells of special organs.

D

2,4-D — *See* 2,4-dichlorophenoxyacetic acid.

D — *See* deuterium; electric displacement.

dactylitis — inflammation of a finger or toe.

daily cover — material that is spread and compacted on the top and sides of solid waste at the end of each operating day.

Dalton — *See* atomic mass unit.

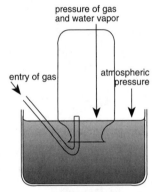

Dalton's Law
in which the total pressure is
equal to the barometric pressure.

Dalton's law — the law that when gases or vapors having no chemical interaction are present as mixture in a given space, the pressure exerted by a component of the gas mixture at a given temperature is the same as it would exert if it filled the whole space alone. Also known as the law of partial pressures.

damage risk criterion — the suggested base line of noise tolerance that should not result in hearing loss.

damp — (*noise*) to reduce the generation of oscillatory or vibrational energy of an electrical or mechanical system; a harmful gas or mixture of gases found in coal mining.

damper — a manually or automatically controlled valve or plate in a breeching duct or stack that is used to regulate a draft or the rate of flow of air or other gases.

damping — the reduction in amplitude of a wave due to the dissipation of wavelength; sound proofing.

dander — small scales from the hair or feathers of animals which may cause an allergic reaction in sensitive persons.

Dano biostabilizer system — an aerobic, thermophilic, composting process in which optimum conditions of moisture, air, and temperature are maintained in a single, slowly revolving cylinder that retains the compostable solid waste for 1–5 days.

DASHO — *See* Designated Agency Safety and Health Official.

data — numerical or qualitative information for analysis used as the basis for decision-making; datum.

database — a collection of interrelated information stored on a magnetic tape or disk. Also known as a tuple.

database management system (DBMS) — a set of computer programs for organizing the information on a database.

data link — the communication lines and related hardware and software systems needed to send data between two or more computers over telephone lines, fiber optics, satellite, or cable.

data set — a minimum aggregation of uniformly defined and classified statistics that describe an element of a situation.

datum — *See* data.

daughter — (*nuclear fission*) a decay product of a radioactive substance. *See also* decay product.

daughter cell — newly formed cells resulting from the division of a previously existing cell called a mother cell. The two daughter cells receive nuclear materials that are identical.

DAVES — *See* desorption and vapor extraction system.

dB — *See* decibel.

dBA — *See* adjusted decibel.

dBC — *See* decibels on C scale.

DBMS — *See* database management system.

dc — *See* direct current.

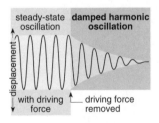

Damping

dcf — notation for dry cubic feet of air.

DD — *See* developmental disability.

DDD — *See* 2,2-bis(para-chlorophenyl)-1,1-dichloroethane.

DDT — $(C_6H_4Cl)_2CHCCl_3$ — the common name for dichlorodiphenyltrichloroethane; colorless crystals or off-white powder with a slight aromatic odor; MW: 354.5, BP: 230°F (decomposes), Sol: insoluble, Fl.P: 162–171°F, sp gr: 0.99. It is used as an insecticide which has been banned in the United States by the Environmental Protection Agency. It is hazardous to the central nervous system, kidneys, liver, skin, and peripheral nervous system; and is toxic by inhalation, absorption, ingestion, and contact. Symptoms of exposure include tremors, apprehension, dizziness, confusion, malaise, headache, fatigue, convulsions, paresis of hands, vomiting, eye and skin irritation, paresthesia of the tongue, lips, and face; carcinogenic. OSHA exposure limit (TWA): 1 mg/m³ [skin].

DE — *See* dose equivalent.

deadly nightshade — *See* belladonna.

deafness — temporary or permanent impairment or loss of hearing.

death rate — the ratio of total deaths to total population usually expressed as 1,000, 10,000, or 100,000 population. *See also* mortality rate.

debilitating — producing weakness, such as a disease, injury, or surgical procedure.

debug — to remove errors from a program or hardware.

debugger — a program that is used to remove programming errors.

debridement — a medical procedure for removing all foreign, contaminated, and devitalized tissues from or adjacent to a traumatic or infected lesion.

debris — the remains of something broken or destroyed.

decaborane — $B_{10}H_{14}$ — a colorless to white crystalline binary compound of boron and hydrogen with an intense, bitter, chocolate-like odor; MW: 122.2, BP: 415°F, Sol: slight, Fl.P: 176°F, sp gr (77°F): 0.94. It is used as a reducing agent in chemical synthesis; as a component or additive for high-energy fuel and rocket propellants; in the manufacture of polymers; as a vulcanizing agent in the manufacture of synthetic rubber. It is hazardous to the central nervous system; and is toxic by inhalation, absorption, ingestion, and contact. Symptoms of exposure include dizziness, headache, nausea, lightheadedness, drowsiness, incoordination, local muscle spasms, tremors, convulsions, fatigue; in animals: dyspnea, weakness, liver and kidney damage. OSHA exposure limit (TWA): 0.3 mg/m³ [skin] or 0.05 ppm.

decalcification — the loss or removal of calcium or calcium compounds from calcified material, such as bone or soil.

decay — the reduction of the organic substances of a plant or animal body to simple inorganic compounds usually by the action of bacteria or fungi; to undergo decomposition; (*radioactivity*) the change in the nuclei of radioactive isotopes that spontaneously emits high-energy electromagnetic radiation and/or subatomic particles, while gradually changing into a different isotope or element.

decay chain — a sequence of radioactive decays (transformations) beginning with one nucleus; that initial nucleus, the parent, decays into a daughter nucleus that differs from the first by whatever particles were emitted during decay. Also known as decay series.

decay constant — the fraction of the number of atoms of the radioactive nuclide which decay in unit time; the "c" in the equation $I=I_0e^{-ct}$, where "I" is the number of disintegrations per unit time.

decay product — a new isotope formed as a result of radioactive decay. Also known as daughter.

decay series — *See* decay chain.

deceleration — the rate at which the velocity of a moving object decreases; the opposite of acceleration.

deci — prefix meaning 1/10 or .1 in the metric system.

decibel (dB) — (*physics*) a unit for measuring the relative intensity of sounds equal to 1/10th of a bel; one decibel equals approximately the smallest difference in acoustic power the human ear can detect where an increase in 10 decibels doubles the loudness of sound; the sound pressure level equal to 20 micropascal.

decibel meter — an instrument for measuring sound waves as they strike a microphone, producing an electric current whose strength depends on the loudness. The current is then amplified and registered on a meter, which is calibrated in decibels.

decibels on C scale (dBC) — the sound level in decibels read on the C-scale of a sound-level meter which discriminates very little against very low frequencies.

deck — any structure that is on top of or adjacent to the outer edges of a pool, spa, or hot tub wall and supports people in a sitting or upright position.

decolonization — (*infection control*) the elimination of methicillin resistant *Staphyloccus*

C
D

aureus from humans by use of infection control measures and/or antibiotics.

decompose — to rot or decay.

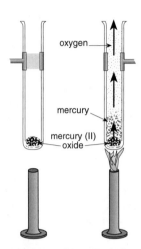

Decomposition Reaction

decomposition — the breaking down of a compound into simple substances or into its constituent elements; (*food*) any food, whole or in part, filthy, putrid, or decomposed and unfit for human consumption; break-down of a material or substance by heat, chemical reaction, electrolysis, decay, or other processes into parts, elements, or simpler compounds.

decomposition product — material produced or generated by the physical or chemical degradation of a parent material.

decontamination — the removing of biological, radiological, or chemical contamination, making the environment or specific object safe to handle.

decontamination room — that area within the hospital where used materials are washed prior to sterilization.

decubitus ulcer — an ulcer of the skin and subcutaneous tissues due to local interference with the circulation. Also known as a bed sore; pressure sore.

dedicated exhaust — an exhaust system serving only the aerator sterilizer and/or the immediate area.

dee — a hollow accelerating cyclotron electrode shaped like the letter D. Also known as cyclotron D.

Deep-Water Port Act — a law passed in 1974 and amended through 1984; it authorizes and regulates location, ownership, construction, and operation of deep-water ports in waters beyond the territorial limits of the United States.

deep well injection — a means of disposal of liquid hazardous waste underground by injecting with force and under pressure the waste into steel and concrete-encased shafts placed deep into the earth.

deer mouse — *Peromyscus maniculatus*; a member of the family Cricetidae, which is native to North America and occupies every type of habitat from forests to grasslands. It may be found in rural, semi-rural, and even urban areas. Characteristics include weight: 2/3 to 1-1/4 ounces; total length including tail: 4-4/5″ to 9″; color: white feet, usually white underside, brownish upper surfaces; gestation: 21–23 days; litter size: 2–4 with 3–5 young each; harborage: nests anywhere, a burrowing animal; range: 0.13 to 1.6 hecacres; food: primarily a seed-eater, but will feed on nuts, acorns, fruits, insects, insect larva, fungi, and green vegetation. It is the reservoir for hantavirus, and the cause of outbreaks of hantavirus pulmonary syndrome.

default decree of condemnation — an order requested of the court when lots under seizure are not claimed or defended; it provides for destruction, donation to charity, sale, or disposed of as the court may elect to decree.

defendant — the opponent of either the plaintiff or prosecution; the party or parties named in an information, indictment, or complaint for injunction, and against whom the government is proceeding.

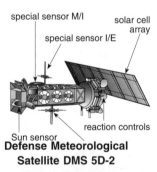

**Defense Meteorological
Satellite DMS 5D-2**

**Defense Meteorological
Satellite DMSP 5D-2**

defense meteorological satellite (DMS) — a satellite using infrared sensors to map global and spectral analysis of the ozone layer.

deflocculation — the action in which groups or clumps of particles are broken up into individual particles and spread out suspended in the solution; dispersion.

defoliant — a pesticide chemical that causes the leaves of a plant to drop off prematurely.

degradation — break-down of a complex chemical by the action of microbes, water, air, sunlight, or other agents; (*protective clothing*) a deleterious change of one or more physical properties of a protective clothing material due to contact with a chemical; the wearing down of a land surface by the process of erosion.

degrease — to use an organic solvent as a surface cleaning agent or to remove grease, oil, or fatty material from a surface.

degree of freedom — statistical term that is a measure of stability related to the number of independent equivalent terms entering into a distribution.

dehydrating agent — a substance which is able to dry another substance by withdrawing water.

dehydration — the removal or loss of water from any substance.

deionizing — the purifying of water of dissolved salts by passing the water through synthetic materials.

DEIS — *See* draft environmental impact statement.

deka — a prefix meaning 10 in the metric system.

Delaney clause — the food additive amendment of 1958 to the Food, Drug, and Cosmetic Act; it states known additives shall be deemed to be unsafe if found to induce cancer when ingested by man and/or animal or found after tests which are appropriate for the evaluation of food additives to induce cancer in man or animal.

delay time — down time; the amount of time which a signal is slowed after transmission.

deleterious — causing harm or injury.

delta — the mass of sediment usually triangular in shape and deposited where rivers lose their velocity as they enter lakes or seas.

delta endotoxin — a toxic molecule produced within the cell of a bacterium.

delta ray — an energetic or swiftly moving electron ejected from an atom during the process of ionization.

demand — the number of funded positions in a given occupation, whether filled or unfilled.

demeton — $C_8H_{19}O_3PS_2$ — an amber, oily liquid with a sulfur-like odor; MW: 258.3, BP: decomposes, Sol: 0.01%, Fl.P: 113°F, sp gr: 1.12. It is used as an insecticide, pesticide, and acaricide. It is hazardous to the respiratory system, cardiovascular system, central nervous system, skin, eyes, and blood cholinesterase; and is toxic by inhalation, absorption, ingestion, and contact. Symptoms of exposure include miosis, eye aches, rhinorrhea, headache, tight chest, wheezing, laryngeal spasms, salivation, cyanosis, anorexia, nausea, vomiting, abdominal cramps, diarrhea, local sweating, muscle fasciculation, weakness, paralysis, giddiness, confusion, ataxia, convulsions, coma, low blood pressure, cardiac irregularities, eye and skin irritation. OSHA exposure limit (TWA): 0.1 mg/m^3 [skin].

de minimus — (*law*) not enough to be considered.

demodulation — the process of extracting the encoded intelligence from a modulated signal.

demography — the statistical study of populations with reference to natality, mortality, migration, age, sex, and other socio, ethno, and economic factors.

denature — the decharacterization of an article through the addition of some foreign substance in sufficient quantity so as to preclude its original purpose for its future use.

dengue fever — (*disease*) an acute, febrile disease with sudden onset of fever for approximately 5–7 days, intense headache, joint muscle pain and rash; incubation time is 3–15 days; caused by the viruses of dengue fever immunological types 1, 2, 3, and 4 which are flaviviruses. It is found in most parts of Asia, Central and South America, Cuba, and Mexico. The reservoir of infection includes people and mosquitos; transmitted by the bite of infected mosquitos, *Aedes aegypti*, as well as other *Aedes*. It is not directly transmitted from person to person; universally susceptible; controlled through community surveys and mosquito control efforts.

denitrification — the biochemical reduction of nitrate or nitrite to gaseous molecular nitrogen or an oxide of nitrogen; brought about by denitrifying bacteria.

density — the weight of a solid or liquid in grams per milliliter; the ratio of the mass of a substance to its volume. Also known as specific weight.

denudation — the wearing away of the land by sun, wind, rain, frost, running water, moving ice, and the intrusion of the sea.

deoxygenation — the removal of oxygen from a substance, such as blood or tissue.

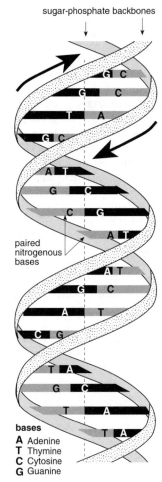

sugar-phosphate backbones

paired
nitrogenous
bases

bases
A Adenine
T Thymine
C Cytosine
G Guanine

Deoxyribonucleic Acid (DNA)
Double Helix

deoxyribonucleic acid (DNA) — a long, double helix molecule in the nucleus of cells that consist of alternating units of deoxyribose sugar, phosphates, and organic bases; the nucleic acid in chromosomes that contains in the arrangement of its units the master code for the hereditary traits of the organism; DNA transmits the hereditary information and controls cellular activities.

Department of Agriculture (USDA) — a department of the United States government which works to improve and maintain farm income and to develop and expand markets abroad for agricultural products; helps to curb and cure poverty, hunger, and malnutrition; works to enhance the environment and to maintain production capacity by helping landowners protect soil, water, forest, and other natural resources.

Department of Commerce (DOC) — a department of the United States government whose function is to encourage, serve, and promote the nation's international trade, economic growth, and technological advancement.

Department of Defense (DOD) — the federal agency responsible for providing the military forces needed to deter war and protect the security of the United States.

Department of Energy (DOE) — the federal agency responsible for providing the framework for a comprehensive and balanced national energy plan through the coordination and administration of the energy functions of the federal government.

Department of Housing and Urban Development (HUD) — the federal agency principally responsible for programs concerned with the nation's housing needs, the development and preservation of the nation's communities, and the provision of equal housing opportunity for every individual.

Department of Interior (DOI) — a department of the United States government whose function is to act as the nation's principle conservation agency; has the responsibility for most of the nationally-owned public lands and natural resources.

Department of Justice (DOJ) — the federal agency which serves as counsel to the citizens of the United States by representing them in enforcing the law in the public interest; it protects citizens against criminals and subversion, ensures healthy competition of business, safeguards the consumer, and works in enforcing drug, immigration, civil rights, and naturalization laws.

Department of Labor (DOL) — the federal agency responsible for fostering, promoting, and developing the welfare of the wage earners of the United States to improve their working conditions, and to advance their opportunities for profitable and safe employment.

Department of Transportation (DOT) — the federal agency responsible for establishing the nation's overall transportation policy; includes highway planning, development, construction, urban mass transit, railroads, and aviation; and the safety of waterways, ports, highways, and oil and gas pipelines.

depleted — being exhausted; reduced.

depletion — the mineral exhaustion of a soil through continued planting of agricultural crops without proper fertilizing; using a resource faster than it can be replenished.

deposit — a coating of solid matter which is gradually laid down on a surface by a natural process or agent.

deposition — the settling of any material, either liquid or solid, on a surface; (*law*) the sworn testimony of a witness in affidavit form; it is obtained outside the courtroom through examination and cross-examination by the attorneys for the government and the defense.

depreciation — the decline in values of capital assets over time with use.

depressant — a substance that reduces a bodily function, activity, or instinctive desire, such as appetite.

depression — (*weather*) *See* low.

derma — *See* corium.

dermal — of or referring to the skin or dermis.

dermal exposure — exposure to a chemical through contact with the skin.

dermal toxicity — adverse effects resulting from skin exposure to a pesticide, chemical, or other substance absorbed through the skin.

dermatitis — inflammation of the skin resulting from various animal, vegetable, and chemical substances, heat, cold, mechanical irritation, malnutrition, infectious diseases, and psychological problems. Symptoms may include itching, redness, crustiness, blisters, water discharges, fissures, or other changes in the normal condition of the skin.

dermatological injury — injury to the skin resulting from instantaneous trauma or brief exposure to toxic agents involving a single incident in the work environment.

dermatophyte — pathogenic fungi that attack the epidermis causing ringworm.

Dermestes lardarius — *See* Larder Beetle.

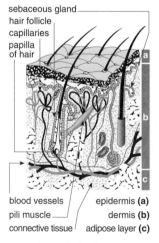

sebaceous gland
hair follicle
capillaries
papilla
of hair

blood vessels   epidermis (a)
pili muscle   dermis (b)
connective tissue   adipose layer (c)

Dermis

dermis — the thick living layer of the skin underneath the thinner epidermis in vertebrates; it contains blood vessels, nerves, and sensory organs.

DES — *See* diethylstilbestrol.

desalination — the purification of saline or brackish water to remove salts by the process of freezing, evaporation, or electrodialysis.

desalinization — the removal of salt from water to make it potable.

descriptive biochemistry — the precise structural chemistry of animate matter.

descriptive epidemiology — (*statistics*) mathematical or graphical techniques to study events, frequency, time, place, and population to describe phenomenon and occurrence of disease.

desertification — the spread of desert to previously fertile areas; a process involving the rapid loss of top soil and depletion of plant life frequently caused by over-grazing of animals and misuse of the land.

desiccant — a chemical agent used to remove moisture and promote drying.

designated health planning areas — geographical areas composed of counties for the purpose of state health planning.

Designated Agency Safety and Health Official (DASHO) — the executive official of a federal department or agency who is responsible for safety and occupational health matters within a federal agency and is designated or appointed by the head of the agency.

design capacity — the number of tons of solid waste that a designer anticipates the incinerator will be able to process in a 24-hour period.

desorption — the removing of a substance from the state of absorption or adsorption.

desorption and vapor extraction system (DAVES) — a low-temperature fluidized bed to remove organic and volatile inorganic compounds from soils, sediments, and sludges.

desquamation — the shedding of the skin in scales or sheets.

destructive distillation — the process of heating an organic substance, such as coal, in the absence of air to break it down into volatile products and a solid char consisting of fixed carbon and ash.

desulfurization — the removal of sulfur from fossil fuels to reduce pollution.

detection — the action of identifying by means of close, continuous or noncontinuous monitoring the appearance of a failure, or the existence of a failing element.

detector — a device that detects the presence of an entity of interest and indicates its magnitude as the deviation from a reference and converts these indications into a signal. Also known as sensor.

detention — the act of detaining an imported product to allow the owner to come forth and

C
D

defend the lot, much in the same manner as a hearing prior to prosecution; the detention impedes entry into the country with any article which appears to be in violation of any federal food or product law.

detention time — the time drinking water or wastewater is kept in a process vessel during treatment.

detergent — a synthetic cleaning agent containing surfactants and added to water to improve its cleaning properties.

detergent disinfectants — disinfectants with cleaning ability.

detergent sanitizers — materials used in highly contaminated areas to first clean the area, and then act as a sanitizer to kill bacteria which may cause disease.

detoxification — the reduction of poison or toxic properties of a substance in the body.

detoxify — to make harmless by removing toxic properties.

detritus — unconsolidated sediments comprised of both inorganic and dead and decaying organic material removed from rocks and minerals by mechanical means; remains of any broken-down tissue.

deuterium (D) — the isotope of the element hydrogen with one neutron and one proton in its nucleus; atomic weight 2.0144. *See also* heavy hydrogen.

deuterium oxide — *See* heavy water.

developmental disability (DD) — federally defined as a severe, chronic disability caused by mental or physical impairment, occurring before the age of twenty-two, and likely to continue indefinitely resulting in substantial functional limitations.

developmental toxicant — an agent affecting growth, development, or acquisition of normal organ function between conception and puberty; chemical, physical, or biological agents that adversely affect the growth or development of an embryo, fetus, or immature individual.

developmental toxicity — the occurrence of adverse effects on a developing organism that may result from an exposure to a chemical in either parent prior to conception during prenatal development or postnatally to the time of sexual maturation.

development toxicology — the field of study dealing with the induction of adverse effects on humans up to the time of puberty.

deviation — failure to meet a required critical limit for a critical control point; the difference between the actual value of a controlled variable and the desired or expected value.

device — a piece of equipment used for a special function.

dew — water vapor condensed on solid surfaces that have cooled below the condensation point of the air in contact with them because of radiation cooling during the night.

dewatering — (*sludge*) the removal of water from sludge so that its physical form is changed from essentially that of a fluid to that of a damp solid.

dew point — the temperature and pressure at which condensation occurs for a given concentration of water vapor in the air.

diabetes mellitus — a metabolic condition resulting from lack of insulin production in the pancreas, causing the body to be unable to store or oxidize sugar efficiently.

diacetone alcohol — $CH_3COCH_2C(CH_3)_2OH$ — a colorless liquid with a faint, minty odor; MW: 116.2, BP: 334°F, Sol: miscible, Fl.P: 125°F, sp gr: 0.94. It is used in the manufacture of artificial silk and leather, quick-drying inks, photographic film, antifreeze preparations, and hydraulic fluids; as a solvent for cellulose esters, epoxy resins, hydrocarbons, oils, fats, resin gums, dyes, tars, cements, and waxes; in paper and textile coatings, wood stains and preservatives. It is hazardous to the eyes, skin, and respiratory system; and is toxic by inhalation, ingestion, and contact. Symptoms of exposure include corneal tissue damage, narcosis, and eye, nose, throat, and skin irritation. OSHA exposure limit (TWA): 50 ppm [air] or 240 mg/m^3.

diagnosis — a concise, technical description of the cause, nature, or manifestations of a condition, situation, or problem by identifying its signs and symptoms.

dialysis — the diffusion of solute molecules through a semipermeable membrane passing from the side of higher concentration to that of the lower concentration; the process of separating a true solution from a colloidal dispersion by means of a membrane.

diameter of a circle — a line segment which passes through the center of a circle and which is equal to 2 × the radius.

diaphragm — (*anatomy*) the thick sheet of muscle that assists in breathing and separates the chest and the abdomen in mammals.

diarrhea — a diffuse and abnormal discharge from the bowels.

diastase — an enzyme controlling the digestion of starch to maltose.

diathermy — the generation of heat in tissues by electric currents for medical, therapeutic, or surgical purposes.

diatomaceous earth — a white, yellow, or light gray powder composed of the fossilized skeletons of one-celled organisms called diatoms; it is highly porous and contains microscopic holes; used as a filler and paint filler, adsorbent, abrasive, and thermal insulator.

diatomaceous earth filter — a filtering system that uses a hydrous form of silica or opal composed of the shells of diatoms; a filter designed to use diatomaceous earth as a filter medium; it may be either pressure or vacuum tight.

diatomite — a shortened name for diatomaceous earth.

diatoms — microscopic, unicellular algae whose silicious skeletons are often found in sea-floor sediments.

diazomethane — CH₂N₂ — a yellow gas with a musty odor; shipped as a liquefied compressed gas; MW: 42.1, BP: –9°F, Sol: reacts. It is used as a methylating agent in chemical analysis and laboratory organic synthesis. It is hazardous to the respiratory system, skin, and eyes; and is toxic by inhalation, ingestion, and contact. Symptoms of exposure include cough, shortness of breath, headache, flushed skin, fever, chest pain, pulmonary edema, pneumonitis, eye irritation, asthma, and fatigue. OSHA exposure limit (TWA): 0.2 ppm [air] or 0.4 mg/m³.

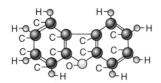

Dibenzofuran

dibenzofuran — an organic compound that contains two benzene rings fused to a central furan ring; chlorodibenzofurans are serious contaminants of some hazardous waste sites. *See* Chlorodibenzofurans.

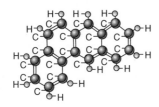

Dibenz (a,h) anthracene

dibenz {a,h} anthracene — C₂₂H₁₄ — one of the polycyclic aromatic hydrocarbon compounds formed when gasoline, garbage, or any animal or plant material burns, usually found in smoke and soot; MW: 278.35, BP: sublimes, Sol: 0.5 mg/L, Fl.P: not known. It is used as a research material; coal tar pitch is used primarily as a binder for electrodes in the aluminum reduction process, to bind the carbon electrodes in the reduction pots; as an adhesive in membrane roofs, as a wood preservative; in clinical treatment of skin disorders such as eczema, dermatitis, and psoriasis; potentially carcinogenic. OSHA exposure limits (TWA): 0.2 mg/m³ [air].

diborane — B₂H₆ — a colorless gas with a repulsive, sweet odor; usually shipped in pressurized cylinders; MW: 27.7, BP: –135°F, Sol: reacts. It is used as a reducing agent in the synthesis of organic chemical intermediates; as a component or additive for high-energy fuels; as a catalyst in olefin polymerization; in the electronics industry to improve crystal growth or impart electrical properties in pure crystals. It is hazardous to the respiratory and central nervous systems; and is toxic by inhalation and contact. Symptoms of exposure include tight chest, precordial pain, shortness of breath, non-productive cough, nausea, headache, lightheadedness, vertigo, chills, fever, fatigue, weakness, tremors, muscle fasciculation; in animals: liver and kidney damage, pulmonary edema, hemorrhage. OSHA exposure limit (TWA): 0.1 ppm [air] or 0.1 mg/m³.

1,2-Dibromo-3-chloropropane
(Dibromochloropropane)

1,2-dibromo-3-chloropropane — CH₂BrCHBrCH₂Cl — a dense yellow or amber liquid with a pungent odor at high concentrations; a solid below 43°F; MW: 236.4, BP: 384°F, Sol (70°F): 0.1%, Fl.P (oc): 170°F, sp gr: 2.05. It is used as a pesticide and fumigant. It is hazardous to the central nervous system, skin, liver, kidneys, spleen, reproductive system, and digestive system; and is toxic by inhalation, absorption, and contact. Symptoms of exposure include drowsiness, nausea, vomiting, irritation of the eyes, nose, skin, and throat, pulmonary edema; carcinogenic. OSHA exposure limit (TWA): 0.001 ppm [air].

C
D

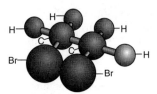

1,2–Dibromoethane
(Ethylene Dibromide)

1,2-dibromoethane — $BrCH_2CH_2Br$ — a colorless liquid with a mild, sweet odor; MW: 187.86, BP: 131–132°C, Sol (20°C): 0.4 g/100, Fl.P: not flammable. It is used as an additive in leaded gasoline where 1,2-dibromoethane acts as a scavenger that converts lead oxides in cars to lead halides; as a soil fumigant to protect against insects, pests, and nematodes, in citrus, vegetable, and grain crops, as a fumigant for turf, especially on golf courses; currently banned as a soil and grain fumigant; the respiratory tract, particularly the nasal cavity, is the point-of-contact target organ affected by inhalation. Symptoms of exposure include pharyngitis, bronchitis, conjunctivitis, anorexia, headache, and depression; the liver is the target organ for toxic effects in experimental animals; renal effects reported also in experimental animals; the reproductive effects are antispermatogenic where there is a change in sperm velocity, and a possible decrease in male fertility; nasal carcinomas are found in laboratory animals; renal lesions have been recorded in humans dying after acute oral exposure; endocrine lesions have also been found; fatal cases of occupational exposure have been reported. OSHA exposure limits (PEL TWA): 20 ppm, ceiling 30 ppm, acceptable maximum peak above ceiling level (5 minutes per 8 hour shift): 50 ppm, STEL (15 minutes) 0.5 ppm [air].

dibutyl phosphate — $(C_4H_9O)_2(OH)PO$ — a pale amber, odorless liquid; MW: 210.2, BP: 212°F (decomposes), Sol: insoluble, Fl.P: not known, sp gr: 1.06. It is used as a catalyst in the manufacture of phenolic and urea resins; in metal separation and extraction. It is hazardous to the respiratory system and skin; and is toxic by inhalation, ingestion, and contact. Symptoms of exposure include headache, irritation of the respiratory system and skin. OSHA exposure limit (TWA): 1 ppm [air] or 5 mg/m³.

dibutylphthalate — $C_6H_4(COOC_4H_9)_2$ — a colorless to faint yellow, oily liquid with a slight, aromatic odor; MW: 278.3, BP: 644°F, Sol (77°F): 0.5%, Fl.P: 315°F, sp gr: 1.05. It is used in the manufacture of nitrile rubber; in the manufacture of polyvinyl acetate surface coatings; in the production of cellulose acetate butyrate and polyvinyl acetate adhesives; in the manufacture of polyester and epoxy resins and nitrocellulose surface coatings; in the manufacture of explosives and propellants. It is hazardous to the respiratory system and gastrointestinal system; and is toxic by inhalation, ingestion, and contact. Symptoms of exposure include upper respiratory tract and stomach irritation. OSHA exposure limit (TWA): 5 mg/m³ [air].

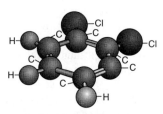

O–Dichlorobenzene
(1,2–Dichlorobenzene)

o-dichlorobenzene — $C_6H_4Cl_2$ — a colorless to pale yellow liquid with a pleasant, aromatic odor; MW: 147.0, BP: 357°F, Sol: 0.01%, Fl.P: 151°F, sp gr: 1.30. It is used in cleaning and degreasing metals, leather, paper, wool, dry cleaning, brick, and upholstery; as a fumigant for poultry houses and stockyards; in organic synthesis in pesticides, herbicides, dyestuffs, and pharmaceuticals; as a chemical intermediate in the manufacture of toluene-diisocynate; as a deodorizing agent; in textile dyeing operation. It is hazardous to the liver, kidneys, skin, and eyes; and is toxic by inhalation, absorption, ingestion, and contact. Symptoms of exposure include eye and nose irritation, liver and kidney damage, skin blistering. OSHA exposure limit (TWA): ceiling 50 ppm [air] or 300 mg/m³.

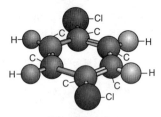

p–Dichlorobenzene
(1,4–Dichlorobenzene)

p-dichlorobenzene — $C_6H_4Cl_2$ — a colorless or white crystalline solid with a mothball-like odor; MW: 147.0, BP: 345°F, Sol: 0.008%, Fl.P: 150°F, sp gr: 1.25. It is used in moth control; as a deodorant for garbage and rest rooms; as an insecticide; in organic synthesis for preparation of dye intermediates. It is hazardous to the liver, respiratory system, eyes, kidneys, and skin; and is toxic by inhalation, ingestion, and contact.

Symptoms of exposure include headache, eye irritation, swelling periorbital, profuse rhinitis, anorexia, nausea, vomiting, low weight, jaundice, cirrhosis; carcinogenic; in animals: liver and kidney damage. OSHA exposure limit (TWA): 75 ppm [air] or 450 mg/m^3.

3,3′-dichlorobenzidine — $C_6H_3ClNH_2C_6H_3ClNH_2$ — a gray to purplish crystalline solid; MW: 253.1, BP: 788°F, Sol: almost insoluble, Fl.P: not known. It is used as an intermediate in the manufacture of dyes, pigments, and isocyanate-containing polymers. It is hazardous to the bladder, liver, lungs, skin, and gastrointestinal tract; and is toxic by inhalation, absorption, ingestion, and contact. Symptoms of exposure include allergic skin reaction, sensitization, dermatitis, headache, dizziness, caustic burns, frequent urination, dysuria, gastrointestinal upsets, upper respiratory infection; carcinogenic. OSHA exposure limit (TWA): regulated use.

dichlorodifluoromethane — CCl_2F_2 — a colorless gas with an ether-like odor at extremely high concentrations; shipped as a liquefied compressed gas; MW: 120.9, BP: –22°F, Sol (77°F): 0.03%. It is used as a propellant and refrigerant; in the manufacture of aerosols for cosmetics, pharmaceuticals, insecticides, paints, adhesives, and cleaners; as a blowing agent for cellular polymers; as a foaming agent in fire extinguishing aerosols; in water purification, copper and aluminum purification; in the manufacture of glass bottles; in regulating devices for leak detection; in thermal expansion valves; in organic synthesis of Freons. It is hazardous to the cardiovascular system and peripheral nervous system; and is toxic by inhalation and contact. Symptoms of exposure include dizziness, tremors, unconsciousness, cardiac arrhythmias, and cardiac arrest. OSHA exposure limit (TWA): 1,000 ppm [air] or 4,950 mg/m^3.

1,3-dichloro-5,5-dimethylhydantion — $C_5H_6Cl_2N_2O_2$ — a white powder with a chlorine-like odor; MW: 197.0, BP: not known, Sol: 0.2%, Fl.P: 346°F, sp gr: 1.5. It is used as a bactericide, sporicide, or sanitizer in swimming pools, dairies, laundries, restaurants, cutting oils, and pharmaceutical industry; as a general cleanser; as a bleaching agent; as a stabilizer, discoloration preventer, and catalyst in the polymer industry; to emboss or texturize resinous sheet preparations. It is hazardous to the respiratory system and eyes; and is toxic by inhalation, ingestion, and contact. Symptoms of exposure include eye, mucous membrane, and respiratory system irritation. OSHA exposure limit (TWA): 0.2 mg/m^3 [air].

dichlorodiphenyltrichloroethane — *See* DDT.

1,1–Dichloroethane

1,1-dichloroethane — $CHCl_2CH_3$ — a colorless, oily liquid with a chloroform-like odor; MW: 99.0, BP: 135°F, Sol: 0.6%, Fl.P (oc): 22°F, sp gr: 1.18. It is used as a dewaxer of mineral oils; an extractant for heat-sensitive substances; as a fumigant; in the manufacture of vinyl chloride by vapor phase cracking; in the manufacture of high vacuum rubber and silicon grease; as a chemical intermediate. It is hazardous to the skin, liver, and kidneys; and is toxic by inhalation, ingestion, and contact. Symptoms of exposure include central nervous system depression, skin irritation, liver and kidney damage. OSHA exposure limit (TWA): 100 ppm [air] or 400 mg/m^3.

1,2-dichloroethene — $C_2H_2Cl_2$ — a highly flammable colorless liquid with a sharp harsh odor; MW: 96.94, BP: 60.3°C; Sol (25°C): 3,500 mg/L, Fl.P: 6°C, sp gr: 1.2837. It is used as a chemical intermediate in the synthesis of chlorinated solvents and compounds, as a low-temperature extraction solvent for organic materials such as dyes, perfumes, lacquers, and thermoplastics. Fatality from inhalation of the vapor in a small enclosure; inhalation of high concentrations depresses the central nervous system causing nausea, drowsiness, fatigue, vertigo, and intracranial pressure; biochemical changes in the liver have been reported in mice and rats. OSHA exposure limits PEL (8 hour TWA): 200 ppm [air].

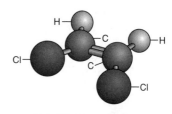

1,2–Dichloroethylene

1,2-dichloroethylene — ClCH=CHCl — a colorless liquid (usually a mixture of cis and trans isomers) with a slightly acrid, chloroform-like odor; MW: 97.0, BP: 118–140°F, Sol: 0.4%, Fl.P: 36°F, sp gr (77°F): 1.27. It is used as a low-temperature solvent for heat-sensitive substances in extraction of caffeine, perfume oils, and fats from the flesh of animals; in rubber and dye industries; as a direct solvent in gums, waxes, oils, camphor, and phenol; in solvent mixtures for esters and ether derivatives, lacquers, resins, thermoplastics, and artificial fibers; in organic synthesis; in miscellaneous applications as liquid dry cleaning agent, cleaning solution for circuit boards, food packaging adhesives, and germicidal fumigants. It is hazardous to the respiratory system, eyes, and central nervous system; and is toxic by inhalation, ingestion, and contact. Symptoms of exposure include central nervous system depression, irritation of the eyes and respiratory system. OSHA exposure limit (TWA): 200 ppm [air] or 790 mg/m^3.

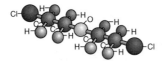

Dichloroethyl Ether
(*Bis* (2-Chloroethyl) ether)

dichloroethyl ether — $(ClCH_2CH_2)_2O$ — a colorless liquid with a chlorinated, solvent-like odor; MW: 143.0, BP: 352°F, Sol: 1%, Fl.P: 131°F, sp gr: 1.22. It is used as a solvent and dewaxing agent in the petroleum industry; in the manufacture of oils, fats, waxes, gums, tars, resins, soaps, ethyl cellulose, paints, varnishes, and lacquers; as a scouring and penetrating agent in the textile industry; as a fumigant; as a chemical intermediate in synthesis during the manufacture of pharmaceuticals, rubber chemicals, resins, plasticizers, and chemicals; as a lead scavenger during production of gasoline engine antiknock compounds. It is hazardous to the respiratory system, skin, and eyes; and is toxic by inhalation, absorption, ingestion, and contact. Symptoms of exposure include lacrimation, nose and throat irritation, cough, nausea, vomiting; carcinogenic; in animals: pulmonary irritation and edema, and liver damage. OSHA exposure limit (TWA): 5 ppm [air] or 30 mg/m^3.

dichloromonofluoromethane — $CHCl_2F$ — a colorless gas with an ether-like odor; a liquid below 48°F; shipped as a liquefied compressed gas; MW: 102.9, BP: 48°F, Sol (86°F): 0.7%. It is used as an aerosol propellant in pharmaceuticals; as a refrigerating agent; as an industrial solvent in the separation of fatty acids, or polymers; as a blowing agent for rigid foams and insulating foams; as a working fluid in Rankine cycle turbogenerators; in water purification, in

organic synthesis in the preparation of other Freons. It is hazardous to the respiratory system and cardiovascular system; and is toxic by inhalation, ingestion, and contact. Symptoms of exposure include asphyxia, cardiac arrhythmias, and cardiac arrest. OSHA exposure limit (TWA): 10 ppm [air] or 40 mg/m^3.

1,1-dichloro-1-nitroethane — $CH_3CCl_2NO_2$ — a colorless liquid with an unpleasant odor; MW: 143.9, BP: 255°F, Sol: 0.3%, Fl.P: 136°F, sp gr: 1.43. It is used as an insecticidal fumigant for grain. It is hazardous to the lungs; and is toxic by inhalation, ingestion, and contact. Symptoms of exposure in animals include eye and skin irritation, liver, heart, and kidney damage, pulmonary edema, hemorrhage. OSHA exposure limit (TWA): 2 ppm [air] or 10 mg/m^3.

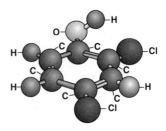

2,4–Dichlorophenol

2,4-dichlorophenol — $C_6H_4Cl_2O$ — a white solid with a medicinal smell; MW: 163.00, BP: 210°C, Sol (20°C): soluble, Fl.P: 93.3°C. It is used mostly to make other organic compounds and as an intermediate in production of herbicides, disinfectants, and mothproofing compounds. Through inhalation or dermal contact, symptoms of exposure include liver injury, chloracne, porphyria, hyperpigmentation, and hirsutism. Little evidence is available to link this chemical with cancer; although human cancer cases have been reported, the individuals were also exposed to pesticides including volatile compounds produced as intermediates. No OSHA standards are available; EPA guidelines 3.09 mg/L [water].

2,4-dichlorophenoxyacetic acid (2,4-D) — $Cl_2C_6H_3OCH_2COOH$ — a white to yellow crystalline, odorless powder; MW: 221.0, BP: decomposes, Sol: 0.05%, sp gr (86°F): 1.57. It is used as a herbicide and plant hormone. It is hazardous to the skin and central nervous system; and is toxic by inhalation, absorption, ingestion, and contact. Symptoms of exposure include weakness, stupor, hyperflexia, muscle twitch, convulsions, dermatitis; in animals: liver and kidney damage. OSHA exposure limit (TWA): 10 mg/m^3 [air].

1,3-dichloropropene — $C_3H_4Cl_2$ — a colorless liquid with a sweet smell; MW: 110.98, BP: 104°C at 1 atmosphere, Sol (25°C): 2,700 ppm, Fl.P: 35°C. It is used predominantly as a component of several formulations in agriculture as soil fumigants for parasitic nematodes; symptoms of exposure include respiratory effects, mucous membrane irritation, chest pain, cough, and breathing difficulties. No OSHA exposure limits; ACGIH guidelines TWA-TLV 1 ppm [air].

dichlorotetrafluoroethane — $CClF_2CClF_2$ — a colorless gas with a faint, ether-like odor at high concentrations; a liquid below 38°F; shipped as a liquefied compressed gas; MW: 170.9, BP: 38°F, Sol: 0.01%. It is used in the manufacture of aerosols with other Freons to lower vapor pressure and produce non-flammable aerosol propellants; as a refrigerant in industrial cooling and air conditioning systems; as a blowing agent for cellular polymers; in the preparation of explosives; in the extraction of volatile substances; as a foaming agent in fire extinguishing and aerosols; in organic synthesis; in strengthening glass bottles; in magnesium refining. It is hazardous to the respiratory system and cardiovascular system; and is toxic by inhalation, ingestion, and contact. Symptoms of exposure include asphyxia, cardiac arrhythmias, cardiac arrest, respiratory system irritation. OSHA exposure limit (TWA): 1,000 ppm [air] or 7,000 mg/m^3.

dichlorvos — $(CH_3O)_2P(O)OCH{=}CCl_2$ — a colorless to amber liquid with a mild, chemical odor; MW: 221.0, BP: decomposes, Sol: 0.5%, Fl.P (oc): >175°F, sp gr (77°F): 1.42. It is used as an insecticide. It is hazardous to the respiratory system, cardiovascular system, central nervous system, eyes, skin, and blood cholinesterase; and is toxic by inhalation, absorption, ingestion, and contact. Symptoms of exposure include miosis, eye aches, rhinorrhea, headache, tight chest, wheezing, laryngeal spasms, salivation, cyanosis, anorexia, nausea, vomiting, diarrhea, sweating, muscle fasciculation, paralysis, giddiness, ataxia, convulsions, low blood pressure, cardiac irregularities, eye, and skin irritation. OSHA exposure limit (TWA): 1 mg/m^3 [skin].

Dick test — a skin test to determine immunity to scarlet fever; *Streptococcus pyogenes* toxin is injected intracutaneously and produces a reaction if there is no circulatory antitoxin.

die — a hard metal or plastic form used to shape material to a particular contour.

dieldrin — $C_{12}H_8Cl_6O$ — colorless to light tan crystals obtained by oxidation of aldrin and with a mild, chemical odor; MW: 380.9, BP: decomposes, Sol: 0.02%, sp gr: 1.75. It is used as an insecticide and in moth-proofing carpets and other furnishings. It is hazardous to the central nervous system, liver, kidneys, and skin; and is toxic by inhalation, absorption, ingestion, and contact. Symptoms of exposure include headache, dizziness, nausea, vomiting, malaise, sweating, myoclonic limb jerks, clonic tonic convulsions, coma; carcinogenic; in animals: liver and kidney damage. OSHA exposure limit (TWA): 0.25 mg/m^3 [skin].

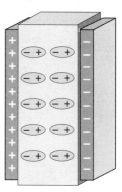

Dielectric
inserted in charged capacitor

dielectric — a nonconductor of direct electric current.

dielectric constant — a property of material which when multiplied by the permitivity of free space determines the electric energy stored per unit volume per unit electric field.

diesel engine — a type of internal-combustion engine that uses a fuel injector and produces combustion temperatures by compression.

diethylamine — $(C_2H_5)_2NH$ — a colorless liquid with a fishy, ammonia-like odor; MW: 73.1, BP: 132°F, Sol: miscible, Fl.P: –15°F, sp gr: 0.71. It is used in the preparation of textile finishing agents, surfactants, rubber processing chemicals, agricultural chemicals, and pharmaceuticals; as a polymerization inhibitor and catalyst; as an intermediate in the dye industry; as a depilatory of animal skins; in electroplating solutions. It is hazardous to the respiratory system, skin, and eyes; and is toxic by inhalation, absorption, ingestion, and contact. Symptoms of exposure include eye, skin, and respiratory system irritation. OSHA exposure limit (TWA): 10 ppm [air] or 30 mg/m^3.

2-diethylaminoethanol — $(C_2H_5)_2NCH_2CH_2OH$ — a colorless liquid with a nauseating, ammonia-like odor; MW: 117.2, BP: 325°F, Sol: miscible, Fl.P: 126°F, sp gr: 0.89. It is used in the preparation of medicinals, pharmaceuticals, pesticides, protective surface coatings for metals, emulsifying agents for polishes, resinous materials for treating fiber surfaces, fluorescent brightening agents, and in polymer production; in organic synthesis to prepare compounds for surfactants, detergents, wetting agents, and yarn-treating; in synthetic fiber dyeing; as a photographic stabilizing solution. It is hazardous to the respiratory system, skin, and eyes; and is toxic by inhalation, absorption, ingestion, and contact. Symptoms of exposure include nausea, vomiting, and irritation of the respiratory system,

skin, and eyes. OSHA exposure limit (TWA): 10 ppm [skin] or 50 mg/m^3.

diethylstilbestrol — $C_{18}H_{20}O_2$ — a white, crystalline, nonsteroid estrogen used as a substitute for natural estrogenic hormones.

differential pressure — the difference in static pressure between two locations.

diffraction — the bending of waves as they pass through an aperture or around the edge of a barrier.

diffused air — (*sewage*) a technique by which air under pressure is forced into sewage in an aeration tank; the air is pumped down into the sewage through a pipe and escapes out through holes in the side of the pipe.

diffuser — a component of the ventilation system that supplies, distributes, and diffuses air to promote air circulation in the occupied space.

diffusion — the tendency of gases or liquids to mix freely, spreading out their component molecules, atoms, and ions in a given space away from the area of greatest concentration.

diffusion badge — a small, clip-on unit worn in the breathing zone using neither pump nor batteries for the testing of a gas or vapor in the air by relying on the tendency of the gas or vapor to disperse based on its density compared to air.

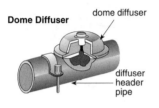

Dome Diffuser
dome diffuser
diffuser header pipe

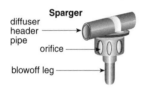

Sparger
diffuser header pipe
orifice
blowoff leg

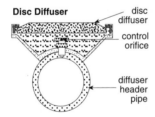

Disc Diffuser
disc diffuser
control orifice
diffuser header pipe

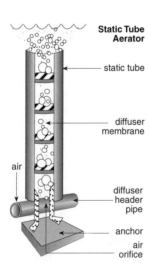

Static Tube Aerator
static tube
diffuser membrane
air
diffuser header pipe
anchor
air orifice

Diffuser

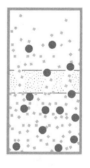

a)

Diffusion

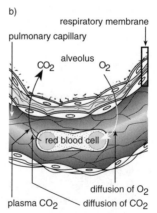

b)
respiratory membrane
pulmonary capillary
alveolus
CO_2 O_2
red blood cell
diffusion of O_2
plasma CO_2 diffusion of CO_2

diffusion rate — a measure of the rate at which one gas or vapor disperses into or mixes with another gas or vapor.

difluorodibromomethane - CBr_2F_2 — a colorless, heavy liquid or gas (above 76°F) with a characteristic odor; MW: 209.8, BP: 76°F, Sol: insoluble, sp gr (59°F): 2.29. It is used as a fire extinguishing agent; in organic synthesis; in processing cutting tools such as razor blades, hypodermic needles, scalpels, culinary knives, and garden tools; as a special solvent for preparation of explosive mixtures. It is hazardous to the respiratory system; and is toxic by inhalation, ingestion, and contact. Symptoms of exposure in animals include respiratory system irritation, central nervous system symptoms, and liver damage. OSHA exposure limit (TWA): 100 ppm [air] or 860 mg/m³.

difluorodichlormethane — CCl_2F_2 — a clear, colorless gas with a sweetish odor at high concentrations and practically odorless at low concentrations; liquid under pressure; MW: 120.92, BP: –29.8°C, Sol: practically insoluble in water, Fl.P: not combustible, sp gr: liquid 1.452 at –17.8°C. It is used as a refrigerant, blowing agent, aerosol propellant, solvent, degreasing agent, monomer for resins, leak detecting agent, and in the preparation of frozen tissue samples. It is hazardous to the central nervous system, lungs, skin, and eyes; and is toxic by inhalation and contact. Symptoms of exposure include mild irritation of the nose, throat, and upper airways, light-headedness, giddiness, dizziness, drowsiness, slurred speech, tingling sensations, humming in the ears, and frostbite. If abused through aerosol sniffing, it can cause an irregular heartbeat leading to death from cardiac arrest. Also known as freon-12. OSHA exposure limit (TWA): 1,000 ppm [air] or 4,950 mg/m³.

digester — (*wastewater*) a closed tank, sometimes heated to 95°F, where sludge is subjected to intensified bacterial action.

digestion — the process by which foods are chemically simplified and made soluble so that they can be used by the cells; (*sewage*) the reduction in volume by decomposition of highly putrescible organic matter to relatively stable or inert, organic, or inorganic compounds carried out by anaerobic organisms in the absence of free oxygen resulting in partial gasification, liquefication, and mineralization.

digestive system — the series of organs taking part in the digestion and absorption of food.

digital — (*computer science*) the ability to represent data in discrete units or digits.

digitize — (*geographic information system*) to encode map coordinates in digital form.

diglycidyl ether — $C_6H_{10}O_3$ — a colorless liquid with a strong, irritating odor; MW: 130.2, BP: 500°F, Sol: not known, Fl.P: 147°F, sp gr (77°F): 1.26. Diglycidyl ether is not generally used outside research laboratories; there are no common industrial uses. It is hazardous to the skin, eyes, and respiratory system; and is toxic by inhalation, ingestion, and contact. Symptoms of exposure include eye and respiratory system irritation, skin burns; carcinogenic; in animals: hematopoietic system, lung, liver, and kidney damage. OSHA exposure limit (TWA): 0.1 ppm [air] or 0.5 mg/m³.

diisobutyl ketone — $((CH_3)_2CHCH_2)_2CO$ — a colorless liquid with a mild, sweet odor; MW: 142.3, BP: 334°F, Sol: 0.05%, Fl.P: 120°F, sp gr: 0.81. It is used as a paint thinner; as a solvent in the production of synthetic coatings, soap, or nitrocellulose; as an intermediate in organic synthesis; an extractant in the pharmaceutical industry; a separating agent in the chemical industry. It is hazardous to the respiratory system, skin, and eyes; and is toxic by inhalation, ingestion, and contact. Symptoms of exposure include eye, nose, and throat irritation, headache, dizziness, dermatitis. OSHA exposure limit (TWA): 25 ppm [air] or 150 mg/m³.

diisopropylamine — $((CH_3)_2CH)_2NH$ — a colorless liquid with an ammonia- or fish-like odor; MW: 101.2, BP: 183°F, Sol: slight, Fl.P: 20°F, sp gr: 0.72. It is used in the synthesis of corrosion inhibitors in iron and steel; in the synthesis of herbicides, delayed action vulcanization accelerator for sulfur-cured rubbers; as a catalyst for chemical synthesis of alkylene cyanohydrin; as a component in gels for cosmetic and medical applications as deodorants and aftershave solutions. It is hazardous to the respiratory system,

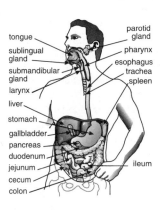

Digestive System

skin, and eyes; and is toxic by inhalation, absorption, ingestion, and contact. Symptoms of exposure include nausea, vomiting, headache, eye irritation, visual disturbance, pulmonary irritation. OSHA exposure limit (TWA): 5 ppm [skin] or 20 mg/m^3.

dike — (*hazardous materials*) a barrier constructed to control or confine hazardous substances and prevent them from entering sewers, ditches, streams, or other flowing waters.

dilapidated — no longer adequate for the purpose or use for which it was originally intended.

diluent — a liquid or dust material used to weaken the concentrated chemical so it can be safely and economically used.

diluent gas — any gas used to dilute the air pollution concentration in an air sample.

dilute — to make less concentrated by adding more solvent, such as water.

dilution rate — the amount of a diluent that must be added to a unit of a chemical to obtain the desired dosage.

dilution ventilation — air flow designed to dilute contaminants to acceptable levels.

dimethyl acetamide — $CH_3CON(CH_3)_2$ — a colorless liquid with a weak ammonia- or fish-like odor; MW: 87.1, BP: 329°F, Sol: miscible, Fl.P (oc): 158°F, sp gr: 0.94. It is used as a spinning solvent for synthetic fibers; as a solvent for film casting and top coating resins; as a reaction medium and catalyst; as an extraction solvent for recovery and purification of butadiene; as a paint stripping solvent. It is hazardous to the liver and skin; and is toxic by inhalation, absorption, ingestion, and contact. Symptoms of exposure include jaundice, liver damage, depression, lethargy, hallucinations, delusions, skin irritation. OSHA exposure limit (TWA): 10 ppm [skin] or 35 mg/m^3.

dimethylamine — $(CH_3)_2NH$ — a colorless gas with an ammonia- or fish-like odor; a liquid below 44°F; shipped as a liquefied compressed gas; MW: 45.1, BP: 44°F, Sol (140°F): 24%, Fl.P (liquid): 20°F, sp gr: 0.67. It is used in the preparation of spinning solvents for acrylic and polymeric fibers; as a raw material in the synthesis of agricultural chemicals; in vulcanization accelerators for sulfur-cured rubber; in textile waterproofing agents; as a photographic chemical; as a stabilizer in natural rubber latex and certain types of resins; as a component of rocket propellants; as an antiknock agent in other fuels. It is hazardous to the respiratory system, skin, and eyes; and is toxic by inhalation, ingestion, and contact. Symptoms of exposure include nose and throat irritation, sneezing, cough, dyspnea, pulmonary edema,

conjunctivitis, skin and mucous membrane burns, dermatitis. OSHA exposure limit (TWA): 10 ppm [air] or 18 mg/m^3.

4-dimethylaminoazobenzene — $C_6H_5NNC_6H_4N(CH_3)_2$ — yellow, leaf-shaped crystals; MW: 25.3, BP: sublimes, Sol: insoluble, Fl.P: not known. It is used in research and laboratory facilities; as a coloring agent for polishes, waxes, polystyrene, petroleum, and soap. It is hazardous to the liver, skin, and bladder; and is toxic by inhalation, absorption, ingestion, and contact. Symptoms of exposure include enlarged liver, hepatic and renal dysfunction, contact dermatitis, cough, wheezing, difficulty breathing, bloody sputum, bronchial secretions, frequent urination, dysuria; carcinogenic. OSHA exposure limit (TWA): regulated use.

dimethylaniline — $C_6H_5N(CH_3)_2$ — a pale yellow, oily liquid with an amine-like odor; a solid below 36°F; MW: 121.2, BP: 378°F, Sol: 2%, Fl.P: 142°F, sp gr: 0.96. It is used in the synthesis of dye and dye intermediates; in the synthesis of explosives; in pharmaceuticals; in intermediates for vanillin; in absorption of sulfur dioxide. It is hazardous to the blood, kidneys, liver, and central nervous system; and is toxic by inhalation, absorption, ingestion, and contact. Symptoms of exposure include anoxia symptoms, cyanosis, weakness, dizziness, ataxia. OSHA exposure limit (TWA): 5 ppm [skin] or 25 mg/m^3.

dimethyl-1,2-dibromo-2,2-dichloroethyl phosphate — $C_4H_7O_4PBr_2Cl_2$ — a colorless to white solid or straw-colored liquid (above 80°F) with a slightly pungent odor; MW: 380.8, BP: decomposes, Sol: insoluble, sp gr (77°F): 1.96. It is used as an agricultural insecticide and acaricide applied to vegetables, fruit, and agronomic crops; in compounding water-base paints and floor polishes. It is hazardous to the respiratory system, central nervous system, cardiovascular system, skin, eyes, and blood cholinesterase; and is toxic by inhalation, absorption, ingestion, and contact. Symptoms of exposure include miosis, lacrimation, headache, tight chest, wheezing, laryngeal spasms, salivation, cyanosis, anorexia, nausea, vomiting, abdominal cramps, diarrhea, weakness, twitching, paralysis, giddiness, ataxia, convulsions, low blood pressure, cardiac irregularities, skin and eye irritation. OSHA exposure limit (TWA): 3 mg/m^3 [skin].

dimethylformamide — $HCON(CH_3)_2$ — a colorless to pale yellow liquid with a faint amine-like odor; MW: 73.1, BP: 307°F, Sol: miscible, Fl.P: 136°F, sp gr: 0.95. It is used as a spinning solvent for acrylic fibers; as a booster solvent in coating, printing, and adhesive formulations; as a chemical intermediate, catalyst, and reaction medium in chemical manufacturing; as a selective absorption and extraction solvent for recovery, purification, absorption, separation, and desulfurization of non-paraffinics from paraffin hydrocarbons; in the manufacture of paint stripper; in the pigment and dye industry; as a crystallization solvent in the pharmaceutical industry; in miscellaneous applications for high-voltage capacitors. It is hazardous to the liver, kidneys, skin, and cardiovascular system; and is toxic by inhalation, absorption, ingestion, and contact. Symptoms of exposure include nausea, vomiting, colic, liver damage, hepatomegaly, high blood pressure, flushed face, dermatitis; in animals: kidney and heart damage. OSHA exposure limit (TWA): 10 ppm [skin] or 30 mg/m^3.

1,1-dimethylhydrazine — $(CH_3)_2NNH_2$ — a colorless liquid with an ammonia- or fish-like odor; MW: 60.1, BP: 147°F, Sol: miscible, Fl.P: 5°F, sp gr: 0.79. It is used in the formulation of jet and rocket propellants; in the chemical synthesis of catalysts, automotive antifreeze, pharmaceuticals, dyestuffs, and stabilizing agents; in the formulation of photographic developers. It is hazardous to the central nervous system, liver, gastrointestinal tract, blood respiratory system, eyes, and skin; and is toxic by inhalation, absorption, ingestion, and contact. Symptoms of exposure include eye and skin irritation, choking, chest pain, dyspnea, lethargy, nausea, anoxia, convulsions, liver injury; carcinogenic. OSHA exposure limit (TWA): 0.5 ppm [skin] or 1 mg/m^3.

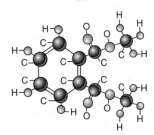

Dimethylphthalate

dimethylphthalate — $C_6H_4(COOCH_3)_2$ — a colorless, oily liquid with a slight, aromatic odor; a solid below 42°F; MW: 194.2, BP: 543°F, Sol: 0.4%, Fl.P: 295°F, sp gr: 1.19. It is used as a plasticizer in the compounding of plastics; in the manufacture of surface coatings containing plasticized resins and polymers including furniture lacquers, printing inks, textile and paper coatings, and moisture-proof coatings for cellophane; in insect repellents. It is hazardous to the respiratory system and gastrointestinal tract; and is toxic by inhalation, ingestion, and contact.

Symptoms of exposure include stomach pain and irritation of the upper respiratory system. OSHA exposure limit (TWA): 5 mg/m^3 [air].

dimethyl sulfate — $(CH_3)_2SO_4$ — a colorless, oily liquid with a faint, onion-like odor; MW: 126.1, BP: 370°F (decomposes), Sol (64°F): 3%, Fl.P: 182°F, sp gr: 1.33. It is used as a methylating agent in the organic chemical industry for the manufacture of esters, ethers, and amines; in the manufacture of dyestuffs, dyes, coloring agents, and perfumes; in the pharmaceutical industry as a solvent for separation and preparation of mineral oils; during analysis of automobile fluids. It is hazardous to the eyes, skin, respiratory system, liver, kidneys, and central nervous system; and is toxic by inhalation, absorption, ingestion, and contact. Symptoms of exposure include eye and nose irritation, headache, giddiness, conjunctivitis, photophobia, periorbital edema, dysphagia, productive cough, chest pain, dyspnea, cyanosis, vomiting, diarrhea, dysuria, analgesia, fever, icterus, albuminuria, skin and eye burns, delirium; carcinogenic. OSHA exposure limit (TWA): 0.1 ppm [skin] or 0.5 mg/m^3.

**Dinitrobenzene
(1,3–Dinitrobenzene)**

dinitrobenzene (ortho-, meta-, and para- isomers) — $C_6H_4(NO_2)_2$ — a pale white or yellow solid; MW: 168.1, BP: 606/572/570°F, Sol: 0.05/0.02/0.01%, Fl.P: 302/302°F/not known, sp gr (64°F): 1.57/1.58/1.63. It is used in the preparation of dyes and dye intermediates; in organic synthesis in the chemical industry as photographic developers and as explosives; in the plastic industry as a substitute for camphor; as a polymerization inhibitor; as an explosive by substituting for TNT; in explosive shells. It is hazardous to the blood, liver, eyes, skin, cardiovascular system, and central nervous system; and is toxic by inhalation, absorption, ingestion, and contact. Symptoms of exposure include anoxia, cyanosis, visual disturbance, central scotomas, bad taste, burning mouth, dry throat, thirst, anemia, liver damage, and yellowing of the hair, eyes, and skin. OSHA exposure limit (TWA): 1 mg/m^3 [skin].

C
D

4,6 Dinitro-o-Cresol

dinitro-o-cresol — CH$_3$C$_6$H$_2$OH(NO$_2$)$_2$ — a yellow, odorless solid; MW: 198.1, BP: 594°F, Sol: 0.01%, sp gr: 1.1 (estimated). It is used as a herbicide, insecticide, and fungicide. It is hazardous to the cardiovascular system, endocrine system, and eyes; and is toxic by inhalation, absorption, ingestion, and contact. Symptoms of exposure include sense of well-being, headache, fever, lassitude, profuse sweating, excessive thirst, tachycardia, hyperpnea, cough, shortness of breath, coma. OSHA exposure limit (TWA): 0.2 mg/m^3 [skin].

dinitrotoluene — C$_6$H$_3$CH$_3$(NO$_2$)$_2$ — an orange-yellow crystalline solid with a characteristic odor; often shipped molten; MW: 182.2, BP: 572°F, Sol: insoluble, Fl.P: 404°F, sp gr: 1.32. It is used in the manufacture of toluene diisocyanate for production of polyurethane plastics. It is hazardous to the blood, liver, and cardiovascular system; and is toxic by inhalation, absorption, ingestion, and contact. Symptoms of exposure include anoxia, cyanosis, anemia, jaundice; carcinogenic. OSHA exposure limit (TWA): 1.5 mg/m^3 [skin].

diode — an electronic tube containing only two electrodes; a cathode and an anode.

diopters — a measure of the power of a lens or prism equal to the reciprocal of its focal length in meters.

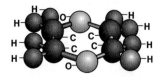

Dioxane
(1,4–Dioxane)

dioxane — C$_4$H$_8$O$_2$ — a colorless liquid or solid (below 53°F) with a mild, ether-like odor; MW: 88.1, BP: 214°F, Sol: miscible, Fl.P: 55°F, sp gr: 1.03. It is used as a solvent for fats, oils, waxes, greases, and natural and synthetic resins; as a wetting agent in textile processing, dye baths, and stain and printing composition; in the manufacture of detergents and cleaning preparations;

as a dehydrating agent in the preparation of histological slides; in the preparation of cosmetics and deodorants; as a working fluid for scintillation counter samples; as a solvent in pulping wood. It is hazardous to the liver, kidneys, skin, and eyes; and is toxic by inhalation, absorption, ingestion, and contact. Symptoms of exposure include drowsiness, headache, nausea, vomiting, liver damage, kidney failure, irritation of the skin, eyes, nose, and throat; carcinogenic. OSHA exposure limit (TWA): 25 ppm [skin] or 90 mg/m^3.

dioxin — *See* 2, 3, 7, 8-tetrachlorodibenzo-p-dioxin.

diphenyl — C$_6$H$_5$C$_6$H$_5$ — a colorless to pale yellow solid with a pleasant, characteristic odor; MW: 154.2, BP: 489°F, Sol: insoluble, Fl.P: 235°F, sp gr: 1.04. It is used as a heat-transfer medium; as a dye carrier for plastics and synthetic resin dyeing; in the impregnation of wrapping and packaging papers. It is hazardous to the liver, skin, central nervous system, upper respiratory system, and eyes; and is toxic by inhalation, absorption, ingestion, and contact. Symptoms of exposure include headache, nausea, fatigue, numb limbs, liver damage, irritation of the throat and eyes. OSHA exposure limit (TWA): 1 mg/m^3 [air].

1,2-diphenylhydrazine — C$_{12}$H$_{12}$N$_2$ — a white solid; MW: 814.24, BP: 309°C, Sol: (20°C): 66.9 mg/L, Fl.P: not known, sp gr: 1.158. It is used in the dye industry for the production of benzidine-based dyes including many of the direct dyes, direct red 28, direct black 4, direct blue 2; it is used as a starting material in the production of benzidine; for the production of the drugs phenylbutazone an anti-inflammatory agent, and sulfinpyrazone a uricosuric agent for the treatment of gouty arthritis. No known health effects in humans by any route of exposure, however in animals, significantly increased incidence of hepatocellular carcinomas, neoplastic, liver nodules, mammary adenocarcinomas, Zymbal's gland carcinomas, and adrenal pheochromocytomas; no OSHA exposure limits. EPA guidelines as an inhalation unit risk 2.2 × 10^{-4} (μg/m^3)$^{-1}$ [air].

diphtheria — (*disease*) an acute bacterial disease of the tonsils, pharynx, larynx, nose, and sometimes mucous membranes of the skin. Incubation time is 2–5 days; caused by *Corynebacterium diphtheriae*. It is found in temperate zones among unimmunized children under 15 years of age or among adults where immunization was not given. The reservoir of infection is people. It is transmitted by contact with a carrier or through raw milk; communicable until virulent bacteria has disappeared in usually two weeks or less; infants and those unprotected by immunization are most susceptible; controlled by proper immunization.

diphyllobothriasis — (*disease*) an intestinal fish tapeworm or broadworm infection of long duration, symptoms range from nonexistent to vitamin B_{12} deficiency anemia with massive infections causing diarrhea, obstructive of the bowel duct or intestine, and toxic symptoms. Incubation time is 3–6 weeks from ingestion of eggs; caused by *Diphyllobothrium latum* cestodes. It is found in lake regions and all areas of consumption of raw or partially cooked fresh water fish. Reservoirs of infection include infected people, dogs, bears, and other fish-eating mammals. It is transmitted by eating raw or inadequately cooked fish; not communicable from person to person; general susceptibility; controlled by proper cooking of all freshwater fish.

diploid number — the full set of chromosomes in a nucleus with both members of each pair present; designated as 2N.

diplopia — the condition of seeing single objects as two; double-vision.

dipole — a positive and a negative charge bound together in such a way that the center of gravity of the positive charge is separated in space from the center of gravity of the negative charge, as with a magnet.

dipropylene glycol methyl ether —
$CH_3OC_3H_6OC_3H_6OH$ — a colorless liquid with a mild, ether-like odor; MW: 148.2, BP: 374°F, Sol: miscible, Fl.P: 180°F, sp gr: 0.95. It is used as a general solvent for oils and greases; as a coupling and dispersing agent in the manufacture and application of printing pastes, dyes, and inks; in the manufacture of cosmetics; in the manufacture of latex paints, lacquers, and leather protective coatings; as a slimicide in food packaging and adhesives in the food industry; as a heat-transfer agent in hydraulic brake fluid; in glass and metal cleaning, and antifogging compounds. It is hazardous to the respiratory system and eyes; and is toxic by inhalation, absorption, ingestion, and contact. Symptoms of exposure include weakness, lightheadedness, headache, irritation of the eyes and nose. OSHA exposure limit (TWA): 100 ppm [skin] or 600 mg/m³.

direct current (dc) — the electric current which flows in one direction only.

direct-dump transfer system — the process of unloading solid waste directly from a collection vehicle to an open-top transfer trailer or container.

direct feed incinerator — an incinerator that accepts solid waste directly into its combustion chamber.

direct-fire afterburner — an afterburner in which the fuel burner supplies the necessary heat; a device whereby a polluted gas stream goes through a fire box before the steam is vented to the atmosphere.

direct-flame incineration — a process by which organic emissions in concentrations well below the lower explosive limit are destroyed under the proper conditions by exposure to temperatures of 900–1,400°F in the presence of a flame.

direct-reading colorimetric devices — an instrument in which the gas reacts directly with a reagent to produce a color whose value is directly measured.

direct-reading instrument — an instrument that gives an immediate indication of the concentration and magnitude of aerosols, gases, vapors, radiation, noise, light, and heat.

disability — a limitation of an individual's physical, mental, or social activity when compared with others of similar age, sex, and occupation; inability to carry on one's normal occupation due to personal injury or illness.

disability benefits — benefits that compensate disabled employees for their loss of incomes or earning capacities due directly to their disabilities.

disability days — the total number of days individuals are disabled during a year.

disaccharide — any sugar, such as $C_{12}H_{22}O_{11}$, which hydrolyzes to form two molecules of simpler sugars or monosaccharides.

disaster plan — a plan outlining individual and departmental responsibilities for natural disasters and multiple casualty incidence.

disc — *See* disk.

discharge — the release of stored charge from a capacitor or a battery.

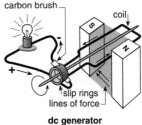

dc generator

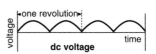

Direct Current

Diskette
(Floppy Disk)

Discharge Monitoring Report — the Environmental Protection Agency's uniform, national form, including any subsequent additions, revisions, or modifications, for the reporting of self-monitoring results by National Pollutant Discharge Elimination System permitees.

discovery — (*law*) the process, through use of interrogatories, of obtaining facts or documents from a party to an action which are in the individual's exclusive knowledge or possession.

disease — a definite pathological process having a characteristic set of signs and symptoms which are detrimental to the well-being of the individual.

di-sec octyl phthalate — $C_{24}H_{38}O_4$ — a colorless, oily liquid with a slight odor; MW: 390.5, BP: 727°F, Sol: insoluble, Fl.P (oc): 420°F, sp gr: 0.99. It is used as a plasticizer in the production of polyvinyl chloride (PVC) and vinyl chloride resins. It is hazardous to the eyes, upper respiratory system, and gastrointestinal tract; and is toxic by inhalation, ingestion, and contact. Symptoms of exposure include eye and mucous membrane irritation; carcinogenic. OSHA exposure limit (TWA): 5 mg/m³ [air].

disinfectant — a product that kills all vegetative bacteria but does not kill spores; a germicide or bactericide.

disintegration — the separation or breaking up of a substance into its parts; (*nuclear*) the transformation or change involving the nuclei in which particles or photons are emitted.

disintegration constant — the fraction of the number of atoms of a radioactive nuclide which decay in unit time.

disintegration rate — the absolute rate at which radiation is emitted from a radioactive source.

disk — (*soils*) an implement used to break up, turn, and loosen the top few inches of the soil; (*computer science*) a storage system for recording digital information. Also spelled disc.

diskette — an inexpensive, low capacity storage medium. Also known as floppy disk.

dismissal — an order or judgment finally disposing of an action, suit, motion, or other proceeding by sending it out of court without a trial of the issues involved.

disperse — to spread out or scatter; (*sewage*) to break up compound particles, such as aggregates, into individual component parts.

dispersing agent — a material which reduces the forces attracting like particles of a chemical mixture to each other in order to get them to mix better with unlike particles (e.g., wetting agents, detergents, etc.).

dispersion — (*light*) the scattering of the values of a frequency distribution from an average; (*physics*) the separation of light or other electromagnetic radiation into its different wavelengths; the separation of nonhomogeneous radiation into components in accordance to some characteristics; a distribution of finely divided particles in a medium; the dilution or removal of a substance by diffusion or turbulence; the distribution of spilled oil into the upper layer of a water column by the natural action of waves or through the action of chemical dispersants. *See also* deflocculation.

dispersion force — *See* van der Waals force.

disposal — the removal of waste material to a site or facility which is specifically designed and permitted to receive such wastes; (*hazardous waste*) the improper discharge, deposit, injection, dumping, spilling, leaking, or placing of any solid waste or hazardous waste into or on any land or water so that any constituent thereof

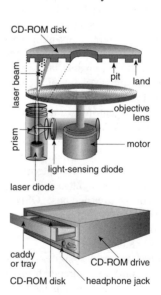

Disk

may enter the environment, be emitted into the air, or discharged into any waters, including groundwater.

disposal container — a plastic or paper sack designed for storing solid waste.

disposal facility — a facility or part of a facility in which hazardous waste is intentionally placed into or on any land or water, and at which waste will remain after closure.

disposal well — a well used for the disposal of waste.

dissipate — to cause to lessen by scattering or spreading out.

dissolution — (*chemistry*) the passing of particles into the liquid phase; dissolving; liquefaction.

dissolve — to cause a substance to pass into solution; turning a solid into a liquid and then going into a solution.

dissolved oxygen (DO) — a measurement of the amount of oxygen dissolved in a given volume of water at a given temperature and atmospheric pressure.

dissolved solids — (*sewage*) a sewage term rather than a technical definition including solids in colloidal as well as those in true solution, both organic and inorganic; theoretically the anhydrous residues of the dissolved constituents in water.

dissolving — a chemical action of detergents that liquefies water-soluble materials or soils.

distal — away from the central axis of the body.

distillate — a liquid product condensed from vapor during distillation.

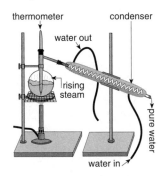

thermometer   condenser

water out

rising steam

pure water

water in

Distillation

distillation — the process of first vaporizing a liquid and then condensing the vapor into a liquid leaving behind nonvolatile impurities used especially for purification, fractionation, or the formation of new substances.

distribution box — a structure designed to distribute effluent by gravity from a septic tank equally into the pipes of an absorption system connected to it.

distributor — the person or business enterprise responsible for delivering goods in commerce.

diuresis — increased excretion of urine.

diuretic — any agent that increases the volume of urine excreted.

diurnal — active during daylight hours; having a daily cycle.

diverted flow position — (*milk*) in high temperature, short time pasteurization, the position in which the diaphragm remains in place across the forward flow line when the temperature is less than 161°F and the microswitch does not activate the solenoid in the flow diversion value.

dizziness — the sensation of unsteadiness and feeling of movement within the head.

DL — *See* ceiling value.

DMS — *See* defense meteorological satellite.

DNA — *See* deoxyribonucleic acid.

DNAPL — acronym for dense non-aqueous phase liquid.

DO — *See* dissolved oxygen.

DOC — *See* Department of Commerce.

docket — a book containing an entry and brief of all the important acts done in court and the conduct of each case from its inception to its conclusion.

documentation — the collection of documents to support a certain element of proof, that is the copying of a bill of lading to demonstrate that substances and goods have gone through interstate commerce.

DOD — *See* Department of Defense.

DOE — *See* Department of Energy.

DOE Energy — a database from 1974 to the present, covering all aspects of energy and related topics including the environment and conservation, indexed journal articles, reports, conference papers, books, patents, dissertations, and translations.

DOI — *See* Department of Interior.

DOJ — *See* Department of Justice.

DOL — *See* Department of Labor.

dolomite — $CaMg(CO_3)_2 = Ca^{2+} + Mg^{2+} + 2CO_3^{2-}$ — a usually white or colorless carbonate mineral consisting of a mixed calcium-magnesium carbonate crystallizing in the rhombohedral system.

domestic waste — wastewater derived principally from plumbing fixture drains in dwellings, business or office buildings, institutions, food service establishments, and similar facilities, and not including industrial or commercial processing waste.

domsat — a domestic satellite regulated by the Federal Communications Commission.

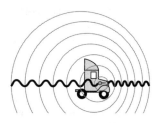

Doppler Effect
(not in proportion)

doppler effect — the apparent change in the frequency of acoustic or electromagnetic waves due to the relative motion of the source of the waves and the observer; the frequency increases as the two approach each other and decreases as they move apart.

dormancy — a period of inactivity.

dosage — the quantity of a given substance associated with a given measurable or observable effect; the amount or rate of a chemical used per unit area, volume, or individual.

dose — (*radiation*) a general term denoting the quantity of radiation or energy absorbed; (*chemical*) the quantity of a chemical absorbed and available for interaction with the metabolic processes.

dose assessment — an estimate of the radiation dose to an individual or a population group usually by means of predictive modeling techniques sometimes supplemented by the results of measurement.

dose-distribution factor — a factor which accounts for modification of the dose effectiveness in cases in which the radionuclei distribution is nonuniform.

dose equivalent (DE) — a quantity used in radiation protection, expressing all radiation on a common scale for calculating the effective absorbed dose. The unit of dose equivalent is the rem, which is numerically equal to the absorbed dose in rads multiplied by certain modifying factors such as the quality factor, the distribution factor, etc.; in SI units, dose equivalent is the sievert, which equals 100 rem.

dose rate — the absorbed dose delivered per unit of time.

dose rate meter — an instrument used to measure radiation dose rates.

dose-response — the mathematical relationship between the dose of the chemical and the physiological response.

dose-response assessment — the relationship between the level of exposure or the dose and the incidence of disease it describes.

dose-response curve — a figure that quantitatively defines the relationship between the degree of exposure to a chemical (dose) and the observed biological effect or response; the extent or incidence of the observed biological effect.

dose-response relationship — the relationship that exists between the degree of exposure to a chemical dose and the magnitude of the effect or response in the exposed organism; projected on a dose-response curve.

dosimeter — an electroscope used in personal monitoring to detect and measure an accumulated dosage of radiation.

dosimetry — scientific determination of amount, rate, and distribution of radiation emitted from a source of ionizing radiation; the process of measuring radiation dose.

DOT — *See* Department of Transportation.

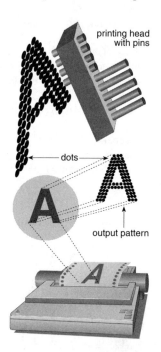

Dot-Matrix Plotter

dot-matrix plotter — a plotter in which the printing head consists of many closely-spaced wire points that can print dots on paper to make a map. Also known as electrostatic plotter.

double-blind design — an experiment in which neither the participants nor the investigator or observer are aware of the allocation of actual and placebo treatments.

downpass — a chamber or gas passage placed between two combustion chambers to carry the products of combustion downward.

downspout — a conduit used to carry water from the gutter to the ground.

dr — *See* dram.

draft — the difference between the pressure in an incinerator or any component part and that in the atmosphere causing air or the product of combustion to flow from the incinerator into the atmosphere.

draft environmental impact statement (DEIS) — the document prepared by the Environmental Protection Agency, or under its guidance, which identifies and analyzes the environmental impacts of a proposed Environmental Protection Agency action and feasible alternatives, and is circulated for public comment prior to preparation of the final environmental impact statement.

drag conveyor — a conveyor that uses vertical steel plates fastened between two continuous chains to drag material across a smooth surface in solid waste disposal.

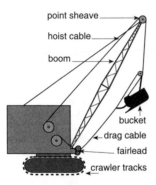

Dragline

dragline — a revolving shovel that carries a bucket attached only by cables and digs by pulling the bucket toward itself.

drainage way — the channel portion of the landscape in which surface water or rain water runoff gathers intermittently to flow to a lower elevation.

dram (dr) — a unit weight formerly used in the avoirdupois system equal to 27.3437 grains, 1/16 ounce; a unit of weight used in the apothecaries system equal to 60 grains, 1/8 ounce.

d-RDF — a refuse-derived fuel that has been compressed to improve certain handling or burning characteristics.

dredge spoils — the material removed from the bottom during dredging operations.

dredging — the removal of earth from the bottom of water bodies using a scooping apparatus.

dressing — any of various materials used for covering and protecting a wound.

drier — any catalytic material which when added to a drying oil accelerates drying or hardening of the film.

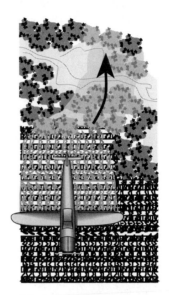

Drift
(from crop dusting)

drift — the movement of droplets, particles, sprays, or dust by air current from the target area to an area not intended for treatment.

drilling mud — a heavy suspension used in drilling an injection well introduced down the drill pipe and through the drill bit to lubricate and cool the bit.

drinking water supply — any raw or finished water source that is or may be used by a public water system or as drinking water.

droplet — a tiny liquid particle suspended in a gas.

droplet infection — a disease spread through contact with cough or sneeze droplets carrying infectious microorganisms.

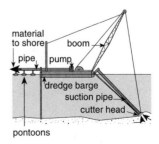

Suction Dredge

dross — a molten or solidified substance that is the non-metallic fused by-product of a furnace operation; a scum that forms on the surface of molten metal. Also known as slag.

drum mill — a long inclined steel drum that rotates and grinds solid waste in its rough interior; smaller material falls through the holes near the end of the drum and the larger material drops out the end.

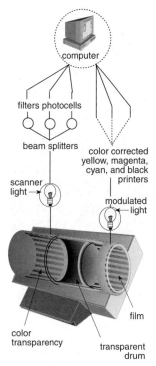

Drum Scanner

drum scanner — (*geographic information system*) a device for converting maps automatically to digital form.

dry acid deposition — the process by which particles such as fly ash or gases such as sulfur dioxide (SO_2) or nitric oxide (NO) are deposited onto surfaces.

dry-adiabatic lapse rate — *See* adiabatic lapse rate.

dry-bulb temperature — the temperature of a gas or mixture of gases shown by a dry bulb thermometer after correcting for radiation.

dry-bulb thermometer — an ordinary thermometer with an unmoistened bulb, not dependent upon atmospheric humidity.

dry cell — an electric cell that has a moist paste instead of a liquid electrolyte.

dry chemical — a powdered, fire extinguishing agent usually composed of sodium bicarbonate or potassium bicarbonate and used especially for electrical fires.

dry cleaning operation — the process by which an organic solvent is used in the commercial cleaning of garments and other fabric materials.

dry gas meter — a device that measures total volume of a gas passed through it without the use of volatile liquids.

dry heat sterilization — a process conducted in a closed vessel at 160°C for one hour to kill all bacteria and spores.

dry ice — solid carbon dioxide usually used as a refrigerant and made into blocks.

dry limestone process — an air pollution control method that uses limestone to absorb the sulfur oxides in furnaces and stack gases.

dry milk — milk from which at least 95% of the water has been removed.

dry soil — any soil from which moisture has been removed.

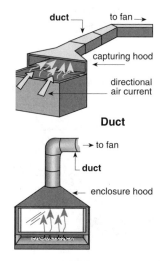

Duct

duct — a passage with well-defined walls for the conveyance of air in a ventilation or other system;

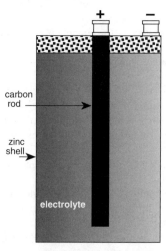

Zinc–Carbon Dry Cell

a tubular structure that delivers glandular secretions to parts of the body through tubes.

ductile — capable of being elongated or deformed without fracture.

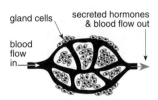

Ductless (Endocrine) Gland

ductless gland — a body organ or tissue that delivers its products (known as hormones) directly into the blood stream. Also known as an endocrine gland.

duct silencers — cylindrical or rectangular structures fitted to the intake or discharge of air moving equipment designed to reduce noise levels.

duct velocity — the velocity of air through the duct cross-section.

dump — a site at which solid waste is disposed in a manner that may contaminate the environment.

dumping — an indiscriminate method of disposing of solid waste.

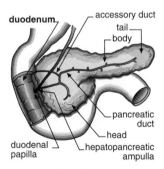

Duodenum

duodenum — the region of the small intestine immediately following the stomach. In humans, it is approximately 10 inches long.

dust — a fine-grain particulate matter that is capable of being temporarily suspended in air, ranging in size from 0.1–25 microns generated by physical processes such as the handling, crushing, or grinding of solids; (*pesticides*) a dry mixture of a finely ground material containing a pesticide chemical.

dust arrester — a piece of equipment for removing dust from air handled by ventilation or exhaust systems.

dust collector — an air-cleaning device used to remove heavy particles from exhaust systems before release to the outside air.

dust diseases of the lungs — *See* pneumoconiosis.

dustfall bucket — (*air pollution*) an open-mouthed container used to collect and roughly estimate the kinds and quantities of materials falling out in a given location.

dust loading — the amount of dust in a gas usually expressed in grains per cubic foot or pounds per thousand pounds of gas.

Dutch liquid — *See* ethylene dichloride.

dwelling — any enclosed space wholly or partially used or intended to be used for living, sleeping, cooking, and eating.

dwelling unit — a room or group of rooms arranged for use by one or more individuals living together as a single household which share living, sleeping, cooking, and eating facilities.

dwell period — that part of the sterilizer cycle during which sterilization takes place.

dye — any of various colored substances containing auxochromes and capable of imparting color to the other materials.

dynamic biochemistry — the chemical or metabolic changes occurring in living systems.

dynamic calibration — the calibration of the complete instrument by means of sampling either a gas of known concentration or an artificial atmosphere containing a pollutant of known concentration.

dynamic head — the sum of all the pressure due to resistance in a complete system in operation.

dynamic imaging — the rapid sequential imaging of vascular blood flow to a body organ or system utilizing an imaging device either in nuclear medicine or by ultrasound.

dynamic loss — the loss of energy in air encountered during air flow resulting from turbulence caused by a change in direction or velocity within a duct.

dynamics — the scientific study of mechanics, which deals with forces in action.

dynamics of stressed ecosystems — the changes that occur in ecosystems that result from human-made stresses.

dyne — the metric unit of force; the amount of force which would, during each second, produce an acceleration of cm per second in a particle of 1 gram in mass.

dynometer — an apparatus used to measure force or work output external to a subject.

dysentery — (*disease*) any of a number of disorders marked by inflammation of the intestine, especially the colon, with abdominal pain and frequent stools often containing blood and mucous.

C
D

dysfunction — a disturbance, impairment, or abnormality of the functioning of an organ.

dysmenorrhea — painful menstruation.

dyspepsia — an impairment of the power or function of digestion.

dysphagia — difficulty in swallowing.

dysplasia — an abnormality of development or growth, especially of the cells.

dyspnea — labored breathing; a symptom of a variety of disorders and an indication of inadequate ventilation or insufficient amounts of oxygen in the circulating blood.

dystrophic — a term related to groundwater, lakes, and streams usually with a low lime content and a high organic content; pertaining to an environment lacking in nutrients.

dysuria — painful or difficult urination.

E

E — *See* electric field.

EAP program — *See* employee assistance program.

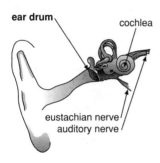

Ear Drum

ear drum — the thin, tympanic membrane which vibrates while three small hinged bones in the middle ear pass the vibration along to the inner ear.

early effects — (*radiation*) effects which appear within the first 60 days of an acute exposure.

ear muffs — devices that fit around the ears and are supported either from a hard hat or from a headband that connects the individual muffs; used to protect the ear from excessive sound.

**Ear Plugs
(3–flange design)**

ear plugs — devices that fit within the ear canal to protect the ear from water or excessive noise.

earth blade — a heavy, broad plate that is connected to the front of a tractor and used to push and spread soil or other materials.

earthquakes — a series of elastic waves traveling through the earth's crust caused by movements along fault lines and emanating from an epicenter.

easily cleanable — a surface that is readily accessible and of such material, finish, and fabrication that residue may be effectively removed by normal cleaning methods.

easily movable — (*food*) weighing 30 pounds or less and mounted on casters, gliders, or rollers, or able to be moved with a force of 30 pounds or less.

eastern equine encephalitis — (*disease*) an acute, mosquito-borne, viral, inflammatory disease of short duration affecting the brain, spinal cord, and meninges with symptoms ranging from mild to severe headaches, high fevers, meningeal signs, stupor, coma, convulsions, and possible death; incubation time is 5–15 days. It is caused by an alpha virus and is found in eastern and north central United States and adjacent Canada, Central and South America, and the Caribbean islands. Possible reservoirs of infection include birds, rodents, bats, reptiles, amphibians, surviving mosquito eggs, or adult mosquitos. It is transmitted by the bite of infected mosquitos to horses as an intermediate, but not person to person; infants and the elderly are most susceptible. It is controlled by the destruction of mosquito larvae, breeding places, adult mosquitos, and the use of mechanical mosquito control techniques.

EBS — *See* Emergency Broadcasting System.

ECG — *See* electrocardiogram, also called EKG.

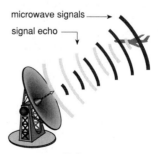

Echo

echo — the reflection of sound waves from a surface or object causing a weaker version to be detected shortly after the original is emitted.

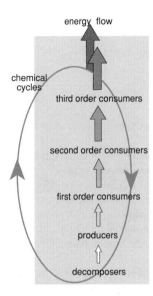

energy flow

chemical cycles

third order consumers

second order consumers

first order consumers

producers

decomposers

Ecosystem

echovirus — any of a group of viruses (genus *Enterovirus*) isolated from people which may produce different types of human disease, especially aseptic meningitis, diarrhea, and various respiratory diseases; the name is derived from the group designation enteric cytopathogenic.

eclampsia — a sudden attack of convulsions, edema, or elevated blood pressure the latter half of pregnancy.

***E. coli* 015: H7** — *See* hemorrhagic colitis.

ecological diversity — the number of different species coexisting in a community of organisms.

ecological marker — an early indicator of potential ecological change, such as the measurement of the addition of phosphates to a stream.

ecological niche — the combination of function and habitat of each of the approximately 1.5 million species of animals and the 1/2 million species of plants on the earth where there are many interactions between species in the ecosystem, and a balance determined by nature.

ecology — the study of the interaction between organisms and their environments.

economic geology — the search for and exploitation of mineral resources and water.

economic poisons — those chemicals used as insecticides, rodenticides, fungicides, herbicides, etc., which may inadvertently or purposely be introduced into the environment.

ecosphere — the layer of earth and troposphere inhabited by or suitable for the existence of living organisms.

ecosystem — the complex of a community and its abiotic and biotic environment functioning as an ecological system in nature.

ectoderm — the outer layer of skin cells.

ectoparasite — a parasite that lives outside the host, such as a tick or mite.

ectopic pregnancy — gestation outside the uterus, often in a fallopian tube.

ecotoxicological study — a measurement of the effects of environmental toxicants on indigenous populations of organisms.

eczema — an inflammatory skin condition usually caused by an allergy with symptoms of redness, itching, and scaly crusts.

ED_{50} — *See* effective dose$_{50}$.

EDB — *See* ethylene dibromide.

eddy — an irregularly patterned motion of turbulent air or fluid.

edema — an excessive, abnormal accumulation of fluid in the intercellular spaces of the body causing a swelling of tissues.

EDF — *See* environmental defense fund.

edible — intended for use as human food.

edit — to remove errors and unwanted material from a work.

EDXA — acronym for energy dispersive X-ray analysis.

EEC — acronym for European Economic Community.

EEG — *See* electroencephalogram.

effective dose$_{50}$ (ED_{50}) — a dose of a chemical that creates a specific effect in 50% of the test subjects.

effective half-life — the time required for a radioisotope in a biological system to diminish 50% as a result of the combined action of radioactive decay and biological elimination where effective half-life = biological half-life × radioactive half-life/biological half-life + radioactive half-life. *See also* biological half-life.

effective homogeneous conductivity — the conductivity of an object viewed as though its dielectric properties were uniform throughout.

effective soil depth — the depth of slightly or moderately limited soil material lying above a restricted soil layer such as clay, hardpan, or bedrock.

effective sound pressure — root-mean-square value of the instantaneous sound pressure over a timed interval at a point during a complete cycle, expressed in dynes per square centimeter.

effective temperature — a combination of globe temperature, wetbulb temperature, air speed, and metabolic rate.

effective temperature index — an empirically determined index of the degree of warmth perceived on exposure to different combinations of temperature, humidity, and air movement.

effervescence — the rapid escape of excess gas bubbles in a liquid by chemical reaction without the application of heat.

efficacy — the probability of benefit to individuals in a defined population from a program applied under ideal conditions.

efficiency — the relationship between the quantity of inputs or resources and the quantity of outputs produced.

effluent — an air flow, discharge, or emission of a liquid or gas that flows outward or away from a main area; the material that comes out of a treatment plant after completion of the treatment process.

egress — an arrangement of exit facilities to ensure a safe means of exit from buildings.

ehv — *See* extra-high-voltage.

EIS — *See* environmental impact statement.

ejector — an air mover consisting of a two-flow system wherein a primary source of compressed gas is passed through a Venturi tube and the vacuum which is developed at the throat of the Venturi tube is used to create a secondary flow of fluid.

EKG — *See* electrocardiogram.

elasticity — the property of matter that causes it to resume its original shape after a distorting force is employed and removed.

elastomer — a synthetic polymer with rubber-like characteristics.

electrical conductivity — a measurement of the ability of a solution to conduct an electrical current.

electrical precipitator — *See* electrostatic precipitator.

electric arc furnace — a steel-making furnace in which the heat is obtained by an electric arc formed between the material to be heated and electrodes.

electric charge — a fundamental quantity required to explain measurable forces and the fundamental unit from which other electromagnetic quantities are developed.

electric circuit — the complete conducting path of an electric current.

electric current — the flow of electrons along a conductor due to potential difference in a circuit.

electric dipole movement — the product of either of the charges and the distance between them.

electric displacement (D) — a field quantity whose source is a net-free charge which has been created by separation of charge.

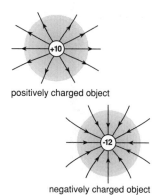

positively charged object

negatively charged object

electric field   field lines

Electric Field

electric field (E) — any space in which force due to an electric charge exists.

electric furnace — a furnace heated by the passage of an electric current, such as in the arc-type, where an electric arc jumps the gap between carbon electrodes.

electric generator — the device that changes mechanical energy to electrical energy by electromagnetic induction.

electric potential — the work required to transport a unit of electric charge from a reference position to another position.

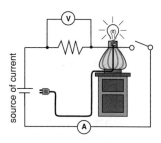

source of current

Electric Circuit
diagram of closed circuit with
battery as emf

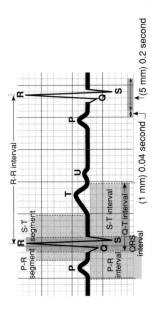

Electrocardiogram

electrocardiogram (EKG) — a graphic tracing and record of the electric current produced by the contraction of the heart muscle.

electrochemical analysis — a system of electrodes and electrochemical techniques used in on-site analysis and continuous monitoring of wastewater effluent.

electrode — an electric conductor for the transfer of charge between an external circuit and a medium; a terminal of an electrolytic cell that emits or collects electrons.

electrode potential — the difference in potential between an electrode and a solution in which it is immersed.

electrodialysis — dialysis accelerated by an electromotive force applied to electrodes adjacent to the membranes; a dialysis which utilizes direct current and the arrangement of permeable-active membranes to achieve separation of the soluble minerals from the water.

electroencephalogram (EEG) — a graphic record of the electric currents developed in the brain as detected by electrodes on the scalp.

electrokinetic soil processing — an *in situ* separation and removal technique for extracting heavy metals and organic contaminants from soils.

electrolysis — the production of chemical changes in matter by means of an electric current passing through an electrolyte.

electrolyte — a nonmetallic chemical compound which will conduct an electric current by the movement of ions when dissolved in certain solvents or when fused by heat; a common example is salt.

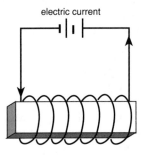

Electromagnet

electromagnet — an iron core that becomes a magnet when current is passed through a coil wrapped around it and becomes demagnetized when the current is interrupted.

electromagnetic field — (*physics*) the field created by the interaction of an electric field and a magnetic field when an electric current passes through a wire.

electromagnetic interference (emi) — a disturbance in the operation of an electronic device which results from the presence of undesired local electric or magnetic fields.

electromagnetic radiation — the propagation of varying electric and magnetic fields through space at the speed of light and exhibits the characteristics of wave motion.

electromagnetic spectrum — a range of electromagnetic radiations including radio, infrared, visible light, ultraviolet, gamma rays, and cosmic rays.

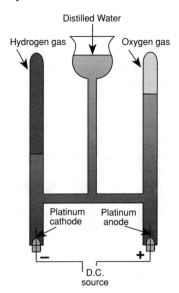

Electrolysis of Water

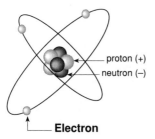

- proton (+)
- neutron (−)

Electron

electromagnetic waves — waves of radiant energy classified according to wave frequency and length.

electromotive force (emf) — the voltage of electricity due to potential difference between two dissimilar electrodes.

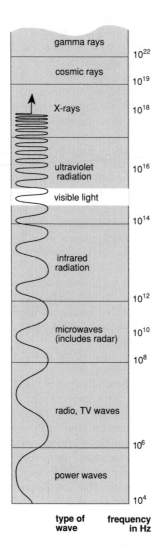

gamma rays	10^{22}
cosmic rays	10^{19}
X-rays	10^{18}
ultraviolet radiation	10^{16}
visible light	
	10^{14}
infrared radiation	
	10^{12}
microwaves (includes radar)	10^{10}
	10^{8}
radio, TV waves	
	10^{6}
power waves	
	10^{4}

| type of wave | frequency in Hz |

Electromagnetic Spectrum
(not to scale)

electron — a negatively charged alimentary particle which is a constituent of the shell of every neutral atom. Its unit of negative electricity equals 4.8 times 10^{-10} electrostatic units or 1.6×10^{-19} coulombs and its mass equals 0.00549 atomic mass units.

electron beam imagers — a device which exposes a view on a photosensitive surface by means of a shutter or similar mechanism and then uses a beam of electrons to provide a means of reading the photosensitive surface after the shutter has closed.

electron beam therapy — the delivery of external beam radiation therapy by means of an electronic beam.

electron beam welding — the welding process used in outer space that produces coalescence of metals with the heat obtained from a concentrated beam composed primarily of high-velocity electrons impinging on the joint to be welded.

electron capture — a mode of radioactive decay involving the capture of an orbital electron by its nucleus where capture from the particular electron shell is designated as "K-electron capture," "L-electron capture," etc.

electron cloud — an average region around the nucleus of an atom where electrons are predicted to be at various states of excitement.

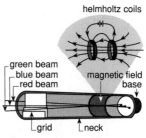

- helmholtz coils
- green beam
- blue beam
- red beam
- magnetic field
- base
- grid
- neck

Electron Gun

electron gun *(electronics)* — the part of a cathode-ray tube that guides the flow and greatly increases the speed of electrons.

electron shell — an energy level surrounding the nucleus of an atom in which an electron may be found.

electronic air sampler — an instrument that measures contaminants through the changes of an electronic signal.

electronic bubble flow meter — an electronic device used for calibrating air-sampling pumps from a specific manufacturer who produces those pumps.

electronic mail (e-mail) — *(computer science)* one-on-one networking between individuals using computers each with some form of network address.

electronic measurement — the detection of pollutants in a gaseous phase through spectroscopic techniques.

electronic product — any manufactured or assembled product which when in operation continues or acts as part of any electronic circuit, emits electronic product radiation, or any manufactured or assembled article which is intended for use as a component, part, or accessory of an electronic product.

electronic product radiation — any ionizing or nonionizing electromagnetic or particulate radiation or any sonic, infrasonic, or ultrasonic wave emitted from an electronic product as a result of the operation of an electronic circuit in such product.

electronic superhighway — *See* communications superhighway.

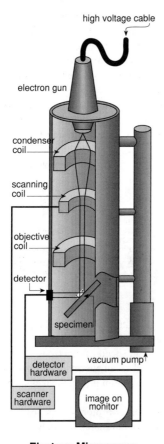

Electron Microscope

electron microscope — a microscope in which high degrees of magnification are produced by an electron beam and magnetic field projecting enlarged images upon a fluorescent surface of photographic plate.

electron-ray tube — *See* cathode-ray tube.

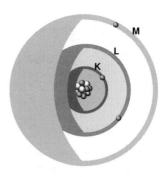

Electron Shells

electron shells — the positions of increasing energy levels occupied by the electrons in an atom. They are designated as K, first next to the nucleus; L, second next to the nucleus; M, third next to the nucleus; N, fourth next to the nucleus, etc.

electron volt (eV) — a unit of electrical energy equal to the amount of energy gained by an electron in passing through a potential difference of one volt which equals 1.6×10^{-19} joules.

electroosmosis — the movement of a charged fluid as a result of the application of an electric field.

electrophoresis — the movement of suspended particles through a fluid under the action of electromotive force especially applying to colloids; an electrochemical process in which colloidal ions or cells with a net electric surface density move in an electric field.

electrophysiology — the science that studies the relationship between living organisms and electricity.

electroplating — the depositing of a thin layer of a metallic element by electrodeposition.

electroscope — an instrument for detecting the presence of electric charges by the deflection of charged bodies.

electrostatic — pertaining to the charge on a substance at rest.

electrostatic field — an electric field such as that produced by a stationary charge.

electrostatic plotter — *See* dot-matrix plotter.

electrostatic precipitator (ESP) — a dry collector that captures particles electrically through the use of electrodes; the particles go to the oppositely charged plate and then either fall or are shaken into a collection hopper. Also known as electrical precipitator.

electrostatic unit of charge (esu) — the quantity of electric charge, which, when placed in a vacuum 1 cm. distance from an equal and like charge, will repel it with a force of one dyne. Also known as a statcoulomb.

electrovalent bond — *See* ionic bond.

element — a molecule composed of atoms of the same atomic number and which cannot be broken down into simpler units by chemical reactions; common examples are hydrogen, gold, and iron; (*geographic information system*) a fundamental geographical unit of information, such as a point, line, or pixel. Also known as entity.

elevated sand mound — a mound of sandy fill material placed on the surface of the ground at a site approved for its use; the fill material serves as a physical and biological medium in which the sewage effluent is filtered and treated before it is absorbed by the natural soil.

ELF — *See* extremely-low-frequency.

ELISA test — *See* Enzyme-Linked Immuno-Sorbent Assay test.

eluate — the solution that results from the elution process.

elution — (*chemistry*) the separation of material by washing.

elutriate — to purify, separate, or remove by washing.

elutriation — purification of a substance by dissolving it in a solvent and pouring off the solution, thereby separating it from the undissolved foreign material; (*sewage*) the process of washing the alkalinity out of anaerobically digested sludge to decrease the demand for acidic chemical conditioning and to improve settling and dewatering characteristics.

e-mail — (*computer science*) *See* electronic mail.

embargo — a legal prohibition on commerce.

EMBASE — a database from 1974 to present abstracting the world's biomedical literature, including environmental health and pollution control.

embolism — the sudden blocking of an artery or vein by a clot or embolus which has been brought to its place by the blood current.

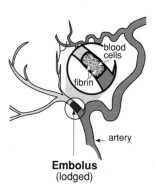

Embolus
(lodged)

embolus — a clot or plug foreign to the bloodstream that is forced into a smaller blood vessel causing obstruction.

embryo — a multicellular organism in the earliest stage of development; from the time of fertilization in the ovum until the beginning of the third month of human pregnancy.

embryotoxicity — any toxic effect on the embryo as a result of prenatal exposure to a chemical.

emergence — the action of a young plant breaking through the surface of the soil; an insect coming out of an egg or pupa; dry land which was part of the ocean floor.

emergency — a situation created by an accidental release or spill of hazardous chemicals and which poses a threat to the safety of workers, residents, the environment, or property.

emergency broadcasting system (EBS) — a system of broadcast stations and interconnecting facilities authorized by the United States Federal Communications Commission to inform the public about the nature of a hazardous state of public peril or national emergency.

emergency level — the level of concentrations of single pollutants or combinations of pollutants and meteorological factors likely to lead to acute illness or death.

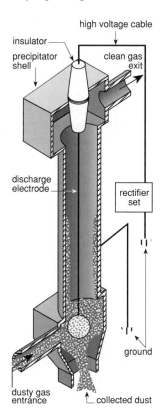

Electrostatic Precipitator

E
F

Emergency Management Institute (EMI) — the part of the federal environmental management agency's National Emergency Training Center that conducts resident and nonresident training activities for federal, state, and local government officials, managers of the private sector, and members of professional and volunteer organizations on subjects from civil nuclear preparedness systems to domestic emergencies caused by natural and technological hazards.

emergency medical service (EMS) — emergency ambulance services and other services used to help individuals in need of immediate medical care to prevent loss of life or aggravation of physiological or psychological illness or injury.

emergency medical system (EMS) — the entire emergency care system in a state or locality.

emergency medical technician (EMT) — an individual certified by the Emergency Medical Services Commission and responsible for the administration of emergency care procedures to emergency patients, and for their handling and transportation.

Emergency Planning and Community-Right-To-Know Act — a federal act passed in 1980 requiring states to establish state emergency response commissions, emergency planning districts, emergency planning committees, comprehensive emergency plans, develop programs to respond to a hazardous chemical release, and also requiring industries to report the presence of hazardous chemicals and the release of hazardous chemicals. Also known as the Community-Right-To-Know-Act.

emergency response personnel — people authorized to respond to fire, rescue, police, and hazardous materials emergencies within a given jurisdiction.

emergent beam diameter — the diameter of the laser beam at the exit aperture of the system.

emery — a natural or synthetic abrasive made of aluminum oxide.

emesis — the act of vomiting.

emetic — any agent that causes vomiting.

emf — *See* electromotive force.

emi — *See* electromagnetic interference.

EMI — *See* Emergency Management Institute.

eminent domain — the authority of the state to seize, appropriate, or limit the use of property in the best interest of the community.

emission — material released into the air either by a primary source or a secondary source, as a result of a photochemical reaction or chain of reactions.

emission factor — a statistical average of the mass of pollutant emitted from each source of pollution per unit, quantity of material handled, processed, or burned.

emission inventory — a list of primary air pollutants emitted into a given community's atmosphere in certain amounts per day by type of source.

emission rate — the amount of pollutant emitted per unit time.

emission spectroscopy — the detection of radiation emissions of characteristic wavelengths from metallic elements (either oxides or salts) when they are excited within an electric arc or spark.

emission standard — the regulations restricting the maximum amounts of pollutants which can be released into the air from a specific polluting source.

emollient — a softening agent, especially for use on the skin.

emphysema — an irreversible, obstructive pulmonary disease in which airways become permanently constricted and alveoli are damaged or destroyed and heart action is frequently impaired.

employee — one who performs services for another under contract of hire, acting under the direction and control of that person, and who performs the service for valuable consideration.

employee assistance program (EAP) — a program set up by a business or agency to help employees with problems such as stress, difficulty at work or home, alcoholism, or drugs; presents various wellness programs to employees.

employer — a person, firm, partnership, association, corporation, or other group that engages the services of another under a contract of hire.

EMS — *See* emergency medical service; emergency medical system.

EMT — *See* emergency medical technician.

EMTD — *See* estimated maximum tolerated dose.

emulsifiable concentrate — a pesticide chemical mixture dissolved in a liquid solvent; the emulsifier allows the dilution of a chemical with oil or water.

emulsification — the process of dispersing one liquid into another liquid with which it is immiscible.

emulsifier — a colloidal chemical that holds in suspension two immiscible liquids by forming a

**alpha₁antitrypsin deficiency
anemia emphysema**
alveoli and bronchioles
permanently distended
with air

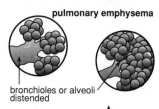

pulmonary emphysema

bronchioles or alveoli
distended

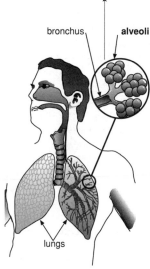

bronchus   **alveoli**

lungs

Emphysema

film around the particles so that they remain
suspended in solution. Also known as emulsifying agent.

emulsifying agent — *See* emulsifier.

emulsion — a stable dispersion of one liquid in
a second immiscible liquid.

enamel — an oil-based, paint-like substance that
produces a glossy finish when applied to a surface.

encapsulate — to seal a substance in an impervious material which will not be degraded physically or chemically.

encephalitis — inflammation of the brain.

encephalopathy — any degenerative disease of
the brain.

endangered species — the animals, birds, fish,
plants, or other living organisms threatened with
extinction by humans or natural changes in the
environment.

endemic — the regular occurrence of a fairly
constant number of human cases of a disease
within an area.

endocarditis — inflammation of the inner lining
of the heart or endocardium.

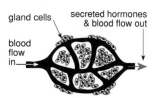

gland cells   secreted hormones
& blood flow out

blood
flow
in

Endocrine Gland

endocrine gland — a ductless gland which secretes hormones directly into the blood. Also
known as ductless gland.

endocrinologist — a physician who specializes
in the diagnoses and treatment of disorders of
the glands of internal secretion such as pituitary,
thyroid, adrenal glands, and pancreas.

endoderm — the layer of cells in which the
lining of the digestive system, the liver, and the
lungs develop.

endometrium — the mucous membrane lining
of the uterus.

endoparasite — a parasite that lives within a
host, such as a tapeworm.

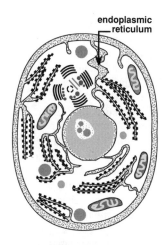

endoplasmic
reticulum

**Endoplasmic Reticulum
in Animal Cell**

endoplasmic reticulum (ER) — a vascular system of the cytoplasm that serves as the circulatory system of cells.

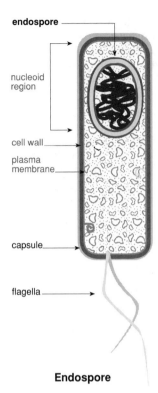

nucleoid region

endospore

cell wall

plasma membrane

capsule

flagella

Endospore

endospore — a thick-walled body formed within the vegetative cells of certain bacteria that is able to withstand adverse environmental conditions for prolonged periods, and then, under favorable conditions, can germinate to form the vegetative bacteria.

endosulfan — $C_9H_6Cl_6O_3S$ — cream-to-brown colored solid crystalline flakes with a distinct terpene-like odor, or odor similar to sulfur dioxide; MW: 406.95, sp gr: (20°C): 1.745. It is used as an insecticide and miticide on vegetable and forage crops, on ornamental flowers, and in controlling termites and tsetse flies. It is hazardous to the nervous system, cardiovascular system, and respiratory system; ingestion can be fatal in humans. Symptoms of exposure include gagging, vomiting, diarrhea, agitation, writhing, loss of consciousness, cyanosis, dyspnea, foaming at the mouth; and noisy breathing; autopsies show edema of the brain and lungs, hemorrhage of the medullary layer of the kidneys, acute lung emphysema, and chromatolysis of the neurons. OSHA exposure limit (PEL TWA): (8 hour skin-designation) 0.1 mg/m³ [air].

endothelium — the epithelial layer of cells that line the serous cavities, lymphatic cavities, joint cavities, and other closed cavities of the body.

endothermic — referring to a chemical change or reaction in which heat is absorbed.

endotoxin — a heat-stable toxin associated with the outer membranes of certain gram-negative bacteria including *Brucella, Neisseria,* and *Vibrio* species. Endotoxins are not secreted and are released only when the cells are disrupted. They are less potent and less specific than the exotoxins, and do not form toxoids.

endpoint — the final result of a health event such as recovery or death; may be either positive or negative; (*titration*) that stage in titration indicating, usually with a color change, that a desired point in titration has been reached.

endrin — $C_{12}H_8CL_6O$ — a colorless to tan crystalline solid with a mild, chemical odor; MW: 380.9, BP: decomposes, Sol: insoluble, sp gr: 1.70. It is used as an insecticide on cotton fields and vegetable crops and as a rodenticide against mice and chipmunks in orchards. It is hazardous to the central nervous system and liver; and is toxic by inhalation, absorption, ingestion, and contact. Symptoms of exposure include epileptiform convulsions, stupor, headache, dizziness, abdominal discomfort, nausea, vomiting, insomnia, aggressiveness, confusion, lethargy, weakness, anorexia; in animals: liver damage. OSHA exposure limit (TWA): 0.1 mg/m³ [skin].

enema — a liquid injected into the rectum to cause movement of the feces.

energy — the capacity to do work; classified as either potential or kinetic energy.

energy density — the intensity of electromagnetic radiation per unit area per pulse; expressed as joules per square centimeter.

Energy Supply and Environmental Act — a law passed in 1974, updated in 1978, providing for a way to assist in meeting the needs of the United States for fuels in a consistent, practical manner and protecting and improving the environment; it allows for coal conversion or coal derivatives in place of oil in power plants.

engineering control — a method of controlling employee exposure to contaminants or hazards by modifying the work environment.

enjoin — (*law*) to direct or impose by court order or with urgent admonition.

enriched flour — flour to which has been added vitamins and other ingredients necessary to conform to the definition standard of identity of enriched flour.

enriched material — material in which the relative amount of one or more isotopes of a constituent has been increased.

Entamoeba — a genus of amebas parasitic in invertebrates and vertebrates, including humans.

enteric — pertaining to the small intestine.

enteric adenovirus — (*disease*) a virus representing serotypes 40 and 41 of the family *Adenoviridae* containing a double-stranded DNA surrounded by a distinctive protein capsid of approximately 70 nanometers in diameter, which may cause 5–20% of viral gastroenteritis in young children with symptoms of diarrhea, nausea, vomiting, malaise, abdominal pain, and fever. The mode of transmission is the oral-fecal route, from person to person, ingestion of contaminated food or water, or may also be transported by the respiratory route. The incubation period is 10–70 hours. The disease has been reported in England and Japan in children in hospitals or day care centers. The reservoir of infection is people. Susceptibility found in young children. Also known as acute nonbacterial infectious gastroenteritis; viral gastroenteritis.

enteritis — inflammation of the intestine, especially the small intestine; a general condition produced by a variety of causes including bacteria and certain viruses, and producing symptoms of abdominal pain, nausea, vomiting, and diarrhea.

pointed posterior end

Enterobius vermicularis
(pinworm)
cause of enterobiasis

enterobiasis — (*disease*) pinworm disease; a common intestinal infestation which may cause no symptoms, anal itching, disturbed sleep, irritability, and sometimes secondary infections.

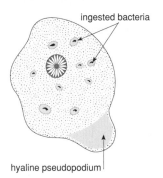

ingested bacteria

hyaline pseudopodium

Entamoeba histolytica
(cause of amebic dysentery)

Incubation time is 4–6 weeks. It is caused by an intestinal nematode, *Enterobius vermicularis*; found worldwide. The reservoir of infection is people. It is transmitted by infected eggs, hands, from anus to mouth, and from clothing, bedding, food, or other contaminated articles, or dust; communicable for approximately 2 weeks; general susceptibility. It is controlled by proper disinfection, cleaning, and good personal hygiene.

enterotoxin — a toxin produced by *Micrococcus pyogenes* var. *aureus* (*Staphylococcus aureus*) which specifically affects cells of the intestinal mucosa causing vomiting and diarrhea.

enterovirus — any of a subgroup of the picornaviruses infecting the gastrointestinal tract and discharged in feces, including coxsackieviruses, echoviruses, and polioviruses; may be involved in respiratory disease, meningitis, and neurological disease.

entity — *See* element.

entomology — the study of insects.

entrainment — a process in which suspended droplets of liquid, particles, or gas are drawn into the flow of a fluid and transported.

entrainment velocity — the gas flow velocity which tends to keep particles suspended and to become airborne.

entrance loss — the loss in static pressure of a fluid that flows from an area into and through a hood or duct opening.

entrapment — to engage in artifice and deceit to afford an individual the opportunity to commit a voluntary criminal act.

entry loss — a loss in static pressure caused by air flowing into a duct or hood opening.

entry loss factor — a factor derived from the coefficients of entry which when multiplied by the velocity pressure at the hood yields the entry loss in inches of water gauged.

entry of decree — an indication that the court has entered a final decree or order of judgment usually signifying the end to a particular proceeding. Also known as entry of judgment.

entry of judgment — *See* entry of decree.

enucleation — the process of removing an organ or tumor in its cavity without cutting into it.

Enviroline — a database from 1971 to present covering all aspects of the environment with coverage of more than 5,000 international sources.

environment — the sum of all surrounding physical conditions in which an organism or cell lives, including the available energy in living and non-living materials.

Environmental Abstract Annual — an index of available environmental articles from 1972 to present; originally known as the Environmental Index.

Environmental Bibliography — a database from 1973 to present indexing periodical articles on general human ecology, land and water resources, pollution, health hazards, solid waste management, and other related topics.

environmental defense fund (EDF) — a national association of individuals concerned about the environment who lobby, bring about legal action, and educate the public as well as legislators on a variety of environmental issues.

environmental design — (*ergonomics*) a design which ensures the lighting, heating, ventilation, noise, vibration, and level of chemical exposure are appropriate to the requirements of the person to prevent disease and injury.

environmental epidemiologist — an environmental health professional who develops, implements, and evaluates environmental epidemiological investigations, including observational and experimental studies of the relationships between etiological, statistical, and mathematical tools, to identify the causes of disease in human populations.

environmental epidemiology — the study of environmental factors that influence the distribution and determinants of disease in human populations; can either identify disease or injury and the characterization of the exposed population, or identify the exposure and determine the health endpoint of the exposure to the population.

environmental factors — conditions other than indoor air contaminants that cause stress, discomfort, and/or health problems to individuals residing in that environment.

environmental fate — the fate of a chemical or biological pollutant following release into the environment based on time, transport, transfer, storage, and transformation.

environmental health — the art and science of the protection of good health, the promotion of aesthetic values, the prevention of disease and injury through the control of positive environmental factors, and the reduction of potential physical, biological, chemical, and radiological hazards.

environmental health practitioner — an applied scientist and educator who uses the knowledge and skills of the natural, behavioral, and environmental sciences to prevent disease and injury and to promote human well-being. Also

known as an environmental health specialist; formerly known as a sanitarian.

environmental health professional — an environmental health practitioner whose educational background includes, at a minimum, a baccalaureate degree in the major components of environmental health and the basic public health sciences of epidemiology, biostatistics, toxicology, and whose primary focus is on protecting human health.

environmental health specialist — *See* environmental health practitioner.

environmental impact statement (EIS) — an analysis required by the National Environmental Policy Act of 1970 before major programs affecting the environment are undertaken.

Environmental Index — *See* Environmental Abstract Annual.

environmental management specialist — *See* occupational safety specialist.

environmental noise — the intensity, duration, and the character of sounds from all sources.

environmental physiology — a measure of the stress due to light, noise, vibration, temperature, radiation, and chemicals on the functions of living organisms, their parts, and their physical and chemical processes.

Environmental Protection Agency (EPA) — an independent agency of the federal government formed in 1970 and responsible for pollution abatement and control programs including air, water, solid and toxic waste management, pesticide control, and noise abatement.

Environmental Protection Agency number — the number assigned to chemicals regulated by the United States Environmental Protection Agency. Also known as EPA number.

environmental response team (ERT) — a group of professionals representing various agencies trained and equipped to determine the extent of an environmental emergency and how to deal with it safely.

environmental safety specialist — *See* occupational safety specialist.

environmental science — the study of the complex interactions of human populations with their environments and energy resources.

environmental tobacco smoke (ETS) — tobacco smoke which poses a risk of lung cancer in nonsmokers and which contains a variety of pollutants including inorganic gases, heavy metals, particulates, volatile organic compounds, and products of incomplete burning, such as polynuclear aromatic hydrocarbons.

environmental toxicologist — an environmental health professional who determines the adverse health effects, and the mechanisms of those effects, resulting from exposure to physical, chemical, and biological aspects in the human environment.

environmental toxicology — a study of the potential poisons in the environment as they relate to individuals affected by exposure, time, dose, and biological effects injuries.

enzyme — any of a group of organic compounds, frequently proteins, found in plants and animals that accelerate by the catalytic action of specific transformations of material without being destroyed or altered, as in the digestion of food, and serves to regulate certain physiological functions.

Enzyme-Linked ImmunoSorbent Assay test (ELISA test) — the diagnostic test that detects the presence of antibodies to HIV in human blood components.

eosin — $C_{20}H_8Br_4O_5$ — a red fluorescent dye obtained by the action of bromine on fluorescein.

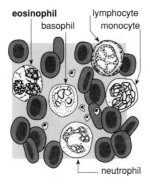

Eosinophile

eosinophile — a leukocyte or other granulocyte with cytoplasmic inclusions readily stained by eosin.

eosinophilia — an increase of certain white cells in the blood or tissues.

EP — *See* extraction procedure.

EPA — *See* Environmental Protection Agency.

EPA Health Advisory — an estimate of acceptable drinking water levels for a chemical substance based on health effects information and not a legally enforceable federal standard.

EPA number — *See* Environmental Protection Agency number.

EPDM — *See* ethylene-propylene terpolymer.

epichlorohydrin — C_3H_5OCl — a colorless liquid with a slightly irritating, chloroform-like odor; MW: 92.5, BP: 242°F, Sol: 7%, Fl.P: 93°F, sp gr: 1.18. It is used as an intermediate in the manufacture of DGEBA resins, epoxy novolac resins, phenoxyl resins, and wet-strength resins for paper; in the manufacture of glycerol; as a stabilizer in the manufacture of chlorinated rubber and chlorinated insecticides; in coating fibers and textile surface coatings. It is hazardous to the respiratory system, skin, and kidneys; and is toxic by inhalation, absorption, ingestion, and contact. Symptoms of exposure include nausea, vomiting, abdominal pain, respiratory distress, coughing, cyanosis, eye and skin irritation with deep pain; carcinogenic. OSHA exposure limit (TWA): 2 ppm [skin] or 8 mg/m³.

epidemic — the occurrence in a community or region of a disease at a rate substantially in excess of normal expectancy and derived from a common or propagated source.

epidemic viral gastroenteritis — *See* Norwalk type disease.

epidemiological reasoning — a three-step process consisting of determining the association between exposure and endpoint, formulating a hypothesis of the relationship, and testing the hypothesis.

epidemiological study — a descriptive study to determine whether correlations exist between the frequency of disease in human populations and some specific factor such as a microorganism or a concentration of a toxic chemical in the environment.

epidemiology — the study of the occurrence and distribution of disease and injury specified by person, place, and time.

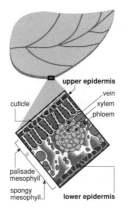

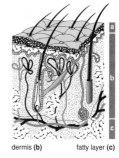

Epidermis

Epidermis layer (a)

epidermis — the outer protective cell layers of an organism.

epiglottis — a thin, leaf-like structure at the opening of the trachea that helps food go from the mouth to the esophagus without blocking the airway.

epihydrin alcohol — *See* glycidol.

epileptiform — resembling epilepsy or its manifestations.

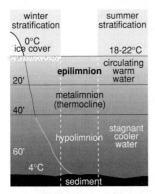

Epilimnion

epilimnion — a fresh-water zone of thermally stratified water which is mixed as a result of wind action and convection currents.

epinephrine — $C_3H_{13}O_3N$ — a hormone produced by the medulla of the adrenal glands that acts to increase blood pressure. Also known as adrenaline.

epiphytic — of or pertaining to a plant which grows nonparasitically on another plant or a nonliving structure.

episode — (*pollution*) an air pollution incident in a given area caused by a concentration of atmospheric pollution reacting with meteorological conditions that may result in a significant increase in illness or death.

epistaxis — hemorrhage from the nose usually due to rupture of small vessels overlying the anterior part of the cartilaginous nasal septum.

epithelial tissue — the layers of cells covering the inside or outside surfaces of an animal's body forming glands and parts of sense organs and serving protective, absorptive, secretory, and sensory functions.

epizootic — affecting many of one kind of animal in a region simultaneously, as with an epidemic disease.

EPN — *See* 0-ethyl-0-para-nitrophenyl phenylphosphonothioate.

epoxide — (*chemistry*) an epoxy compound.

epoxy — a large group of compounds containing a triangular structure in which oxygen acts as a bridge between two atoms already bound together.

epoxy resin — a synthetic resin composed of epoxy polymers and characterized by adhesiveness, flexibility, and resistance to chemical actions.

$\in$ — *See* permittivity.

EPTC — acronym for extraction procedure toxicity characteristic.

EP toxicity test — an extraction procedure designed to identify waste likely to leach hazardous concentrations of particular toxic constituents into the groundwater as a result of improper management.

equilibrium fraction (F) — a measure of the equilibrium/disequilibrium in radon-radon daughters where the parents and daughters have equal radioactivity, that is as many decay into a specific nuclide as decay out.

simple squamous

stratified squamous

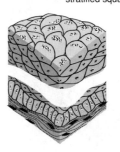

Epithelial Tissue

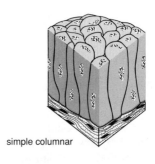

simple columnar

simple cuboidal

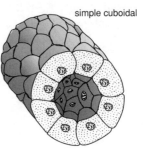

equine — pertaining to, characteristic of, or derived from the horse.

equipment — (*food*) items other than utensils used in the storage, preparation, display, and transportation of food such as stoves, ovens, hoods, slicers, grinders, mixers, scales, meat blocks, tables, shelving, reach-in refrigerators and freezers, sinks, icemakers, and similar equipment in retail food stores.

equipment room — (*swimming pools*) an area provided for storage of safety and deck equipment.

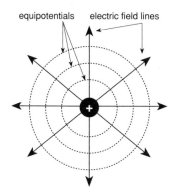

Equipotential

equipotential — a line or surface characterized by a single electric potential or voltage.

equity law — a decision made by a judge in a given situation where rules and regulations have been violated and the judge decides the penalty rather than deciding if the law is valid.

equivalent method — a method of sampling and analyzing ambient air for an air pollutant that has been demonstrated as equivalent to a reference sampling technique in accordance with the Code of Federal Regulations.

ER — notation for emergency response; emergency room. *See* endoplasmic reticulum.

eradication — the elimination of pests from an area; the elimination of infections from a facility.

erg — a unit of work equal to a force of one dyne acting through a distance of 1 centimeter in the direction of the force; 10^{-7} joule, or 2.4×10^{-11} kcal.

ergonomics — a multidisciplinary activity dealing with interactions between people and the total working environment and stresses related to such environmental elements as atmosphere, heat, light, and sound, as well as all tools and equipment in the workplace; literally the study of measurement; organization of work by making purposeful human activities more effective.

ergotism — a rare, chronic poisoning associated with the consumption of rye contaminated with the fungus *Claviceps purpurea* or ergot fungus.

Glacial Erosion

Stream Erosion

Wave Erosion

Erosion

erosion — the removal of soil and rock fragments by natural agents such as wind, water, organisms, and gravity.

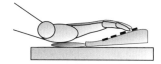

Ergonomic Keyboard

E F

erosion-corrosion — deterioration which occurs from movement of a corrosive over a surface, increasing the rate of attack due to mechanical wear and corrosion.

ERT — *See* environmental response team.

erysipeloid — (*disease*) a bacterial cellulitis with symptoms of fever, leukocytosis, and red, tender edematous spreading lesions of the skin with a definite raised border. Three forms exist: a usually self-limited, mild, localized form manifested by an erythemous and painful swelling at the site of inoculation which spreads peripherally with central clearing; a generalized or diffuse form which may be accompanied by fever and arthritis symptoms and resolves spontaneously; and a rare and sometimes fatal systemic form associated with endocarditis. Incubation time is 1–7 days, longer if the infection enters through an unobserved breech of the skin. It is caused by *Erysipelothrix rhusiopathiae* found in occupational food processing settings. The reservoir of infection is meat, poultry, and fish. It is transmitted by contact with the food source; not communicable with unknown susceptibility. It is controlled by avoiding direct contact between the food source and skin.

erythema — an abnormal redness of the skin due to the distention of the capillaries with blood caused by physical or chemical agents due to skin injury, infection, or inflammation.

erythemal region — the ultraviolet light radiation between 280–320 millimicrons as absorbed by the eye.

erythremia — an increase in the number of circulating erythrocytes.

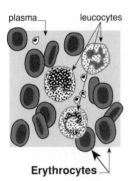

Erythrocytes

erythrocyte — red blood cell or corpuscle.

eschar — the crust formed after injury by a caustic chemical or heat.

Escheria coli diarrhea — (*disease*) three types exist; invasive strains cause disease localized primarily in the colon with fever, mucoid, and occasional bloody diarrhea. Enterotoxigenic (toxin-producing) strains are like *Vibrio cholerae*, producing profuse watery diarrhea without blood or mucus, abdominal cramping, vomiting, acidosis, prostration, and dehydration with symptoms lasting 3–5 days. Enteropathogenic strains are usually associated with outbreaks of acute diarrheal disease in newborn nurseries. Incubation time is 12–72 hours; caused by enterotoxigenic (heat-labile) toxin-LT, heat stabile toxin-ST or both -LT/-ST invasive or enteropathogenic strains of *E.coli*; found in nurseries, institutions, as outbreaks of foodborne or waterborne disease in communities, and especially where there is poor water treatment and handling. The reservoir of infection is infected people. It is transmitted through fecal contamination of food and water, from mites, and person to person. It is communicable probably for the duration of the excretion of feces, but may last several weeks; infants and malnourished children are most susceptible; travelers may also be affected, sometimes called "travelers diarrhea." It is controlled through the proper handling of food, water, fecal material, and breaking the oral fecal route of disease; also important to have appropriate hand washing and other preventative procedures in institutions.

Escherichia coli — *See* coliform.

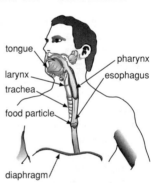

Esophagus

esophagus — the tube connecting the mouth and the stomach in the digestive system.

ESP — *See* electrostatic precipitator.

essential oil — any of a class of volatile, odoriferous oils found in plants and imparting to the plant's odor and often other characteristic properties; used in perfume.

ester — the organic compound formed by the action between an alcohol and organic or inorganic acid; a compound of the type RCOOR, where R and R are hydrocarbon groups.

estimated maximum tolerated dose (EMTD) — the estimated maximum dose of a chemical that individuals can be exposed to without causing disease or injury.

estrogen — the female sex hormone including estradiol, estriol, and estrone formed in the ovary, adrenal cortex, testis, and fetoplacental unit; responsible for secondary sex characteristic development and during the menstrual cycle, acts on the female genitalia to produce an environment suitable for fertilization, implantation, and nutrition in the early embryo.

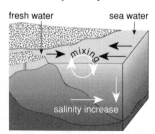

Estuary

estuary — the enclosed or semi-enclosed shallow body of water that forms where a river enters the ocean and creates an area that mixes fresh water and ocean water.

esu — *See* electrostatic unit of charge.

etch — to cut or eat away material with acid or other corrosive substance.

ethamine — *See* ethyl amine.

ethane — $CH_3·CH_3$ — a colorless, odorless gas used as a fuel and refrigerant and forms an explosive mixture with air.

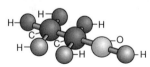

Ethanol

ethanol — C_2H_5OH — a colorless liquid used as a reagent and solvent and derived either through a fermentation process or synthesis from petroleum or natural gas. Also known as ethyl alcohol; grain alcohol.

ethanolamine — $NH_2CH_2CH_2OH$ — a colorless, viscous liquid or solid (below 51°F) with an unpleasant, ammonia-like odor; MW: 61.1, BP: 339°F, Sol: miscible, Fl.P: 185°F, sp gr: 1.02. It is used in the production of monoalkanolamides, in fuel oil additives, pharmaceuticals, agricultural chemicals, cosmetics, emulsion paints, polishers, and cleansers; in the manufacture of inks, paper, glues, textiles, and polishes; in the synthesis of 2-mercaptothiazole in rubber vulcanization acceleration; in the recovery and removal of acid gases from natural, fuel, and process gas in the synthesis of ammonia; in the manufacture of dry ice. It is hazardous to the

skin, eyes, and respiratory system; and is toxic by inhalation, ingestion, and contact. Symptoms of exposure include lethargy, and skin, eye, and respiratory system irritation. OSHA exposure limit (TWA): 3 ppm [air] or 8 mg/m³.

ethene — *See* ethylene.

ether — *See* ethyl ether.

2-ethoxyethanol — $C_2H_5OCH_2CH_2OH$ — colorless liquid with a sweet, pleasant, ether-like odor; MW: 90.1, BP: 313°F, Sol: miscible, Fl.P: 110°F, sp gr: 0.93. It is used as a solvent in the manufacture of lacquers, lacquer thinners, nitrocellulose lacquers, alkyd resins, printing ink solvents, varnish removers, cleaning compounds, soaps, cosmetics, pesticides, pharmaceuticals, adhesives, and detergents; in textile dyeing and printing; in the manufacture of leather finishes; as an anti-icing additive, in the manufacture of brake fluids, aviation fuels, and automotive anti-stall additives. In animals, it is hazardous to the lungs, eyes, blood, kidneys, and liver; and is toxic by inhalation, absorption, ingestion, and contact. Symptoms of exposure in animals include pulmonary irritation, hematologic effects, eye irritation, and damage to the liver, kidneys and lungs. Also known as cellosolve. OSHA exposure limit (TWA): 200 ppm [skin] or 740 mg/m³.

2-ethoxyethyl acetate — $CH_3COOCH_2CH_2OC_2H_5$ — a colorless liquid with a mild odor; MW: 132.2, BP: 313°F, Sol: 23%, Fl.P: 124°F, sp gr: 0.98. It is used as a solvent vehicle during hot and cold spray applications of lacquers and wood stains; in the manufacture of varnishes, thinners, lacquers, wood stains, adhesives, varnish removers, fabric and paper coatings, and textile sizing; in the manufacture of cellulose acetate films; as a solvent for many oils and gums. It is hazardous to the respiratory system, eyes, and gastrointestinal tract; and is toxic by inhalation, absorption, ingestion, and contact. Symptoms of exposure include vomiting, kidney damage, paralysis, irritation of the eyes and nose. OSHA exposure limit (TWA): 100 ppm [skin] or 540 mg/m³.

ethyl acetate — $CH_3COOC_2H_5$ — a colorless liquid with an ether-like, fruity odor; MW: 88.1, BP: 171°F, Sol (77°F): 10%, Fl.P: 24°F, sp gr: 0.90. It is used in the manufacture of smokeless powder and artificial leather; in the preparation of photographic films; as a cleaning agent in the textile industry; in the manufacture of linoleum and plastic wood; in the manufacture of dyes and drug intermediates. It is hazardous to the eyes, skin, and respiratory system; and is toxic by inhalation, ingestion, and contact. Symptoms of exposure include narcosis, dermatitis, and

E
F

irritation of the eyes, nose, and throat. OSHA exposure limit (TWA): 400 ppm [air] or 1,400 mg/m³.

ethyl acrylate — $CH_2=CHCOOC_2H_5$ — a colorless liquid with an acrid odor; MW: 100.1, BP: 211°F, Sol: 2%, Fl.P: 48°F, sp gr: 0.92. It is used in the manufacture of acrylic resins for use in paint formulations, industrial coatings, and latexes; in the manufacture of plastics such as ethylene ethyl acrylate; in the manufacture of polyacrylate elastomers and acrylic rubber; in forming denture materials. It is hazardous to the respiratory system, eyes, and skin; and is toxic by inhalation, absorption, ingestion, and contact. Symptoms of exposure include irritation of the eyes, respiratory system, and skin; carcinogenic. OSHA exposure limit (TWA): 5 ppm [skin] or 20 mg/m³.

ethyl alcohol — *See* ethanol.

ethyl amine — $CH_3CH_2NH_2$ — a colorless gas or water-white liquid (below 62°F) with an ammonia-like odor; shipped as a liquefied compressed gas; MW: 45.1, BP: 62°F, Sol: miscible, Fl.P: 1°F, sp gr (liquid): 0.69. It is used in the synthesis of agricultural chemicals for herbicides and of chemical intermediates and solvents; as a dyestuff intermediate; as a solvent for dyes, resins, and oils; as a catalyst for polyurethane foams and for curing epoxy resins; in pharmaceuticals, emulsifying agents, and as a vulcanization accelerator for sulfur-cured rubber; as a stabilizer for rubber latex; as a selective solvent in the refining petroleum and vegetable oils; in the synthesis of rhodamine dyes; as a deflocculating agent in the ceramics industry; in the manufacture of detergents. It is hazardous to the respiratory system, eyes, and skin; and is toxic by inhalation, absorption, ingestion, and contact. Symptoms of exposure include skin burns, dermatitis, eye and respiratory irritation. Also known as aminoethane; ethamine. OSHA exposure limit (TWA): 10 ppm [air] or 18 mg/m³.

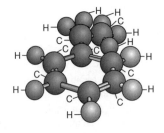

Ethyl benzene

ethyl benzene — $CH_3CH_2C_6H_5$ — a colorless liquid with an aromatic odor; MW: 106.2, BP: 277°F, Sol: 0.01%, Fl.P: 55°F, sp gr: 0.87. It is

used in the manufacture and application of rubber adhesives; during electroplating of aluminum on copper or steel; as a heat-transfer medium and as a dielectric; as an intermediate in dye manufacture. It is hazardous to the eyes, upper respiratory system, skin, and central nervous system; and is toxic by inhalation, ingestion, and contact. Symptoms of exposure include headache, dermatitis, narcosis, coma, and irritation of the eyes and mucous membrane. OSHA exposure limit (TWA): 100 ppm [air] or 435 mg/m³.

ethyl bromide — CH_3CH_2Br — a colorless to yellow liquid with an ether-like odor; a gas above 101°F; MW: 109.0, BP: 101°F, Sol: 0.9%, Fl.P: <4%°F, sp gr: 0.87. It is used as an alkylating or ethylating agent in the manufacture of pharmaceuticals, and to a lesser extent of chemicals, dyes, and perfumes; as a solvent for fats, waxes, and resins; as a local surface anesthetic for minor surgery and neuralgic pain; as a liquid fumigant for grain and fruit; as a refrigerant. It is hazardous to the skin, liver, kidneys, respiratory system, central nervous system, and cardiovascular system; and is toxic by inhalation, ingestion, and contact. Symptoms of exposure include central nervous system depression, pulmonary edema, liver and kidney disease, cardiac arrhythmias, cardiac arrest. OSHA exposure limit (TWA): 200 ppm [air] or 890 mg/m³.

ethyl butyl ketone — $CH_3CH_2CO(CH_2)_3CH_3$ — a colorless liquid with a powerful, fruity odor; MW: 114.2, BP: 298°F, Sol: 1%, Fl.P (open cup): 115°F, sp gr: 0.82. It is used in lacquers, varnishes, epoxies, vinyl coatings, finishes, and adhesives; in cellulosic, acetate, and rubber adhesives in shoe manufacture, book binding, and packaging; in mastics and other natural gum-based adhesives for floor and wall tiles, other floor coverings, automobile silencer, and lining pads. It is hazardous to the eyes, skin, and respiratory system; and is toxic by inhalation, ingestion, and contact. Symptoms of exposure include headache, narcosis, coma, dermatitis, irritation of the eyes and mucous membrane. Also known as 3-heptanone. OSHA exposure limit (TWA): 50 ppm [air] or 230 mg/m³.

Ethyl Chloride
(Chloroethane)

ethyl chloride — CH_3CH_2Cl — a colorless gas or liquid (below 54°F) with a pungent, ether-like odor; shipped as a liquefied compressed gas;

MW: 64.5, BP: 54°F, Sol: 0.6%, sp gr: 0.92 (liquid at 32°F). It is used in the production of tetraethyl lead and ethyl cellulose; as a local general anesthetic; as a refrigeration compound; as a solvent for fats, oils, waxes, phosphorus, acetylene, and many resins; in organic synthesis of perchloroethane, esters, and Grignard reagents; in the manufacture of dyes, drugs, and perfumes; as a propellant in aerosols. It is hazardous to the liver, kidneys, respiratory system, and cardiovascular system; and is toxic by inhalation, absorption, ingestion, and contact. Symptoms of exposure include incoordination, inebriation, abdominal cramps, cardiac arrhythmias, cardiac arrest, damage to the liver and kidneys. Also known as chloroethane. OSHA exposure limit (TWA): 1,000 ppm [air] or 2,600 mg/m³.

ethyl chlorohydrin — CH_2ClCH_2OH — a colorless liquid with a faint, ether-like odor; MW: 80.5, BP: 262°F, Sol: miscible, Fl.P: 140°F, sp gr: 1.20. It is used as a fumigant and in treatment of seed potatoes; in dyeing operations during color printing of textiles; in cleaning, degreasing, and dewaxing operations; in organic synthesis and synthesis of dyes, pharmaceuticals, and resin processing; in the extraction of pine lignin and oil of rose. It is hazardous to the respiratory system, liver, kidneys, skin, cardiovascular system, and central nervous system; and is toxic by inhalation, absorption, ingestion, and contact. Symptoms of exposure include nausea, vomiting, vertigo, incoordination, numbness, visual disturbance, headache, thirst, delirium, low blood pressure, collapse, shock, coma, irritation of the mucous membrane. OSHA exposure limit (TWA): ceiling 1 ppm [skin] or 3 mg/m³.

ethylene — C_2H_4 — a colorless, flammable, gaseous organic compound resulting from the cracking of either selective petroleum fractures or natural gas. It is widely used as a petrochemical base for numerous chemical reactions, especially for the preparation of polyethylene and related plastics. Also known as ethene; olefiant gas.

ethylene bromide — *See* ethylene dibromide.
ethylene chloride — *See* ethylene dichloride.
ethylenediamine — $NH_2CH_2CH_2NH_2$ — a colorless, viscous liquid with an ammonia-like odor; a solid below 47°F; MW: 60.1, BP: 241°F, Sol: miscible, Fl.P: 93°F, sp gr: 0.91. It is used in the manufacture of carbamate fungicides, pesticides, and weed killers; in the manufacture of chelating agents in dyes, soaps, and cleaning compounds, water treatment, in agriculture, in rubber, pulp, and paper processing; as an activator for epoxy resin menders; in the preparation of

spandex, adhesives, inks, surface coatings, crosslinking agents, plasticizers, resins curing compounds, and as a dye assist; in the manufacture of surfactants, emulsifying agents, dispersants, corrosion inhibitors, detergents, and textile surface treatments; in organic synthesis for preparation of dye intermediates, heterocyclic compounds, pharmaceuticals, and salts. It is hazardous to the respiratory system, liver, kidneys, and skin; and is toxic by inhalation, absorption, ingestion, and contact. Symptoms of exposure include nasal irritation, primary irritation, sensitization dermatitis, asthma, damage to the liver and kidneys, irritation of the respiratory system. OSHA exposure limit (TWA): 10 ppm [air] or 25 mg/m³.

ethylene dibromide — $BrCH_2CH_2Br$ — a colorless liquid or solid (below 50°F) with a sweet odor; MW: 187.9, BP: 268°F, Sol: 0.4%, sp gr: 2.17. It is used in fumigation operations in preplanting and on grains, fruits, and vegetables; in the production of antiknock fluids in fuels; in the production of water-proofing agents, fire extinguishing agents, and gauge fluids during the manufacture of measuring instruments; in organic synthesis in the production of dyes, pharmaceuticals, and ethylene oxide; as a specialty solvent for resins, gums, and waxes. It is hazardous to the respiratory system, liver, kidneys, skin, and eyes; and is toxic by inhalation, absorption, ingestion, and contact. Symptoms of exposure include dermatitis with vesiculation; irritation of the eyes and respiratory system; carcinogenic. Also known as ethylene bromide. OSHA exposure limit (TWA): 20 ppm [air].

**Ethylene Dichloride
(1,2–Dichloroethane)**

ethylene dichloride — $ClCH_2CH_2Cl$ — a colorless liquid with a pleasant, chloroform-like odor; decomposes slowly, becomes acidic, and darkens in color; MW: 99.0, BP: 182°F, Sol: 0.9%, Fl.P: 56°F, sp gr: 1.24. It is used as a chemical intermediate in the manufacture of vinyl chloride; as an intermediate in the production of chlorinated solvents and ethyleneamines; in the production of gasoline using tetraethyl lead as an antiknock agent and ethylene dichloride as a lead scavenger; as a fumigant or industrial solvent. It is hazardous to the kidneys, liver, eyes,

E
F

skin, and central nervous system; and is toxic by inhalation, absorption, ingestion, and contact. Symptoms of exposure include central nervous system depression, nausea, vomiting, dermatitis, corneal opacity, irritation of the eyes; carcinogenic. Also known as ethylene chloride; Dutch liquid. OSHA exposure limit (TWA): 1 ppm [air] or 4 mg/m^3.

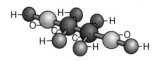

Ethylene Glycol

ethylene glycol — HOCH$_2$CH$_2$OH — a colorless organic liquid used as an absorbent since it absorbs twice its weight of water at room temperature and 100% humidity. It is used as an antifreeze, solvent for dyes, resins, drugs, or other organic chemicals, and in the manufacture of dynamites. Also known as glycol.

ethylene glycol dinitrate — O$_2$NOCH$_2$CH$_2$ONO$_2$ — a colorless to yellow, oily, odorless liquid; MW: 152.1, BP: 387°F, Sol: insoluble, Fl.P: 419°F, sp gr: 1.49. It is used as an explosive ingredient (60–80%) in dynamite along with nitroglycerine (20–40%). It is hazardous to the cardiovascular system, blood, and skin; and is toxic by inhalation, absorption, ingestion, and contact. Symptoms of exposure include throbbing headache, dizziness, nausea, vomiting, abdominal pain, hypotension, flush, palpitations, methemoglobinemia, delirium, depressed central nervous system, angina, skin irritation; in animals, anemia, and mild liver and kidney damage. OSHA exposure limit (TWA): 0.1 mg/m^3 [skin].

ethyleneimine — C$_2$H$_5$N — a colorless liquid with an ammonia-like odor; MW: 43.1, BP: 133°F, Sol: miscible, Fl.P: 12°F, sp gr: 0.83. It is used in the manufacture of paper, textiles, adhesives, binders, petroleum refining products, rocket and jet fuels, lubricants, chemo-sterilant chemicals, chemotherapeutic agents, coating resins, varnishes, lacquers, agricultural chemicals, cosmetics, ion-exchange resins, photographic chemicals, colloid flocculants, and surfactants. It is hazardous to the eyes, lungs, skin, liver, and kidneys; and is toxic by inhalation, absorption, ingestion, and contact. Symptoms of exposure include nausea, vomiting, headache, dizziness, pulmonary edema, damage to the liver and kidneys, eye burns, skin sensitization, irritation of the nose and throat; carcinogenic. Also known as aziridine. OSHA exposure limit (TWA): regulated use.

ethylene oxide (ETO) — C$_2$H$_4$O — a colorless, flammable gas with a somewhat sweet odor; MW: 44.05, BP: 11°C, Sol (20°C): 1 × 1^{-6} mg/ L, Fl.P: < –18°C. It is used in organic synthesis, for sterilizing, and for fumigating. It is hazardous to the respiratory system, nervous system, and reproductive system; and is toxic by inhalation, ingestion, absorption, and contact. Symptoms of exposure include bronchitis, pulmonary edema, emphysema, nausea and vomiting, cataracts, headache, peripheral neuropathy, impaired eye coordination, memory loss, lethargy, and numbness and weakness in the extremities. Also known as oxirane. OSHA exposure limit (PEL TWA): 1 ppm [air].

ethylene-propylene terpolymer (EPDM) — a high strength, flexible compound designed especially for contact with potable water.

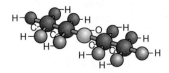

**Ethyl Ether
(Diethyl Ether)**

ethyl ether — C$_2$H$_5$OC$_2$H$_5$ — a colorless liquid with a pungent, sweetish odor; a gas above 84°F; MW: 74.1, BP: 94°F, Sol: 8%, Fl.P: –49°F, sp gr: 0.71. It is used as a solvent in the manufacture of smokeless powder; as a laboratory solvent and chemical extractant; as a solvent and cleaning agent in the shoe and textile industry; in the manufacture of warm-process and cold-process pharmaceuticals; as an additive in motor fuels, perfumes, and denatured alcohol; as an anesthetic by animal handlers. It is hazardous to the central nervous system, skin, respiratory system, and eyes; and is toxic by inhalation, ingestion, and contact. Symptoms of exposure include dizziness, drowsiness, headache, excitedness, narcosis, nausea, vomiting, irritation of the eyes, skin, and upper respiratory system. Also known as ether. OSHA exposure limit (TWA): 400 ppm [air] or 1,200 mg/m^3.

ethyl formate — CH$_3$CH$_2$OCHO — a colorless liquid with a fruity odor; MW: 74.1, BP: 130°F, Sol (64°F): 9%, Fl.P: –4°F, sp gr: 0.92. It is used in the shoe industry to dissolve celluloid heel coverings; as a solvent in the manufacture of artificial leather and in the manufacture of cellulose acetate for artificial silk; in the manufacture of safety glass; as a fumigant and larvacide for tobacco, cereals, and dried fruits; as an intermediate in organic synthesis; during the formulation of synthetic flavors. It is hazardous to the eyes and respiratory system; and is toxic by inhalation, ingestion, and contact. Symptoms of exposure include narcosis, and irritation of the eyes and upper respiratory system. OSHA exposure limit (TWA): 100 ppm [air] or 300 mg/m³.

di (2-ethylhexyl) phthalate — $C_{24}H_{38}O_4$ — a colorless liquid with almost no odor; MW: 390.57, BP: 385°C, Sol (24°C): 0.285 mg/L, Fl.P: 196°C. It is used principally as a plasticizer in the production of polyvinyl chloride (PVC) and vinyl chloride resins; PVC is used in toys, vinyl, upholstery, shower curtains, adhesives, coatings, components of paper and paper board, disposable medical examination and surgical gloves, flexible tubing used to administer parenteral solutions, and in hemodialysis treatment. Single oral doses up to 10 grams are not lethal to humans and no cases of death in humans after oral exposure are known; higher doses are lethal to rats, rabbits, and guinea pigs; acute exposures to large oral doses can cause gastrointestinal distress with mild abdominal pain and diarrhea; oral exposures to rats and mice result in marked increase in liver mass; hepatic hyperplasia appears to be the initial physiological response in rats; an autopsy of the liver had fat deposits, a decline in centrilovular glycogen deposits and structural changes in the bile ducts; developmental toxicity including teratogenic effects have occurred in both rats and mice. OSHA exposure limits (PEL TWA): 5 mg/m³ [air], STL 10 milligrams/m³ [air].

ethyl mercaptan — CH_3CH_2SH — a colorless liquid with a strong skunk-like odor; a gas above 95°F; MW: 62.1, BP: 95°F, Sol: 0.7%, Fl.P: –55°F, sp gr: 0.84. It is used as a fuel gas odorant; as a chemical intermediate, or as raw material in the production of insecticides, antioxidants, plasticizers, and pharmaceuticals; as a stabilizer; as a solvent for elastomeric polymers and oil soluble dyes in industrial stains and quick-setting rubber cements. It is hazardous to the respiratory system; in animals: liver and kidneys; and is toxic by inhalation, ingestion, and contact. Symptoms of exposure include headache, nausea, irritation of the mucous membrane; in animals: incoordination, weakness, pulmonary irritation, cyanosis, liver and kidney damage. OSHA exposure limit (TWA): 0.5 ppm [air] or 1 mg/m³.

n-ethylmorpholine — $C_4H_8ONCH_2CH_3$ — a colorless liquid with an ammonia-like odor; MW: 115.2, BP: 281°F, Sol: miscible, Fl.P (open cup): 90°F, sp gr: 0.90. It is used as a catalyst for flexible, semi-flexible, and rigid polyurethane foam production; in polymer technology as a promoter for resin surface curing, and as a stabilizer for fiber spinning solutions; in the manufacture of vat dyes; in the manufacture of pharmaceuticals in the purification of Penicillin G; in organic synthesis as a special solvent, pH regulator, and for the preparation of chemical intermediates. It is hazardous to the respiratory system, skin, and eyes; and is toxic by inhalation, absorption, ingestion, and contact. Symptoms of exposure include visual disturbance, irritation of the eyes, nose, and throat, and severe eye irritation from splashes. OSHA exposure limit (TWA): 5 ppm [skin] or 23 mg/m³.

0-ethyl-0-para-nitrophenyl phenylphosphonothioate (EPN) — $C_2H_5O(C_6H_5)P(S)OC_6H_4NO_2$ — a yellow solid with an aromatic odor; a brown liquid above 97°F; MW: 323.3, BP: not known, Sol: insoluble, sp gr (77°F): 1.27. It is used as an insecticide and acaricide. It is hazardous to the respiratory system, cardiovascular system, central nervous system, eyes, skin, and blood cholinesterase; and is toxic by inhalation, absorption, ingestion, and contact. Symptoms of exposure include miosis, lacrimation, rhinorrhea, headache, tight chest, wheezing, laryngeal spasms, salivation, cyanosis, anorexia, nausea, abdominal cramps, diarrhea, paralysis, convulsions, low blood pressure, cardiac irregularities, eye and skin irritation. OSHA exposure limit (TWA): 0.5 mg/m³ [skin].

ethyl pyrophosphate — *See* TEPP.

ethyl silicate — $(C_2H_5)_4SiO_4$ — a colorless liquid with a sharp, alcohol-like odor; MW: 208.3, BP: 336°F, Sol: reacts, Fl.P: 99°F. It is used as a bonding agent for investment castings, ceramic shells, crucibles, and types of refractory shapes; as a protective coating for paints, lacquers, and films; in the impregnation of porous materials to increase strength, hardness, stiffness, and abrasion; as a waterproofing and weatherproofing agent in porous rocks; as a chemical intermediate; as a gelling agent in organic liquids; as a coating agent inside electric lamp bulbs; in the synthesis of fused quartz; in the textile industry. It is hazardous to the respiratory system, liver, kidneys, blood, and skin; and is toxic by inhalation, ingestion, and contact. Symptoms of exposure include eye and nose irritation; in animals: lacrimation, dyspnea, pulmonary edema, tremors, narcosis, liver and kidney damage, anemia. OSHA exposure limit (TWA): 10 ppm [air] or 85 mg/m³.

etiological agent — a pathogenic organism or chemical causing a specific disease in a living body.

etiology — the science dealing with causes of disease.

ETO — *See* ethylene oxide. Also known as oxirane.

ETS — acronym for emergency temporary standards. *See* environmental tobacco smoke.

eugenics — the science of human heredity.

eukaryotic cells — (*genetics*) the cells of higher animals and plants which contain genetic material in the nuclei.

euphoria — the absence of pain or distress.

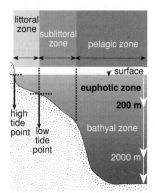

Euphotic Zone

euphotic zone — the lighted region that extends vertically from the water surface to the level at which photosynthesis fails to occur because of ineffective light penetration.

eustachian tube — a tube connecting the nasopharynx with the middle ear cavity.

eutrophic — pertaining to a lake abundant in nutrients and having high rates of productivity frequently resulting in oxygen depletion below the surface layer.

eutrophication — a process by which pollutants cause a body of water to become overly rich in organic and mineral nutrients so that algae grows rapidly and depletes the oxygen supply.

eV — *See* electron volt.

evacuation — removal of residents from an area of danger.

evacuation/exhaust — that part of the sterilizer cycle when the vacuum pump runs to remove the bulk of the ethylene oxide from the chamber followed by the opening of a valve to admit filtered air into the sterilizer chamber returning it to atmospheric pressure.

evaluation — a critical appraisal or assessment.

evaporation — the physical transformation of a liquid to a gaseous state at any temperature below its boiling point.

evaporation ponds — areas where sewage, sludge, or wastes are dumped and allowed to dry.

evaporation rate — the rate at which a material will vaporize or evaporate compared to the known rate of vaporization of a standard material.

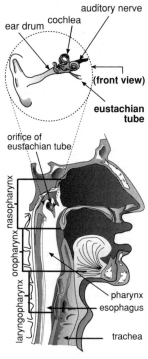

Eustachian Tube

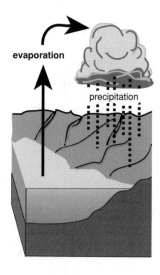

Evaporation

evapotranspiration — the combined loss of water in a given area during a specified period of time by evaporation from the soil or water surface and by transpiration from plants.

evolution — the slow process of change by which organisms have acquired their distinguishing characteristics and come to differ from their ancestors.

exacerbation — an increase in severity.

excess combustion air — air supplied in excess of theoretical air, usually expressed as a percentage of the theoretical air.

exchange capacity — the total ionic charge of the adsorption complex active in the adsorption of ions.

excitation — the process by which the addition of energy to a system thereby transfers it from its ground state to an excited state; an act of irritation or stimulation.

excitation energy — the minimum energy required to change a system from its ground state to an excited state.

excreta — waste material eliminated from the body, including feces and urine.

excretion — the process by which waste materials are removed by living cells from the body.

excretory system — the series of organs adapted to removing waste from an organism.

execution of decree — a decree ordering an action that when completed the decree is considered to have been executed, e.g., the destruction of goods under seizure by the United States Marshall in response to a default decree of condemnation.

exhalation valve — a device that allows exhaled air to leave a respirator while it prevents outside air from entering.

exhaust emissions — substances emitted to the atmosphere from any opening downstream from the exhaust port of a motor vehicle.

exhaust rate — the volumetric rate at which air is removed.

exhaust ventilation — mechanical removal of air from a portion of a building.

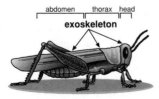

Grasshopper Exoskeleton

exoskeleton — the stiff, tough, external covering on the body of certain invertebrates, especially arthropods, providing support and protection to the softer inner parts.

exothermic — referring to a chemical reaction in which heat is given off, such as in the burning of materials.

exothermic reaction — a chemical reaction in which the energy released is greater than the energy required to start the reaction.

exotoxin — a soluble toxin excreted by certain microbacteria and absorbed by the tissues of host organisms.

expansion chamber — *See* cloud chamber.

expansion joint — a joint which is purposely left open to allow for the expansion of its parts.

ex parte — (*law*) on one side only; referring to cases in which only one side is represented.

expected deaths — an estimation of the number of deaths occurring from a particular disease.

expenses — all payments transferred or paid to entities outside the operating organization for costs incurred for and related to the plant operation.

experience — a record of premiums and losses relating to insurance and used as the basis upon which future rates or costs are predicted.

experience modification — a provision designed to recognize the merits or demerits of individual risks as it relates to insurance.

experimental animal — any animal used in any research investigation involving the feeding or other administration of, or subjection to an experimental or nonexperimental biological product, drug, or chemical used in a manner for which it was not intended.

experimental study — research in which the investigator has complete control over allocation of subjects to treatment groups.

expiration — the discharge of air from the lungs.

expiration date — the date beyond which a drug, chemical, or food should not be used.

expiratory flow rate — the maximum rate at which air can be expelled from the lungs.

explosive — any chemical compound, mixture, or device whose primary or common purpose is to function by explosion with substantial instantaneous release of gas or heat when subjected to a sudden shock, pressure, or high temperature.

explosive limit — *See* flammable limit.

explosive range — *See* flammable range.

exposure — a measure of the ionization produced in air by x- or gamma-radiation. It is the sum of the electrical charges of all ions of one sign produced in air when all electrons liberated by photons in a volume element of air are completely stopped and divided by the mass of air in the volume element. This unit measurement of exposure is the roentgen; the act or condition of coming in contact with, but not necessarily being infected by, a disease-causing agent; (*hazardous materials*) the state of being open and vulnerable to a hazardous chemical in the course of employment by inhalation, ingestion, skin contact, absorption or any other method, including potential and accidental exposure.

exposure assessment — the determination or estimation, either qualitative and/or quantitative, of the magnitude, frequency, duration, and route of human exposure to a particular agent; a combination of field monitoring, mathematical modeling, measurement of actual concentrations of chemicals in tissue and laboratory monitoring data; the process of determining how much of a chemical is in the environment and capable of contact with organisms.

exposure period — the length of time an object or individual is exposed to a chemical or microorganism.

exposure point — any hypothetical event that would lead to human contact with a toxin.

exposure unit — 1 electrostatic unit (ESU) of charge in 1 cm^3 of air at standard temperature and pressure; formerly called the roentgen.

extended aeration — the modification of the activated sludge process which provides for aerobic sludge digestion within the aeration system.

extermination — the control and elimination of insects, rodents, and other pests by eliminating harborage, food, and the use of poisoning or trapping techniques.

external radiation — radiation that penetrates the skin from a source outside the body.

external respiration — the exchange of oxygen and carbon dioxide between the alveolar air and blood.

extinguishing media — the fire fighting substances used to control a material during a fire.

extraction procedure (EP) — a procedure used to identify waste likely to leach hazardous concentrations of particular toxic constituents into the groundwater from landfills.

extra-high-voltage (ehv) — a voltage above 345 kilovolts used for power transmission.

extrapolation — estimation of unknown values by extension or projection from known values; to project, extend, or expand known data or experience into an area not known or experienced to arrive at a projection of knowledge.

extrasystole — a premature cardiac contraction that is independent of the normal rhythm and arises in response to a sinoatrial node.

extremely-low-frequency (ELF) — the frequency range below 300 hertz in the radio spectrum.

extrusion — the forcing of raw material through a die or a form in either a hot or cold state in a solid or partial solution.

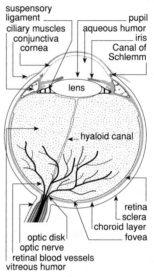

Eye

eye — one of a pair of organs of sight, contained in a bony orbit at the front of the skull, embedded in orbital fat and innervated by one of a pair of optic nerves from the forebrain.

eye hazards — chemicals which affect the eye or visual capacity with conjunctivitis and corneal damage.

eyepiece — a gas-tight, transparent window in a full facepiece through which the person can see.

eye protection — recommended safety glasses, chemical splash goggles, and face shields to be utilized when handling a hazardous material.

eye toxins — chemicals, including organic solvents and acids, which produce symptoms of conjunctivitis and corneal damage.

f — *See* frequency.

F — *See* fahrenheit scale; equilibrium fraction; farad.

F₁ — notation for the first filial generation resulting from a cross.

FAA — *See* Federal Aviation Administration.

fabric filter — a device designed to remove particles from a carrier gas by passage of the gas through a porous fabric medium.

facepiece — that portion of a respirator that covers the wearer's nose and mouth in a half-mask facepiece or nose, mouth, and eyes in a full facepiece.

face velocity — the average air velocity into the exhaust system measured at the opening into the hood or booth.

facilitate — the act of easing understanding or action.

FACOSH — *See* Federal Advisory Council for Occupational Safety and Health.

facultative aerobe — a microorganism that lives in the presence of oxygen but does not require it.

facultative anaerobe — an organism that grows best under aerobic conditions but may also grow under anaerobic conditions.

fahrenheit scale (F) — temperature scale ranging from 32° at freezing to 212° at boiling.

faint — *See* syncope.

fallout — the dust particles that contain radioactive fission products resulting from a nuclear explosion.

false galena — *See* sphalerite.

false negative — the result of a test whose subject does have the disease tested for.

false positive — the result of a test whose subject does not have the disease tested for.

FAM — *See* fibrous aerosol monitor.

fan rating table — the data describing the volumetric output of a fan at different static pressures.

fan static pressure — the pressure added to a system by the fan equal to the sum of the pressure lost in the system minus the velocity pressure in the air at the fan outlet.

FAO — *See* Food and Agricultural Organization of the United Nations.

farad (F) — the unit of capacitance in the meter-kilogram-second system.

faraday — a unit of electric charge required to deposit by electrolysis one equivalent weight of

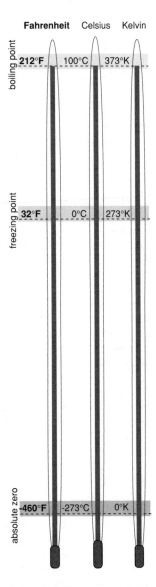

Fahrenheit Temperature Scale
compared to K and C scales
(not proportionate)

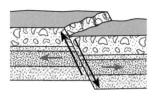

normal thrust fault

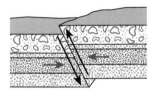

reverse thrust fault

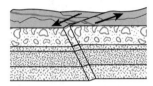

strike-slip fault

Faults

an element; the measurement equal to 96,500 coulombs. Also known as Faraday Constant.

Faraday Constant — *See* faraday.

Faraday's law — the law which states that during electrolysis, the weight of an element liberated is proportional to the quantity of electricity passing through the cell and equivalent to the weight of the element.

fasciculation — a small, local, involuntary, muscular contraction visible under the skin representing spontaneous discharge of a number of fibers innervated by a single motor nerve filament.

fat — any glycerol ester of a saturated fatty acid which form a class of neutral organic compounds.

fatal — having the capacity to cause death.

fatality rate — the number of persons diagnosed as having a specific disease who die as a result of the illness.

fate — the combined transport and transformation of a pollutant.

fat free — *See* non-fat.

fatigue — a state of increased discomfort and decreased efficiency resulting from prolonged exertion; a generalized feeling of tiredness or exhaustion; a phenomenon due to overexertion leading to a decrease in physical performance and a build-up of lactic acid in the body.

fatty acid — one of the soluble products into which fats are broken down in the body by digestion.

faults — fractures where once-continuous rocks have suffered relative displacement.

fauna — the animals characteristic of region, period, or special environment.

favism — a poisoning from eating the fava bean Viciafava or being around the plants which are in blossom.

F/B — *See* freight bill.

FBC — *See* fluidized-bed combustion.

fc — *See* footcandle.

f/cc — *See* fibers per cubic centimeter.

FCC — *See* Federal Communications Commission.

FDA — *See* Food and Drug Administration.

FDCA — *See* Federal Food, Drug, and Cosmetic Act.

FDV — *See* flow diversion valve.

feasibility study — the acquisition of information to determine if a proposed project is practical and/or capable of being accomplished from the standpoint of advantages versus disadvantages.

febrile — associated or characterized by fever.

fecal coliform bacteria — bacteria of the coliform group of fecal origin from the intestines of warm blooded animals as opposed to coliforms of non-fecal sources.

feces — solid intestinal waste materials.

Federal Advisory Council for Occupational Safety and Health (FACOSH) — a joint management-labor council that advises the Secretary of Labor on matters relating to the occupational safety and health of federal employees.

Federal Aviation Administration (FAA) — the division of the United States Department of Transportation that inspects and rates civilian aircraft and airmen and enforces the rules of air safety.

Federal Communications Commission (FCC) — a federal agency which regulates interstate and foreign communications by radio, television, wire, and cable.

Federal Emergency Management Agency (FEMA) — a federal agency serving as a single point of accountability for all federal emergency preparedness mitigation and response activities and chartered to enhance the multiple use of emergency preparedness and response resources at the federal, state, and local levels.

Federal Energy Regulatory Commission (FERC) — an independent, five-member commission within the Department of Energy which sets rates and charges for transportation and sale of natural gas, for the transmission and sale of electricity, and the licensing of hydroelectric power projects.

Federal Food, Drug, and Cosmetic Act (FDCA) — an amendment to the Food and Drug Act of 1906, passed in 1938, and amended through the 1980s, establishing the authority of the Food and Drug Administration to regulate food additives, contaminants, drugs, cosmetics, and potential carcinogens related to food, drugs, and cosmetics.

Federal Housing Administration (FHA) — an agency of the United States government whose function has been transferred to the Department of Housing and Urban Development.

Federal Insecticide, Fungicide, and Rodenticide Act (FIFRA) — a 1947 law updated through 1988, authorizing the Environmental Protection Agency to register and reregister pesticides, review, cancel, and suspend registered pesticides and regulate the production, storage, transportation, use, and disposal of pesticides, including areas of research and monitoring.

Federal Register — a bulletin published daily by the federal government containing executive branch proposed or final regulations, proclamation, schedule of hearings before Congress, and federal agency committees.

Federal Water Pollution Control Act (FWPCA) — a 1956 law funding water pollution research and training, establishing construction grants for municipalities, and establishing a three-stage enforcement process.

Federal Water Pollution Control Act Amendments (FWPCAA) — a 1972 amendment to the Federal Water Pollution Control Act, setting national water quality goals, establishment of a zero discharge goal by 1985, setting technology-based effluent limitations, establishing National Pollution Discharge Elimination System permits, providing federal enforcement based on permit violations, and increasing federal funding to water treatment facilities.

feed — animal food; the act of giving food to a plant or animal.

feedback — providing valuative or corrective information concerning an action or process.

feedlot — a concentrated, confined, animal- or poultry-growing operation for meat, milk, or egg production or stabling in pens or houses wherein the animals or poultry are fed at the place of confinement and the waste material is concentrated.

feedstock — the raw materials supplied to manufacturing or processing plants for use in the production of goods or for treatment.

fee for service — the fee charged by a governmental agency to license or approve an environmental service, such as a food establishment, septic tank system, waste site, etc.; the fee is comprised of the time, materials, and other expenses needed to perform the service; can be partially subsidized by taxes.

feelers — *See* antenna.

FEF — notation for forced expiratory flow rate.

felony — a crime punishable by more than one year in prison.

FEMA — *See* Federal Emergency Management Agency.

FEPCA — acronym for Federal Environmental Pesticide Control Act.

ferbam — $((CH_3)_2NCS_2)_3Fe$ — a dark brown to black, odorless solid; MW: 416.5, BP: decomposes, Sol: 0.01%, Fl.P: not known. It is used as a fungicide for the control of scab and rust diseases. It is hazardous to the skin, respiratory system, gastrointestinal tract; and is toxic by inhalation, ingestion, and contact. Symptoms of exposure include gastrointestinal distress, dermatitis, irritation of the eyes and respiratory system. OSHA exposure limit (TWA): 10 mg/m^3 [air].

FERC — *See* Federal Energy Regulatory Commission.

fermentation — the glucose oxidation process in which an anaerobic chemical change is brought about by enzymes, bacteria, or other living organisms resulting in the formation of lactic acid or alcohol and generally accompanied by the evolution of gas. Also known as zymosis.

ferric iron — a compound of iron which is insoluble in water and will precipitate.

ferrihemoglobin — *See* methemoglobin.

ferromagnetism — a property exhibited by certain metals, alloys, and compounds in which the internal magnetic moments spontaneously organize in a common direction; gives rise to permeability somewhat greater than air.

ferrous iron — a compound of iron which is soluble in water and which will impart a clear green color.

ferrovanadium dust (FeV) — a dark, odorless particulate dispersed in air; ferrovanadium metal is an alloy usually containing 35–80% vanadium; MW: 106.8, BP: not known, Sol: insoluble. It is used in the production of steel as an additive to produce refinement and hardenability. It is hazardous to the respiratory system and eyes; and is toxic by inhalation and contact. Symptoms of exposure include irritation of the eyes and respiratory system; in animals: bronchitis and pneumonitis. OSHA exposure limit (TWA): 1 mg/m^3 [air].

fertilization — the physiochemical union of sperm and egg to form a zygote.

fertilizer — a plant food usually sold in mixed formula, containing basic plant nutrients including nitrogen, potassium, soluble phosphorous, sulfur, and sometimes mineral compounds.

fetal — pertaining to the fetus.

fetal death — the death of a product of conception prior to the complete expulsion or extraction from the mother in at least the twentieth week of gestation.

fetal death rate — the number of fetal deaths per 1,000 total births for a specified time period.

fetotoxicity — any toxic or damaging effect on a fetus as a result of prenatal exposure to a chemical.

fetus — a mammalian embryo after the main body features are present; in humans, the embryo from 7–8 weeks after fertilization to birth.

FeV — *See* ferrovandium dust.

FEV₁ — *See* forced expiratory volume measure at 1 second.

fever — an elevation of the central body temperature caused by abnormal functioning of the thermoregulatory mechanisms.

FFSHC — *See* Field, Federal, Safety, and Health Councils.

FGD — *See* flue gas desulfurization.

FHA — *See* Federal Housing Administration.

FHCP — acronym for Federal Hazard Communication Program.

FHSA — acronym for Federal Hazardous Substances Act.

fiat — a command or order directing some legal act be done.

fiber — (*EPA definition*) a structure having a minimum length equal to 0.5 micrometers and an aspect ratio (length to width) of 5:1 or greater with substantially parallel sides; (*NIOSH definition*) "a rules" — thread-like structure longer than 5 micrometers measured along the curve if applicable with a length-to-width ratio equal to or greater than 3:1. "b rules" — thread-like structure longer than 5 micrometers and less than 3 micrometers in diameter with a length-to-width ratio equal or greater than 5:1.

fiber optic — an optical fiber composed of thin glass wire designed for light transmission; capable of transmitting billions of bits per second which are not affected by random radiation in the environment.

fibers per cubic centimeter (f/cc) — the number of fibers present in a cubic centimeter of air.

fibers per cubic meter (f/m³) — the number of fibers present in a cubic meter of air.

fibrillation — *See* atrial fibrillation; ventricular fibrillation.

fibrin — an insoluble protein formed during blood clotting by the union of thrombin and fibrinogen.

fibrinogen — a plasma protein involved in clotting.

fibrosis — an abnormal thickening of fibrous connecting tissue, usually in the lungs.

fibrous aerosol monitor (FAM) — a direct-reading instrument used to measure airborne fibers, such as asbestos, by use of a helium-neon laser and electro-optical sensor that detects fiber oscillations as the aerosol passes through a rapidly oscillating, high-intensity electrical field.

FID — *See* flame ionization detector.

field blank — a clean filter cassette assembly which is taken to the sampling site, handled in every way as the air samples, except that no air is drawn through it.

Field, Federal, Safety, and Health Councils (FFSHC) — councils organized throughout the country to improve the federal safety and health programs at the field level and within a geographic location.

FIFRA — *See* Federal Insecticide, Fungicide, and Rodenticide Act.

filaria — a parasitic nematode worm of the superfamily *Filarioidea*.

Filariasis
(*Wuchereria bancrofti,* cause of filariasis and elephantiasis)

filariasis — (*disease*) a human infection by a nematode which normally develops in the lymph glands; the female worms produce microfilariae which reach the blood stream and circulate. Symptoms of exposure include fever, lymph node disease, swelling of the breasts and genitalia. The incubation time is 1–12 months depending on the organism; caused by *Wuchereria bancrofti, Brugia malayi* and *B. timori*; found in Latin America, Africa, Asia, the Pacific, and among some American veterans of foreign wars.

The reservoir of infection is humans. It is transmitted by the bite of a mosquito and is not communicable from person to person; general susceptibility; controlled by elimination of mosquitos.

filiform — thread-like.

fill — (*solid waste*) soil transported and deposited intentionally as well as soil transported and deposited by natural forces and used as backfill for sanitary landfill operations.

film badge — a personal radiation monitor where radiation interacts with the silver atoms in a photographic film to expose the film the same way as light rays do; the amount of darkening of a film is then compared to a control film not exposed to radiation to determine the amount of radiation exposure.

filter — a porous mesh of spun glass fibers, asbestos cellulose, or other materials that allow air or liquid to pass through but retain solid particles; a mechanical device for straining suspended particles from water; (*swimming pools*) any of a variety of systems in which water continuously flows in order to remove the particulate matter from the water; a device for removing extraneous signals or radiation.

filterable virus — a virus that passes through the extremely small pores of unglazed porcelain filters used in separating bacteria from fluid.

filter aid — a powder-like substance, such as diatomaceous earth or volcanic ash, used to coat a septum pipe filter; can also be used as an aid on a sand filter.

filter background level — the concentration of structures per square millimeter of filter that is considered indistinguishable from the concentration measured on a blank filter through which no air has been drawn.

filter cartridge — a disposal element, usually a fibrous material, used as a filter septum in some pool filters.

filter collector — a mechanical filtration system for removing particulate matter from a gas stream for measurement, analysis, or control.

filter cycle — the time of filter operation between backwash procedures.

filter efficiency — the efficiency of a filter based on the amount of particles collected compared to the amount of particles which could be collected.

filter element — a filter cartridge or that part of a diatomite filter on which the filter aid is deposited.

filter media — any fine grain material carefully graded to size which entraps suspended particles as water passes through.

filter rate — the rate of flow of water through a filter during the filtering cycle expressed in gallons per square of effective filter area.

filter rock — graded, rounded rock, and/or gravel used to support filter media.

filter room — the area which houses the recirculation and filtration equipment, and chemical treatment units with the exception of gas chlorinators, chemicals, and supplies for repairs.

filter run — the time of filter operation between backwash procedures.

filter sand — a type of filter media composed of hard, sharp silica, quartz, or similar particles with proper grading for size and uniformity.

filter septum — the part of the filter in which diatomaceous earth or filter media is deposited.

filtrate — the liquid which is passed through a filter.

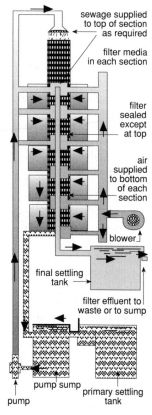

Filtration
(showing controlled filtration)

filtration — the process in which suspended matter is removed from a liquid by passing the liquid through a porous material.

final cover — material at a landfill that is used to permanently seal the surface.

final order — (*law*) a court order stating the final judicial action.

final sedimentation — a technique for thickening of sludge.

Final Soil Cover

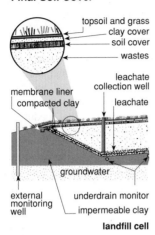

landfill cell

final soil cover — the upper two feet of compacted, graded earth on top of the buried solid waste, used to prevent insect and rodent problems, or as a base for top soil and grass.

financial impact analysis — a study of the probable changes in the financial structure of a system which may result from a proposed or impending action.

financing — the source and method of acquiring monies for a specified use.

fines — fine particulates.

fine texture — the texture exhibited by soils having clay as a part of their textural class name.

finfish — that portion of the aquatic community made up of the true fishes as opposed to invertebrate shellfish.

firebox — a chamber that contains a fire; a furnace.

firebrick — refractory brick made from fireclay able to withstand high temperatures.

firedamp — the accumulation of an explosive gas, usually methane, in a mine.

fire point — the lowest temperature at which a volatile, combustible material can evolve enough vapors to support combustion.

fire retardant — a treatment or coating applied to interior-finished building materials that reduces the rate of flame spread from that of the material in its natural state.

firing rate — the rate of fuel usage in boilers or other direct-fired process equipment expressed in terms of energy equivalent as BTU/hr.

first law of thermodynamics — the law stating that energy cannot be created or destroyed, only converted from one form into another; mechanical energy may be converted into heat or heat into mechanical energy. Also known as conservation of energy.

first responder — the individual who is first to arrive at the scene of a medical emergency and is capable of rendering basic life support.

fiscal year — any twelve-month period for which annual accounts are kept.

fish — any edible, commercially distributed fresh or salt water member of the animal kingdom classed as fish (*Pisces*).

fish processing plant — any food establishment or portion thereof in which fish or shellfish are handled, processed, or packed for market or shipment as fresh fish or fresh or perishable shellfish.

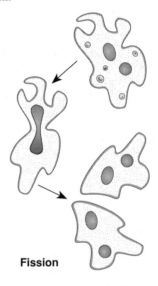

Fission

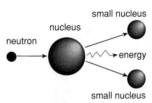

Nuclear Fission

fission — (*biology*) a method of asexual reproduction in single-celled organisms in which the cell divides in half and each half develops into a new organism exactly like the original organism; (*physics*) the splitting of an atom nuclei accompanied by the release of great quantities of energy.

fixative — an agent used in preserving a histological or pathologic specimen so as to maintain the normal structure of its constituent elements.

fixed carbon — the ash-free carbonaceous material that remains after volatile matter is driven off during the proximate analysis of a dry solid waste sample.

fixed grate — a grate that has no moving parts. Also known as a stationary grate.

fixed location monitoring — the sampling of an environmental or ambient medium for pollutants at one location continuously or repeatedly over time.

fixed noise source — a stationary device which creates sounds while fixed or motionless, including but not limited to residential, agricultural, industrial, and commercial machinery and equipment, pumps, fans, compressor air conditioners, and refrigeration equipment.

Flagella →

flagellum — a flexible whip-like appendage on motile cells used as an organ of locomotion.

flame ionization detector (FID) — a direct-reading physical instrument where gas is fed into a stainless steel burner mixed with hydrogen and then burned with air or oxygen; loop of platinum is set about 6 millimeters above the tip of the burner; the current carried across the electrode gap is proportional to the number of ions generated during the burning of the gas.

flame photometry — the spectrochemical measurement made of light emitted from certain chemicals excited by a flame.

flameproofing material — a chemical that catalytically controls the decomposition of material at the flame temperature.

flame reactor process — a hydrocarbon-fueled flash smelting system that treats residues and wastes containing metals.

flammable — the capability of a substance to be set on fire or support combustion easily.

flammable aerosol — an aerosol that when tested by federally regulated methods yields a flame projection exceeding 18 inches at full valve opening or a flashback at any degree of valve opening.

flammable chemical — any aerosol, liquid, gas, or solid capable of combustion.

flammable gas — a gas that at ambient temperature and pressure forms a flammable mixture with air at a concentration of 13% or less by volume.

flammable limit — a minimum concentration of a substance below which propagation of flame does not occur on contact with a source of ignition. Also known as explosive limit.

flammable liquid — any liquid having a flash point below 100°F except any mixture having components with flash points of 100°F or higher, the total of which make up 99% or more of the total volume of mixture; a liquid which gives off combustible vapors.

flammable range — the difference between the lower and upper flammable limits, expressed in terms of percentage of vapor or gas by volume range. Also known as explosive range.

flammable solid — a solid other than a federally defined blasting agent or explosive that is liable to cause fire through friction, absorption of moisture, spontaneous chemical change, retention of heat from manufacturing or processing, or which can be ignited readily and burns so vigorously and consistently as to create a serious hazard.

flange — a projecting rim of an organism or mechanical part.

flare — a fire primarily used to control hydrocarbon gases released intermittently in certain refinery and petrochemical processes; causes smokeless combustion of gases.

flashback — a flame from a torch which burns back into the tip, the torch, or the hose. Also known as backfire.

flash blindness — temporary loss of sight caused by an intense light source.

flashover — the arcing between conductors when the local electric fields exceed the dielectric's strength of air.

E
F

flash point (Fl.P) — the minimum temperature at which a liquid gives off a vapor in sufficient concentration to ignite when tested by specific tests established by the American National Standards' Methods; test conditions can be either open- or closed-cup.

flash point minimum — the lowest temperature at which a chemical will give off fumes able to burst into flames.

flat — a tin can with both ends concave; the can remains in this condition even when it is brought down sharply on its end on a solid, flat surface; possibly caused by a loss of vacuum or in storage.

Flat Grain Beetle (*Cryptolestes pusillius*) — one of the smallest beetles found in stored grain; a tiny reddish-brown beetle about 1/16″ long with antennae nearly as long as the insect.

flatulence — distention of the stomach or intestines with gas.

flatware — eating and serving utensils that are more or less flat such as forks, knives, and spoons.

flatworm — *See* fluke.

flavivirus — an arbovirus spread by mosquitos and ticks which may cause fever, encephalitis, aseptic meningitis, rashes, hemorrhagic fever, and hepatitis.

flavored milk — a beverage consisting of milk to which has been added syrup or flavor made from wholesome-like ingredients.

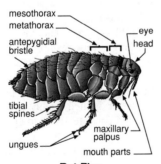

Rat Flea
(*Xenopsylla cheopis*)

flea — a small, wingless, blood sucking ectoparasite that may act as a disease carrier.

flipper — a tin can that normally appears flat, though when brought down sharply on its end on a flat surface, flips out one end; when pressure is applied to this end, it flips in again and can appear flat.

floc — (*sewage*) a clump of solid formed in sewage by biological or chemical action. Also known as flocculent.

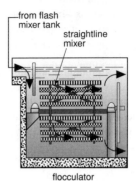

Flocculation

flocculation — the process by which suspended colloidal or very fine particles are assembled into large masses or floccules which eventually settle out of suspension; the stirring of water after coagulant chemicals have been added to promote the formation of particles that will settle; (*sewage*) the process by which clumps of solids in sewage are made to increase in size by chemical, physical, or biological action.

flocculent — a compound, usually some type of alum, used with sand-type filters to form a thin layer of gelatinous substance on the top of the sand. Also known as floc.

flood base elevation — the elevation in terms of the United States Geological Survey mean sea-level that establishes the 100-year flood limit.

flood hazard area — any area composed of flood plane land.

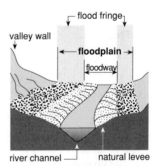

Flood Plain

flood plain — the area contiguous to a lake, water course, stream or stream bed, depressional pocket or area, the elevation of which is greater than the normal water level or pool elevation but equal to or lower than the flood base elevation; the part of a stream valley which is covered with water during the flood stage.

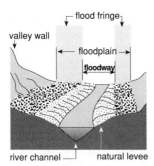

Floodway

floodway — a channel built to carry excess water from a stream.

floppy disk — *See* diskette.

flora — plant life present in a given environment.

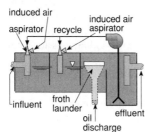

induced air flotation

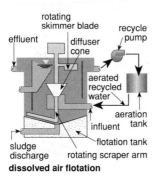

dissolved air flotation

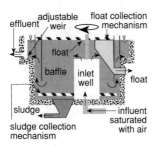

circular dissolved air flotation unit

Flotation

flotation — the process used to separate certain ores from rock by wetting.

flow coefficient — a correction factor used for calculating volume flow rate of a fluid through an orifice, including the effects of contraction and turbulence loss, the compressibility effect, and the effect of an upstream velocity other than zero.

flow diversion valve (FDV) — (*milk*) a three-way valve automatically controlling product flow where forward flow position occurs when recorder controller microswitch preset to operate at 161°F energizes a solenoid-operated air valve, permitting compressed air to flow across a diaphragm, forcing it to sit securely on diverted flow line.

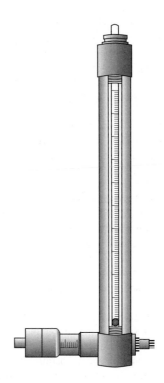

Flowmeter

flowmeter — an apparatus for measuring the rate of flow of liquids or gases.

flow rate — the volume per unit time given to the flow of a gas or other fluid substance which emerges from an orifice, pump, turbine, or passes along a conduit or channel.

flow technique — the flow of a substance from the point of delivery to the point of production of the product and by-products, as well as waste products and ultimate disposal; importance of the flow technique is to determine the potential problems that may occur at each point of the flow of the process and how to deal with them in an appropriate manner.

Fl.P — *See* flash point.

flue — a conduit made of masonry material or other approved non-combustible heat-resisting material and that is used to remove the products of combustion from solid, liquid, or gaseous fuel.

flue fed incinerator — an incinerator that is charged through a shaft that functions as a chute for waste and has a flue to carry away the products of combustion.

flue gas — waste gas from a combustion process.

flue gas desulfurization (FGD) — any pollution control process which treats stationary source combustion flue gas to remove sulfur oxides.

flue gas scrubber — a piece of equipment that removes fly ash and other objectionable materials from flue gas by the use of sprays, wet baffles, or other means that require water as the primary separation medium.

fluid — the gas or liquid phase of matter in which molecules are able to flow past each other without limit.

fluid balance — a state in which the volume of body water and its solutes, both electrolytes and non-electrolytes, is within normal limits and there is normal distribution of fluids within the intracellular and extracellular compartments.

fluidized-bed combustion — a method of burning particulate fuel, such as coal, in which high pressure air is forced through a mixture of crushed coal and limestone particles, causing the burning fuel to move like a boiling liquid, where the limestone combines with the sulfur oxide released in combustion to form calcium sulfate.

fluidized bed technique — a combustion process in which heat is transferred from finely divided particles, such as sand, to noncombustible materials in a combustion chamber; the materials are supported and fluidized by calm or moving air.

fluke — the common name for more than 40,000 parasitic organisms of the class *Trematoda*, characterized by a body that is usually flat and often leaf-like; can infect the blood, liver, intestines, and lungs. Also known as parasitic flatworm.

flume — a natural or synthetic channel that diverts water from a natural course for power, industry, and so on.

fluorescein — $C_{20}H_{12}O_5$ — a yellow or red crystalline dye with a bright yellow-green fluorescence in alkaline solution.

fluorescence — the emission of electromagnetic radiation, usually visible, as the immediate result of absorption of energy from another source.

fluorescent screen — a screen coated with a fluorescent substance that emits light when irradiated with X-rays.

fluoridation — the addition of fluoride to the municipal water supply at a rate of one part per million as part of the public health program to prevent or reduce the incidence of dental caries and strengthen brittle bones.

fluoride — NaF and Na_3AlF_6 — NaF is an odorless white powder or colorless crystals; pesticide grade is often dyed blue; Na_3AlF_6 is a colorless to dark odorless solid which loses color on heating; MW: 42.0/209.9, BP: 3099°F/decomposes, Sol: 4/0.4%, sp gr: 2.78/2.90. It is used in the manufacture of cryolite from phosphate rock processing; in the formulation of insecticides; in the chemical industry; in the treatment of water; in the manufacture of fiberglass, optical glass, lenses, and pottery; in the manufacture of welding rods and welding fluxes. It is hazardous to the eyes, respiratory system, central nervous system, skeleton, kidneys, and skin; and is toxic by inhalation, ingestion, and contact. Symptoms of exposure include nausea, abdominal pain, diarrhea, excessive salivation, thirst, sweating, stiff spine, dermatitis, calcification of the ligaments of the ribs and pelvis, irritation of the eyes and respiratory system. OSHA exposure limit (TWA): 2.5 mg/m³ [air].

fluorine — F_2 — pale yellow to water-white nearly odorless liquid or gas (above 75°F) and member of the halide family; MW: 38.0, BP: –307°F, Sol:

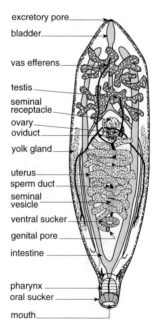

Fluke

reacts, atomic weight: 9. It is used during organic and inorganic synthesis in the production of fluorine compounds, refrigerants, plastics, incendiary devices, and electrolytic solvents; as a rocket fuel oxidizer for ammonia, hydrazine, hydrogen, JP-4, and diborane fuels. It is hazardous to the respiratory system, skin, and eyes; in animals: the liver and kidneys; and is toxic by inhalation and contact. Symptoms of exposure include laryngeal spasms, bronchial spasms, pulmonary edema, eye and skin burns, irritation of the eyes, nose, and respiratory system; in animals: liver and kidney damage. OSHA exposure limit (TWA): 0.1 ppm [air] or 0.2 mg/m³.

fluoroacetamide — (*rodenticides*) causes convulsions, cyanosis, ventricular fibrillation, and death. Also known as 1081.

fluorocarbon — a hydrocarbon, such as Freon, in which all or part of the hydrogenations have been replaced by fluorine; used as a propellant in aerosols, refrigerant, and solvent; thought to be modifying the ozone layer by allowing increased levels of harmful solar radiation to reach the earth's surface.

fluorography — the photographic examination by means of a device which exposes deep body structures by using X-rays produced on a fluorescent screen.

fluorophotometry — the measurement of light given off by fluorescent substances.

fluoroscope — a fluorescent screen mounted with respect to an X-ray tube for ease of direct observation of internal organs or other internal structures in materials; a device for examining deep tissues by means of roentgen rays.

fluorotrichloromethane — CCl_3F — a colorless to water-white, nearly odorless liquid or gas (above 75°F); MW: 137.4, BP: 75°F, Sol (77°F): 0.1%, sp gr: 1.47. It is used as a propellant in aerosols for insecticides, floor waxes, paint, cosmetics, and perfumes; as a refrigerant; as a blowing agent in foam plastics, as a solvent and degreaser; in the production of polymeric resins; as a dielectric fluid in bubble chambers and in wind tunnels; as a sulfonation solvent in chemical synthesis. It is hazardous to the cardiovascular system and skin; and is toxic by inhalation and contact. Symptoms of exposure include incoordination, tremors, dermatitis, frostbite, cardiac arrhythmia, and cardiac arrest. OSHA exposure limit (TWA): ceiling 1,000 ppm [air] or 0.2 mg/m³.

flushometer valve — a device activated by direct water pressure that discharges a predetermined quantity of water to fixtures for flushing purposes.

flush valve — a device located at the bottom of the tank of flushing water closets and similar fixtures.

flush water closet — a toilet bowl that is flushed with water supplied under pressure and equipped with a water sealed trap above floor level.

flux — a material or mixture of materials used in soldering and welding that causes other compounds with which it comes into contact to fuse at a temperature lower than their normal fusion temperatures and to prevent the formation of oxides; a substance that helps to melt and remove the solid impurities as a slag.

flux-cored arc welding — an arc welding process that produces coalescence of metals by heating them with an arc between a continuous filler metal electrode and the work.

flux density — (*neutron*) an expression of the number of neutrons entering a sphere of unit cross-sectional area in unit time.

fly — *See musca domestica.*

fly ash — fine, solid particulate refuse including ash, charred paper, cinders, dust, soot, or other partially incinerated matter carried in a gas stream from a furnace.

f/m³ — *See* fibers per cubic meter.

F/M — notation for food to microorganism ratio.

FM — *See* frequency modulation.

FML — acronym for flexible membrane liner.

foam — a colloidal dispersion of gas bubbles on the surface of a liquid.

foam resins — plastic materials rendered light and porous by the inclusion of minute gas bubbles.

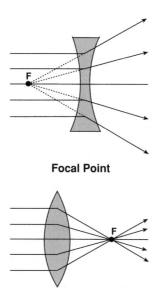

Focal Point

focal point — the point to which a lens or mirror converges parallel rays of light.

focus — the site of a disease process; (*optics*) to move an optical lens toward or away from an object to obtain the sharpest image.

fog — suspended liquid particles formed by condensation of vapor in sufficient quantities to reduce visibility.

fog chamber — *See* cloud chamber.

fogging — (*pesticides*) the application of a pesticide chemical by rapidly heating the liquid form, resulting in very fine droplets with the appearance of smoke.

foliage — the leaves, needles, and blades of plants.

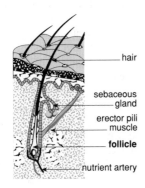

Hair Follicle

follicle — a sac or pouch-like depression or cavity.

folliculitis — an infection of a hair follicle often caused by a natural or industrial oil obstruction.

fomite — an inanimate object that can harbor or transmit pathogenic organisms.

food — any substance ingested by an organism which yields material for energy, growth, and repair of tissue, and regulation of the life processes.

food additive — a substance, usually a chemical, added to the production, processing, packaging, or cooking of food to preserve, protect, or enhance flavor, nutritive content, or appearance.

Food and Agricultural Organization (FAO) — the agency of the United Nations that institutes and administers programs, especially in underdeveloped countries, for improving farming methods and increasing food production.

Food and Drug Administration (FDA) — a federal agency within the Department of Health and Human Services authorized to regulate food additives, contaminants, drugs, cosmetics, and potential carcinogens related to food additives, drugs, and cosmetics.

foodborne bacterial poisoning — a bacterial poisoning resulting from the ingestion of food on or in which the organisms have grown and produced toxins.

foodborne chemical poisoning — a chemical poisoning characterized by a short interval varying from ten minutes to two hours after ingestion of the food containing a specific chemical or chemicals and the onset of symptoms.

foodborne disease outbreak — an incident in which 2 or more persons have a similar illness after ingesting a common food when the epidemiological analysis implicates the food as the source of the illness; a single case of illness, such as one case of botulism or chemical food poisoning.

foodborne infection — a disease caused by the ingestion of food in which large numbers of specific microorganisms have grown.

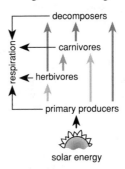

Food Chain

food chain — the feeding of organisms in an ecological community based on trophic levels which result in the transfer of food energy from plants or organic detritus organisms.

food contact surfaces — those surfaces, equipment, and utensils which normally come in contact with food.

food handler — any individual who comes into contact with food or utensils used in connection with the storage, manufacture, preparation, handling, sale, or service of food.

food infection — the introduction of infectious organisms into the body by means of food.

food poisoning — a group of acute illnesses due to ingestion of contaminated food from toxins inherent in the food or formed by bacteria; usually causes inflammation of the gastrointestinal tract, pain, cramps in the abdomen, nausea, vomiting, diarrhea, weakness, and dizziness; the disease is relieved when the toxin leaves the system.

food preservation — processing designed to keep food in a clean, fresh, and wholesome manner in such a way that it will not lose nutritive value or acquire injurious properties.

food processing establishment — a commercial establishment in which food is manufactured or packaged for human consumption.

food processing waste — the combination of wastes resulting from the growing, harvesting, processing, and packaging of fruits, vegetables, and other food crops.

food protection — a technique used to prevent food-borne disease consisting of proper plans review, development, and use of adequate facilities, procurement and use of correct equipment, necessary cleanliness, proper storage, preparation and serving of food, removal of infected individuals, and adequate disposal of food remnants.

Food, Science, and Technology Abstracts — a database from 1969 to the present which provides access to research and new development literature; also covers allied disciplines such as agriculture, chemistry, biochemistry and physics, some related disciplines such as engineering and home economics when relevant to food science; coverage of over 1,200 journals from over 50 countries, patents from 20 countries, and books in any language.

food service establishment — any place where food is prepared and intended for individual portion service.

food spoilage — deterioration of the color, texture, odor, or flavor of a food caused by chemicals, including enzymes, bacteria, molds, and yeast.

food-vending vehicle — any conveyance used by a food vendor offering food for human consumption.

food vendor — any person who sells, offers for sale, or gives away from any vehicle, container, or public market any article of food for human consumption.

food web — the many connected food chains by which the organisms of a particular ecological community obtain resources of energy.

fool's gold — *See* pyrite.

foot (ft) — the unit of measure equivalent to 0.3048 meters or 12 inches.

foot bath — a shallow water area between bath house showers and pool deck through which patrons must walk.

footcandle (fc) — the unit of illumination equal to one lumen/ft.^{2}; it is the brightness of light one foot from the light source of one candela.

footlambert (ft-l) — a unit of brightness, one lumen uniformly emitted or reflected by an area of one square foot.

foot-pound (ft-lb.) — the unit of work done by a force of one pound acting through a distance of one foot; equal to approximately 1.355818 joule.

foot-pound of torque — a measurement of the physiological stress exerted upon any joint during the performance of a task. Also known as pound foot.

foot spray — a device for spraying bathers' feet with water or a disinfectant in public bathing areas.

forage — plants grown to feed dairy animals, beef cattle, horses, sheep, and goats.

force (*physics*) — any influence that causes a body to move or accelerate.

forced draft — the positive pressure created by the action of the fan or blower which supplies the primary or secondary combustion air in an incinerator.

forced expiratory volume measure at 1 second (FEV$_1$) — a test of pulmonary function used for routine monitoring.

forced vital capacity (FVC) — the maximum gas volume which can be expired immediately after a maximum expiration; used as a test of pulmonary function.

forecast year — any twelve-month period in the future for which an estimation or calculation is made.

foreseeable emergency — any potential occurrence such as, but not limited to, equipment failure, rupture of containers, or failure of control equipment, which could result in an uncontrolled release of a hazardous chemical into the workplace.

forfeiture of bond — a failure to perform the conditions upon which an obligor was to be excused of the penalty in the bond.

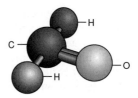

Formaldehyde

formaldehyde — HCHO — a nearly colorless gas with a pungent, suffocating odor: often used in an aqueous solution; MW: 30.0, BP: 207–214°F, Sol: miscible, Fl.P: 140–185°F, sp gr (77°F): 1.08–1.10. It is used in the synthesis of chelating agents and dyes; in textile manufacturing and handling and in tanning operations; in the manufacture of particle board, soft wood plywood, sandpaper, and grinding wheels; as an embalming fluid. It is hazardous to the respiratory system, skin, and eyes; and is toxic by inhalation, ingestion, and contact. Children, the elderly, and individuals with lung disease or poorly functioning immune systems are particularly affected.

E
F

Symptoms of exposure include lacrimation, burning nose and cough, bronchial spasms, pulmonary irritation, dermatitis, irritation of the eyes, nose, and throat; carcinogenic. OSHA exposure limit (TWA): 1 ppm [air].

format — the arrangement of data in a systematic manner.

formic acid — HCOOH — a colorless liquid with a pungent, penetrating odor; often used in an aqueous solution; the 90% solution freezes at 20°F; MW: 46.0, BP: 224°F, Sol: miscible, Fl.P (open cup): 122°F, sp gr: 1.22. It is used in the textile dyeing and finishing industries as a dye exhausting agent; in chrome dyeing, to impart finishes on cotton; as an acidifying agent; as a shrink- and wrinkle-proofing agent; as a chemical intermediate for acids, salts, dyes, fumigants, refrigerants, pharmaceuticals, and solvents; in the leather processing industry as a deliming agent and neutralizer; in the rubber industry as a coagulant for natural rubber latex; in electroplating to control particle size and plating thickness; as an antiseptic in wine and beer brewing; as a preservative in animal feed additives; in miscellaneous operations. It is hazardous to the respiratory system, skin, kidneys, liver, and eyes; and is toxic by inhalation, ingestion, and contact. Symptoms of exposure include eye and throat irritation, lacrimation, nasal discharge, cough, dyspnea, nausea, skin burns, dermatitis. Also known as methanoic acid. OSHA exposure limit (TWA): 5 ppm [air] or 9 mg/m^3.

formulation — (*pesticides*) a mixture of one or more pesticides plus other materials needed to make it safe and easy to store, dilute, and apply.

formula translation — *See* FORTRAN.

FORTRAN (formula translation) — (*computer science*) a high-level programming language used in computer graphics and computer-aided design/computer-aided manufacturing.

fossil fuels — coal, oil, and natural gas; remnants of ancient plants and animals having high carbon or hydrogen contents and typically containing varying levels of sulfur as a contaminant.

fouling — the impedance to the flow of gas or heat that results when material accumulates in gas passages or on heat-absorbing surfaces in an incinerator or other combustion chamber; (*naval architecture*) the adhesion of marine organisms to the underwater parts of ships, causing ships to lose speed.

foundation drain — the portion of a residential drainage system provided to drain only ground-water from outside the foundation of the house or from under the basement floor.

Fourier analysis — (*mathematics*) a method of dissociating time series or spatial data into sets of sine and cosine waves.

fp — *See* freezing point.

FPLA — acronym for Fair Packaging and Labeling Act.

fps — acronym for feet per second.

fractional distillation — the process of separating a mixture of liquids having different boiling points.

fractional efficiency — the percentage of particles of a specific size which are removed and retained by a particular type of collector or sampler.

fractionation — the separation of a mixture into two different portions.

fractionation dose — a method of administering radiation in which relatively small doses are given daily or at longer intervals.

fracto clouds — fragmented or wind-blown clouds.

franchise collection — collection made by a private firm that is given the right by a governmental agency to collect solid waste for a fee.

FRC — *See* functional residual capacity.

free field — (*noise*) the distance from the noise source where the sound pressure level decreases 6 decibels on the A scale for each doubling of distance.

free groundwater — groundwater in aquifers that are not bounded by or confined in impervious strata.

free radical — an electrically neutral fragment of a molecule having at least one impaired electron or an unused chemical bond.

free residual chlorination — chlorination that maintains the presence of hypochlorous acid (HOCl) or hypochlorite ion (OCl$^-$) in water.

freeze-out sampling — the process by which an air sample is drawn through an extremely cold chamber, and condensed contaminants such as hydrocarbons are collected.

freezing point (fp) — the temperature at which a liquid changes to a solid at normal pressure.

freight bill (F/B) — the document stating the transportation charges by the carrier for a shipment; next to the bill of lading, the freight bill is the most authentic document that supports movement of a shipment in interstate commerce.

Freon — the commercial name for chlorofluorocarbon.

freon-12 — *See* difluorodichloromethane.

frequency (f) — the number of waves, vibrations, or cycles that pass any given point in a certain period of time, usually one second expressed in hertz.

frequency modulation (FM) — (*electronics*) a deliberate modulation of the frequency of the transmitting wave in broadcasting in order to agree with the sounds or images being transmitted to reduce static.

fresh meat — regular retail cuts of beef, veal, lamb, and pork.

friable — easily crumbled or pulverized.

friction — the force opposing the motion of a body relative to another with which it is in contact.

friction factor — a factor used in calculating loss of pressure due to friction of a fluid flowing through a pipe or duct.

friction head — pressure lost by the flow in a stream.

friction loss — the loss in energy of a substance when it encounters resistance along any surface.

front — (*weather*) the boundary at which air masses of different temperature and density collide to produce very unstable weather.

front-end loader — a waste collection vehicle with forward arms that engage a detachable container, move it up over the cab, enter it into the vehicle's body, and return it to the ground.

frost — a covering of ice caused by the sublimation of water on objects colder than 32°F.

frostbite — injury to tissues due to exposure to extreme cold.

frozen dessert — any clean frozen or partially frozen combination of two or more milk or milk products, eggs or egg products, sugar, water, fruit or fruit juices, candy, nut meats, flavors, color, and harmless stabilizers; includes ice cream, frozen custard, ice milk, milk sherbet, ices, and other products.

frozen food processing plant — any food establishment or portion thereof where food is prepared, processed, packed, stored for sale or shipment, or held in storage as a frozen product.

FRP — acronym for fiberglass-reinforced plastic.

fructose — $C_6H_{12}O_6$ — a sugar whose molecules contain the same number and kind of atoms as the glucose molecule, but in different arrangements; the sweetest of sugars, found in the free state in fruit juices, honey, and nectar.

fruit sherbets — foods which are prepared by freezing while stirring a mix composed of one or more of the optional fruit ingredients, and one or more of the pasteurized dairy ingredients, sweetened with one or more of the optional saccharin ingredients; sugar, dextrose, or corn syrup, with or without added water.

ft — *See* foot.

FTE — *See* full-time equivalency.

ft-l — *See* footlambert.

ft-lb. — *See* foot-pound.

fuel — any material which is capable of releasing energy or power by combustion or other chemical or physical reaction.

fuel cell — a device for converting chemical energy into electrical energy.

fugitive dust — particulate matter composed of soil which is uncontaminated by pollutants resulting from industrial activity.

fugitive emissions — any air pollutants emitted to the atmosphere other than from a stack.

fuller's earth — a hydrated silica-alumina compound used as a filter medium.

full-time equivalency (FTE) — a standard of hours per year used in determining the standard number of employees; the actual time delivering the direct services in the field by one person over the course of a year.

fully-closed position — the closed position of the valve seat which requires the maximum movement of the valve to reach the fully-open position.

fully-open position — the position of the valve seat which permits the maximum flow into or out of the holder.

fulminate — to come suddenly and follow a severe, intense, or rapid course.

fume fever — *See* metal fume fever.

fumes — fine, solid particles deriving from the condensing of the volatization of molten metals from the gaseous state under one micrometer in diameter; they cause metal fume fever, allergic responses, and other respiratory-type problems.

fumigant — a pesticide chemical compound which burns or evaporates to form a gas or vapor that destroys pests.

fumigation — the use of a fumigant to destroy pests.

functional antagonism — an effect occurring when two chemicals counter-balance each other by producing opposite reactions to the same physiological function.

functional group — a relatively reactive group of atoms responsible for the characteristic reactions of a compound such as –COOH, or –OH, or $–NH_2$. These may be attached to a relatively inert hydrocarbon group and their presence imparts characteristic properties to the molecule.

functional residual capacity (FRC) — the volume of gas remaining in the lungs at the end of a normal expiration.

E
F

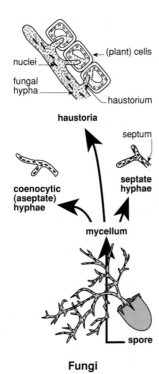

Fungi

fungi — spore-bearing plants lacking chlorophyll, such as mushrooms, molds, and yeast.

fungicide — a chemical that kills the non-green microscopic plants, fungi.

fungus — singular of fungi.

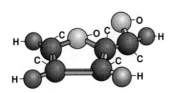

Furfural
(2–Furancarboxaldehyde)

furfural — C_4H_3OCHO — a colorless to amber liquid with an almond-like odor; darkens in light and air; MW: 96.1, BP: 323°F, Sol: 8%, Fl.P: 140°F, sp gr: 1.16. It is used in the manufacture of reinforced plastic products and surface coatings; during *in situ* desludging and decarbonizing of internal combustion engines; during cold molding of abrasive grinding wheels; during the molding of friction materials such as brake linings and clutch facings; as an accelerator in the vulcanization of rubber. It is hazardous to the eyes, respiratory system, and skin; and is toxic by inhalation, absorption, ingestion, and contact. Symptoms of exposure include headache, dermatitis, irritation of the eyes and upper respiratory system. OSHA exposure limit (TWA): 2 ppm [skin] or 8 mg/m³.

furfuryl alcohol — $C_4H_3OCH_2OH$ — an amber liquid with a faint, burning odor; MW: 98.1, BP: 338°F, Sol: miscible, Fl.P: 149°F, sp gr: 1.13. It is used during the manufacture of cements, molded high-density carbon and graphite articles; in sand consolidation for oil and gas recovery operations. It is hazardous to the respiratory system; and is toxic by inhalation, absorption, ingestion, and contact. Symptoms of exposure include dizziness, nausea, diarrhea, diuresis, respiratory and body temperature depression, vomiting; in animals: drowsiness and eye irritation. OSHA exposure limit (TWA): 10 ppm [skin] or 40 mg/m³.

furnace — an enclosed combustion chamber in which heat is produced for the purpose of affecting a physical or chemical change.

furnace brazing — a process in which the parts to be joined are placed in a furnace heated to a suitable temperature.

furuncle — a localized inflammation by a bacterium in a hair follicle or skin gland that discharges pus and has a central core of dead tissue. Also known as a boil.

fusion — the changing of the state of a substance from solid to liquid by heating. Also known as melting.

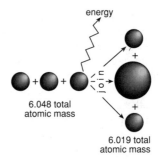

Nuclear Fusion

fusion point — the temperature of a plasma above which the rate of energy generation by nuclear fusion reactions exceeds the rate of energy loss from the plasma, so that the fusion reaction can be self-sustaining.

FVC — *See* forced vital capacity.

FWPCA — *See* Federal Water Pollution Control Act.

G

g — *See* gram.

G — *See* gauss.

gabbro — (*geology*) any of a group of fine- to course-grained, greenish-gray to black, plutonic, igneous rock containing pyroxene.

gal — *See* gallon.

galaxy — a large-scale concentration of stars, gas, and dust.

gamete — a reproductive cell that fuses with another gamete to form a zygote.

gamma — one nanotesla; a unit of magnetic field strength.

gamma globulin — a blood protein sometimes used to impart temporary immunity to people for certain infectious diseases.

Galena

galena (PbS) — a bluish-gray lead sulfide mineral which is heavy, brittle, and breaks into cubes.

Galileo spacecraft — a spacecraft launched October 18, 1989, sent deep into space and designed to orbit the planet Jupiter in December 1995, while sending probes into its atmosphere to determine the nature and compostition of the planet and atmosphere.

gallon — the liquid measurement equal to 231 cubic inches, 128 fluid ounces, 3.7853 liters, or 4 quarts.

galvanic action — the creation of an electrical current by electro-chemical action.

galvanic cell — an electrolytic cell capable of producing electrical energy by electrochemical action.

galvanic corrosion — electrochemical corrosion of metals which occurs when two or more dissimilar metals are immersed in an electrolyte.

galvanizing — the deposition of a protective coating for metals by dipping them into a bath of molten zinc.

galvanometer — an instrument used to detect small electric currents by means of a coil of wire which pivots between the poles of a magnet.

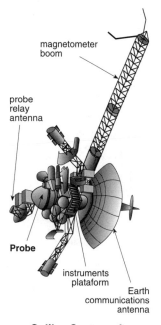

Galileo Spacecraft

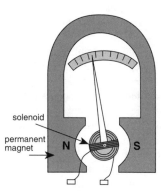

Weston Galvanometer

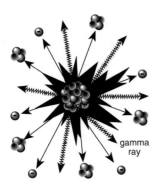

Gamma Ray

gamma ray — radiant electromagnetic energy of a very short wavelength (about 10^{-8} to 10^{-11} cm) capable of penetrating most substances and given off by radioactive substances and during fission and fusion reactions.

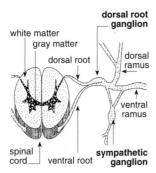

Ganglion

ganglion — a mass of nerve tissue containing cell bodies and synapses lying outside of the central nervous system in vertebrates.

gangrene — the death of body tissue, usually in considerable mass, caused by loss of blood supply and followed by bacterial invasion and putrefaction.

GAO — *See* General Accounting Office.

gap — (*geographic information system*) the distance between two graphic entities on a digitized map.

garbage — all putrescible animal and vegetable waste, except sewage and body waste.

gas — (*physics*) a substance characterized by very low density and viscosity compared with liquids and solids; a fluid that has neither independent shape nor volume, but tends to expand indefinitely. *See also* gasoline.

gas absorption — (*chemistry*) the solution of one component of a gaseous mixture in a liquid or the absorption onto a surface; selective absorption of gases from a mixture is a means of separating them for testing or collection.

gas barrier — any device or material used to divert the flow of gases produced in a sanitary landfill, or by other land disposal techniques.

gas chromatography (GC) — a type of automated chromatography in which the mobile phase is separated in columns and measured by a detector; a direct-reading physical instrument principle where the components of a complex mixture are separated physically because of the varying infinities of the components for different packing materials.

gas flammable — a gas that at ambient temperatures and pressures forms a flammable mixture with air at a concentration of 13% by volume or less.

gas-forming bacteria — (*milk*) microorganisms which ferment lactose, producing both acid and gas, and a smooth, gelatinous curd which often contains gas bubbles.

gasification — any chemical or heat process of converting a solid or liquid fuel into a gaseous fuel.

gas ionization instruments — instruments based on the principle of collecting ions formed by the action of ionizing radiation in a gas.

gas-liquid chromatography (GLC) — a form of gas chromatography in which the substances to be separated are moved by an inert gas along a tube filled with a finely divided inert solid coated with a nonvolatile oil. Each component migrates at a rate determined by its solubility in oil and its vapor pressure.

gas metal arc welding — a process that produces coalescence of metals by heating them with an arc between a continuous filler metal electrode and the work, using an inert gas to prevent oxidation of the weld.

gasohol — mixture of gasoline and ethanol used as a fuel.

gasoline — any petroleum distillate having a Reid vapor pressure of 4 pounds or greater which is produced for use as a motor fuel; derived from crude petroleum. Commonly known as gas.

gasometric method — an analytic technique in which a substance is isolated in gaseous form or is adsorbed from a gaseous mixture.

gas sorption — a piece of equipment used to reduce levels of airborne gaseous compounds by passing the air through materials that extract the gases.

gas stream — the air, clean or polluted, that is present during a production process or combustion and is eventually vented to the atmosphere.

gastric acid — the digestive acid in the stomach secreted by glands in the stomach wall.

gastritis — inflammation of the lining of the stomach.

gastroenteritis — inflammation of the lining of the stomach and intestine due to psychological or emotional upset, allergic reactions to certain foods, irritation to alcohol, and microorganisms.

gastroenterologist — an internist specializing in the diagnoses and treatment of the stomach and gastrointestinal tract.

gastrointestinal (GI) — pertaining to the stomach and intestine.

gas-tungsten arc welding — a process that produces coalescence of metals by heating them with an arc between a tungsten electrode and the work.

gauge pressure (psig) — the pressure measured within a piece of equipment compared to the atmospheric pressure.

gauss (G) — the cgs unit of magnetic field replaced by the m.k.s. tesla with one tesla equal to 10^4 gauss.

gavage — forced feeding, especially through a tube passed into the stomach. Also known as tube feeding.

Gay-Lussac law — *See* Charles' law.

GC — *See* gas chromatography; General Council.

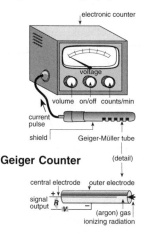

Geiger Counter

geiger counter — a radiation counter using a gas-filled tube that indicates the presence of ionizing particles.

Geiger-Müeller counter (GM counter) — a highly sensitive, gas-filled, radiation-measuring device which operates at voltages sufficiently high to produce ionization.

gel — a colloidal dispersion of a solid and a liquid, whose dispersed substance coalesces in thin films forming a jelly-like material.

gene — a segment of deoxyribonucleic acid nucleus of a cell that contains information for the synthesis of a specific protein or for determining the inheritance of a particular trait.

General Accounting Office (GAO) — a federal accounting office established by the Budget and Accounting Act of 1921 to independently audit government offices; under the control and direction of the Comptroller General of the United States.

General Council (GC) — a group of attorneys employed by the Department of Health and Human Services and assigned to handle litigation of lawsuits brought by the Food and Drug Administration to assure the legality of proposed regulations.

general exhaust — a system for exhausting contaminated air in a general work area.

general obligation bond — a bond issued by a municipality backed by their basic taxing authority for the purpose of raising capital for programs, such as wastewater treatment; there is a fixed rate of interest and fixed maturities.

General Science Index — a bibliographical index available for journals on general science papers from 1978 to the present.

general ventilation — mechanical ventilation applied to a room or an area for the purpose of climate control and dilution to safe levels of hazardous chemical concentration.

generation rate — the quantity of solid waste that originates from a defined activity.

generation time — the interval between receipt of infection by a host and maximal communicability of a host; an important factor in determining the rapidity of person-to-person spread of a disease.

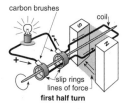

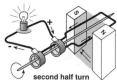

Alternating Current Generator

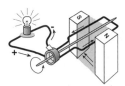

Direct Current Generator

generator — (*electricity*) a device that changes mechanical energy into electrical energy; (*waste*) a facility whose act or process produces hazardous waste.

G
H

generic — a term relating to or descriptive of an entire group, class, or general category.

genetically significant dose (GSD) — the dose of radiation which if received by every member of the population would be expected to produce the same total genetic injury to the population as would the actual dose received by an individual producing a genetic injury.

genetic code — the sequential arrangement of the bases in the deoxyribonucleic acid molecules which control the traits of an organism.

genetic diversity — the amount of genetic variability among individuals of a single species.

genetic effect of radiation — inheritable change, chiefly mutations, produced by the absorption of ionizing radiation by male and female germ cells.

genetics — a branch of biology dealing with the phenomenon of heredity and variation.

genetic screening — testing to identify individuals at risk of having children with specific genetic disorders.

genome — the complete set of hereditary factors contained in the haploid set of chromosomes.

genotoxic — pertaining to a chemical that causes adverse effects in the genetic material of living organisms.

genotoxic carcinogens — those carcinogens which affect genetic material only.

genotoxicity — the kind of toxin specifically causing chromosomal change.

genotype — all or part of the genetic constituents of an organism.

genus — (*classification system*) a taxonomic subgrouping of plants or animals of related species.

geocoding — (*geographic information system*) the activity of defining the position of geographic objects relative to a standard reference grid.

Geographic Information Manipulation and Mapping Systems (GIMMS) — a low-cost polygon mapping geographic information system.

geographic information system (GIS) — a computer-based system used to store and manipulate geographic information linking the location of objects to the characteristics of objects such as soil types, hazardous waste disposal sites, and ground water supplies.

geomagnetism — (*geophysics*) the science concerned with the magnetism of the earth.

geometric isomers — molecules having the same types of bond in the same order but having atoms with different spatial relationships. Also known as CIS-trans isomers.

geophysics — the study of the earth's physical properties, including the air and space.

Geostationary Operational Environmental Satellite (GOES) — a satellite that remains fixed at 22,300 miles above a point on the equator, takes photographs once every 30 minutes day and night, and provides up-to-the-minute data used by weather forecasters around the world; it moves at a speed synchronous with the earth's rotation.

geothermal energy — an energy produced by tapping the earth's internal heat. At present, the only available technologies to do this are those that extract heat from hydrothermal convection systems, where water or steam transfer the heat from the deeper part of the earth to the areas where the energy can be tapped. The amounts of pollutants found in geothermal vary from area to area but may contain arsenic, boron, selenium, lead, cadmium, and fluorides. They also may contain hydrogen sulphide, mercury, ammonia, radon, carbon dioxide, and methane. They may cause contamination of surface water from the wastewater, elliptical dish shaped depression in the ground, enhance seismic activity, and create noise levels as high as 120 dBA.

geothermal heat pump system — a system which uses a condenser circuit that is either in the earth, in the case of a closed-loop system, or a water supply, in the case of an open-loop system, to provide or remove heat from the dwelling.

g-eq — *See* gram equivalent.

geriatrics — the study and treatment of the biological and physical changes and the diseases of old age.

germ — a pathogenic microorganism.

germicide — an agent capable of killing microorganisms.

germination — the process of development of a seed or spore.

gerontology — the scientific study of the aging process.

gestation period — the period of intrauterine development in mammals from fertilization to birth.

geyser — a natural, thermal spring which periodically discharges its water or steam with enormous power.

GI — *See* gastrointestinal.

Giardia — a genus of flagellate protozoa parasitic in the intestines of people and animals which may cause protracted, intermittent diarrhea with symptoms suggesting malabsorption.

giardiasis — (*disease*) a protozoan infection usually of the upper small intestine which may be asymptomatic or associated with intestinal symptoms such as chronic diarrhea, abdominal cramps, fatigue, weight loss, poor absorption of fats or fat-soluble vitamins; incubation time is 5–25 days or longer, usually 7–10 days. It is caused by *Giardia lamblia*, a flagellate protozoan; found worldwide with children being infected more frequently than adults, especially in areas of poor environmental control and in institutions including day-care centers; may be the cause of waterborne outbreaks of disease in the United States. The reservoir of infection includes people and possibly wild or domestic animals. It is transmitted by ingestion of cysts in water or food contaminated with feces or by person-to-person transmission through the oral-fecal route; communicable for the entire period of the infection, the asymptomatic carrier rate is high and therefore susceptibility may be sizeable. It is controlled by the proper protection and processing of water and food, the proper sanitary disposal of feces, and good personal hygiene.

GIMMS — *See* Geographic Information Manipulation and Mapping Systems.

gingivitis — an inflammation of the gums.

GIS — *See* geographic information system.

g/kg — *See* grams per kilogram.

glacial drift — the unconsolidated mixture of gravel and partly weathered rock fragments left by glaciers.

glacial drift aquifers — geological formations found in those parts of the United States which were covered by the advance of glaciers during the ice age. As the climate cooled and the glaciers advanced, the ice packed aerobic soil and bedrock. When temperatures warmed, the glaciers melted, the material was redeposited, and some of the water was trapped from the melted glacier.

Glacial Erosion

glacial erosion — the covering or alteration of an area by an ice-sheet or by glaciers. Also known as glaciation.

glacial outwash — the partially sorted mixture of gravel, sand, and silt left from the ice melting in a preexisting valley bottom or broadcast in a form similar to an alluvial plain.

glacial till — the unsorted mixture of clay, silt, sand, gravel, and boulders deposited by the ice sheet.

glaciation — *See* glacial erosion.

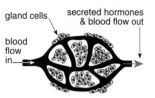

ductless (endocrine) gland

gland cells

secreted hormones & blood flow out

blood flow in

Gland

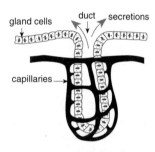

gland cells   duct   secretions

capillaries

Duct (Exocrine Gland)

gland — an organ in animals secreting a particular substance or substances vital to existence.

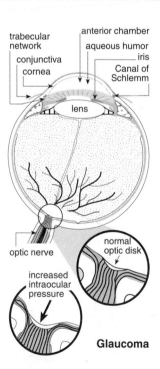

trabecular network

conjunctiva

cornea

anterior chamber

aqueous humor

iris

Canal of Schlemm

lens

optic nerve

increased intraocular pressure

normal optic disk

Glaucoma

glaucoma — a condition of the eye where the fluid that normally fills the eyeball fails to drain properly, thereby causing excess pressure to build and damage the optic nerve; caused by aging, infection, injury, and congenital defect; a leading cause of blindness.

GLC — *See* ground level concentration; gas-liquid chromatography.

globe thermometer — a thermometer placed in the center of a metal sphere that has been painted black to measure radiant heat.

globulin — a general term for proteins that are insoluble in water and highly-soluble concentrated salt solutions but are soluble in moderately concentrated salt solutions; all plasma proteins except albumin and prealbumin are globulins.

glove box — a sealed enclosure in which long impervious gloves sealed to the openings are used to handle all items in the box.

glucose — $C_6H_{12}O_6$ — a white, crystalline, monosaccharide; a product of photosynthesis and an end-product of digestion; the most common sugar.

glycerin — *See* glycerol.

glycerol — $CH_2OHCHOHCH_2OH$ — a clear, sweet, colorless, syrup-like liquid which may be solid at lower temperatures. It is used in the manufacture of medicines, explosives, soaps, antifreeze, inks, perfumes, cosmetics, finishes, or as a solvent; one of the soluble products into which fats are broken down in the body by digestion. Also known as glycerin; glycyl alcohol.

glycidol — $C_3H_6O_2$ — colorless liquid; MW: 74.1, BP: 320°F (decomposes), Sol: miscible, Fl.P: 162°F, sp gr: 1.12. It is used in surface coatings; in chemical synthesis; as a chemical stabilizer for natural oils and vinyl polymers; as a demulsifying agent; as a dye leveling agent. It is hazardous to the eyes, skin, respiratory system, and central nervous system; and is toxic by inhalation, ingestion, and contact. Symptoms of exposure include narcosis, irritation of the eyes, nose, skin, and throat. Also known as epihydrin alcohol. OSHA exposure limit (TWA): 25 ppm [air] or 75 mg/m³.

glycogen — a starch-like reserve carbohydrate formed and stored in the liver cells and muscles of all higher animals.

glycol — *See* ethylene glycol.

glycyl alcohol — *See* glycerol.

gm — *See* gram.

GM counter — *See* Geiger-Müeller counter.

GMP — acronym for good manufacturing practice.

goal — an articulation of community expectations of desired obtainable levels of program and system performance; the goal differs from the objective in that it does not have a deadline, and it is usually for a longer term.

goat milk — the lacteal secretion practically free from colostrum obtained by the complete milking of healthy goats; the generic term milk includes goat milk.

GOES — *See* Geostationary Operational Environmental Satellite.

goitrogen — a natural product found in plant foods causing hypothyroidism.

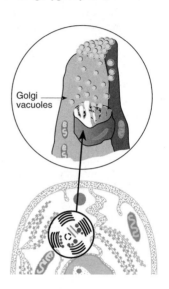

**Golgi Apparatus
in Animal Cell**

Golgi apparatus — a complex cellular organella consisting mainly of a number of flattened sacs and associated vesicles, which is involved in the synthesis of glycoproteins, lipoproteins, membrane-bound proteins, and lysosomal enzymes.

gonads — the male and female primary sex glands.

good soil suitability — the quality of cover soil characterized by favorable properties for use.

gouging — the removal of material by the forming of a bevel or groove.

gpd — acronym for gallons per day.

gpm — acronym for gallons per minute.

GPO Monthly Catalog — a database published by the Government Printing Office from 1976 to the present that covers all types of federal government publications.

gr — *See* grain.

grab sample — a sample representing a larger gross sample taken at a particular place in a brief period of time and produced when a container emptied of liquid or air is used to quickly draw in a small representative sample of a gas.

gradient — the degree of slope or a rate of descent or ascent.

grading — any excavating, filling, or leveling of earth material conducted at a site in preparation of construction, landfilling, or improvements.

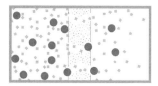

Graham's Law

Graham's law of diffusion — the law that the rate of diffusion of a gas through porous membranes varies inversely with the square root of its density.

grain (gr) — a unit of weight equal to 1/7,000 of a pound, or to 6.479891×10^{-5} kilogram.

grain alcohol — *See* ethanol.

grain per gallon — a measure of solution concentration equal to 17.1 parts per million.

gram (g or gm) — the basic unit of mass (weight) in the metric system equivalent to 15.432 grains or 0.0353 ounces; 454 grams equal one pound.

gram-calorie — the amount of heat necessary to raise the temperature of one gram of water 1°C at the range of 14.5°C to 15.5°C. Also known as calorie.

gram equivalent (g-eq) — the equivalent weight of a substance expressed in grams.

gram formula — *See* mole.

grams per kilogram (g/kg) — a unit used in the measurement of toxicological effects on humans or animals.

Gram's stain — a differential bacteriological staining procedure in which a smear is stained with crystal violet treated with strong iodine solution decolorized with ethanol or ethanol-acetone and counterstained with a contrasting dye. Those retaining the primary stain are gram positive and those with the counterstain are gram negative.

Granary Weevil (*Sitophilus granarius*) — a stored food product insect similar to the Rice Weevil in appearance, but with oval pits on the thorax and with winged covers uniformly dark brown.

grand jury — a jury of inquiry whose duty is to receive complaints and accusations in criminal cases, hear the evidence on the part of the federal or state government, and find bills of indictment in cases where it is satisfied a trial should commence.

grand name — *See* trade name.

granite — light-colored crystalline, coarse, grain, and rock containing approximately 30% quartz and 60% alkali feldspar with plagioclase, biotile, and hornblende.

granular pesticide — a pesticide chemical mixed with or coated with small pellets or a sand-like material.

granular soil — a soil structure in which the aggregates are small, non-porous, and are strongly held together; nearly spherical with many irregular surfaces, usually found in surface soil or A horizon.

granuloma — a mass or nodule of chronically-inflamed tissue with granulations, usually associated with an infection.

graphite (natural) — C — steel gray to black, greasy-feeling, odorless mineral; MW: 12.0, BP: not known, Sol: insoluble, Fl.P: not known, sp gr: 2.0–2.25. It is used in foundry facings; in the manufacture of refractories in crucibles and retorts; in the manufacture of lubricants and adhesives in bearings, slides, gears, engines, and mold release agents; in the manufacture of writing and drafting agents; in the manufacture of industrial paints, ink, and polishes; in the manufacture of motor and generator brushes, batteries, carbon resistors, and electrodes; in the manufacture of tungsten carbide cutting tools and friction materials; in electroplating; in the manufacture of ammunition; miscellaneous uses in hard rubber, engine packing, cord, rope, twine, floor coverings, and coatings for fertilizers. It is hazardous to the respiratory system and cardiovascular system; and is toxic by inhalation, and contact. Symptoms of exposure include cough, dyspnea, black sputum, pulmonary function impairment, lung fibrosis. Also known as black lead; plumbago. OSHA exposure limit (TWA): 2.5 mg/m³ [resp].

GRAS — (*food additives*) acronym for generally recognized as safe.

grass — the common name for all plants with jointed stems; narrow leaves having parallel veins that encircle the stem and produce a seed-like grain as their fruit.

grate — a device used to support the solid fuel or solid waste in a furnace during drying, ignition, or combustion; openings permit combustion air to pass through it.

gravel — rock fragments from 2 millimeters to 64 millimeters in diameter; an unconsolidated mix with sand, cobbles, and boulders.

gravimetric method — a method of chemical analysis in which a substance is isolated in the pure state or as one of its compounds, and its weight is determined with an analytical balance.

G H

gravimetry — a procedure dependent upon the formation or use of a precipitate or residue which is weighed to determine a concentration of a specific contaminant in a previously collected sample.

gravity spray tower — an air pollution control device in which particles in the gas stream are trapped by much larger drops of water sprayed from above causing the particles to be removed from the gas stream.

gray (Gy) — the International System unit of absorbed dose; replaces the term rad. *See also* absorbed dose; 1 Gy = 1 joule/kg — 100 rads.

graywater — wastewater generated by water-using fixtures and appliances, excluding the toilet and possibly the garbage disposal.

grease skimmer — a device for removing floating grease or scum from the surface of wastewater in a tank.

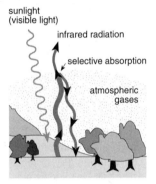

Greenhouse Effect

greenhouse effect — the phenomenon in which the sun's energy in the form of light waves passes through the atmosphere and is absorbed by the earth, then reradiated as heat waves which are absorbed by the air. The air then acts in a similar manner as glass in a greenhouse.

Grey scale — (*geographic information system*) the level of brightness used to display information on a monochrome display device.

grid — an open lattice for mounting a sample to aid in its examination by transmission electron microscope or microscopy; the term is used by the Environmental Protection Agency to denote a 200-mesh copper lattice approximately 3 millimeters in diameter; a network of uniformly spaced points or lines on the visual display device for locating positions; a set of regularly spaced sample points; (*cartography*) an exact set of reference lines over the earth's surface; the distribution network of utility resources.

Grignard reagents — compounds of the type RMgX where R represents an alkyl group and X an aryl halide, especially bromides and iodides; formed from the reaction of alkyl and aryl halides with magnesium in the presence of dry ether.

grinding — the mechanical pulverization of solid waste.

gross heating value — the total heat obtained from the complete combustion of fuel or waste at 60°F when combustion starts and the combustion products of which are cooled to 60°F before the quantity of heat released is measured.

ground based inversion — temperature inversion based at the ground's surface that forms during the night when there are radiated losses.

ground cover — grasses and low-growing plants that keep soil from being blown or washed away.

grounding — the procedure used to carry an electrical charge to the ground through a conductive path.

ground level concentration (GLC) — the concentration of air contaminants at the ground surface.

groundwater — water which adheres to the surface of soil particles in thin films and prevents the soil from ever becoming totally dry; the water is held in deposits of gravel, sand, or porous rock below the earth's surface in the saturation zone.

groundwater contamination — the pollution of springs and wells from their source underground, potentially by landfills, agriculture, industry, and urban areas.

groundwater infiltration — water which enters treatment plants or pipes from natural springs, aquifers, or run-off from land.

groundwater runoff — that part of the groundwater that is discharged into a stream channel as a spring or seepage water.

groundwater supply — the water beneath the ground tapped by wells and springs.

group — *See* radical.

grout — a sealing mixture of cement and water to which sand, sawdust, and other fillers may be added; neat cement, inert natural materials, concrete, heavy drilling mud, heavy bentonite, water slurry impervious to and capable of preventing movement of water.

growth regulator — a chemical used to increase, decrease, or change the normal growth of a plant or animal; does not include fertilizers and other nutrients.

grubs — the larvae of certain beetles, wasps, bees, and ants.

GSD — *See* genetically significant dose.

guaranty — a formal and signed agreement between buyers and sellers in which the latter verifies that the items for sale are not in violation of the Food, Drug, and Cosmetic Act when shipped.

gutter — a trough under an eave to carry off rain water.

Gy — *See* gray.

gynecology — the study of the reproductive organs in women.

gypsum — $CaSO_4 \cdot 2H_2O$ — a monoclinic mineral form of hydrated calcium sulfate; used in the building industry and in the manufacture of cement, rubber, paper, and plaster of paris.

gypsum

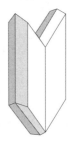

G
H

H

h — *See* Planck's constant.

H — *See* magnetic field intensity; magnetic field strength.

H₂O — the scientific notation for water.

H₂O₂ — the scientific notation for hydrogen peroxide.

H₂S — the scientific notation for hydrogen sulfide.

H₂SO₄ — the scientific notation for sulfuric acid.

habitat — the sum of environmental conditions in a specific place that is occupied by an organism, population, or community.

HACCP — *See* Hazard Analysis/Critical Control Point Inspection.

hafnium and compounds — Hf — a highly lustrous, ductile, grayish metal; MW: 178.5, BP: 8316°F, Sol: insoluble, sp gr: 13.31. It is used in the manufacture and fabrication of special alloys; as control rods in water-cooled nuclear reactors; in gas-filled tubes and incandescent lamps as a scavenger for oxygen and nitrogen; in the manufacture of photographic flash bulbs, light bulb filaments, and in electronic equipment as cathodes and capacitors. It is hazardous to the eyes, skin, and mucous membrane; and is toxic by inhalation, ingestion, and contact. Symptoms of exposure in animals include liver damage, and eye, skin, and mucous membrane irritation. OSHA exposure limit (TWA): 0.5 mg/m³ [air].

hail — precipitation in the form of frozen rain drops produced high in convective clouds usually in the latitudes between 30° and 60°.

hair catcher — a piece of equipment used for straining out lint and other materials before the water from a swimming pool precedes to the motor in the filter operation.

half-and-half — a product consisting of a mixture of milk and cream containing not less than 11.5% milk-fat.

half-life — the time required for one-half the atoms of a given amount of radioactive material to undergo radioactive decay.

half-value layer (HVL) — the thickness of absorbing material required to reduce intensity of a beam of radiation to one half its original intensity.

halocarbon — (*chemistry*) any of a group of compounds of carbon and one or more halogens; used as refrigerants, propellant gases, and

so on; may be anticipated to contribute to reductions in the concentration of ozone in the stratosphere.

halogen — any member of the chlorine family containing the elements fluorine, chlorine, bromine, iodine, and astatine.

halogenated hydrocarbon — a chemical substance containing carbon plus one or more of the halogens chlorine, fluorine, bromine, or iodine.

halophile — an organism that requires a high salt concentration for growth and maintenance.

hammermill — a device used for reducing the size of bulk material by means of hammers usually placed on a rotating axle inside a steel cylinder.

hand protection — specific types of gloves or other hand protectors required to prevent harmful exposure to hazardous materials.

hantavirus — a genus in the family *Bunyaviridae* which are lipid-enveloped viruses with a negative-stranded RNA genome composed of three unique segments; susceptible to most disinfectants including dilute hypochlorite solutions, phenolics, detergents, 70% alcohol, and other general household disinfectants. The survival time of the virus in the environment in liquids, aerosols, or the dried state is not known. It is spread by deer mice in the United States, and is the cause of hantavirus pulmonary syndrome.

hantavirus infection — *See* hantavirus pulmonary syndrome.

hantavirus pulmonary syndrome — (*disease*) an acute, viral, flu-like disease with high fever, muscle aches, cough, and headaches; respiratory problems worsen rapidly as lungs fill with fluid and death may occur because of respiratory failure (over 60% of the victims die). Incubation time is 1–2 weeks, but may range from a few days to 6 weeks. It is caused by hantaviruses; found worldwide. The reservoir of infection in the United States is the deer mouse, while worldwide it may be in other field rodents. It is transmitted through the inhalation of aerosols from infective saliva or excreta or bites of rodents. The possibility of this being a foodborne infection has not been evaluated. There is no evidence of person to person transmission. It is controlled through the elimination of rodents,

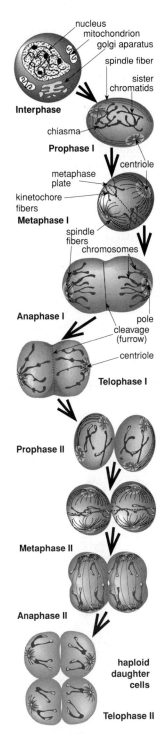

Interphase

nucleus
mitochondrion
golgi aparatus
spindle fiber
sister
chromatids

chiasma

Prophase I

centriole

metaphase
plate

kinetochore
fibers

Metaphase I

spindle
fibers

chromosomes

Anaphase I

pole
cleavage
(furrow)

centriole

Telophase I

Prophase II

Metaphase II

Anaphase II

haploid
daughter
cells

Telophase II

Haploid Cells

especially deer mice, in structures where people live or work; by the cleaning of all surfaces with good household disinfectants; and removal of dead rodents and contaminated materials. Workers must wear rubber or plastic gloves, and serious infestations require special clothing and breathing apparatus.

HAP — *See* hazardous air pollution.

haploid — having half the number of chromosomes found in the somatic cells of an organism.

hard coal — *See* anthracite.

hard-copy — (*computer science*) a copy on paper of information shown on the visual display unit.

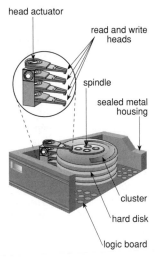

head actuator

read and write
heads

spindle

sealed metal
housing

cluster

hard disk

logic board

Hard Disk

hard disk — a magnetic disk made of rigid material for high-capacity random-access storage.

hardpan — a hardened soil layer in the lower A or in the B horizon caused by cementation of soil particles with organic matter or with materials such as silica sesquioxides or calcium carbonate; found in arid and semi-arid regions of the Southwestern United States. Also known as caliche; kanker.

hard swell — (*food*) a tin can bulged so tightly at both ends that no indentation can be made with thumb pressure.

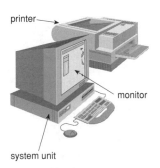

printer

monitor

system unit

Hardware

hardware — the physical components of a computer.

hard water — water containing dissolved minerals such as calcium, iron, and magnesium which do not lather with soap easily and which form insoluble deposits in boilers. Also known as water hardness.

harmonic — a series of sounds, each with a frequency which is an integral multiple of a fundamental frequency.

harvest — to pick or gather a crop; to gather in shellfish.

haustorium — a food-absorbing outgrowth of a plant organ such as a hypha or stem.

Haverhill fever — (*disease*) *See* streptobacillary fever.

hazard — any biological, chemical, or physical risk to a worker, consumer, or other individual or environment; the probability that a concentration of a substance will cause a significant physiological response.

Hazard Analysis/Critical Control Point Inspection (HACCP) — an in-depth inspection process developed by the Food and Drug Administration used to resolve or prevent disease outbreaks in restaurants and other food establishments by identifying the foods at greatest risk and the critical control points.

Hazard and Operability Study (HAZOP) — a systematic technique for identifying hazards or operable problems throughout an entire facility.

Hazard Communication Standard (HCS) — an OSHA regulation issued under federal regulations to prevent disease and injury in working people.

hazard identification — the gathering of data on a specific substance including information about the link between the substance and adverse health effects.

hazard information transmission (HIT) — a digital transmission of CHEMTREC emergency data.

hazardous air pollutant — an air pollutant to which no ambient air quality standard is applicable and which causes or contributes to air pollution resulting in an increase in mortality, or an increase in serious irreversible or incapacitating reversible illnesses.

hazardous air pollutants engineering program — a program designed to assess various industrial and combustion sources of hazardous air pollutants to determine the magnitude of emissions and the capability of technologies to reduce or eliminate them.

hazardous air pollution — *See* airborne toxics.

Hazardous and Solid Waste Amendments (HSWA) — a 1984 law banning liquid waste from land disposal; prohibiting certain land disposal practices; establishing minimum technology requirements for landfills, surface impoundments, and incinerators; expanding requirements for groundwater monitoring and clean-up; requiring double-liners in landfills; and expediting permits for new treatment technologies.

hazardous chemical — any chemical which is explosive, caustic, flammable, poisonous, corrosive, reactive, or radioactive and which requires special care when handling because its presence or use is a physical or health hazard.

hazardous material — any substance or compound capable of producing adverse health and/or safety effects.

Hazardous Material Information System (HMIS) — a Material Safety Data Sheets file maintained by the Department of Defense, containing material safety data sheets and transportation data for products purchased by the Department of Defense and the General Services Administration.

hazardous materials manager — an environmental health practitioner responsible for the control and management of materials that are potentially hazardous, from the point of extraction of raw products to their elemental destruction, transformation into non-hazardous materials, or disposal in controlled facilities.

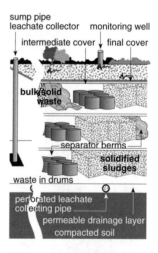

Hazardous Waste Landfill

hazardous waste (HW) — the legal designation for discarded solid or liquid waste containing substances known to be mutagenic, synergetic, or teratogenic to humans or animals and considered to be ignitible, corrosive, reactive, or toxic as defined by Title 40 of the Code of Federal Regulations; excludes domestic, agricultural, and mining wastes.

hazardous waste discharge — the accidental or intentional spilling, leaking, pumping, pouring, emitting, emptying, or dumping of hazardous waste into any land or water.

hazardous waste management facility (HWM facility) — all contiguous land and structures or other appurtenances and improvements on the land used for treating, storing, or disposing of hazardous waste.

hazard warning — words, pictures, symbols, or a combination thereof presented on a label or other appropriate medium to inform of the presence of various hazardous materials.

HAZMAT — acronym for hazardous materials.

HAZOP — *See* Hazard and Operability Study.

HBIG — *See* hepatitis B immune globulin.

HCB — *See* hexachlorobenzene.

HCH — *See* hexachlorocyclohexane.

HCS — *See* Hazard Communication Standard.

HDPE — *See* high-density polyethylene.

head — (*physics*) a basic measurement of pressure or resistance in a hydraulic system which is equivalent to the height of a column of water capable of exerting a given pressure.

health — the avoidance of disease and injury and the promotion of normalcy through efficient use of the environment, a properly functioning society, and an inner sense of well-being.

health and safety guides (HSG) — basic information for employers and employees for a safe and healthful work environment.

Health and Safety Science Abstract — an abstract service available of health and safety science articles from 1977 to the present.

health care facility — a facility or location where medical, dental, surgical, or nursing attention or treatment is provided to humans or animals.

health care institution — any hospital, convalescent home, or other facility which provides health care, medical treatment, room, board, or other services for the ill, physically or mentally disabled.

health hazard — a chemical for which there is significant evidence that acute or chronic health effects may occur in exposed individuals based on at least one study conducted in accordance with established scientific principles; the health hazards may be chemicals which are carcinogens, toxic or highly toxic agents, reproductive toxins, irritants, corrosives, sensitizers, hepatotoxins, nephrotoxins, neurotoxins, agents which act on the hematopoietic system, and agents which damage the lungs, skin, eyes, and mucous membranes of systems of the body.

hearing — the general perceptual behavior and the specific responses made in relation to sound stimuli; (*law*) the opportunity for a party to present views respondent to a charge.

hearing level — the amount in decibels by which the threshold of audibility for an ear differs from a standard audiometric threshold.

hearing loss — partial or complete loss of the sense of hearing; conductive hearing loss is associated with impaired transmission of sound waves through the external ear canal to the bones of the middle ear; sensorineural hearing loss is associated with some pathological change in structures within the inner ear or in the acoustic nerve; central hearing loss occurs when there is a pathological condition above the junction of the acoustic nerve and the brain stem.

heart disease — a disease of the heart having a definite pathological process with a characteristic set of signs and symptoms which may affect the heart itself or any of its parts; may be of unknown etiology with an unknown prognosis.

hearth — the bottom of a furnace on which waste materials or fuels are exposed to the flame.

heat — a form of energy that raises the temperature of a body or substance. It is the kinetic energy of its molecules related to the capacity of a substance to do work. Heat will flow from the hot area to the cold area. Heat can be created by mechanical, chemical, electric, and nuclear energy; causes expansion in gases, liquids, and solids.

heat balance — an accounting of the distribution of the heat input and output of an incinerator, usually on an hourly basis.

heat capacity — the quantity of heat in calories needed to raise the temperature of one gram of a substance 1°C. Also known as thermal capacity.

heat content — the sum total of the latent and sensible heat present in a gas, liquid, or solid minus that contained at an arbitrary set of conditions chosen as the base or zero point.

heat cramps — a condition marked by sudden development of cramps in skeletal muscles resulting from prolonged work in high temperatures, accompanied by profuse perspiration and loss of salt from the body.

heat drying — (*sludge*) the application of heat to evaporate sufficient moisture and render the sludge dry to the touch and relatively free-flowing.

heat exchanger — any device that transfers heat from one fluid to another without allowing them to mix.

heat exhaustion — a condition marked by weakness, nausea, dizziness, and profuse sweating, usually precipitated by physical exertion in a hot environment.

G
H

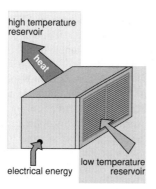

Heat Pump

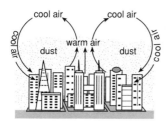

Heat Island

heat island — a reservoir for storing heat through the absorption of the sun's rays during the day and the release of energy at night; typically occurring in cities.

heat island effect — the phenomenon of air circulation found in urban areas with tall buildings; heat and pollution create a haze dome that prevents warm air from rising and being cooled at a normal rate.

heat of combustion — (*chemistry*) the quantity of heat released per unit mass or unit volume of 1 mole of a substance when the substance is completely burned at constant pressure.

heat of formation — (*chemistry*) the quantity of heat released when 1 mole of a compound is formed from its elements in the phases in which they exist at ordinary temperatures.

heat of fusion — the heat absorbed and used in the melting process of one unit mass of a solid to a liquid at constant temperature and pressure.

heat of neutralization — the amount of heat evolved when 1 gram equivalent of an acid is neutralized by 1 gram equivalent of a base; for strong acids and bases in dilute solution, the only reaction is the production of water.

heat of reaction (*chemistry*) the quantity of heat released by a chemical reaction.

heat of vaporization — the heat required to change a liquid into a gas (e.g., water at 100°C does not change into steam until 540 calories of heat have been absorbed by each gram of the boiling water).

heat pump — a device used to bring heat from outside a structure during the winter months and to remove heat from inside during the summer months.

heat rash — an inflammatory skin condition that occurs in hot, humid environments where sweat is not easily removed from the surface of the skin by evaporation and the skin stays moist; may cause a tingling or prickling sensation. Also known as prickly heat; miliaria.

heat release rate — the amount of heat liberated during complete combustion per second divided by its volume.

heatsink — (*physics*) any substance, body, or region that absorbs or dissipates heat.

heat stress — relative amount of thermal stress from the environment.

heat stress index (HSI) — the relation of the amount of evaporation or perspiration required for particular job conditions as related to the maximum evaporative capacity of an average person.

heatstroke — a condition marked by the cessation of sweating, extremely high body temperature, and collapse that results from prolonged exposure to high temperature.

heat survey — a study conducted to determine the measurement of ambient temperatures, air motion, humidity, and radiant heat sources.

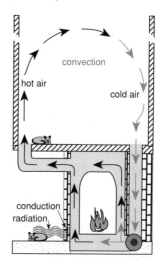

Heat Transfer

heat transfer — the movement of heat from an area of higher temperature to one of lower temperature until there is an equilibrium; transferred by conduction, convection, or radiation.

heat treatment — (*sewage*) the pressure cooking of sludges in such a manner that little sludge oxidation occurs; any of several processes of metal modification, such as annealing, which occurs.

heat wave — a period of at least one day of abnormally hot weather.

heavy-duty cleaners — liquid products formulated to emulsify and soften wax and synthetic floor finishes for easy removal; also used to remove heavy soil from surfaces.

heavy hydrogen — the hydrogen isotope having an atom weight greater than one. Also known as deuterium; tritium.

heavy metal — a metallic element such as mercury, chromium, cadmium, arsenic, and lead with a high molecular weight; can damage living organisms at low concentrations and tends to accumulate in the food chain.

heavy water — the water in which molecule combinations of heavy hydrogen and oxygen are used as a moderator in a nuclear reactor. Also known as deuterium oxide.

hectare — a square kilometer which is 0.3861 of a square mile.

hecto (h) — prefix representing 100 in the metric system.

HEEP — acronym for Health and Environmental Effects Profile.

helmet — a device that protects the eyes, face, neck, and other parts of the head.

helmholtz coils — a pair of flat, circular coils with an equal number of turns and equal diameters arranged with a common axis and connected in series to have a common current. The purpose of the arrangement is to obtain a magnetic field that is more nearly uniform than that of a simple coil.

helminth — any parasitic worm; a nematode or trematode.

helper/suppressor T-cells — white blood cells that are part of the immune system.

hematology — the study of anatomy, physiology, pathology, symptomatology, and therapeutics related to the blood and blood-forming tissues.

hematoma — a localized blood clot under the surface of the skin.

hematopoietic system — the blood-forming mechanism of the human body. Also known as reticuloendothelial system.

hematopoietic toxin — any chemical, including carbon monoxide and cyanide, which decreases hemoglobin function and deprives the body tissues of oxygen causing symptoms of cyanosis and loss of consciousness.

hematuria — blood in the urine.

heme — $C_{34}H_{32}O_4N_4Fe$ — the nonprotein, insoluble, iron proroporphyrin constituent of hemoglobin.

hemimetabolism — *See* incomplete metamorphosis.

hemoglobin — the iron-containing, oxygen-carrying protein compound in red blood cells that gives them their color.

hemolysin — a substance that liberates hemoglobin from erythrocytes by interrupting their structural integrity.

hemolysis — rupture of erythrocytes with release of hemoglobin into the plasma.

hemolytic — pertaining to, characterized by, or producing hemolysis.

hemolytic staphylococci — strings of cocci that cause the red blood cells to leave the hemoglobin.

hemoptysis — bleeding from the lungs, larynx, trachea, or bronchi.

hemorrhage — the escape of blood from the vascular system.

hemorrhagic colitis — an acute disease caused by *E. coli* 015: H7 (enterohemorrhagic strain); generally symptoms include severe abdominal cramping, watery diarrhea which may become grossly bloody, occasional vomiting, possible low grade fever. Incubation time is 2–9 days; the disease has been found in the pacific northwest, Canada, and Michigan. The reservoir of infection is cattle, and it is transmitted by raw or improperly cooked ground beef or raw milk. It is not communicable from person to person; general susceptibility but the very young may develop renal failure and the very old may develop fever and neurologic symptoms which can lead to death. It is controlled by proper refrigeration and cooking of ground beef and using only pasteurized milk. Also known as enterohemorrhagic colitis.

hemotoxin — a poison which destroys red blood cells and breaks down the walls of small blood vessels.

Henry's law — the law stating that at a constant temperature the solubility of a gas dissolved in a liquid at equilibrium is directly proportional to the partial pressure of the gas.

HEPA — *See* high efficiency particulate air filters.

hepatic — pertaining to the liver.

hepatitis — inflammation of the liver, commonly of viral origin.

hepatitis B immune globulin (HBIG) — a preparation that provides some temporary protection following exposure to the hepatitis B virus if given within seven days after exposure.

hepatolenticular degeneration — *See* Wilson's disease.

hepatomegaly — enlargement of the liver.

hepatosis — any functional disorder of the liver.

hepatotoxin — an agent capable of damaging the liver, causing symptoms such as jaundice and

G
H

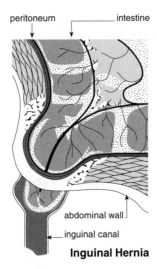

Inguinal Hernia

liver enlargement; chemicals include carbon tetrachloride and nitrosamines.

heptachlor — $C_{10}H_5Cl_7$ — white to light tan crystals with a camphor-like odor; MW: 373.4, BP: 293°F (decomposes), Sol: insoluble, sp gr: 1.66. It is used in the formulation of pesticides; as an insecticide in seed treatment. In animals, it is hazardous to the liver and central nervous system; and is toxic by inhalation, absorption, ingestion, and contact. Symptoms of exposure in animals include tremors, convulsions, liver damage; carcinogenic. OSHA exposure limit (TWA): 0.5 mg/m³ [skin].

n-heptane — $CH_3(CH_2)_5CH_3$ — a colorless liquid with a gasoline-like odor; MW: 100.2, BP: 209°F, Sol (60°F): 0.005%, Fl.P: 25°F, sp gr: 0.68. It is used as a carrier and penetrant solvent for adhesives, in azeotrophic distillations, and in rubber tire manufacture; as a solvent in rubber cements; as an ink solvent in gravure printing; as a solvent in the polymer industry as a swelling and blowing agent for plastic foams; as a reference fuel for testing gasoline engine knock and pollution and combustion studies; as a diluent solvent for lacquers during preparation and application; in organic chemical synthesis; as a laboratory solvent for scientific testing. It is hazardous to the skin, respiratory system, and peripheral nervous system; and is toxic by inhalation, ingestion, and contact. Symptoms of exposure include lightheadedness, giddiness, stupor, no appetite, nausea, dermatitis, chemical pneumonia, unconsciousness. OSHA exposure limit (TWA): 400 ppm [air] or 1,600 mg/m³.

2-heptanone — *See* methyl (n-amyl) ketone.

3-heptanone — *See* ethyl butyl ketone.

herbicide — a pesticide chemical used to prevent or destroy the growth of plants.

herd immunity — the resistance of a group to invasion and spread of an infectious agent based on the immunity of a high proportion of individual members of the group.

heredity — the sum of the transmission of traits from parent to offspring.

hermetically-sealed container — an air-tight container designed and intended to be secure against the entry of microorganisms and to maintain the commercial sterility of its contents after processing.

hernia — an abnormal protrusion of tissue through its containing wall.

hertz (Hz) — the unit for frequency equal to one cycle per second.

heterotroph — an organism that feeds upon organic compounds produced by other organisms in order to synthesize its own food. *See also* consumer.

heterotrophic reducers — bacteria or fungi which return the complex organic compounds to their original abiotic state and release the remaining chemical energy.

HEW — acronym for Department of Health, Education, and Welfare; now known as HHS or Health and Human Services.

Hexachlorobenzene

hexachlorobenzene (HCB) — C_6Cl_6 — white or colorless needle-like crystals which occur as a by-product during the manufacture of chlorine containing compounds; MW: 284.79, BP: 322°C (sublimes), Sol: 0.006 mg/L, Fl.P: 242°C. It is not currently in commercial use in the United States. Prior to 1985, it was used as a pesticide and in the production of synthetic rubber and of pyrotechnic and ordinance materials for the military. It is hazardous to the liver, digestive system, central nervous system, and reproductive system; and is toxic through ingestion. Symptoms of exposure include skin lesions; weakness; convulsions; osteoporosis, especially of the bones of the hands; muscle damage. OSHA does not regulate exposure limits. World Health Organization drinking water guidelines: 0.01 µg/L.

Hexachlorobutadiene

hexachlorobutadiene — C_4Cl_6 — a colorless liquid with a turpentine-like odor; MW: 260.76, BP: 215°C, Sol: 2–2.55 mg/L at 20°C in water, density: 1.55 g/cm³ at 20°C. It is used as a chemical intermediate in the manufacture of rubber compounds, as a solvent, a fluid for gyroscopes, a heat transfer liquid, hydraulic fluid, and as a chemical intermediate in the production of chlorofluorocarbons and lubricants. It is hazardous to the respiratory system, kidneys, and liver in animals; may cause death; and is toxic by inhalation, ingestion, and contact. Symptoms of exposure in animals include nasal irritation, labored and slow breathing, damaged cortical and proximal renal tubules. OSHA exposure limit (TWA): 0.002 ppm [air].

hexachlorocyclohexane (HCH) — $C_6H_6Cl_6$ — a white, solid, synthetic chemical existing in eight different isomer forms (gamma-HCH is called lindane); MW: 290.83, BP: 60°C at 0.36 mm Hg to 323.4°C at 760 mm Hg depending on the isomer, Sol: 5 ppm to 17 ppm, density: 1.89 to 1.891 at 19°C. It is used as an insecticide, therapeutic scabicide, pediculicide, and ectoparasiticide for humans and animals. It is hazardous to the respiratory system, neurological system, cardiovascular system, gastrointestinal system, musculoskeletal system; and is toxic by inhalation, ingestion, and contact. Symptoms of exposure include mucous membrane irritation of the nose and throat, electrocardiogram abnormalities, hypochromic anemia, generalized urticaria, paresthesia of the face and extremities, headache, and occasional deaths. Formerly known as benzene hexachloride (BHC). OSHA exposure limit (TWA): 0.5 mg/m³ [air].

Hexachloroethane

hexachloroethane — Cl_3CCCl_3 — colorless crystals with a camphor-like odor; MW: 236.7, BP:

sublimes, Sol (72°F): 0.005%, sp gr: 2.09. It is used in the manufacture of pyrotechnics; in lubricants in foundries for mold treatment; in preparation of nitrocellulose esters and camphor substitutes; as a fumigant, insecticide, fungicide; in the manufacture of synthetic rubber; as a chemical additive, and as an additive to fire extinguishers. It is hazardous to the eyes; and is toxic by inhalation, absorption, ingestion, and contact. Symptoms of exposure include eye irritation; carcinogenic. OSHA exposure limit (TWA): 1 ppm [skin] or 10 mg/m³.

hexachloronaphthalene — $C_{10}H_2Cl_6$ — a white to light-yellow solid with an aromatic odor; MW: 334.9, BP: 650–730°F, Sol: insoluble, sp gr: 1.78. It is used in the manufacture of electric equipment as insulating material; as an inert component of resins or polymers for coating and impregnating textiles, wood, paper for flame- and water-proofing, and fungicidal and insecticidal properties; as an additive to special lubricants and cutting oils; in polymer manufacture as fillers. It is hazardous to the liver and skin; and is toxic by inhalation, absorption, ingestion, and contact. Symptoms of exposure include acne-forming dermatitis, nausea, confusion, jaundice, coma. OSHA exposure limit (TWA): 0.2 mg/m³ [skin].

n-hexane — $CH_3(CH_2)_4CH_3$ — a colorless liquid with a gasoline-like odor; MW: 86.2, BP: 156°F, Sol: 0.002%, Fl.P: –7°F, sp gr: 0.66. It is used as an extractant of agricultural products and animal fat; in the manufacture of polyolefins and certain elastomers as a catalyst carrier and assist in controlling molecular weight; as a solvent in adhesives; in the pharmaceutical industry as a reaction medium, immiscible solvent, and extraction ergot; in compounding printing inks, lacquers, or stains; as a laboratory reagent and general solvent; in the manufacture of low-temperature thermometers. It is hazardous to the skin, eyes, and respiratory system; and is toxic by inhalation, ingestion, and contact. Symptoms of exposure include lightheadedness, nausea, headache, numbness in the extremities, muscle weakness, dermatitis, chemical pneumonia, giddiness, irritation of the eyes and nose. OSHA exposure limit (TWA): 50 ppm [air] or 180 mg/m³.

2-hexanone — $CH_3CO(CH_2)_3CH_3$ — a colorless liquid with an acetone-like odor; MW: 100.2, BP: 262°F, Sol: 2%, Fl.P: 77°F, sp gr: 0.81. It is used as a commercial solvent for nitrocellulose, natural and synthetic resins, oils, waxes, vinyl polymers, and cellulose acetates; as a solvent in the manufacture of varnish removers,

G
H

vinyl lacquers, and nitrate wood lacquers; as an extractive solvent for paraffin wax; in the separation and purification of certain metals. It is hazardous to the central nervous system, skin, and respiratory system; and is toxic by inhalation, absorption, ingestion, and contact. Symptoms of exposure include peripheral neuropathy, weakness, paresthesia, dermatitis, headache, drowsiness, irritation of the eyes and nose. OSHA exposure limit (TWA): 5 ppm [air] or 20 mg/m³.

hexone — $CH_3COCH_2CH(CH_3)_2$ — a colorless liquid with a pleasant odor; MW: 100.2, BP: 242°F, Sol: 2%, Fl.P: 64°F, sp gr: 0.80. It is used as a separating agent for certain inorganic salts; in the extraction and the manufacture of antibiotics and purification of petroleum products; in the manufacture of dry cleaning preparations, germicides, fungicides, and electroplating solutions; in blending raw materials for molded plastics; in cleaning and maintaining ketone processing equipment; in the application and drying of lacquers, varnishes, epoxy, acrylic, vinyl, or other cellulose- or resin-based coatings, finishes, and adhesives. It is hazardous to the respiratory system, eyes, skin, and central nervous system; and is toxic by inhalation, ingestion, and contact. Symptoms of exposure include headache, narcosis, coma, dermatitis, irritation of the eyes and mucous membrane. OSHA exposure limit (TWA): 50 ppm [air] or 205 mg/m³.

sec-hexyl acetate — $C_8H_{16}O_2$ — a colorless liquid with a mild, pleasant, fruity odor; MW: 144.2, BP: 236°F, Sol: 0.08%, Fl.P: 113°F, sp gr: 0.86. It is used in the application of nitrocellulose lacquers and other lacquers; as an inhibitor in handling diacyl peroxide solutions. It is hazardous to the eyes and central nervous system; and is toxic by inhalation, ingestion, and contact. Symptoms of exposure include headache; in animals: narcosis, irritation of the eyes, nose, and throat. OSHA exposure limit (TWA): 50 ppm [air] or 300 mg/m³.

HFID — acronym for heated flame ionization detector.

Hg — *See* mercury.

HHS — acronym for Department of Health and Human Services; previously known as HEW or Health, Education, and Welfare.

high — (*weather*) an area of high pressure areas which may develop any place that air cools, compresses, and sinks; wind movement that follows a normal flow from high pressure to low pressure.

high altitude — any elevation over 4,000 feet.

high clouds — types of clouds composed almost entirely of tiny ice crystals with bases averaging 2,000 feet above the earth.

high-density polyethylene (HDPE) — a material used to make plastic bottles that produces toxic fumes when burned.

high efficiency particulate air filters (HEPA) — a designation for a type of filter that is 99.97% efficient in removing particles from a body of air with a diameter of 0.3 micrometers.

high-end exposure — a reasonable estimate of an individual's risk to an exposure that is greater than the 90th percentile.

high-flow pump — a pump used for sampling particulates, gases, and vapors where the flow rate is two liters per minute of air.

high frequency loss — a hearing deficit starting with 2,000 hertz and greater.

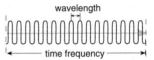

wavelength

time frequency

High Frequency Sound Waves

high-frequency sound waves — a frequency of 3–30 megahertz which produce high-pitched or shrill tones.

high heat value — the BTUs liberated when a pound of solid waste is burned completely and the products of combustion are cooled to the initial temperature of the solid waste.

high level — having a high level of radioactivity.

high-level radioactive waste — aqueous waste from the operation of a nuclear reactor.

highly alkaline detergent — a heavy-duty cleaning material used to remove wax and baked-on grease in ranges, ovens, dish washing machines, and power sprays.

highly susceptible population — a group of persons who are more likely than other populations to experience disease because their immune systems have been compromised as with the elderly, very young, or those with chronic diseases.

highly toxic — relating to a chemical with a median lethal dose (LD_{50}) or concentration (LC_{50}) which is fatal to laboratory animals when administered by physical contact, ingestion, or inhalation; relating to a chemical with a median lethal dose (LD_{50}) of 50 milligrams or less per kilogram of body weight when administered orally to albino rats weighing between 200–300 grams each; a chemical with a median lethal dose (LD_{50}) of 200 milligrams or less per kilogram of body weight when administered by continuous contact for 24 hours, or less if death occurs within 24 hours, with the bare skin of albino rabbits weighing between 200–300 grams each; a chemical that has a median lethal concentration (LC_{50}) in air of 200 ppm by volume or less of gas or vapor, or 2 milligrams per liter or less of mist, fume, or dust when administered by continuous inhalation for one hour, or less if death occurs within one hour, to albino rats weighing between 200–300 grams each.

highly volatile — a liquid that quickly forms a gas or vapor at room temperature.

high-pressure cell — *See* air mass.

high-resolution infra-red radiation sounder (HIRS) — a satellite instrument used in determining vertical temperature profiles.

high-temperature short-time pasteurization (HTST) — pasteurization of milk that occurs at 161°F for 15 seconds.

high-to-low extrapolation — (*animals*) a mathematical manipulation of data which allows for chemicals to be expressed in terms of dose and cancer risk.

high velocity air filter (HVAF) — an air pollution control filtration device for the removal of sticky, oily, or liquid aerosol particulate matter from exhaust gas streams.

high-volume sampler (Hi-Vol) — a device operating somewhat like a vacuum cleaner, used in the measurement and analysis of suspended particulate pollution, and operated at an average sampling rate of 72,000 ft^3 per 24-hour period.

Hill-Burton Act — legislation and programs operated under the legislation for federal support to construct and modernize hospitals and other health facilities beginning with public law 79-725, the Hospital Surveying Construction Act of 1946.

HIRS — *See* high-resolution infra-red radiation sounder.

histamine — C_5HgN_3 — a powerful dilator of the capillaries; also thought to be responsible for many human allergies.

histogram — (*statistics*) a diagram showing the number of samples and their distribution by category.

histology — the study of the structure of cells and tissues as related to their function.

histopathology — a branch of pathology dealing with tissue changes associated with disease.

histoplasmosis — (*disease*) a systemic infectious mycosis of varying severity with the primary lesion usually in the human lungs; ranging from asymptomatic symptoms to chronic pulmonary symptoms. Incubation time is 5–18 days, commonly 10 days. It is caused by *Histoplasma capsulatum*, a fungus with two different forms,

growing as a mold in soil and as a yeast in animal and human hosts; found in many parts of the world including the Americas. The reservoirs of infection include soil around old chicken houses, in caves, around starling roosts, soils with high organic content, and decaying trees. It is transmitted through inhalation; not communicable from person to person; general susceptibility. It is controlled by disinfection of sputum and articles soiled with sputum, and pharmaceutical treatment.

historical geology — the study of the geologic history of the earth and its inhabitants.

HIT — *See* hazard information transmission.

hit space theory — *See* target theory.

HIV — *See* human immunodeficiency virus.

hives — a skin condition with elevations and itching resulting from an allergic reaction. Also known as urticaria.

Hi-Vol — *See* high-volume sampler.

HMIS — *See* Hazardous Material Information System.

HMTC — acronym for Hazardous Materials Technical Center.

holding tank — a water-tight receptacle designed and constructed to receive and retain sewage for ultimate disposal at another site.

holding time — (*milk pasteurization*) the flow time of the fastest particle of milk at or above 161°F through the holder section, which is that portion of the system outside the influence of the heating medium and sloping continuously upward in the down stream direction; located upstream from the flow-diversion valve.

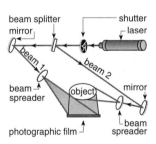

Laser System
in the making of a hologram

hologram — a realistic, three-dimensional image produced by a laser and photographic plates.

homeostasis — a steady state which an organism maintains by self-regulating adjustments to its internal environment.

homogeneous — pertaining to a substance having uniform composition and structure.

homogenized milk — milk which has been treated in such a manner as to ensure break-up of the fat globules to such an extent that after 48 hours of quiescent storage, no visible cream separation occurs on the milk, and the fat percentage of the top 100 milliliters of milk in a quart bottle does not differ by more than 10% from the fat percentage of the remaining milk.

homogenizing — the process of making a substance more uniform.

homoiothermal — (*animals*) referring to an organism which maintains a relatively constant internal temperature regardless of the environmental temperature. Also known as warm-blooded.

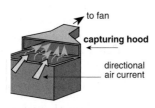

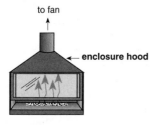

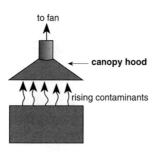

Hoods

hood — the point of entry into a local exhaust system.

hood entry loss — the pressure loss from turbulence and friction as air enters the ventilation system.

hood slot — a hood consisting of a narrow slot leading into a plenum chamber under suction to distribute air velocity along the length of the slot.

hood static pressure — the suction or static pressure in a duct near a hood which represents the suction that is available to draw air into the hood.

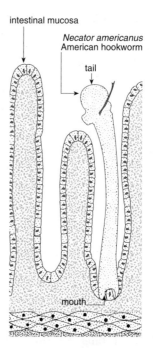

Hookworm Disease

hookworm disease — (*disease*) a common chronic parasitic infection with various symptoms including anemia. Incubation time is a few weeks to many months; caused by *Necator americanus*, *Ancylostoma duodenale* and *A. ceylanicum*; found in tropical and subtropical climates. The reservoirs of infection include people for *N. americanus* and *A. duodenale*, and cats and dogs for *A. ceylanicum*; transmitted by the eggs in feces deposited on the ground and hatched. Larvae develop and become infective in 7–10 days, penetrate the skin, usually at the foot, and pass through the lymphatics and blood stream to the lungs to enter the alveoli, where they migrate up the trachea to the pharynx, are swallowed and reach the small intestine; they attach to the intestinal wall. It is not transmitted from person to person, but infected people can contaminate soil for several years; general susceptibility. It is controlled through public education, wearing shoes, and proper disposal of sewage. Also known as *Ancylostomiasis*.

hormone — a chemical secretion from one of the ductless glands of the body which acts as a catalyst in regulating specific activities of certain cells or organs.

horsepower (hp) — a unit of power equal to 745.7 watts.

host — (*parasitic relationship*) the organism from which the parasite derives its food supply and shelter.

hot — a relatively high temperature; highly radioactive.

hot tub — a hydrotherapy spa constructed of wood with separately formed sides and bottoms and a hull shaped to join together by pressure from the surrounding hoops, veins, or rods.

hot water — water at 180°F at a pressure of 16–25 lb/in² for 30 seconds which is an effective physical, sanitizing agent for nonselective sanitization of food-contact surfaces where spores may survive.

hot work — any work involving burning, welding, riveting, or similar fire-producing operations, as well as work which produces a source of ignition such as drilling, abrasive blasting, and space heating.

household — one or more individuals living together in a single dwelling unit and sharing common living, sleeping, cooking, and eating facilities.

housing maintenance specialist — an environmental health practitioner involved in the regulation of housing, including zoning and occupancy approvals, elimination of nuisances, and regular inspection to assure that minimum standards of health and safety are met.

hp — *See* horsepower.

HPLC — acronym for high performance liquid chromatography.

hr — notation for hour or hours.

HRS — acronym for Hazardous Ranking System.

HSG — *See* health and safety guides.

HSI — *See* heat stress index.

HSWA — *See* Hazardous and Solid Waste Amendments.

HTGR — acronym for high temperature gas-cooled reactor.

HTST — *See* high-temperature short-time pasteurization.

HUD — *See* Department of Housing and Urban Development.

Human Factors Society — a professional society of psychologists, engineers, physiologists, and other related scientists concerned with the use of human factors in the development of systems and devices.

human immunodeficiency virus (HIV) — a retrovirus designated as either human T-lymphotropic virus type III (HTLV-III) or lymphadenopathy associated virus (LAV) both of which compromise the body's ability to deal with various infections and rare forms of cancer by damaging the autoimmune system; may lie dormant for several years; causes acquired immune deficiency syndrome.

human threadworm — *See* pinworm.

human waste — normal excretory waste of the human body.

humidity — atmospheric water vapor content.

humpback — *See* kyphosis.

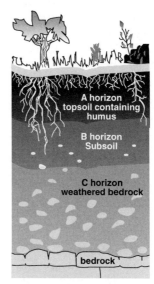

Humus
in A Horizon

humus — the normally dark-colored amorphous colloid matter in the soil formed by decomposition of plant and animal remains.

hunchback — *See* kyphosis.

HVAC — acronym for heating, ventilation, and air conditioning.

HVAF — *See* high velocity air filter.

HVL — *See* half-value layer.

HW — *See* hazardous waste.

HWM facility — *See* hazardous waste management facility.

hydration — the combination of a substance with water.

hydraulic conductivity — (*soils*) the ability of the soil to transmit water in liquid form through pores.

hydraulic fracturing — the physical process that creates fractures in soils to enhance fluid or vapor flow in the subsurface; used especially in petroleum drilling.

hydraulic loading — the flow rate of liquid applied to the granular bed expressed as gallons per minute per ft².

hydrazine — N_2H_4 — a colorless liquid with an ammonia-like odor; a solid below 36°F; MW: 32.1, BP: 236°F, Sol: miscible, Fl.P: 99°F, sp gr: 1.01. It is used in the synthesis and handling of high-energy fuels, agricultural chemicals, pharmaceuticals, chemicals for plastics and

G
H

rubber manufacturing, textile agents and dye intermediates, and photographic chemicals; as an anticorrosion agent. It is hazardous to the skin, eyes, respiratory system, and central nervous system; and is toxic by inhalation, absorption, ingestion, and contact. Symptoms of exposure include temporary blindness, dizziness, nausea, dermatitis, skin and eye burns, irritation of the eyes, nose, and throat; carcinogenic; in animals: bronchitis, pulmonary edema, liver and kidney damage, convulsions. OSHA exposure limit (TWA): 0.1 ppm [skin] or 0.1 mg/m^3.

hydrocarbon — one of a very large group of compounds containing hydrogen and carbon atoms which may be saturated or unsaturated. Saturated hydrocarbons have no double or triple bonds. Unsaturated hydrocarbons contain double and/or triple bonds. Aromatic hydrocarbons contain benzene or benzene-like rings. Cyclic hydrocarbons are saturated or unsaturated but contain a ring of carbon atoms.

hydrocarbon radical (R) — (*chemistry*) a group of atoms of C and H with one or more free bonds.

hydrochinone — *See* hydroquinone.

hydrochloric acid — *See* hydrogen chloride.

hydrogenation — the process in which hydrogen is made to combine with another substance, usually an organic one, in the presence of a catalyst, such as nickel.

hydrogen bromide — HBr — a colorless gas with a sharp, irritating odor; often used in an aqueous solution; MW: 80.9, BP: –88°F, Sol: 49%. It is used in the manufacture of organic bromides for use in photography, pharmaceuticals, industrial drying, textile finishing, engraving and lithography, chemical synthesis, and fire retardants; in the manufacture of brominated fluorocarbons for fire extinguishing, refrigeration, and aerosols; in the organic synthesis as intermediates for barbiturate manufacture, manufacture of synthetic hormones, catalyst for alkylations, controlled oxidations, isomerizations, and polymerizations; as a reagent in analytical chemistry; in the etching of germanium crystals, silicon discs, and metal alloys; as a solvent for ore minerals. It is hazardous to the respiratory system, eyes, and skin; and is toxic by inhalation, ingestion, and contact. Symptoms of exposure include skin and eye burns, irritation of the eyes, nose, and throat. OSHA exposure limit (TWA): ceiling 3 ppm [air] or 10 mg/m^3.

hydrogen chloride — HCl — a colorless to slightly yellow gas with a pungent, irritating odor; often used in an aqueous solution; MW: 36.5, BP: –121°F, Sol (86°F): 67%. It is used during pickling of metals including stainless steel, iron, and nickel; as a catalyst or chlorinat-ing agent in chemical synthesis; in food processing and manufacture including sugar cane refining, glucose, and corn sugar, and brewing operations; in industrial chemical cleaning operations; in the production of plastics and resins; in rubber manufacture including synthesis of chloroprene; as a deliner of hides in leather manufacture; in activation of petroleum wells; in waste treatment operations for neutralization of alkaline waste streams; in the production of chlorine; in swimming pools. It is hazardous to the respiratory system, skin, and eyes; toxic by inhalation, ingestion, and contact. Symptoms of exposure include cough, burning throat, choking, burning eyes and skin, dermatitis, inflammation of the nose, throat, and larynx; in animals: laryngeal spasms, pulmonary edema. Also known as hydrochloric acid. OSHA exposure limit (TWA): ceiling 5 ppm [air] or 7 mg/m^3.

hydrogen cyanide — HCN — a colorless to pale blue liquid or gas (above 78°F) with a bitter, almond-like odor; MW: 27.0, BP: 78°F, Sol: miscible, Fl.P: 0°F, sp gr: 0.69. It is used in fumigation of structures and agricultural crops; in the production of intermediates in synthesis of acrylic plastics, nylon 66, chelating agents, dyes, pharmaceuticals, and specialty chemicals. It is hazardous to the central nervous system, cardiovascular system, liver, and kidneys; and is toxic by inhalation, absorption, ingestion, and contact. Symptoms of exposure include asphyxia and death at high levels, weakness, headache, confusion, nausea, vomiting, increased rate and depth of respiration or slow respiration and gasping. OSHA exposure limit (TWA): 4.7 ppm [skin] or 5 mg/m^3.

hydrogen fluoride — HF — a colorless gas or fuming liquid (below 67°F) with a strong, irritating odor; shipped in cylinders; MW: 20.0, BP: 286°F, Sol: miscible, sp gr (67°): 1.00. It is used in the manufacture of chlorofluorohydrocarbons for application as refrigeration fluids, aerosol propellants, specialty solvents, high-performance plastics, and foaming agents; in cleaning sandstone and marble; as a pickling agent for stainless steel and other metals; as a cleaner in meat packing; in anhydrous acid in the manufacture of aluminum fluoride and synthetic cryolite and as a catalyst in alkylation of petroleum fractions to produce high-octane fuels, in separation of uranium isotopes, in the manufacture of pharmaceuticals and special dyes; in aqueous acid in etching, frosting, and polishing glassware and ceramics; in the manufacture of insecticides, laundry sours, and stain removers. It is hazardous to the eyes, skin, and respiratory system; and is toxic by inhalation, absorption, ingestion, and contact. Symptoms of exposure include pulmonary edema, skin and eye burns, nasal congestion, bronchitis, irritation of the eyes, nose, and throat. OSHA exposure limit (TWA): 3 ppm [air] or 1.4 mg/m^3.

hydrogen ion balance — *See* acid-base balance.

hydrogen peroxide — H_2O_2 — a colorless liquid with a slightly sharp odor; the pure compound is a crystalline solid below 12°F; often used in aqueous solution; MW: 34.0, BP: 286°F, Sol: miscible, sp gr: 1.39. It is used in the manufacture of propellants for military and space programs; as a component of explosives; in chemical synthesis as an oxidant in organic and inorganic synthesis; as a polymerization promoter; as a bleaching agent for oils, waxes, and fats, and discolored concentrated acids. It is hazardous to the eyes, skin, and respiratory system; and is toxic by inhalation, ingestion, and contact. Symptoms of exposure include corneal ulcers, erythema, vesicles on the skin, bleaching hair, irritation of the eyes, nose, and throat. OSHA exposure limit (TWA): 1 ppm [air] or 1.4 mg/m^3.

hydrogen selenide — H_2Se — a colorless gas with an odor resembling decayed horseradish; MW: 81.0, BP: –42°F, Sol (73°F): 0.9%. It is used in the preparation of semiconductor materials; in chemical synthesis for metal selenides and organoselenium, lasers, and emulsions. It is hazardous to the respiratory system and eyes; toxic by inhalation and contact. Symptoms of exposure include nausea, vomiting, diarrhea, metallic taste, garlic breath, dizziness, lassitude, fatigue, irritation of the eyes, nose, and throat. OSHA exposure limit (TWA): 0.5 ppm [air] or 0.2 mg/m^3.

hydrogen sulfide — H_2S — a colorless gas with a strong odor of rotten eggs; MW: 34.1, BP: –77°F, Sol: 0.4%; sense of smell becomes rapidly fatigued and cannot be relied on to warn of its continuous presence; shipped as a liquefied compressed gas. It is used in underground mining operations; in refining of high-sulfur petroleum; in tanneries, glue factories, fat-rendering plants, and fertilizer plants; in the manufacture of viscose rayon; in the production of sulfur dyes, carbon disulfide, sulfur, oleum, and thioprene; in the vulcanization of rubber; in the manufacture of coke from coal with high gypsum content. It is hazardous to the respiratory system and eyes; and is toxic by inhalation, ingestion, and contact. Symptoms of exposure include apnea, coma, convulsions, eye irritation, conjunctivitis, pain, lacrimation, photophobia, corneal vesiculation, dizziness, headache, fatigue, irritability, insomnia, gastrointestinal dis-

tress, respiratory system irritation. OSHA exposure limit (TWA): 10 ppm [air] or 14 mg/m^3.

hydrojet — a device which blends air and water to create a high velocity turbulent stream of air-enriched water.

hydrokinetic — of or relating to the motion of fluids or the forces which produce or affect such motions.

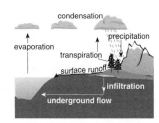

Hydrologic Cycle

hydrologic cycle — the continuous process of water being evaporated from the sea, precipitated over the land, and eventually returned to the sea. Also known as water cycle.

hydrology — the science dealing with the properties, distribution, and flow of water on or in the earth.

hydrolysis — a chemical reaction in which complex molecules are broken down into simpler molecules by combining with water, for instance fats into fatty acids and glycerol; (*carbohydrates*) the action of water in the presence of a catalyst upon one carbohydrate to form simpler carbohydrates; (*salts*) a reaction involving the splitting of water into its ions by the formation of a weak acid, base, or both.

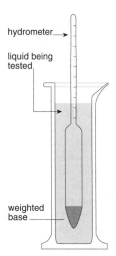

Hydrometer

hydrometer — a direct-reading instrument for indicating the density, specific gravity, or some similar characteristic of liquids.

hydrophilic — characterized by a tendency to hydrate easily and become quite stable.

hydrophobic — characterized by an inability to combine with or dissolve in water.

hydrophytic — characterized by growth in or in close proximity to water.

hydroponics — a commercial technique for the growing of plants without the use of soil by the use of nutrient solutions.

hydroquinone — $C_6H_4(OH)_2$ — light tan, light gray, or colorless crystals; MW: 110.1, BP: 545°F, Sol: 7%, Fl.P: 329°F (molten), sp gr: 1.33. It is used in the preparation and use of photographic developers; in the dyeing and fur processing industries; as an antioxidant; as a chemical stabilizer; in the synthesis of hydroquinone ethers and vitamin E. It is hazardous to the eyes, respiratory system, skin, and central nervous system; and is toxic by inhalation, ingestion, and contact. Symptoms of exposure include conjunctivitis, keratitis, central nervous system excitement, colored urine, nausea, dizziness, suffocation, rapid breathing, muscle twitches, delirium, collapse, irritation of the eyes. Also known as hydrochinone. OSHA exposure limit (TWA): 2 mg/m³ [air].

hydrosphere — the aqueous or water envelope of the earth, including all bodies of water (oceans, lakes, streams, underground waters, and vapor in the atmosphere).

hydrostatic — of or relating to fluids at rest or to the pressures they exert or transmit.

hydroxyl — a chemical prefix indicating the –OH radical in a chemical compound.

hyetometry — the measurement of precipitation; the study of the origin, structure, and other features of all forms of condensation, sublimation, and precipitation.

hygrograph — an instrument for recording relative humidity.

hygrometer — an instrument to measure humidity in the atmosphere.

hygrometry — the science of moisture measurements of evaporation, evapotranspiration, condensation, precipitation, and the water vapor content of the atmosphere.

hygroscopic — characterized by the capacity to take up and retain moisture readily.

hygroscopic particulates — particles which collect water from the atmosphere and form a mist that may limit visibility.

hymenolepiasis — (*disease*) an intestinal infection with dwarf tapeworms; ranges from asymptomatic to causing enteritis with or without diarrhea, abdominal pain, loss of weight, and weakness. Incubation time is approximately 2 weeks; caused by *Hymenolepis nana*; found more commonly in warm than cold areas as common tapeworm in southeastern United States. Reservoirs of infection include infected people and possibly mice. It is transmitted by the eggs of *H. nana* passed into the feces and contaminating food or water which is ingested by the individual; communicable as long as eggs are passed in the feces, which may be for several years; general susceptibility. It is controlled by good personal hygiene and proper protection of food and water from contamination with human and rodent feces.

hypalon — a widely used synthetic rubber providing exceptional weather, ozone, and sunlight resistance.

hyperbaric conditions — pressure conditions in excess of surface pressure.

hyperkeratosis — hypertrophy of the horny layer of the skin; hypertrophy of the cornea.

hypermenorrhea — *See* menorrhagia.

hyperpigmentation — a darkening of the skin which may be a postinflammatory response to chemical photosensitizers such as tar, pitch, and drugs; physical agents such as ultraviolet light, thermal radiation, and ionizing radiation; trauma; and chemicals, such as arsenic and certain aromatic hydrocarbons.

hyperplasia — increase in volume of a tissue or organ caused by the growth of new cells.

hypersensitivity — a state of altered reactivity in which the body reacts with an exaggerated immune response to a foreign agent such as a chemical or antigen; responsible for allergic reactions.

hypersensitivity diseases — diseases characterized by allergic responses to animal antigens; those most clearly associated with indoor air quality are asthma, rhinitis, and hypersensitivity pneumonitis.

hypertension — abnormally elevated blood pressure; in adults generally regarded at 140 mm Hg in the systolic pressure and 90 mm Hg in the diastolic pressure; a diagnosis of hypertension should be based on several readings, since a single reading can be influenced by emotional state or physical activity.

hypertrophy — nontumorous increase in the size of an organ as a result of enlargement of constituent cells without increase in the numbers.

hyperventilation — an increase of air in the lungs above the normal amount; leads to fainting, impaired consciousness, tightness of the chest, and a sensation of smothering and apprehension.

hypha — the tread-like structure of the mycelium in a fungus.

hypochlorinator — a chemical feeder through which liquid solutions of chlorine bearing chemicals are fed into the pool water at a controllable rate.

hypochlorite — any organic compound containing a metal and the (OCL) radical.

hypochlorous acid — HOCl — an unstable acid with excellent bactericidal and algicidal properties.

hypoglycemia — low levels of sugar in the blood.

hypolimnion — a region of a body of water that extends from below the thermocline to the bottom of the lake; it is removed from much of the surface influence and is characterized by uniform temperature that is generally cooler than that of other strata.

hypophysis — *See* pituitary gland.

hypopigmentation — a loss of pigment in the skin caused by a postinflammatory response due to exposure to thermal radiation, ultraviolet radiation, or chemical burns.

hypotension — abnormally low blood pressure, generally regarded at levels below 100 diastolic and 40 systolic.

hypothermia — subnormal temperature of the body.

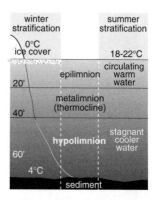

Hypolimnion

hypothyroidism — a deficiency of thyroid gland activity with underproduction of thyroxin or the condition resulting from it; may be caused by goitrogens.

hypoxemia — deficient oxygenation of the blood.

hypoxia — a deficiency of oxygen reaching the tissues of the body.

hypsometry — (*geography*) the measurement of the elevation of the earth's surface with respect to sea-level.

Hz — *See* hertz.

G
H

I

IAQ — *See* indoor air quality.

IARC — acronym for International Agency for Research on Cancer.

IBP — acronym for initial boiling point.

ICB — *See* immobilized cell bioreactor.

ICC — *See* Interstate Commerce Commission.

ICD — *See* International Statistical Classification of Diseases, Injuries, and Causes of Death.

ice fog — an atmospheric suspension of highly reflective ice crystals which affect visibility.

ICRP — acronym for International Commission on Radiological Protection.

ID — notation for inside diameter.

ideal gas — gas that obeys Boyle's law $PV = RT$ strictly at all temperatures and pressures. Also known as perfect gas.

idiopathic — a morbid state arising without a known cause.

idiosyncratic response — an unusual physiologic or metabolic reaction to a chemical that is consistently reproducible; the response observed is qualitatively similar to that observed in all individuals, but may take the form of extreme sensitivity to low doses or extreme insensitivity to high doses of the chemical.

IDLH — *See* immediately dangerous to life or health.

IEMS — acronym for Integrated Emergency Management System.

Igneous Rock

igneous rocks — rocks formed by the solidification of magma (molten silicate liquid); classified as plutonic, hypabyssal, and volcanic depending on the depth at which they were formed.

ignitable — characterized by the capacity to be set on fire.

ignition temperature — the lowest temperature of a fuel at which combustion becomes self-sustaining.

IGR — *See* insect growth regulator.

ileum — the third and longest region of the small intestine in mammals; it is the site of digestion and absorption.

illness — a condition marked by a pronounced deviation from the normal health state.

illness prevention — the art devoted to the recognition, evaluation, and control of health hazards.

ILO — *See* International Labor Organization.

Imhoff tank — (*sewage*) a two-story tank used for the settling of solids from sewage in which the liquid flows to the upper story, settling occurs, and the solids fall through a slot into the sludge compartment.

immediately dangerous to life or health (IDLH) — the maximum environmental concentration of a contaminant for which one could escape within thirty minutes without developing escape-impairing symptoms or irreversible health effects.

imminent health hazard — a significant threat or danger to health that is considered to exist when sufficient evidence shows that a product, practice, circumstance, or event creates a situation that requires immediate correction or cessation of an operation to prevent potential injuries and/or severe injuries.

immiscible — referring to liquids which are not soluble in each other or do not mix.

immobilized cell bioreactor (ICB) — an aerobic fixed-film bioreactor system designed to remove organic contaminants including nitrogen-containing compounds and chlorinated solvents from processed wastewater, contaminated groundwater, and other aqueous systems.

immune — resistant to disease.

immune status — the state of the body's immune system in regard to heredity, age, diet, and physical and mental health.

immune system — a body system that helps resist disease-causing microorganisms.

immunity — the ability of an organism to resist disease or toxins by natural or artificial means.

immunization — the process of rendering a subject immune. Also known as inoculation; vaccination.

immunoassay — any of several methods for the quantitative determination of chemical substances such as hormones, drugs, and certain proteins that utilize the highly specific binding between an antigen and an antibody.

immunodeficiency — a deficiency of immune response or a disorder characterized by deficient immune response.

immunologic toxicity — the occurrence of adverse effects on the immune system that may result from exposure to environmental agents such as chemicals.

immunology — the science that deals with the phenomenon and causes of immunity and immune responses.

immunosuppression — suppression of an immune response by the use of drugs or radiation.

impact injury — deformation of human body tissues beyond their failure limits which results in damage of anatomic structures or alteration in normal function.

impaction — the forcible contact of particles of matter on a surface.

impact mill — a machine that grinds material by throwing it against heavy metal projections rigidly attached to a rapidly rotating shaft.

impact tube — *See* pitot tube.

impedance (Z+) — the opposition to the flow of alternating electrical current measured in ohms; the rate at which a substance absorbs and transmits sound.

impeller — the rotating veins of a centripetal pump, turbine, blower, fan, or mixing apparatus.

impermeable — *See* impervious.

im personan — against the person.

impervious — not permitting water or other fluid to pass through. Also known as impermeable.

impingement — removal of liquid droplets from a flowing gas or vapor stream by causing it to collide with a baffle plate at high velocity so that the droplets fall away from the stream. Also known as liquid knock-out.

impinger — a device used to sample dust in the air that draws in air and directs it through a jet and onto a wetted glass plate for counting.

impoundment — a body of water confined by a dam, dike, floodgate, or other barrier.

impulse — a nerve message triggered by a stimulus or a response carried by nerve cells through the nervous system.

impulse noise — a noise of short duration, usually less than one second, and of high intensity with an abrupt onset and rapid decay.

impurity — a chemical substance which is unintentionally and undesirably present in a pure substance.

IMS — *See* ion mobility spectrometry.

in. — *See* inch.

inactivation — a reaction between two chemicals to produce a less toxic product.

incandescent — pertaining to an object that has been heated to the point where it becomes hot enough to radiate visible light.

inch (in.) — a unit of length equal to 1/12 foot or 2.54 centimeters.

incidence — (*epidemiology*) the number of cases of disease, infection, or some other event having an onset during a prescribed period of time in relation to the unit of population in which they occur.

incidence rate — the rate at which new cases of a disease occur in a population at risk during a

I
J

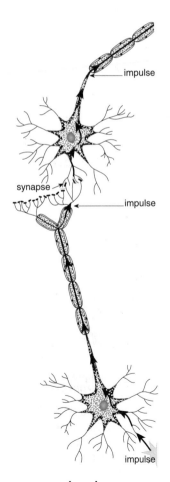

Impulse

specified period of time; most useful in determining factors associated with the etiology of disease and in evaluating programs of prevention. Also known as incidence ratio.

incidence ratio — *See* incidence rate.

incineration — the controlled combustion of solid, liquid, or gaseous combustible wastes which are ignited and burned to form carbon dioxide, water vapor, and other gaseous products, and from which the solid residues contain little or no combustible material.

incinerator — an engineered piece of equipment used to burn waste substances and of which all of the factors of combustion, including temperature, retention time, turbulence, and combustion air, can be controlled.

incinerator residue — all solid material remaining after an incineration process is completed

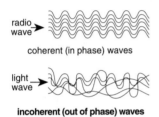

Incoherent Waves

incoherent waves — waves whose crests and troughs are not synchronized.

incompatible — materials that can cause dangerous reactions by direct contact with one another.

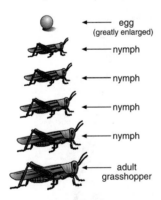

Incomplete Metamorphosis

incomplete metamorphosis — (*zoology*) an incomplete form of insect metamorphosis in which the pupal stage is lacking, as in roaches. Also known as hemimetabolism.

incubation — the provision and maintenance of ideal conditions for growth and development, as in the incubation and growth of bacteria.

incubation period — the time period of infection required before the appearance or development of disease symptoms.

incubator — an apparatus for maintaining optimal conditions of temperature and humidity for growth and development of organisms.

Index Medicus — a monthly index of the world's leading biomedical literature which is indexed by subject and author and published by the National Library of Medicine.

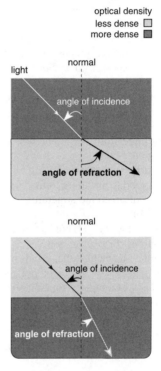

Index of Refraction

index of refraction — the refracted or bending ability of light as it passes through various substances.

Index to United States Government Periodicals — an index of government periodicals from 1974 to the present.

Indian Meal Moth (*Plodia interpunctella*) — a moth with a wing expansion of 1/2–3/4″ with the basal 1/3 of each forewing dull white to cream and the other two thirds brown to copper; a stored food product insect.

indicator compounds — chemical compounds, such as carbon dioxide, whose presence at certain concentrations may be used to estimate air quality conditions in buildings.

indicator paper test — a rapid test of limited accuracy in which test papers usually impregnated with starch iodide are immersed and the developed color compared with a standard.

indicator tube — a simple and useful field test for detecting a gas or vapor with limited accuracy utilizing instantaneous samples of air.

indictment — the formal statement of the charges against an individual presented by a grand jury, usually utilized in felony cases; the United States Attorney or State Attorney (with witnesses deemed necessary) appears before a grand jury and attempts to convince this body that an offense has been committed. If the grand jury concurs, it approves, signs, and returns the true bill of indictment to the court, and the matter is then entered on the court docket.

indirect association — a valid, but non-causal association between a factor and a disease due to some common underlying condition.

indirect waste pipe — a waste pipe that does not connect directly with the drainage system, but discharges into the drainage system through an air break or air gap into a trap, fixture, receptacle, or interceptor.

individual sewage system — a single system of piping, tanks, or other facilities collecting and disposing of sewage in whole or part into the soil.

individual sewage treatment system — a system or part thereof serving a dwelling, other establishment or group thereof, which utilizes subsurface soil treatment and disposal.

indoor air pollutants — pollutants within a structure that may contribute to discomfort or disease; these pollutants include radon, environmental tobacco smoke, biological contaminants, contaminants from stoves, heaters, fireplaces and chimneys, household products, pesticides, formaldehyde, asbestos, lead, and solvents.

indoor air pollution strategies — pollution elimination strategies include source control, the removal of the material before it enters the structure, the elimination of the individual sources of pollution, or ventilation improvement, which includes greater flow of air and the use of air cleaners either table top models or whole-house systems.

indoor air quality (IAQ) — a constantly changing interaction of complex factors in the indoor environment including odors; biological, chemical, or physical contaminants; design and/or operation and maintenance of the heating, ventilation, and air conditioning system; the building occupants; and the potential for healthful living conditions or disease and discomfort.

indoor air research — the development and testing of monitoring devices, and the design and implementation of field studies to identify and quantitate indoor sources of pollutants.

induced currents — currents that flow in an object as a result of exposure to an electric or magnetic field.

induced draft — the negative pressure created by the action of the fan, blower, or ejector located between an incinerator and a stack, or at a point where air or gases leave a unit.

induced draft fan — a fan that exhausts hot gases from heat-absorbing equipment, dust collectors, or scrubbers.

induced radioactivity — radioactivity produced in a substance after bombardment with neutrons or other particles. Also known as artificial radioactivity.

induction — the interval between an exposure to a carcinogen and the transformation of cells to occur.

induction period — the time interval during which a sufficient cause is complete to indicate disease.

induration — the process of hardening.

industrial gases — gases produced by industry and classified as industrial, organic solvent vapors, upper respiratory irritants, pulmonary irritants, chemical asphyxiants, and simple asphyxiants. They enter the body through the respiratory system, digestive system, and the skin; the gases are primarily inhaled, dissolved in the alveoli, and diffused into the blood; may form toxic substances causing acute poisoning, rapid injury, possible death, or chronic poisoning; gases enter the digestive tract when absorbed in food and saliva; some gases can enter the unbroken skin, or may cause dermatitis if they do not penetrate the skin.

industrial hygiene — the science and art of the recognition, evaluation, and control of factors or stresses which may lead to disease and injury in the workplace.

industrial hygienist — an environmental health professional who identifies and solves potentially hazardous problems by measuring and assessing harmful chemical, physical, and biological agents in the work environment.

industrial incinerator — an incinerator designed to burn a particular industrial waste.

industrial property — a parcel of property which is developed and used either in part or whole for manufacturing purposes.

industrial site selection process — an environmental study conducted to determine the location of emissions, quantity and quality of pollutants, control of pollutants, impact on the environment, and compliance with regulations prior to purchase of the property.

industrial solid waste — the solid waste that results from industrial processes in manufacturing.

industrial waste — any liquid, gaseous, or solid waste substance or combination thereof resulting from any process of industry, manufacturing, trade or business, or from the development, processing, or recovery of any natural resources.

inedible — adulterated or not intended for use as human food.

inert — lacking activity, reactivity, or effect.

inert gas — a rare gas that does not react with other substances under ordinary conditions; in the group O of the periodic table of elements. Also known as noble gas.

inertia — the property of matter which resists change in motion.

inertial collector — a dry collector mechanism in which a rapidly flowing gas is forced to change direction on contact with an obstacle, causing the particles in the gas to continue in the initial direction; an example is the cyclone collector where the gas stream moves rapidly into a cylindrical chamber through a tangential inlet duct (one set at an angle to the chamber wall) at the top of the cylinder; the gas stream whirls downward with increasing rapidity toward a cone-shaped base; the centripetal force throws the entrained particles out of the spinning gas stream onto the wall of the chamber, where they fall into a collecting hopper.

inertial separator — air pollution control equipment that uses the principle of inertia to remove particulate matter from an air or gas stream.

inerting — displacement of the atmosphere by a nonreactive gas, such as nitrogen, to such an extent that resulting atmosphere is noncombustible.

inert ingredient — inactive ingredient used as a filler or binder.

infant — a live-borne child from the moment of birth through the completion of the first year of life.

infant botulism — (*disease*) a distinct clinical form of botulism found only in infants under one year of age. Symptoms of exposure start with constipation, then lethargy, listlessness, poor feeding, sagging or prolapse of an organ or part thereof, difficulty in swallowing, weakness, and sometimes respiratory insufficiency, and respiratory arrest; may result in death. It results from colonization of the intestine by the botulimun bacillus with production of a toxin in the body. Incubation time not known. It is caused by *Clostridiun botulinum* types A, B or F; found worldwide. The reservoir of infection is spores in the soil; transmitted by ingestion of the spores on food or dust; not communicable from person to person. Those 2 weeks to 9 months of age are susceptible; controlled by proper handling of soiled diapers, feces, and soiled articles.

infant death — any death at any time from birth up to but not including one year of age.

infantile gastroenteritis — *See* rotavirus.

infant mortality rate — the number of infant deaths per 1,000 live-borne infants through a given period of time.

infarct — the death of a portion of tissue because of lack of blood supply resulting from obstruction of circulation in the area. Also known as infarction.

infarction — *See* infarct.

infection — the entry and development or multiplication of an infectious agent in the body of a living organism; pathogenic condition resulting from invasion of a pathogen; colonization occurring when an infectious agent has established itself on the host and propagates at a rate sufficient to maintain its numbers with or without disease manifestation.

infectious dose — the number of microbial cells required to initiate an infection.

infectious hepatitis — (*disease*) *See* viral hepatitis A.

infectious waste — equipment, instruments, utensils, and other objects capable of transmitting infectious material from areas or rooms where persons or animals with suspected or diagnosed communicable disease are housed or treated. They are not intended for re-use and are designated for disposal; human and animal specimens and disposal objects derived from surgical or other medical procedures on persons or animals with suspected or diagnosed communicable disease capable of transmitting infectious materials; laboratory wastes which may harbor or transmit infectious material.

infestation — the state or condition of having insects or other pests in a place where they may cause damage, disease, illness, injury, death, or annoyance.

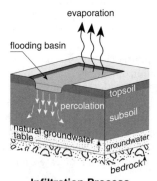

Infiltration Process

infiltration — (*water*) water entering a sewer system and service connections from the ground through such means as, but not limited to, defective pipes, pipe joints, connections, or man-hole walls; (*hydrology*) the movement of water into soil.

infiltration air — air that leaks into the chambers or ducts of an incinerator.

infirmity — an abnormal, disabling condition of the mind or body.

inflammation — a localized protective response caused by injury or destruction of tissues which serves to destroy or dilute the injurious agent and protect the injured tissue.

inflow — water discharged into a sewer system including service connections from such sources as roof leaders, cellar, yard and area drains, foundation drains, pooling water discharges, drains from springs and swampy areas, manhole covers, cross connections from storm sewers and combined sewers, catch basins, storm sewers, surface run-off, street wash waters, or drainage.

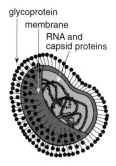

glycoprotein
membrane
RNA and
capsid proteins

Influenza Virus

influenza — (*disease*) an acute viral disease of the respiratory tract characterized by fever, chills, headache, myalgia, prostration, mild sore throat, and cough. Incubation time is 24–72 hours; caused by three types of influenza viruses: A, B, and C, with type A associated with wide spread epidemics and pandemics; B localized epidemics; and type C, sporadic cases; found in all areas. Reservoirs of infection include humans and possibly animals. It is transmitted by direct contact through droplet infection and airborne spread; communicable for 3 days; general susceptibility. It is controlled by active immunization, pharmaceutical treatment, and good personal hygiene.

information — (*law*) the formal statement of the charges against an individual in a misdemeanor; the government may proceed through furnishing the United States Attorney or State's Attorney the information which is then filed with the clerk of the court.

information letter — (*law*) a letter to responsible officials that affirms warning of the existence of a violative practice.

information superhighway — *See* communications superhighway.

informed consent — consent by an individual to become the subject of a human subject study after the individual understands the nature and implications of the study and its objectives.

infrared — the long rays of radiation beyond the red end of the visible spectrum.

infrared absorption spectrophotometry — molecular absorption spectrometric technique involving the dispersing of a polychromatic infrared beam of light using a suitable prism or diffraction grading.

infrared light — the light with lower frequencies than red light, which is less than about 4.3×10^{14} sec^{-1}.

infrared radiation (IR) — electromagnetic radiation below the visible light spectrum with wavelengths longer than those of red light but shorter than radio waves; from 0.75 or 0.8 micrometers to 1,000 micrometers.

infrared thermal destruction — a mobile thermal processing system that uses electrically-powered silicon carbide rods to heat organic waste to combustion temperatures.

ingestion — the act or process of taking food and other substances into the body by the mouth; a route of entry into the body along with food or water, or through inhalation and then ingested; the materials may cross through the gastrointestinal tract and into the blood.

ingot — a block of iron, steel, or other metal cast in a mold for ease in handling before processing.

ingredients — (*pesticides*) the active pesticide chemical plus other inactive chemicals in a pesticide formulation or mixture listed in the ingredient statement on the container or package.

ingredient statement — the part of the label on a pesticide container which gives the name and amount by weight of each pesticide chemical, plus the amount by weight of the inactive material in the container.

inhabit — to occupy as a place of settled residence.

inhalation — the act or instance of breathing in a substance in the form of a gas, vapor, fume, mist, or dust; a route of entry into the body of microorganisms, chemicals, or physical agents during breathing; the microorganisms can cause infections or chronic disease; the chemicals can be absorbed into the lungs, rapidly pass into the bloodstream, and ultimately reach the vital organs; the physical agents can cause a mechanical blocking or irritation of the airway.

inhalation LC$_{50}$ — a concentration of a substance expressed as milligrams per liter or parts per million parts of air that is lethal to 50% of the test population under specified test conditions.

I
J

inhalation therapy — the therapeutic use of gases or aerosols by inhalation into the respiratory system.

inhalation valve — a device that allows respirable air to enter the facepiece and prevents exhaled air from leaving the facepiece through the intake opening.

inhalator — a device providing a mixture of oxygen and carbon dioxide for breathing that is used especially in conjunction with artificial respiration.

medication

mouthpiece

Pump Style Inhaler

inhaler — a device containing a medication which is drawn into the air passages by inhalation.

in. Hg — notation for inches of Mercury.

inhibitor — any substance that interferes with a chemical reaction, growth, or other biological activity; also used to prevent or retard rust or corrosion.

in. H_2O — notation for inches of water.

initial review — (*environmental impact statement*) a preliminary public discussion of the information to be developed, alternatives to be considered, and issues to be discussed.

initiation — (*cancer*) the transformation of a benign or harmless cell to one with the potential for malignant growth.

initiator — a chemical or substance that can cause the initial step in a chain reaction or in the process of carcinogenesis.

injection — the forcing of a liquid into the skin, vessels, muscles, subcutaneous tissue, or any cavity of the body; the forcing of a chemical or other substance into a plant, animal, soil material, or enclosure.

injection well — a well into which fluids are being injected to drive remaining oil into the vicinity of production wells.

injunction — an order issued by the court requiring the defendant to do or refrain from doing a specified act.

injurious — harmful.

injury — a stress upon an organism that disrupts the structure or function and results in a pathological process.

injury controller (*nonworkplace*) — an environmental health practitioner concerned with the identification and elimination of potentially hazardous construction or conditions, and the education of the public in order to prevent unintentional, non-occupational, and non-motor vehicle accidents or injuries.

injury prevention — the art of the recognition, evaluation, and control of safety hazards.

in lieu of — instead of; in place of.

innage — space occupied in a product container.

inner ear — a cavity in the vertebrate ear filled with a liquid containing many end-fibers of the auditory nerve. As the sound waves spread through the liquid, they vibrate the nerve endings, and the auditory nerve picks up the vibrations, sensing electric disturbances to the auditory center of the brain which produces the sensation sound.

subcutaneous injection

hypodermic needle

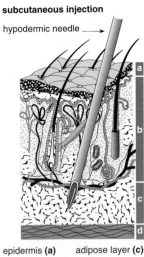

epidermis **(a)**　　　adipose layer **(c)**
dermis **(b)**　　　muscle layer **(d)**

Injection

hypodermic needle

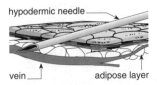

vein　　　　　　　adipose layer

intravenous injection

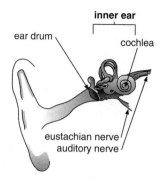

Inner Ear

inner liner — a continuous layer of material placed inside a tank or container which protects the construction materials from the contained waste reagents used to treat the waste.

innocuous — harmless.

innovative control technology — any system of pollution control technology that has not been adequately demonstrated in practice, but would have a substantial likelihood of achieving better results in reducing pollutants than standard control technology.

inoculate — to inject, introduce, or place fungi, bacteria, nematodes, viruses, or chemicals into a plant or animal to produce a mild form of disease and render the individual immune. *See also* immunization.

inoculum — bacteria placed in compost to start a biological action.

inorganic acid — compounds of hydrogen and one or more other elements that break down in water and other solvents to produce hydrogen ions; examples are chromic, hydrochloric, nitric, and sulfuric acids; in high concentrations they will destroy body tissue and cause chemical burns. They are especially hazardous to the upper respiratory tract, mucous membrane, and the teeth.

inorganic compound (IOC) — (*chemistry*) compounds that do not contain carbon as the principal element (except carbon dioxide, carbonates, cyanides, and cyanates) as distinguished from organic compounds; organic compounds may be limited to those that contain hydrocarbon groups, often along with oxygen and other elements.

inorganic gases — oxides of nitrogen, such as nitrogen dioxide, nitric oxide, and nitrous oxide; oxides of sulfur, such as sulfur dioxide and sulfur trioxide; oxides of carbon, such as carbon monoxide and carbon dioxide; and other inorganics, such as hydrogen sulfide, ammonia, and chlorine, which are constituents of nonliving things and are not characteristically biological.

inorganic soil material — inorganic components consisting of rock fragments and minerals ranging in size from stones and gravel to microscopic pieces of clay.

inorganic solids — (*sewage*) inert materials not subject to decomposition including sand, gravel, silt, and mineral salts.

in-process control technology — the conservation of chemicals and water throughout production operations to reduce the amount of wastewater discharged.

input — the process of entering data.

input bias current — current that flows into the inputs of a nonideal amplifier due to leakage current, gate current, or transistor bias current.

in rem — (*law*) against the defendant.

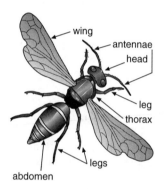

Insect
(Potter Wasp)

insect — an arthropod with three pairs of legs and three body regions: head, thorax, abdomen, and often when adult, wings.

insect growth regulator (IGR) — a chemical substance, such as methoprene; a true analog of the mosquito's own juvenile hormone which causes the death of the mosquito by maintaining high levels of the hormone during the latter instar stages and therefore prohibiting the development of the physical features necessary for adult emergence from the pupal stage.

insecticide — a pesticide chemical that prevents damage from or destroys insects.

in situ — in the natural or normal place.

in situ **air sampler** — a sensor which makes constant contact with the air sample directly with or without the aid of an air mover.

in situ **vitrification** (ISV) — a technique using an electric current to melt soil or sludge at extremely high temperatures of 1,600–2,000°C to destroy organic pollutants by pyrolysis.

insol — *See* insoluble.

I
J

insoluble (insol) — incapable of being dissolved in another material.

insomnia — the inability to sleep.

inspection — (*insurance*) the investigation of certain risks which may be made by independent inspection firms or by the company before issuance or during the term of the policy; an evaluation of a problem or area by visual means.

inspection criteria — the maximum and minimum acceptable values associated with a particular sampling plan.

inspirable particulate mass threshold limit value — those materials which are hazardous when deposited anywhere in the respiratory tract.

instar — a stage in the life of an arthropod, such as a mosquito, between two successive molts.

institutional and research laboratory waste — the combination of garbage, refuse, disposables, and dressings which are considered to be contaminated.

institutional environmental health manager — an environmental health professional responsible for the control of biological, physical, and chemical factors which affect the institutional setting.

institutional solid waste — the solid waste originating from educational, health care, and research facilities.

instrument — a device for making observations and measurements of a quantity under investigation.

instrument inlet — the opening at the instrument through which the sample enters the analyzer, excluding all external sample lines, probes, and manifolds.

insulating brick — a firebrick with low thermal conductivity and a bulk density of less than 70 pounds per cubic foot; suitable for insulating industrial furnaces.

insulator — a material through which electric charges, heat, or sound are not readily conducted.

insulin — a hormone secretion of the islet of Langerhans in the pancreas which regulates the oxidation of sugar in the tissues and controls its concentration in the blood.

intake area — *See* recharge zone.

integer — a number without a decimal component.

integrated pest management (IPM) — a systematic, comprehensive approach to pest control that uses the insect's or rodent's own biology and behavior to find the least toxic control methods at the lowest cost; it includes monitoring, biological control, chemical control, removal of harborage, removal of food, and insect- and rodent-proofing.

integrated sample — the sum of a series of small samples or a continuous flow of samples collected over a defined time period so as to create a large, average sample.

integrating nephelometer — an instrument for remote sensing of the vertical and horizontal profile of particle concentration in air in terms of the light-scattering coefficient.

integration — the operation of computing the area within mathematically defined limits; the arrangement of components in a system so they function together in a logical and efficient way.

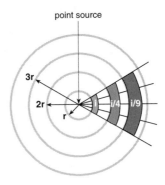

Sound Intensity

intensity — amount of energy per unit time passing through a unit area perpendicular to the line of propagation at the point in question.

interactive — (*computer science*) a system which allows the operator to receive information from the computer and initiate or modify the program.

interactive graphics system — a computer system used for the preparation of graphics consisting of a central computer and workstation allowing input from all areas.

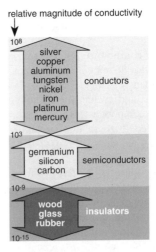

Insulator

interceptor — a device designed and installed so as to separate and retain hazardous or undesirable material from normal waste while permitting normal sewage or liquid waste to discharge into the drainage system by gravity.

interceptor sewers — a means of control of flow of the sewage to the treatment plant; in the event of a storm, they allow some of the sewage to flow directly into a receiving stream, which protects the treatment plant from being overloaded in case of a sudden surge of water.

interface design — (*ergonomics*) the exchange of information between person, machine, and environment to provide necessary data for an efficient operation and safe work place.

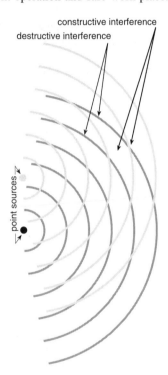

constructive interference
destructive interference
point sources

Interference

interference — any undesired energy that tends to disrupt the reception of desired signals; an effect resulting when two series of waves merge into each other.

interferon — a natural glycoprotein released by cells invaded by viruses.

interleukin — one of several proteins important for lymphocyte proliferation.

intermediate — an organic chemical formed as a step in a series of chemical reactions; a precursor to a desired product.

intermediate exposure — occupational exposure to a chemical for a duration of 15–364 days.

intermedia transport — movement from one environmental medium to another.

intermittent — characterized by alternating periods of activity and inactivity.

intermittent filter — a natural or artificial bed of sand or other fine grain material to the surface of which wastewaters are applied intermittently in flooding doses giving an opportunity for filtration and the maintenance of an aerobic condition.

intermittent positive-pressure breathing (IPPB) — a form of respiratory therapy using a ventilator for the treatment of patients with inadequate pulmonary activity. The treatment involves application of pressure only during the inspiratory phase forcing adequate oxygen delivery and carbon dioxide removal from the lungs.

intermittent sand filter — specially prepared beds of sand from which effluent from primary treatment, trickling filters, or secondary settling tanks are applied intermittently; the filter bed is a true filter for the solids present, although oxidation may occur at or near the surface of the sand.

intermittent stream — a stream that flows during parts of the year.

internal combustion engine — an engine in which the fuel burns inside the cylinders rather than in a separate furnace.

internal conversion — one of the possible mechanisms of decay from the metastable state (isomeric transition) in which the transition energy is transferred to an orbital electron causing its ejection from the atom.

internal dose — the amount of a substance penetrating across absorption barriers by physical or biological means.

internal housing environment — heat, light, ventilation, electrical facilities, plumbing, structural soundness, and physical structure of dwellings as they affect people.

internal radiation — exposure to ionizing radiation from radioisotopes deposited inside body tissues through inhalation, ingestion, absorption, or injection.

internal respiration — the exchange of gas between the blood and tissues of an organism.

internal soil drainage — the quality of a soil that permits the downward flow of excess water through the soil.

International Labor Organization (ILO) — a permanent international organization created by the Paris Peace Conference of 1919 to promote the improvement of working and living conditions as an essential contribution to safeguarding peace throughout the world.

I
J

International Nursing Index — an international index of over 260 nursing journals and selected topics from 2,700 allied health and biomedical journals from 1967 to the present.

International Statistical Classification of Diseases, Injuries, and Causes of Death (ICD) — an internationally recognized system for coding problems of health; revised every ten years by the World Health Organization.

International System of Units (SI units) — the international system of physical units recommended by the General Conference on Weights and Measures (1960); fundamental quantities are length, time, mass, electric current, temperature, luminous intensity, and amount of substance; corresponding units are the meter, second, kilogram, ampere, kelvin, candela, and mole. Also known as Système International d'Unités.

internet — (*computer science*) an international network of networks; an electronic communications vehicle allowing computers to talk to other computers thereby sending and receiving mail, database searching, file transfer transmissions, etc.

interphase — the phase of the cell cycle between cell divisions during which much of the synthesis of cellular constituents occurs.

interpolate — to estimate that value of an attribute at an unsampled point from measurements made at surrounding sites.

interrogatories — (*law*) written questions propounded by one party and served to an advisor who must present written answers thereto under oath.

intersection — nonparallel touching or crossing of fibers with the projection having an aspect ratio of 5:1 or greater.

interspike interval (ISI) — the time between two successive repetitions of a motor unit action potential.

interstate (IS) — between or among states.

Interstate Commerce Commission (ICC) — a federal agency regulating interstate surface transportation including trains, trucks, buses, water carriers, freight forwarders, transportation brokers, and coal slurry pipelines.

interstitial — pertaining to or situated in the space between two things.

interstitial water — the moisture which collects in voids between soil particles. It is relatively mobile, dissolves solids, and moves through the soil.

intertrigo — an erythemous skin eruption occurring on surfaces of skin such as the creases of the neck, folds of the groin, and armpits.

interval — period of time separating two events or the distance between two objects.

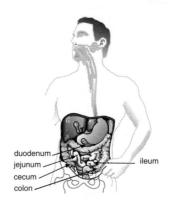

duodenum
jejunum
cecum
colon
ileum

Intestine

intestine — the portion of the alimentary canal extending beyond the stomach. It is the site of final digestion of food matter from the stomach, the absorption of soluble food matter and water, and the production of feces.

intima — the innermost structure.

intoxication — the state of being poisoned; the state produced by overindulgence in alcohol.

intracellular — within the cell.

intracutaneous — within the substance of a skin.

intramedia transport — movement within an environmental medium.

intraperitoneal — within the abdominal cavity.

intrastate — within a state.

intrauterine — within the uterus.

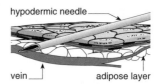

hypodermic needle

vein   adipose layer

Intravenous Injection

intravenous — within a vein.

intravenous drug — a drug injected by needle directly into a vein.

intruding noise level — the total sound level in decibels created, caused, maintained, or originated from an alleged offensive source at a specified location while the alleged offensive source is in operation.

invasive — the tendency to spread, especially the tendency to invade healthy tissues.

inventory — a specific list of problems or equipment-related problems determined by inspections, field visits, or use of specialized forms prepared by personnel at the source of the potential pollution.

inverse-square law — a relationship in which the strength of a physical quantity is proportional to the reciprocal of the square of the distance from the source of that property; applies to sound, gravity, heat, light, and radio waves.

inversion — the phenomenon of a layer of cool air trapped below a layer of warmer air so that the bottom layer cannot rise as usual; an increase in an area's temperature with height, whereas usually there is a decrease of temperature with height. Also known as temperature inversion.

invertebrate — an animal without a backbone or internal skeleton.

invert emulsion — an emulsion having the water suspended as small droplets in oil.

in vitro — pertaining to a biological reaction taking place outside the host organism in an artificial environment.

in vivo — pertaining to a biological reaction taking place within a living organism.

invoice — paperwork which shows the vendor's intent to sell an article but does not prove actual movement.

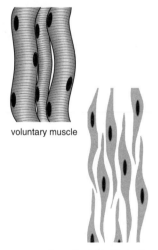

voluntary muscle

Involuntary Muscle

involuntary muscle — any muscle which cannot be controlled at will. Also known as smooth muscle.

IOC — *See* inorganic compound.

iodide — a chemical compound containing iodine; when potassium or sodium iodide is used with a super oxidizing agent, such as chlorine, iodine will be released in pool water.

iodine — I_2 — a violet solid with a sharp, characteristic odor; MW: 253.8, BP: 365°F, Sol: 0.01%, sp gr: 4.93. It is used in the synthesis of chemical intermediates, pharmaceuticals, photographic chemicals, antiseptics, disinfectants, detergent sanitizers, and organic and inorganic compounds; as a reagent in analytical chemistry; as a catalyst in organic synthesis, synthesis of food dyes, food additives, and coloring agents; in the production of intermediates in purification of minerals; as a catalyst in the modification of selenium during the manufacture of photoelectric cells and rectifiers; during the manufacture of specialty lubricants and in the production of stereospecific polymers. It is hazardous to the respiratory system, eyes, skin, central nervous system, and cardiovascular system; and is toxic by inhalation, ingestion, and contact. Symptoms of exposure include lacrimation, headache, tight chest, skin burns, rash, cutaneous hypersensitivity, irritation of the eyes and nose. OSHA exposure limit (TWA): ceiling 0.1 ppm [air] or 1 mg/m^3.

iodophore — a soluble complex of iodine and a nonionic surface-active agent that releases iodine gradually and creates a bacteriocidal action in cold or hot water.

ion — an atomic particle, atom, or chemical radical bearing an electrical charge, which is either negative or positive, depending on whether it has gained or lost one or more electrons; its migration affects the transport of electricity through an electrolyte, or to a certain extent, through a gas.

ion exchange — a process by which certain ions of given charge may be absorbed from solution and replaced in the solution by other ions with a similar charge from the absorbent; a chemical process involving reversible interchange of ions between a liquid and a solid, but no radical change in structure of the solid.

ion exchange resin — a special type of molecule, such as zeolite, or a synthetic resin containing active groups that give the resin the property of combining with or exchanging ions between the resin and the solution, as in the process of softening hard water.

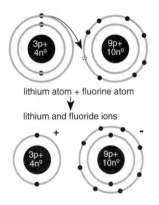

lithium atom + fluorine atom

lithium and fluoride ions

Ionic Bond

ionic bond — a type of chemical force holding atoms together resulting from differences in electrical charges. Also known as electrovalent bond.

ionization — the process by which a neutral atom or molecule acquires a positive or negative charge.

I
J

ionization chamber — an instrument designed to measure the quantity of ionizing radiation in terms of the charge of electricity associated with ions produced within a defined volume.

ionization density — the number of ion pairs per unit volume.

ionization path — the trail of ion pairs produced by ionizing radiation and their passage through matter.

ionizing energy — the average energy lost by ionizing radiation producing an ion parent in a gas.

ionizing radiation — high-energy electromagnetic or particulate radiation released by nuclear decay, and capable of producing ions directly or indirectly in its passage through matter.

ion mobility spectrometry (IMS) — a technique used to detect and characterize organic vapors in air involving the ionization of molecules and their subsequent temporal drift through an electric field.

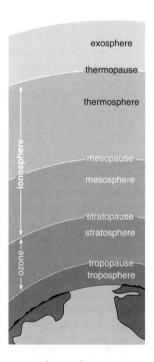

Ionosphere

ionosphere — the layer of the atmosphere which reflects radio waves and which is located directly above the stratosphere at 40 or 50 miles, and it extends indefinitely.

ion pair — two particles of opposite charge, usually referring to the electron and positive atomic or molecular residue resulting after the interaction of ionizing radiation with the orbital electrons of atoms.

ipecac — the dried rhizome and root of several tropical South American shrubs used as an emetic or expectorant.

IPM — *See* integrated pest management.

IPM-TLVs — *See* inspirable particulate mass threshold limit value.

IPPB — *See* intermittent positive-pressure breathing.

ipso facto — by the act itself.

IR — *See* infrared radiation.

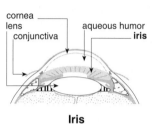

Iris

iris — the muscular colored portion of the eye behind the cornea and surrounding the pupil which regulates the amount of light reaching the retina in vertebrates and some cephalopod molluscs.

IRLG — acronym for International Regulatory Liaison Group.

iron bacteria — those bacteria able to obtain their energy for living by oxidizing iron compounds.

iron oxide dust and fume — Fe_2O_3 — a reddish-brown solid; exposure to fume may occur during the arc-welding of iron; MW: 159.7, BP: not known, Sol: insoluble, sp gr: 5.24. It is liberated in the production of steel ingots; processing of iron ore and pig iron; heating and pouring of molten metal in foundry operations; hot rolling sheet and strip steel; during the forging of metal items containing iron and steel; pressing of metal items, grinding and polishing of glass, precious metals, stones, and gem stones. It is hazardous to the respiratory system; and is toxic by inhalation. Symptoms of exposure include benign pneumoconiosis with X-ray shadows indistinguishable from fibrotic pneumoconiosis. OSHA exposure limit (TWA): 10 mg/m³ [air].

iron sulfide pyrite — FeS — an ore commonly found in rocks. Also known as fool's gold.

irradiation — exposure to any form of radiation.

irritability — the ability to detect, interpret, and respond to a stimulus in the environment; a condition of abnormal excitability to slight stimuli.

irritant — a chemical which is not corrosive, but which causes a reversible inflammatory effect on a living tissue by chemical action at the site of contact; a liquid or solid substance which upon contact with fire or when exposed to air gives off dangerous or intensely irritating fumes.

irritation — an aggravation of a tissue at the point of contact with a material.

IS — *See* interstate.

ISC — acronym for Intersociety Committee on Methods for Air Sampling and Analysis.

ischemia — a local and temporary deficiency of blood as a result of obstruction of the blood supply.

ISI — *See* interspike interval.

islets of Langerhans — the special cells in the pancreas that secrete insulin.

isoamyl acetate — $CH_3COOCH_2CH_2CH(CH_3)_2$ — a colorless liquid with a banana-like odor; MW: 130.2, BP: 288°F, Sol: 0.3%, Fl.P: 77°F, sp gr: 0.87. It is used in treating natural leathers by tanning; as an extractant in purification of pharmaceuticals including penicillin; in the manufacture of artificial leather; in dry cleaning preparations; during the manufacture of artificial silk, rayon, and pearls; liberated during application of varnishes and nitrocellulose lacquers as protective and finish coatings for wood, paper, metal, leather, and other surfaces; during application of cellulosic adhesives in shoe manufacturing, book binding, packaging, leather and paper processing; during the manufacture of shoe and furniture polishes; during fermentation of whiskey grains; during the manufacture of cellulosic photographic film; during the prepara-

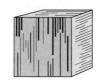

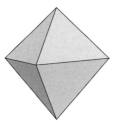

Iron Sulfide Pyrite

tion of perfumes, foods, and other materials for use as a flavoring and odorant; during the manufacture of bath sponges; during the cleaning and maintenance of acetate-processing equipment. It is hazardous to the eyes, skin, and respiratory system; and is toxic by inhalation, ingestion, and contact. Symptoms of exposure include narcosis, dermatitis, irritation of the eyes, nose, and throat. OSHA exposure limit (TWA): 100 ppm [air] or 525 mg/m³.

isoamyl alcohol—primary—$(CH_3)_2CHCH_2CH_2OH$; secondary — $(CH_3)_2CHCH(OH)CH_3$ — colorless liquids with a disagreeable odor; MW: 130.2, BP: 270/234°F, Sol: 2%/Not known (57°), Fl.P (open cup): 109/95°F, sp gr: 0.81/0.82. It is used as a vehicle, latent, or diluent solvent during the application of paints, lacquers, varnishes, thinners, and paint removers; in the synthesis of drugs and medicinals, as a solvent for alkaloids, and as an extractant of antibiotics in the pharmaceutical industry; in the manufacture of lacquers, paints, varnishes, thinners, and paint removers; as a chemical intermediate in organic synthesis of photographic chemicals and esters; in the manufacture of printing inks; as a solvent for resins, gums, waxes, and oils, perfumes, explosives, shoe cement, analytical determination for fat in milk, and artificial rubber; in the manufacture of antifoaming agents; in the mining industry as a frothing agent for flotation of non-ferrous ores; as a vehicle solvent for celloidin solutions in microscopy. It is hazardous to the eyes, skin, and respiratory system; and is toxic by inhalation, ingestion, and contact. Symptoms of exposure include narcosis, headache, dizziness, dyspnea, nausea, vomiting, diarrhea, skin cracking, irritation of the eyes, nose, and throat. OSHA exposure limit (TWA): 100 ppm [air] or 360 mg/m³.

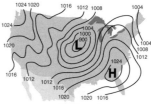

Isobar

isobar — (*physiology*) one of two or more chemical species with the same atomic weight, but different atomic numbers; (*meteorology*) a line drawn on a map or chart connecting locations of identical atmospheric pressure.

isobutyl acetate — $CH_3COOCH_2CH(CH_3)_2$ — a colorless liquid with a fruity, floral odor; MW: 116.2, BP: 243°F, Sol (77°F): 0.6%, Fl.P: 64°F, sp gr: 0.87. It is used in the manufacture of some perfumes, cosmetics, and flavoring agents; in shoe manufacturing, book binding, packaging, leather processing, photographic film manufacturing, and paper processing; liberated during the application of varnishes and nitrocellulose lacquers as protective coatings for wood, plastic, metal, leather, and other surfaces; during oven baking of phenolic and epoxy coatings; during the cleaning and maintenance of acetate-processing equipment. It is hazardous to the skin, eyes, and respiratory system; and is toxic by inhalation, ingestion, and contact. Symptoms of exposure include headache, drowsiness, anesthesia, irritation of the eyes, skin, and upper respiratory system. OSHA exposure limit (TWA): 150 ppm [air] or 700 mg/m^3.

isobutyl alcohol — $(CH_3)_2CHCH_2OH$ — a colorless, oily liquid with a sweet, musty odor; MW: 74.1, BP: 227°F, Sol: 10%, Fl.P: 82°F, sp gr: 0.80. It is used in the manufacture of industrial cleaners, nitrocellulose lacquers, paint removers, lubricating oil and hydraulic fluids, amino resins, and plasticizers; during the distillation of whiskey. It is hazardous to the eyes, skin, and respiratory system; and is toxic by inhalation, ingestion, and contact. Symptoms of exposure include headache, drowsiness, skin irritation and cracking, irritation of the eyes and throat. OSHA exposure limit (TWA): 50 ppm [air] or 150 mg/m^3.

isokinetic sampling — any technique for collecting airborne particulate matter in which the linear velocity of the gas entering the sampling nozzle is equal to that of the undisturbed gas stream at the sample point.

isolation — the separation of infected individuals from those uninfected for the period of communicability of a particular disease; a technique used in environmental control where potential pollutants are removed from substantial population centers.

isolation room — a separate room containing the sterilizer and the sterilizer loading area as it relates to the use of ethylene oxide; a room in which an infected patient is treated and housed.

isomer — one of several nuclides having the same number of neutrons and protons but capable of existing for a certain period of time in different states with different energies, and physical and chemical properties.

isomeric transition — the process by which a nuclide decays to an isomeric nuclide of lower quantum energy.

isometric — maintaining or pertaining to the same length.

isometric work — a state of muscular contraction without movement.

isophorone — $C_9H_{14}O$ — a colorless to white liquid with a peppermint-like odor; MW: 138.2, BP: 419°F, Sol: 1%, Fl.P: 184°F, sp gr: 0.92. It is used as a solvent in the manufacture of vinyl resins, nitrocellulose, fats, chlorinated rubber, herbicides, coatings, roll-coating finishes, paints, adhesives, and inks; in organic synthesis in the manufacture of lubricating oil additives, fungicides, and tetramethylquanidine. It is hazardous to the respiratory system and skin; and is toxic by inhalation, ingestion, and contact. Symptoms of exposure include narcosis, dermatitis, irritation of the eyes, nose, and throat. OSHA exposure limit (TWA): 4 ppm [air] or 23 mg/m^3.

isopropyl acetate — $CH_3COOCH(CH_3)_2$ — a colorless liquid with a fruity odor; MW: 102.2, BP: 194°F, Sol: 3%, Fl.P: 36°F, sp gr: 0.87. It is used as a vehicle solvent during spray applications of cellulose nitrate and ethyl cellulose lacquers; as a solvent in synthetic resins in the plastic industry; in the manufacture of printing and litho inks, perfumes, and flavoring agents; as a general solvent for waxes, gums, and oils; in organic synthesis. It is hazardous to the eyes, skin, and respiratory system; and is toxic by inhalation, ingestion, and contact. Symptoms of exposure include narcosis, dermatitis, irritation of the eyes, nose, and skin. OSHA exposure limit (TWA): 250 ppm [air] or 950 mg/m^3.

isopropyl alcohol — $(CH_3)_2CHOH$ — a colorless liquid with the odor of rubbing alcohol; MW: 60.1, BP: 181°F, Sol: miscible, Fl.P: 53°F, sp gr: 0.79. It is used as a solvent in spray and heat applications of surface coatings, including stain, varnish, nitrocellulose lacquers, and quick-drying inks and paints; in the manufacture of acetone; as a solvent in the manufacture of surface coatings and thinners; in organic synthesis for isopropyl derivatives; in the manufacture of cosmetics including liniments, skin lotions, permanent wave lotions, and color hair rinses; as a disinfectant and sanitizer; in cleaning and degreasing operations; in the extraction and purification of alkaloids, proteins, chlorophyll, vitamins, kelp, pectin, resins, waxes, and gums; in the manufacture of rubber products and adhesives; as an additive in anti-stalling gasoline, lubricants, denatured ethyl alcohol, hydraulic brake fluids, and rocket fuel. It is hazardous to the eyes, skin, and respiratory system; and is toxic by inhalation, ingestion, and contact. Symptoms of exposure include drowsiness, dizziness, headache, dry and cracking skin, mild irritation of the eyes, nose, and throat. OSHA exposure limit (TWA): 400 ppm [air] or 980 mg/m³.

isopropylamine — $(CH_3)_2CHNH_2$ — a colorless liquid with an ammonia-like odor; a gas above 91°F; MW: 59.1, BP: 91°F, Sol: miscible, Fl.P (open cup): –35°F, sp gr: 0.69. It is used as an intermediate in the synthesis of agricultural chemicals, insecticides, and bactericides; as an intermediate in the synthesis of vulcanization accelerators for sulfur-cured rubbers; in the manufacture of ore flotation agents; as a depilatory on skins and hides in leather manufacture; in the manufacture of emulsion-type floor polish; as a general solvent; in the manufacture of medicinals in the purification of penicillin and streptomycin; as an intermediate in the synthesis of some dyes; as a solubilizer for certain herbicides in hard water. It is hazardous to the respiratory system, skin, and eyes; and is toxic by inhalation, absorption, ingestion, and contact. Symptoms of exposure include pulmonary edema, visual disturbances, skin and eye burns, dermatitis, irritation of the eyes, nose, and throat. OSHA exposure limit (TWA): 5 ppm [air] or 12 mg/m³.

isopropyl ether — $(CH_3)_2CHOCH(CH_3)_2$ — a colorless liquid with a sharp, sweet, ether-like odor; MW: 102.2, BP: 154°F, Sol: 0.2%, Fl.P: 92°F, sp gr: 0.73. It is used as a solvent in extraction processes, rubber adhesives, lacquers, resins, oils, cellulose, pharmaceutical manufacture, smokeless gunpowder, and textile spot cleaning; in organic synthesis as an alkylation agent; as an emulsion breaker in the petroleum industry; as a blending agent for gasoline. It is hazardous to the respiratory system and skin; and is toxic by inhalation, ingestion, and contact. Symptoms of exposure include respiratory discomfort, dermatitis, irritation of the eyes and nose; in animals: drowsiness, dizziness, unconsciousness, narcosis. OSHA exposure limit (TWA): 500 ppm [air] or 2,100 mg/m³.

isopropyl glycidyl ether — $C_6H_{12}O_2$ — a colorless liquid; MW: 116.2, BP: 279°F, Sol: 19%, Fl.P: 92°F, sp gr: 0.92. It is used as a reactive diluent for epoxy resins; as a chemical intermediate for the synthesis of esters and ethers; as a stabilizing agent for organic chemicals. It is hazardous to the eyes, skin, and respiratory system; and is toxic by inhalation, ingestion, and contact. Symptoms of exposure include skin sensitization, irritation of the eyes, skin, and upper respiratory system. OSHA exposure limit (TWA): 50 ppm [air] or 240 mg/m³.

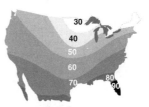

Isotherm

isotherm — a line on a map or chart joining locations recording identical temperature.

isothermal process — any process that occurs at constant temperature.

isotone — one of two or more atoms having the same number of neutrons in their nuclei.

isotope — one of two or more forms of the same element differing in atomic weight but having the same atomic number and the same chemical properties.

isotopic tracer — the isotope or non-natural mixture of isotopes of an element which may be incorporated into a sample, making possible the observation of the course of that element, alone or in combination, through a chemical, biological, or physical process. Also known as label.

ISV — *See in situ* vitrification.

ITC — acronym for International Trade Commission.

J

J — *See* joule.

JCAH — *See* Joint Commission on Accreditation of Hospitals.

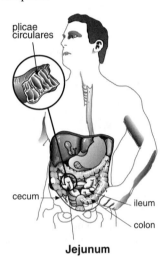

Jejunum

jejunum — a section of the small intestine lying between the duodenum and the ileum; its primary function is the absorption of digested material.

Jet Stream

jet stream — a swift, high altitude wind or current of air at the edges of the circumpolar whirls flowing toward the east.

jigging — a process for separating presized, solid materials of different densities by using the periodic pulsation of a liquid, usually water, through a bed of the mixed material to float the lighter solids.

join — (*geographic information system*) to connect two or more separately digitized maps.

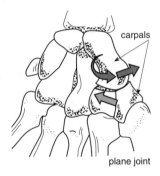

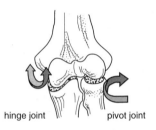

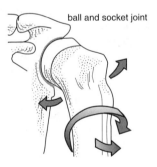

Joints

joint — (*anatomy*) the contact point at which two separate bones are joined by ligaments; (*geology*) a fracture or crack in which the rock on either side of the fracture does not exhibit evidence of relative movement.

Joint Commission on Accreditation of Hospitals (JCAH) — a private, nonprofit organization whose purpose is to encourage the attainment of uniformly high standards of institutional medical care; comprised of representatives of the American Hospital Association, American Medical Association, American College of Physicians, and American College of Surgeons.

joule (J) — the standard international unit of work and energy equal to the work done when the point of application of a force of one newton moves a distance of one meter; one calorie equals 4.184 joules; the mks unit of work; a unit of energy equivalent to 10^7 ergs. Also known as newton-meter of energy.

judicial district — physical perimeters set by Congress or a state legislature designating the counties under the jurisdiction of a specific court.

judicial review — a constitutional doctrine that gives a court system the power to annul legislative or executive acts which the judge declares to be unconstitutional.

junk — unprocessed materials suitable for reuse or recycling.

jurat — that part of an affidavit or other document where an officer certifies witnessing the sworn testimony.

I
J

k — *See* kilo-.

kanker — *See* hardpan.

kb — *See* kilobase.

kcal — *See* kilocalorie.

Kd — *See* adsorption ratio.

k-electron capture — electron capture from a *k* shell by the nucleus of the atom. *See also* electron capture.

Kelvin temperature scale — absolute temperature scale.

keratin — the horny substance present in the epidermis, hair, and all horny tissues, such as feathers and nails.

keratitis — an inflammation of the cornea of the eye.

kerogen — the hydrocarbon liquid found in sedimentary rock. *See also* oil shale.

kerosene — a flammable hydrocarbon oil usually obtained by distillation of petroleum and used as a fuel and a solvent.

ketene — $CH_2{=}CO$ — a colorless gas with a penetrating odor; MW: 42.0, BP: –69°F, Sol: decomposes. It is used as an intermediate in the production of acetic anhydride, cellulose and vinyl acetate resins and plastics, acrylic resins, dyes, pigments, and pharmaceuticals; in industrial organic synthesis; in laboratory operations as an acetylating agent. It is hazardous to the eyes, skin, and respiratory system; and toxic by inhalation and contact. Symptoms of exposure include pulmonary edema, irritation of the eyes, nose, lungs, and throat. OSHA exposure limit (TWA): 0.5 ppm [air] or 0.9 mg/m^3.

ketone — one of a class of compounds containing a carbonyl group, –CO–, in the molecule attached to two hydrocarbon radicals; ketones are important intermediates in the synthesis of organic compounds; the simplest ketone is acetone.

keV — *See* kiloelectronvolts.

keyboard — (*computer science*) a device for typing alphanumeric characters into the computer.

K-Factor — the thermal conductivity of a material expressed as BTUs per square foot per hour in degrees Fahrenheit.

kg — *See* kilogram.

kidney — either of a pair of main glandular organs in mammals which excrete urine.

kilkenny coal — *See* anthracite.

kiln — a furnace in which the heating operations stop just before complete fusion; used for drying, burning, or firing materials.

kilo- (k) — a prefix meaning 1,000 in the metric system, as in kilogram, which is 1,000 grams.

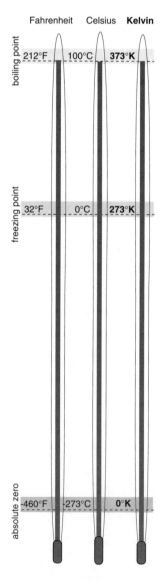

Kelvin Temperature Scale
compared to C and F scales
(not proportionate)

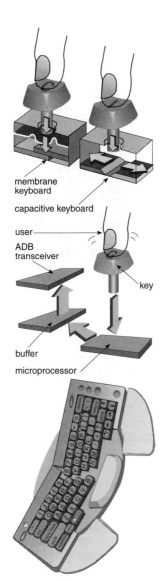

membrane keyboard

capacitive keyboard

user

ADB transceiver

key

buffer

microprocessor

Keyboard
(Showing an
Ergonomic Design)

kilobase (kb) — a unit used in designating the length of a nucleic acid sequence where 1 kilobase equals a sequence of 1,000 purine bases.

kilocalorie (kcal) — a unit of heat energy equal to 1,000 calories. Also known as kilogram calorie.

kilocurie — 1,000 curies.

kiloelectronvolts (keV) — 1,000 electron volts.

kilogram (kg) — a unit of measurement equal to 2.2046 pounds.

kilogram calorie — *See* kilocalorie.

kilometer (km) — a unit of measurement equal to 0.6214 miles.

kilopascals (kpa) — 1,000 pascals. *See also* pascal.

kilovolt meter (kV/m) — a unit of measurement equal to 1,000 volts per meter.

kilowatt (kW) — a unit of electrical power equal to 1,000 watts.

kilowatt-hour (kWh) — a unit of electrical energy equivalent to using 1,000 watts of power for one hour.

kinematics — a branch of dynamics that deals with aspects of motion apart from consideration of mass and force.

kinesiology — the study of the principles, mechanics, and anatomy of human movement.

kinetic energy — the energy possessed by a mass because of its motion; equal to one-half the body's mass times the square of its speed, or $1/2 \, mv^2$.

kinetochore — *See* centromere.

kitchenware — all multi-use utensils other than tableware.

kjeldahl method — an analytical method for the determination of nitrogen content, particularly in organic materials.

km — *See* kilometer.

knapsack sprayer — a sprayer that can be strapped to one's back and used to apply liquid pesticide chemicals; the attached hose has a nozzle at the tip that can be aimed at the spots to be treated.

knot — (*wind*) a speed unit of one nautical mile per hour, equal to approximately 1.15 statute miles per hour.

Koc — *See* adsorption coefficient.

Koch's law — a law stating that to determine the cause of a given disease by an organism, the following conditions must be met: (1) a microorganism must be present in every case of the disease, (2) it must be isolated and cultivated in pure culture, (3) inoculation of the culture must produce the disease in susceptible animals, (4) it must be observed in and recovered from the experimentally diseased animals. Also known as Koch's postulates; law of specificity of bacteria.

Koch's postulates — *See* Koch's law.

Kow — *See* octanol-water partition coefficient.

kpa — *See* kilopascals.

kraft processing — (*chemicals*) a method of producing wood pulp by digesting wood chips in an alkaline liquor consisting chiefly of caustic soda together with sodium sulfate.

kV/m — *See* kilovolt meter.

kW — *See* kilowatt.

kWh — *See* kilowatt-hour.

kyphosis — an abnormal curvature of the spine of the upper back in the anteroposterior plane. Also known as humpback; hunchback.

K
L

L

l — *See* liter.

label — notice attached to a container bearing information concerning its contents, proper use, manufacturer, and any cautions or hazards of use; (*nuclearphysics*) *See* isotopic tracer.

labelled compound — a compound consisting in part of labelled molecules; that is, molecules, including radionuclides, in their structure.

labile — unstable and likely to change under certain influences.

laboratory — any place organized and operated for the performance of any microbiological, biochemical, hematological, microscopical, immunological, parasitological, or other tests, examinations, or evaluations; those facilities that work with pathogens or animals, or that use various biotechnologies which generate infectious waste.

laboratory acquired infection — any infection resulting from exposure to biohazardous material in a laboratory environment.

laboratory waste — discarded materials generated by research and analytical activities.

labradorite — (*mineralogy*) a triclinic feldspar that is an essential constituent of basalts and gabbros.

lacquer — a colloidal dispersion or solution of nitrocellulose or similar film-forming compounds used as a glossy, protective, and decorative coating for surfaces.

lacrimation — secretion and discharge of tears.

lacrosse encephalitis — (*disease*) an acute inflammatory disease of short duration involving parts of the brain, spinal cord, and meninges ranging from mild symptoms to high fevers, meningeal signs, disorientation, coma, tremors, seizures, and possibly death. Incubation time 5–15 days; caused by a bunyavirus; found in the United States and Canada. The reservoir of infection is *Aedes* eggs; transmitted by the bite of a mosquito but not from person to person; usually a disease of children but can affect others. It is controlled by destroying larvae, eggs, and adult mosquitos, and using mechanical means of mosquito control.

lactase — a digestive enzyme of the intestinal fluid which changes lactose to glucose.

lactating animal — any animal that is producing milk.

Label
examples of warning placards
required on trucks

lactation — the secretion of milk by the mammary glands.

lactic acid — $C_3H_6O_3$ — a hygroscopic acid produced by anaerobic metabolism of glucose.

lading — *See* bill of lading.

LAER — acronym for lowest achievable emission rates.

lagoon — a shallow pond generally near but separated from a larger body of water; often created by humans using rigid specifications in which sunlight, algae, and oxygen interact to restore water to a reasonable state of purity.

lag time — time between the occurrence of a primary event and the occurrence of the symptoms of the event.

laminar flow — a flow in which no turbulence is present; diffusion in such a flow occurs only by molecular processes; (*air*) the introduction of large volumes of clean air through a very large diffuser or perforated panel which reduces the velocity of the incoming air, thereby preventing agitation and reintroduction of settled contaminants.

laminate — a sheet of material made of many different layers.

landfill — a disposal facility or part of a facility where waste is buried in layers below ground, compacted, and covered.

landfill blade — a U-blade with an extension on top that increases the volume of solid waste that can be pushed and spread, and protects the operator from any debris thrown out of the solid waste.

landfill gas — a gas produced in sanitary landfills during anaerobic digestion of the organic contents containing 40–60% methane.

land planner — an environmental health professional who reviews land-use proposals to assess environmental impacts, evaluate health risks, recommend actions that will protect the public from exposure to disease and other health and safety hazards, and will protect the environment from degradation.

landslide — mass movement of earth or rock along a definite plane under the influence of gravity.

LANDSTAT — the generic name for a series of earth resource scanning satellites launched by the United States beginning in 1972.

Lantz process — a destructive distillation technique in which the combustible components of solid waste are converted into combustible gases, charcoal, and a variety of distillates.

Larder Beetle (*Dermestes lardarius*) — a stored food product insect about 1/3″ long, dark brown with a wide yellow band across the front part of the wing cover.

large dyne — *See* newton.

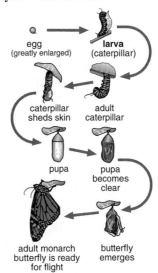

Larval Stage
of *Danaus plexippus*
(Monarch butterfly)

larva — the immature worm-like stage of an insect that undergoes complete metamorphosis before assuming characteristic features of the parent.

larvacide — a pesticide chemical that will prevent damage from or destroy larvae.

laryngitis — an inflammation of the mucous membrane of the larynx characterized by dryness and soreness of the throat, hoarseness, cough, and dysphagia.

larynx — the anterior portion of the trachea made of cartilages and related structures which contains the vocal chords in amphibians, reptiles, and mammals.

LAS — *See* linear alkylate sulfonate.

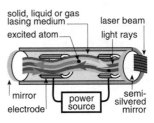

Laser

laser (light amplification by stimulated emission of radiation) — a device that utilizes the natural oscillations of atoms or molecules between energy levels for generating coherent electromagnetic radiation in the ultraviolet visible or infrared regions of the spectrum. Lasers produce monochromatic coherent radiation. It travels in a plane wave front with very little spreading. It is amplified, meaning a chain reaction of energy

K
L

is produced when the highly excitable photon travels between polished or mirror-like ends of the laser, and releases an enormous amount of energy simultaneously. Lasers are used in industry, medicine, and sometimes defense.

laser air pollution probe — a device used for measuring the concentration and location of a particulate pollution by use of a laser.

laser beam welding — a process that produces coalescence of materials with the heat obtained in the application of a concentrated coherent light beam impinging on the members to be joined.

laser doppler velocimeter (LDV) — a mechanism used to measure velocity and concentration levels of pollution by means of a carbon dioxide (CO_2) laser beam.

laser light region — a portion of the electromagnetic spectrum which includes ultraviolet, visible, and infrared light.

laser plotter — a plotter for mapping in which the information is written onto a light-sensitive material using a laser.

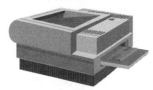

Laser Printer

laser printer — an extremely fast printer that uses a laser to form areas of static electric charge which attract metallic powder to paper.

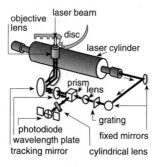

Laser System
in a compact disc player

laser system — a device of electrical, mechanical, and optical components including a laser.

Lasioderma serricorne — *See* cigarette beetle.

lassitude — weariness; exhaustion.

late effects (*radiation*) — effects which appear 60 days or more following an acute exposure to radiation.

latency period — the presymptomatic interval between initiation of a disease and its becoming clinically apparent.

latent energy — the heat released when water vapor changes back to liquid water.

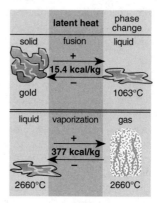

Latent Heat

latent heat — the quantity of heat absorbed or given off per unit weight of material during a change of state, such as ice to water or water to steam.

latent heat of vaporization — the quantity of energy required to evaporate a unit mass of liquid at constant temperature and pressure.

latent period — the time between exposure and the first manifestation of disease or injury.

lateral sewer — a sewer that runs under city streets and into which empty the sewers from homes or businesses.

latex — a water emulsion of natural or synthetic rubbers or resins; used in emulsion and paints.

latrine — a receptacle in the earth used as a toilet.

laughing gas — *See* nitrous oxide.

law — the rule of civil conduct prescribed by the supreme power in a state demanding what is right and forbidding what is wrong.

law of conservation of energy — *See* conservation of energy.

law of conservation of matter — *See* conservation of matter.

law of definite composition — *See* law of definite proportion.

law of definite proportion — every compound always contains the same proportions by weight of the elements that compose it. Also known as the law of definite composition.

law of electrostatic attraction — *See* Coulomb's law.

law of limiting factors — the biological law that a minimum quantity of essentials, such as nutrients, light, heat, moisture, and space, must be available within the ecosystem for survival of the organisms. The ecosystem can be affected by pesticides or other environmental changes.

law of partial pressures — *See* Dalton's law.

law of specificity of bacteria — *See* Koch's law.

layer — (*geographic information system*) a logical separation of mapped information according to the theme of the information required.

lb — *See* pound.

lb-ft — *See* pound foot.

LC — *See* lethal concentration.

LC$_{50}$ — *See* lethal concentration$_{50}$.

LCCA — *See* Lead Contamination Control Act.

LCL — *See* low lethal concentration.

LC$_{lo}$ — *See* lethal concentration$_{lo}$.

LD — *See* lethal dose.

LD$_{50}$ — *See* lethal dose$_{50}$.

LDL — *See* low lethal dose.

LD$_{lo}$ — *See* lethal dose$_{lo}$.

LDV — *See* laser doppler velocimeter.

leachate — liquid that has percolated through solid waste or other medium and has extracted, dissolved, or suspended materials from solid waste; a solution formed by leaching.

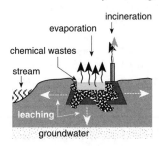

Leaching

leaching — the loss of soluble soil minerals as the result of the movement of groundwater; the operation by which color is partially or wholly removed form a colored material; the dissolving by a liquid solvent of soluble material from its mixture with an insoluble solid; movement of a chemical downward through the soil.

lead — Pb — a naturally occurring bluish-gray metal found in small amounts in the earth's crust; has no characteristic taste or smell; MW: 207.20, BP: 1,740°C, Sol: insoluble, Fl.P: not known, density 11.34 g/cm^3 at 20°C. It is used in the production of batteries, ammunition, sheet lead, solder, pipes, in the tetraethyl lead and tetramethyl lead forms as gasoline additives, paint, ceramic products, roofing, caulking, electrical applications, tubes, or containers. Lead exposure may be due to air, water, food, or soil. Lead in the air is primarily due to lead-based fuels and the combustion of solid waste, coal, oils, and emissions from alkyl lead manufacturers, wind blown dust volcanoes, the burning of lead-painted surfaces, and cigarette smoke. Lead in drinking water comes from leaching from lead pipes, connectors, and solder in both the distribution system and household plumbing. In surface water, lead comes from lead-containing dust which has settled from the atmosphere, wastewater from industries, handling lead, primarily steel, iron, lead producers, and from urban runoff. Lead in food comes from dust from the atmosphere, on crops during fruit processing, and from lead-soldered food cans. Lead in the soil comes from dust from the atmosphere, solid waste disposal, paint chips, deterioration of lead-based paints, and fertilizers which become part of sewage sludge. It is hazardous to the digestive system, reproductive system, blood, central nervous system, kidneys, and the mental and physical development of children exposed. Mortality occurs in individuals in the working environment in battery plants and lead production plants as well as in the ambient environment, especially in children. Symptoms of exposure include cerebrovascular disease, hypertension, nephritis, nephrosis, cancer of the digestive tract, colic, hepatic and renal effects, dullness, irritability, poor attention span, headaches, muscular tremors, memory loss, hallucinations, paralysis, coma, death; in animals, decreased leukocyte counts, circulating antibodies and antibody forming cells. A suspected causative agent in sudden infant death syndrome; carcinogenic. Respiratory effects appear to be rare but may occur; colic is a consistent, early gastrointestinal symptom of lead poisoning in occupationally exposed individuals at acutely high levels of lead, especially during the removal of lead-base paint; colic symptoms include abdominal pain, constipation, cramps, nausea, vomiting, anorexia, weight loss, and decreased appetite; colic is also a symptom of lead poisoning in children; lead has a profound affect on hemabiosynthesis and inhibits the activity of certain enzymes involved in hemabiosynthesis; children have hepatic effects with a decreased activity of hepatic mix-function oxygenases; renal effects of early or acute lead-induced nephropathy include nuclear inclusion bodies, mitochondrial changes and cytomegaly of the proximal tubular epithelial cells, disfunction of the proximal tubules manifested as aminoaciduria glucosuria phosphaturia with hypophosphatemia; thyroid function is adversely affected; lead interferes with the conversion of vitamin D to its hormonal form, 1,25-dihydroxy-vitamin D; there is an association between bloodlead levels and growth in children; it may have an affect on the cellular component of the immune system; neurological effects are related to brain function, dullness, irritability, poor attention span, headaches, muscular tremors, loss

K
L

of memory, and hallucinations; these may worsen to become delirium, convulsion, paralysis, coma, and death; behavioral changes occur in children with a drop of 5 IQ points; developmental effects include reduced birth rate and gestational age and neurobehavioral deficits or delays; adverse reproductive effects include miscarriages and still births; because of the indirect effects of lead on maternal nutrition or the hormonal status of the pregnant woman, lead may have an effect on chromosomes; immunological effects include in animals decrease in leukocyte counts, circulating antibodies and antibody forming cells. OSHA exposure limits (TWA): 50 micrograms/m^3 [air].

lead additive — any substance containing lead or lead compounds.

lead-based paint — any paint with a lead content exceeding 0.06% by weight of the total non-volatile content of the paint or the weight of the dry paint film; any applied paint film that contains 0.7 mg/cm^2 or more of lead as measured by an *in situ* analyzer device; paint that contains more lead as determined by chemical analysis than the appropriate authority considers to be safe.

Lead Contamination Control Act (LCCA) — legislation passed in 1988 requiring identification of water coolers that are not lead-free, repair or removal of water coolers with lead-lined tanks, identification and resolution of lead problems in schools' drinking water, and banning the manufacture and sale of water coolers that are not lead-free.

leader — a conduit used to carry rain water.

lead marcasite — *See* sphalerite.

lead peroxide candle — a method for determining the amount of sulfur dioxide in the atmosphere when exposing lead oxide paste to air in which gaseous sulfur dioxide is present, and sulphation occurs at a rate proportional to the amount of sulfur dioxide present.

lead poisoning — a form of poisoning caused by the presence of lead or lead salts in the body with symptoms of weight loss, anemia, stomach cramps, mental depression, irritation, and convulsions affecting the brain, nervous system, blood, and digestive system.

leakage — the undesired entrance or escape of fluid.

leaking underground storage tank (LUST) — older tanks placed underground as a fire prevention measure that leak into the ground and potentially into the ground water supply; leaking tanks used to store gasoline, other oil products, or hazardous substances.

leak-protector valve — a valve equipped with a leak-diverting device which will prevent leakage of the fluid past the valve in any closed position.

lectins — any of a variety of plants containing substances that agglutinate red blood cells.

legal status — the legal constitution of a business enterprise as to whether it is a corporation, partnership, cooperative, or sole ownership.

legend — (*mapping*) the part of the drawn map explaining the meaning of the symbols used to code the geographical information.

legionellosis — (*disease*) an acute bacterial disease with two distinct manifestations; Legionnaire's disease and Pontiac fever. Symptoms include malaise, anorexia, myalgia, headache, cough, and high temperature. Incubation time for Legionnaire's disease is 2–10 days, usually 5–6 days. Incubation time for Pontiac fever is 5–66 hours, usually 24–48 hours. Both are caused by *Legionella pneumophila*, a poor-staining, gram-negative rod that is difficult to grow outside of the body; found in a variety of places for at least the last 45 years. Reservoirs of infection include possibly water and soil; transmitted possibly by air; not communicable from person to person; general susceptibility but rarely found under age 20. It is controlled by use of antibiotics, appropriate disinfection of cooling tower waters, and proper treatment of water supplies.

Legionnaire's disease — *See* legionellosis.

legislation — the exercise of power and function of making laws.

LEL — *See* lower explosive limit.

L-electron capture — *See* electron capture.

lens — the transparent disc which directs light rays to the retina of the eye of many invertebrates.

lentic environment — standing water and its various intergrades such as lakes, ponds, and swamps.

LEPC — *See* Local Emergency Planning Committee.

leptospirosis — (*disease*) zoonotic diseases with the ability to assume different forms including fever, headache, chills, severe malaise, vomiting, sometimes jaundice, renal insufficiency, and hemorrhage. Incubation time is 10 days, with a range of 4–19 days; caused by *Leptospira*, especially in the United States, and *icterohaemorrhagie*, found worldwide. Reservoirs of infection include farm and pet animals as well as rats; transmitted through contact with the skin, especially if abraded; or contact with the mucous membranes in water. Rarely transmitted from person to person; may be communicable for 1–11 months;

general susceptibility. It is controlled by protective clothing, boots for workers, and proper rodent control.

lesion — any pathological or traumatic discontinuity of tissue or loss of function due to injury or disease.

Lesser Grain Borer (*Rhyzopertha dominica*) — a brown or black slender cylindrical beetle with numerous course elevations on the pronatum, about 1/8″ long and commonly found in the Gulf states, but may occur elsewhere in the country; a stored food product insect.

LET — *See* linear energy transfer.

lethal — capable of causing death.

lethal concentration (LC) — a calculated concentration of a chemical in air to which exposure for a specific length of time is expected to cause a death rate of 100% in members of the population affected.

lethal concentration$_{50}$ (LC$_{50}$) — a calculated concentration of a chemical in air to which exposure for a specific length of time is expected to cause death in 50% of a defined experimental animal population.

lethal concentration$_{lo}$ (LC$_{lo}$) — the lowest concentration of a chemical in air which has been reported to cause death in humans or animals.

lethal dose (LD) — the quantity of a substance being tested that will kill 100% of the defined experimental population.

lethal dose$_{50}$ (LD$_{50}$) — the dose of a chemical which has been calculated to cause death in 50% of a defined animal laboratory population.

lethal dose$_{lo}$ (LD$_{lo}$) — the lowest dose of a chemical introduced by a route other than inhalation that is expected to cause death in humans or animals.

lethal time$_{50}$ (LT$_{50}$) — a calculated period of time within which a specific concentration of a chemical is expected to cause death in 50% of a defined experimental animal population.

lethargy — a state of deep drowsiness usually of mental origin but also caused by illness or fatigue.

leukemia — a progressive, malignant disease of the blood-forming organs; a distorted proliferation and development of leukocytes and their precursors in the blood and bone marrow.

leukocyte — a colorless cell with a nucleus found in blood and lymph. Also known as white blood cell.

leukocytosis — a transient increase in the number of white blood cells in the blood.

leukopenia — reduction of the number of leukocytes in the blood; 5,000 or less.

LEV — *See* local exhaust ventilation.

LFEM — acronym for low-frequency electromagnetic field.

LFL — *See* lower flammable limit.

Lfm — *See* linear feet per minute.

libel of information — *See* complaint for forfeiture.

Library of Congress (LOC) — the national library of the United States offering diverse materials for research, including the world's most extensive collection in many areas, such as American history, music, and art.

lice — the plural of louse.

license — a legal document granting the bearer the right to carry forth a given business dependent upon adherence to the rules, regulations, and applicable laws.

licensure — the process by which a state government agency grants permission to a person meeting predetermined qualifications to engage in a given occupation.

lidar — a ruby laser radar used to measure atmospheric backscattering and/or attenuation as a function of the slant range, the primary and secondary haze layers, the smoke-plume capacity, and particulate clouds; derived from laser infrared radar.

lidar ground sensor — a mechanism used to determine the opacity of gases coming from smoke stacks; accurate within 3% opacity over opacities of less than 50%.

lidar radar backscatter — a device which measures cross-sections of the plume for particulate emissions by using a laser beam through the plume and comparing the relative backscattering intensity of the beam in front and behind the plume; used in stack transmissometer measurements.

life cycle — the stages an organism passes through during its existence.

life expectancy — the average length of time an individual may live based on an analysis of

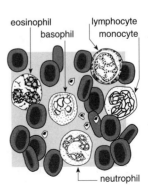

Leukocytes

actual time until the death of a large group of similar people.

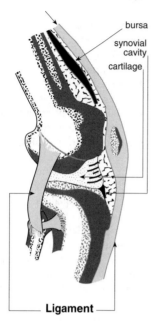

Ligament

ligament — a flexible, tough strand of connective tissue which holds bones together at a joint.

light — a form of radiant energy coming from atoms when they are violently disturbed by heat or electricity; electromagnetic waves that travel freely through space in a vacuum at a speed of 299,792 kilometers per second where the oscillations are at right angles to the direction in which the light travels; an electromagnetic radiation in the wavelength range, includes infrared, visible, ultraviolet, and X-rays.

light amplification by stimulated emission of radiation — *See* laser.

light cream — coffee cream or cream which contains less than 30% milk-fat.

light meter — a small, portable photometer with light-sensitive, barrier-layer cells used to measure illumination directly; usually calibrated in foot candles.

lightning — the large spark produced by attraction of unlike electrical charges within a thunder cloud or between a thunder cloud and the earth.

light quantum — *See* photon.

light trap — a device used for collecting or destroying insects that consists of a bright light in association with a trapping or killing medium.

lignin — an amorphous polymeric substance that together with cellulose forms the woody cell walls of plants.

lignite — a brownish-black solid fuel in the second stage in the development of coal. It has a little over half the heating value of bituminous or anthracite coal.

lignocellulose — any of several closely related substances constituting the essential part of woody plant cell walls and consisting of cellulose and lignin.

lime — *See* calcium oxide.

lime scrubbing process — a process used on flue gases for control of sulfur dioxide; dried limestone is injected into the fire box where it becomes reactive lime; this combines with the gas stream and passes into a wet scrubber where the lime reacts with the water and together with the fly ash and forms a slurry; the slurry and the sulfur dioxide react to form sulphite and sulfate salts, which are then disposed of.

limestone — a sedimentary rock composed mainly of calcite and concentrated shale, coral, algae, and other debris.

lime treatment — (*sewage*) a means of control of odor and reduction of pathogens through the use of lime without significantly reducing sludge solids.

limiting factor — a condition whose absence or excessive concentration exerts some restraining influence upon a population through incompatibility with a species' requirements or tolerance.

limit of detection — the minimum concentration of a substance which can be detected and identified quantitatively or qualitatively 99% of the time.

Lightning into ground

limnetic zone — the open-water region of a lake supporting plankton and fish as the principle inhabitants.

limnology — the study of biological, chemical, geographical, and physical features of fresh

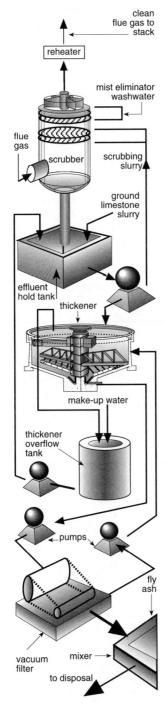

Lime Scrubbing System

waters, especially lakes and ponds; used in understanding aquatic ecology.

limonite — a group of hydrated ferric oxide minerals ($Fe_2O_3H_2O$) which occur commonly in many types of rocks; it is generally rusty or blackish with a dull earthy luster and a yellow-brown streak.

lindane — $C_6H_6CL_6$ — a white to yellow, crystalline powder with a slight musty odor; the pure gamma isomer of benzene hexachloride; an insecticide; MW: 290.8, BP: 614°F, Sol: 0.001%, sp gr: 1.85. It is used in the formulation of insecticides, scabicides, pediculicides, and vermifuges. It is hazardous to the eyes, central nervous system, blood, liver, kidneys, and skin; and is toxic by inhalation, absorption, ingestion, and contact. Symptoms of exposure include headache, nausea, clonic convulsions, respiratory difficulty, cyanosis, aplastic anemia, skin irritation, muscle spasms, irritation of the eyes, nose, and throat; in animals: liver and kidney damage. OSHA exposure limit (TWA): 0.5 mg/m^3 [skin].

line — (*mapping*) one of the basic geographical elements defined by at least two pairs of XY coordinates.

linear — a single, straight line or lines.

Linear Accelerator

linear accelerator — a device for accelerating charged particles in a straight line through a vacuum tube or series of tubes by means of alternating negative and positive impulses from electric fields; a machine with a rating of 1,000,000 electron volts (1 mev) or greater which accelerates particles, electrons, or protons with high velocities along a straight line.

linear alkylate sulfonate (LAS) — a rapidly biodegradable variety of alkylbenzene sulfonate.

linear energy transfer (LET) — the average amount of energy transferred locally to the medium per unit of particle track length.

linear envelope detection — an electrical circuit assembly designed to detect the shape of the envelope of an amplitude modulated carrier or signal.

linear feet per minute (Lfm) — a unit of measurement for air velocity.

linear hypothesis — (*radiation*) the assumption that a dose-effect curve in the high dose and high dose-rate ranges may be extrapolated through the low dose and low dose range to zero implying that theoretically any amount of radiation will cause some damage.

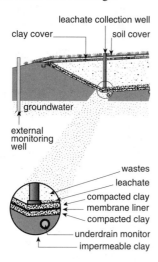

Landfill Liner

liner — (*swimming pools*) the membrane that acts as a container for the water; material used on the inside of a furnace wall to ensure a chamber impervious to escaping gases; (*hazardous waste*) a continuous layer of natural or people-made materials beneath or on the sides of a surface impoundment, landfill, or landfill cell which restricts the downward or lateral movement of hazardous waste or leachate.

lingual — pertaining to the tongue.

lining — the material used on the inside of a furnace wall; usually a high-grade refractory tile or brick, or plastic refractory material; a material used to protect inner surfaces.

lipase — a digestive enzyme of the pancreatic fluid that changes fats to glycerol and fatty acids.

lipid — one of a class of nonpolar organic compounds that is insoluble in water but soluble in certain solvents including alcohol and ether; includes fatty acids, greasy, oily, and waxy compounds.

lipophilic — having an affinity for fats or other lipids.

liquefaction — the change in phase from solid into a liquid form.

liquefied petroleum gas (LPG) — $C_3H_8/C_3H_6/C_4H_{10}/C_4H_8$ — a colorless, noncorrosive, odorless gas when pure; a foul-smelling odorant is usually added; shipped as a liquefied compressed gas; MW: 42 to 58, BP: >−44°F, Sol: not known.

It is used in the chemical industry as feedstock in catalytic cracking in the manufacture of petrochemicals and polymers; as an internal combustion engine fuel; in the manufacture of synthetic rubber; as a fuel for industrial space and wall heating, air conditioning, and water heating; as a fuel for drying agricultural products; as a low-sulfur fuel for cutting, soldering, and brazing; as an aerosol propellant for shaving lather, window cleaners, starch sprays, room deodorants, insecticides, and pharmaceuticals; as an agricultural tool in non-selective cultivation, and weed control. It is hazardous to the respiratory system and central nervous system; and is toxic by inhalation and contact. Symptoms of exposure include lightheadedness and drowsiness. OSHA exposure limit (TWA): 1,000 ppm [air] or 1,800 mg/m^3.

liquid — the condensed phase of matter between solid and gas in which molecules are very close together but still have enough energy to move over one another and liquids can flow; it conforms to the shape of a confining vessel.

liquid air — air cooled and compressed until it liquefies; used as a refrigerant.

liquid all-purpose cleaners — anionic and nonionic synthetic detergents that are low sudsing and will not damage or dull synthetic or wax floor finishes when used on floors.

liquid and solid biological treatment (LST) — a process that can be used to remediate soils and sludges contaminated with biodegradable organics.

liquid capacity — (*sewage*) the internal volume of a tank below the invert of the outlet line.

liquid carriage — (*air pollution*) an air control system in which dust particles are forced to separate from a gas stream by striking a liquid surface and are then entrained and removed from the area.

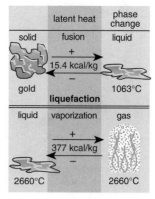

Liquefaction

liquid knock-out — *See* impingement.

liquid-liquid extractions — a chemical technique in which two mutually immersible liquids redistribute the desired solute on the basis of differences in solubility.

liquid-media sampler — a device using an absorbing liquid to capture gaseous substances from air for sampling and analysis.

liquid oxygen — an intensely cold, transparent liquid formed by putting oxygen under very great pressure and cooling it to –182.97°C.

Listeria — a genus of gram-negative bacteria (*Corynebacterium* family) found chiefly in the feces of lower animals; produces upper respiratory disease, septicemia, and encephalitic disease in people.

listeriosis — (*disease*) a bacterial disease usually found in the very young or old, during pregnancy, or people with immunosuppressed diseases. Symptoms include septicemia, fever, meningo-encephalitis, intense headaches, nausea, vomiting, delirium, coma, shock, endocarditis, lesions in the liver and other organs, and infections of the fetus. Incubation time is not known, probably a few days to 3 weeks, the fetus is usually infected within several days of the maternal disease; caused by the bacterium *Listeria monocytogenes*, types I_a, I_b, IV_a, IV_b; unusual infection, sporadic, occurring in all seasons; found in institutions and elsewhere. Reservoirs of infection include infected domestic and wild mammals, fowl, and people; transmitted from mother to fetus in the passage through the infected birth canal, and direct contact from lesions on hands and arms, infectious material, or contaminated soil; associated with ingestion of contaminated vegetables and dairy products, also through person-to-person transmission; communicable for mothers of infected newborn infants for 7–10 days in vaginal discharges or urine, other person-to-person communicability not known. Fetuses and newborn infants are highly susceptible, children and young adults generally are resistant; older people are more susceptible. It is controlled by avoidance of contact with infected materials or aborted animal fetuses, and pharmaceutical treatment.

liter (l) — the metric unit of volume which is equal to 1.057 quarts, 2.2046 pounds of pure water at 4°C, 0.0353 cubic feet, or 0.2642 gallons.

liters per minute — a measurement of air flow where 1 liter of air flows per minute past a structure or into a piece of equipment.

lithium hydride — LiH — an odorless, off-white to gray translucent, crystalline mass or white powder; MW: 7.95, BP: decomposes, Sol: reacts, Fl.P: not known, sp gr: 0.78. It is used in the manufacture of buoyancy devices, fuel cells, and portable field generators; in the manufacture of reducing agents and propellants; in powder metallurgy; as a shielding material for thermal neutrons in the nuclear industry; as a desiccant; as a condensing agent in organic synthesis; in a condensation polymerization; in the manufacture of electronic tubes and in ceramics. It is hazardous to respiratory system, skin, and eyes; and is toxic by inhalation, ingestion, and contact. Symptoms of exposure include nausea, muscle twitches, mental confusion, blurred vision, eye and skin burns; if ingested, burns of the mouth and esophagus. OSHA exposure limit (TWA): 0.025 mg/m^3 [air].

lithology — the description of rocks on the basis of their physical and chemical characteristics as determined by the eye or a low-power microscope.

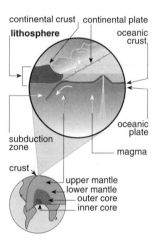

Lithosphere
in a whole earth context

lithosphere — the outer part of the solid earth composed of rock, essentially like that exposed at the surface; considered to be 60 miles thick.

litigation — the act or process of carrying on a lawsuit.

litmus — a water-soluble organic substance obtained from the lichen plant and used as an indicator to determine if a solution is acidic by turning red, or basic by turning blue.

litter — unwanted, discarded material which is solid waste that exists outside of a collection system.

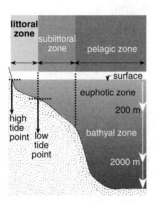

Littoral Zone

littoral zone — the shallow, shoreward, or coastal region of a body of water having light penetration to the bottom; frequently occupied by rooted plants; the zone between the high- and low-watermarks.

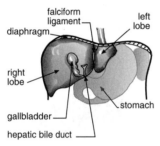

Liver

liver — the large, dark red gland located in the upper right portion of the abdomen in the human body associated with several vital activities including digestion, sugar metabolism, and removal of toxins.

LLD — *See* low lethal dose.

lm — *See* lumen.

loading rate — the allowable rate of application of septic tank effluent to the soil, expressed in gallons per day per square foot.

LOAEL — *See* lowest-observed-adverse-effect-level.

loam — a soil in which there is an even mixture of sand, silt, clay, and humus.

LOC — *See* Library of Congress.

Local Emergency Planning Committee — an organization of state and local officials; police, fire, civil defense, public health professionals; environmental, hospital, and transportation officials; as well as representatives of facilities on community groups, and the media, whose primary responsibility is the development of an emergency response plan for a local area;

established under the Superfund Amendments and Reauthorization Act of 1988, Title III.

local exhaust ventilation (LEV) — mechanical ventilation applied at or close to the source of an emission for the purpose of drawing clean, uncontaminated air past the worker, capturing the emission, collecting it in an exhaust hood, and removing it from the building.

localized — limited to a given area or part.

lockjaw — *See* tetanus.

locular pneumonia — *See* bronchopneumonia.

LOD — acronym for limit of detection.

loess — a silty, wind-deposited material consisting almost entirely of silt but may include small amounts of very fine sand, clay, or fossils.

logarithm — (*mathematics*) the power to which a base (usually 10) must be raised to produce a given number; base 10 raised by a logarithm of 3 equals 1,000.

longitudinal study — (*epidemiology*) *See* prospective study.

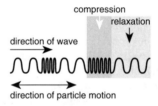

Longitudinal Wave

longitudinal wave — a wave in which the particles of a medium move back and forth in the same direction as the wave itself moves.

long-path monitoring system — a field-deployable, long-path, Fourier-transformed infrared spectrometer that measures the absorption caused by infrared-active molecules.

long-term animal bioassays — laboratory studies in which animals are exposed to a suspected hazard. The animals are examined for the presence of tumors and other signs of disease throughout the study. At the end of the study, the surviving animals are sacrificed, and their tissues are examined.

long-term exposure — the prolonged condition of being exposed to infectious, chemical, or physical agents which may have an adverse effect on the body.

long ton — measurement equal to 2,240 pounds.

loop — *See* antinode.

LOQ — acronym for limit of quantification.

lot — the product produced during a period of time indicated by a specific code.

lotic environment — running waters, such as streams or rivers.

loudness — the physiological perception of sound intensity; loudness is dependent on frequency.

louse — the common name for a grayish, wingless, and dorsoventrally flattened insect varying in length from 1/6–1/16 of an inch; a human parasite.

low — (*weather*) an area of low pressure. Also known as a depression.

low-acid foods — any food, other than alcoholic beverages, with a finished equilibrium pH value greater than 4.6 and a water activity greater than 0.85, and also includes any normally low-acid fruits, vegetables, or vegetable products in which, for the purpose of thermal processing, the pH value is reduced by acidification.

low altitude — any elevation less than 4,000 feet above sea level.

low clouds — types of clouds whose bases range in height from near the earth's surface to 6,500 feet; principle clouds in this group include stratocumulus, stratus, and nimbostratus.

lower atmosphere — the part of the atmosphere in which most weather occurs; from the ground to approximately 60 kilometers up in the air.

lower explosive limit (LEL) — the lowest concentration of a vapor or gas at ordinary ambient temperatures expressed in percent of the gas or vapor that will produce a flash of fire when an ignition source is present.

lower flammable limit (LFL) — *See* lower explosive limit.

lowest-observed-adverse-effect-level (LOAEL) — the lowest dose producing an observable adverse effect.

low-flow pump — a pump used for gas and vapor sampling where the flow rate of air is 200 milliliters per minute.

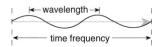

Low Frequency Sound Waves

low frequency sound waves — of or having to do with a frequency of sound ranging from 30 to 300 kilohertz.

low heat value — the high heat value minus the latent heat of vaporization of water as formed by burning the hydrogen in the fuel. Also known as net heating value.

low-LET — *See* low-linear energy transfer.

low lethal concentration (LCL) — the lowest concentration of a gas or vapor capable of killing a specified species.

low lethal dose (LDL) — the lowest administered dose of a material capable of killing a specified species.

low-linear energy transfer (low-LET) — radiation characteristic of electrons, X-rays, and gamma rays.

low-volatile — referring to a liquid or solid that does not evaporate quickly at normal temperatures.

LOX — *See* liquid oxygen.

LPG — *See* liquefied petroleum gas.

LPM — *See* liters per minute.

LSE — acronym for levels of significant exposure.

LST — *See* liquid and solid biological treatment.

LT$_{50}$ — *See* lethal time$_{50}$.

lumen (lm) — (*anatomy*) a canal or cavity in a tubular organ; (*optics*) the unit for the measurement of the rate at which light energy is radiated from a source. It is the amount of light striking one square foot of a surface one foot from the light source. One candle power is equal to about 12-1/2 lumens.

luminous — referring to the emission of a steady, suffused light.

luminous flux — the rate or flow of light measured in lumens.

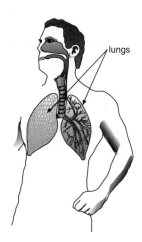

Lungs

lung — either of a pair of the respiratory organs for breathing air in vertebrates.

lung toxins — chemicals which irritate or damage the pulmonary tissue with symptoms of cough, tightness in the chest, and shortness of breath.

LUST — *See* leaking underground storage tank.

lye — a common name for sodium hydroxide or potassium hydroxide; used as a strong alkaline solution in industry.

Lyme disease — (*disease*) a tickborne spirochetal zoonotic disease with a distinctive skin lesion, systemic symptoms, polyarthritis, neurological, and cardiac involvement. Incubation time is 3–32 days; caused by a spirochet *Borrelia burgdorferi*; found in the United States, along the Atlantic coast and also along the Pacific coast and in the south. The reservoir of infection is Ixodid ticks through transstadial transmission by ticks; not communicable; general susceptibility. It is controlled by tick repellent, proper clothing, and avoiding areas containing ticks.

lymph — the clear, liquid part of blood which enters the tissue spaces and lymph vessels of the lymphatic system.

lymph node — glandular centers in which lymphatics converge.

lymphoid — resembling lymph tissue.

lyse — the cause or produced disintegration of a compound, substance, or cell.

lysimeter — a device to measure the quantity or rate of water movement or leaching loss through or from a block of soil usually undisturbed and *in situ*, or to collect such percolated water for quality analysis.

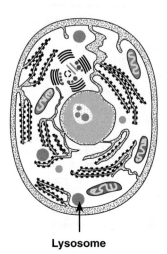

Lysosome

lysis — destruction or decomposition especially by enzymatic digestion.

lysosome — the digestive system of a cell.

lysozyme — an enzyme that dissolves the cell walls of many bacteria; present in many body secretions, such as tears.

K
L

M

m — *See* meter; *See* milli-.

M — *See* mega-.

mA — *See* milliampere.

MAC — *See* maximum allowable concentration.

maceration — the softening of a solid by soaking.

machine location — the room, enclosure, space, or area where one or more vending machines are installed and operated.

macronutrient — a chemical element necessary in large amounts, usually greater than 1 ppm, for the growth and development of plants.

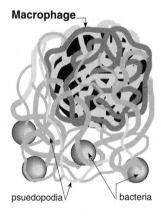

Macrophage

psuedopodia   bacteria

macrophage — any of the large mononuclear, high phagocytic cells derived by monocytes that occur in the walls of blood vessels and in loose connective tissue.

macrophyte — (*botany*) the larger aquatic plants.

macroscopic — visible without the aid of a microscope.

macula — a discolored spot on the skin not elevated above the surface.

made land — areas in which the natural soil profile has been worked or disturbed to such an extent that the soil cannot be classified in accordance with the standards established for soil classification by the National Cooperative Soil Survey.

maggot — the larval stage of a fly.

magistrate — a public civil officer having power to issue a warrant for arrest and charge an individual with a public offense.

magma — the molten silicate source material from which igneous rocks are formed.

magnesium oxide fume — MgO — a finely divided white particulate dispersed in the air; exposure may occur when magnesium is burned,

thermally cut, or welded upon; MW: 40.3, BP: 6512°F, Sol (86°F): 0.009%, sp gr: 3.58. It is liberated from the fabrication of alloys for aircraft, ships, automobiles, boats, tools, machinery, and military equipment; from casting of metal and alloys. It is hazardous to the respiratory system and eyes; and is toxic by inhalation and contact. Symptoms of exposure include metal fume fever, cough, chest pain, flu-like fever, irritation of the eyes and nose. OSHA exposure limit (TWA): 10 mg/m³ [air].

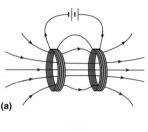

(a)

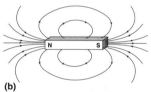

(b)

Magnetic Field
(a) Helmholtz coils
(b) bar magnet

magnetic field (B) — the space around a magnet or electric current in which magnetic force is felt.

Magma

magnetic field intensity — *See* magnetic field strength.

magnetic field strength (H) — vector quantity given by the ratio B/μ where B is the magnetic flux density and μ the permeability of the medium; its magnitude gives the strength of the magnetic field at a point in the direction of the line of force at that point. Also known as magnetic field intensity.

magnetic flux (Φ) — the product of a particular area under consideration and the component, normal to the area, of the average magnetic flux density over it.

magnetic resonance imaging (MRI) — a diagnostic imaging technique that employs magnetic and radio frequency fields to image body tissues and monitor body chemistry non-invasively; produces cross-sectional images; formerly known as nuclear magnetic resonance (NMR).

magnetic separator — any device that removes ferrous metals by means of magnets.

magnetometer — an instrument to measure the local intensity of a magnetic field, such as the earth's.

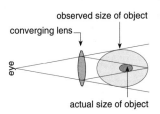

Magnification

magnification — the apparent enlargement of an object by an optical instrument.

mainframe — (*computer science*) a large computer supporting many users.

maintenance — the combination of technical actions and corresponding administrative actions undertaken to maintain or reestablish a piece of equipment in the state necessary to fulfill its required function.

makeup air — clean, temperature-adjusted, outdoor air supplied to a workplace to replace air removed by exhaust ventilation.

makeup tank — balance tank; (*swimming pools*) a tank provided for a continuous flow of makeup water for the pool; may be used as a receiving point for overflow; may be used as a receiving point for chemicals in the event they have to be added on an emergency basis.

makeup water — fresh water used to fill or refill the pool; water added to boiler cooling towers, water tanks, or other systems to maintain the volume of water required.

malaise — a vague feeling of discomfort, ill-being, and poor health.

malaria — (*disease*) four different human diseases: Falciparum malaria, the most serious form; causes fever, chills, sweats, headaches leading to jaundice, coagulation defects, shock, renal, and liver failure and coma; case fatality is 10% among children and untreated adults. Incubation time is 12 days; Vivax, Malariae, and Ovale malaria are usually not life threatening except in the very young and very old. Symptoms include malaise, followed by a shaking chill and rapidly rising temperature, headache, nausea, and profuse sweating; the cycle of chills, fever, and sweating is repeated daily, every other day, or every third day with relapses occurring irregularly for 2–5 years; with Malariae, relapses may persist for 50 years. Incubation time is 14 days for *Plasmodium vivax, P. ovale,* and *P. malariae.* Although not typically in temperate climates any longer, malaria is still found in tropical and subtropical areas. The reservoir of infection is people; transmitted by the bite of an infective female anopheline mosquito where the mosquito ingests blood containing the sexual stages of the parasite, the gametocytes. The male and female gametes are free in the mosquito's stomach, unite, and enter the stomach wall to form a cyst in which thousands of sporozoites develop in about a period of 8–35 days, depending on the parasite and temperature; the sporozoites migrate to various organs of the infected mosquito and some reach the salivary glands, mature, and are infective when injected into humans. In humans, the sporozoites enter hepatocytes and develop into exoerythrocytic schizonts; the hepatocytes rupture and asexual parasites, called tissue merozoites, appear in the blood stream and invade the erythrocytes, grow, and multiply. Some develop into asexual forms to mature blood schizonts, which rupture and liberate merozoites and invade other erythrocytes. The clinical symptoms occur with the rupture of the erythrocytic schizonts. It is communicable for as long as there are infective gametocytes in the blood of patients, which can be 3 years in malariae, 1–2 years in vivax, and 1 year in falciparum; may also be spread by transfusions; general susceptibility. It is controlled by removal of mosquito breeding sites, larvaciding and adultaciding, using residual insecticide, mechanical screening of all sleeping areas, use of insect repellents, and avoidance of blood donors who have a history of malaria.

malathion — $C_{10}H_{19}O_6PS_2$ — a deep brown to yellow liquid with a garlic-like odor; a solid below 37°F; MW: 330.4, BP: 140°F (decomposes),

Sol: 0.02%, Fl.P (open cup): >325°F, sp gr: 1.21. It is used in the formulation of pesticide products; as an insecticide for treatment of grain, nut, fruit, and fiber crops. It is hazardous to the respiratory system, liver, blood cholinesterase, the central nervous system, cardiovascular system, and gastrointestinal tract; and is toxic by inhalation, absorption, ingestion, and contact. Symptoms of exposure include miosis, aching eyes, blurred vision, lacrimation, salivation, anorexia, nausea, vomiting, abdominal cramps, diarrhea, giddiness, confusion, ataxia, rhinorrhea, headache, tight chest, wheezing, laryngeal spasms, irritation of the eyes and skin. OSHA exposure limit (TWA): 10 mg/m³ [skin].

maleic anhydride — $C_4H_2O_3$ — colorless needles, white lumps, or pellets with an irritating, choking odor; MW: 98.1, BP: 396°F, Sol: reacts, Fl.P: 218°F, sp gr: 1.48. It is used in the manufacture of polyester resins for automobile bodies, structural building panels, molded boats, and chemical storage; during the manufacture of fumaric acid and polyester; in the manufacture of alkyd resins as enamels, interior flat finishes, automotive finishes, printing inks, reactive plasticizers, and marine paints and varnishes; in the manufacture of detergents and lubricating additives as dispersant and wetting agent; in the manufacture of drying oils, oil resins as general coating and industrial dip coating; in the manufacture of terpene resins as shellac substitutes, in aniline inks, as protective coatings on paper, in metal foil, cellulose films, and natural synthetic fibers; in the manufacture of chlorendic anhydride for fire-retardant polyester resins; in the manufacture of extreme pressure lubricants; in organic synthesis and in production of chemical intermediates. It is hazardous to the eyes, skin, and respiratory system; and is toxic by inhalation, ingestion, and contact. Symptoms of exposure include conjunctivitis, photophobia, double vision, bronchial asthma, dermatitis, nasal and upper respiratory system irritation. OSHA exposure limit (TWA): 0.25 ppm [air] or 1 mg/m³.

malfeasance — inappropriate action or official misconduct.

malformation — permanent structural change that may adversely affect survival, development, or function.

malfunction — any unanticipated and unavoidable failure of pollution control equipment or process.

malignant — an abnormal growth that endangers the life or health of an individual and is resistant to treatment; having the potential to be transmitted from an original site to one or more sites

elsewhere in the body if not checked by treatment.

malinger — to pretend or exaggerate illness to avoid work.

malingerer — a person who pretends to be ill to avoid work.

malleable — capable of being hammered or pounded into thin sheets without structural failure.

malnutrition — poor nourishment resulting in improper diet or from some defect in metabolism that prevents the body from using its food properly.

malpractice — failure of a professional person to render proper services through reprehensible ignorance, negligence, or criminal intent.

mammary gland — a highly modified sebaceous gland found in female mammals that secretes milk.

mandate — a command, order or direction, written or oral, which the court is authorized to give and a person is bound to obey; this is done upon the decision of an appeal or writ of error.

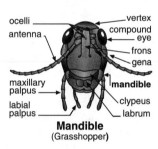

Mandible
(Grasshopper)

mandible — the strong, cutting mouthparts of arthropods used for crushing foods; the lower jaw.

manganese — Mn — a lustrous, brittle, silvery metal; MW: 54.9, BP: 3564°F, Sol: insoluble, Fl.P: not known, sp gr: 7.20. It is liberated during welding operations; during the casting of

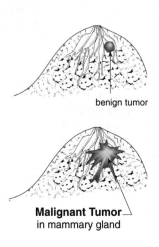

benign tumor

Malignant Tumor
in mammary gland

molten ferro-manganese; during the mixing and pressing of dry battery depolarization; during dumping, weighing, and mixing operations in ceramics and glass manufacture for pigmentation and coloration purposes; during the manufacture of manganese soap and wood preservatives; in the manufacture of safety matches, signal flares, and fire works. It is hazardous to the respiratory system, central nervous system, blood, and kidneys; and is toxic by inhalation and ingestion. Symptoms of exposure include Parkinson's disease, asthenia, insomnia, mental confusion, metal fume fever, dry throat, cough, tight chest, dyspnea, râles, flu-like fever, low-back pain, vomiting, malaise, fatigue. OSHA exposure limit (TWA): ceiling 5 mg/m^3 [air].

manifest — an itemized listing of a vessel's cargo with other particulars for the facility of customs' officers.

manifold — a pipefitting with numerous branches that convey fluids between a large pipe and several smaller pipes, or to permit choice of diverting flow from one of several sources or to one of several discharge points.

manometer — a device for measuring pressure using common mercury or another liquid to compare the difference in height of two liquid columns; an instrument used for measuring pressure; U-tube-shaped device partially filled with a liquid (usually water, mercury, or a light oil), so constructed that the amount of displacement of the liquid indicates the pressure being exerted on the instrument.

manual immersion cleaners — a blend of anionic and nonionic organic liquids or powders containing special additives for corrosion inhibition and skin protection.

manure — the excrete of animals collected from stables and barnyards; may also contain some spilled feed or bedding.

map — *See* cartography.

mapping unit — a soil or combination of soils delineated on a map or showing the taxonomic unit or units included; depicts soil types, phases, associations, or complexes.

margin of safety (MOS) — a factor estimated by dividing the experimental no observed adverse effect level (NOAEL) by the estimated daily human dose; the difference between an allowable level for a given pollutant and the criteria level at which adverse effects have been noted, assuming the allowable level is lower.

marine — referring to life in salt water.

Marine Protection and Sanctuaries Act — a 1972 law amended through 1988 stating that unregulated dumping of material into the ocean waters endangers human health, welfare, amenities, the marine environment, ecological systems, and economic potential; the policy of the United States is to regulate dumping of all types of materials into ocean waters and prevent or strictly limit dumping of any material that would adversely affect health, welfare, and ecosystem; also regulates transportation of material and dumping at sea.

Mariotte's law — *See* Boyle's law.

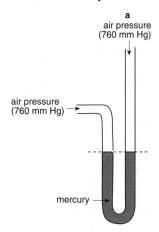

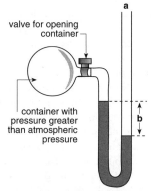

a + b = pressure in container

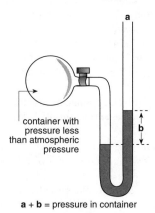

a + b = pressure in container

Manometer

marl — an earthy, unconsolidated deposit formed at freshwater lakes consisting chiefly of calcium carbonate mixed with clay or other impurities in varying proportions; used as fertilizer.

marrow — the soft tissue in the central cavity of large bones.

marsh — wet, soft, low-lying land that provides a habitat for many plants and animals and is covered at least part of the time by estuarine or coastal waters.

marsh gas — *See* methane.

maser (microwave amplification by stimulated emission of radiation) — an instrument used to amplify radio microwaves.

mask — a protective device usually fitting over the face or mouth and nose to filter poisonous chemicals in a gas or vapor form.

masking — the blocking out of a sight, sound, or smell with another.

mass — the definite quantity of matter a substance possesses depending on the gravitational force acting on it; an assigned unit of weight.

Massive Soil

massive soil — a soil structure in which the material clings together in large uniform masses, usually found in sub-stratum of C horizon.

mass number — the sum of the numbers of protons and neutrons in the nucleus of an atom; the integer nearest to the atomic mass. Also known as nucleon number.

mass spectrograph (MS) — a device used to determine the relative atomic masses or weights of electrically charged particles by separating them into distinct streams by means of magnetic deflection.

mass spectrometer — a device similar to the mass spectrograph designed in such a way that the beam constituents of a given mass-to-charge ratio are focused on an electrode and detected or measured electrically.

mastitis — (*disease*) inflammation of the breast occurring in a variety of forms and in varying degrees of severity.

Material Safety Data Sheet (MSDS) — a form and format used by industry and government to meet the specifications of the Occupational Safety and Health Administration Hazard Communication Standard. It contains information on the manufacture, identity of hazardous ingredients, physical/chemical characteristics, fire and explosion hazard data, reactivity data, precautions for safe handling and use, and control measures.

maternal death — the death of a woman while pregnant or within 42 days of termination of the pregnancy irrespective of the duration and site of the pregnancy.

maternal death rate — the number of direct and indirect maternal deaths per 100,000 live births for a specified time period.

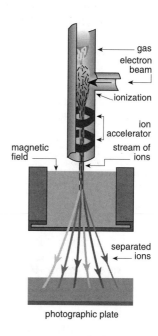

Mass Spectrograph

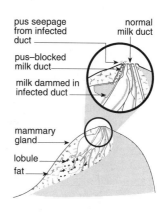

Mastitis

maticulism — a combination of catabolism and anabolism; the chemical change in living cells by which energy is provided for bioprocesses and activities and new material is assimilated.

matrix — (*EPA definition*) fiber or fibers with one end free and the other end imbedded in or hidden by a particulate; the exposed fiber must meet the fiber definition; (*NIOSH definition*) one or more fibers attached to or imbedded in a nonasbestos particle.

matte — a mixture of sulfides formed in smelting sulphide ores.

matter — anything that occupies space and has mass; occurs in three states — solid, liquid, and gas. Solid matter has both a definite size and shape. Liquid matter has a definite volume that takes the shape of its container. Gas has neither definite shape nor definite volume.

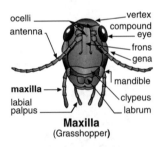

ocelli — vertex — compound eye — antenna — frons — gena — maxilla — mandible — labial palpus — clypeus — labrum

Maxilla
(Grasshopper)

maxilla — a mouth part of an arthropod; the upper jaw.

maximum allowable concentration (MAC) — the maximum exposure to a physical or chemical agent allowed in an 8-hour work day to prevent disease or injury.

maximum contaminant level — the maximum permissible level of a contaminant in water which is delivered to the free-flowing outlet of the consumer.

maximum dosage — the largest amount of a chemical that is safe to use without resultant excess residues or damage.

maximum evaporative capacity — the amount of sweat from a person which can be accepted into the surrounding air.

maximum permissible concentration (MPC) — the recommended maximum average concentration of radionuclides or chemicals a worker can be exposed to 8 hours a day, 5 days a week, 50 weeks per year.

maximum permissible dose (MPD) — the maximum dose of radiation which may be received by persons working with ionized radiation producing no detectable damage over a normal life span.

maximum permissible dose equivalent — the greatest dose equivalent that a person or specified

part thereof shall be allowed to receive in a given period of time.

maximum permissible exposure (MPE) — the maximum quantity of a chemical a person can be exposed to without causing short-term or long-term disease or injury.

maximum permissible level (MPL) — the tolerable dose rate of nuclear radiation humans can be exposed to.

maximum permissible limit (MPL) — *See* maximum permissible level.

mb — *See* millibar.

mCi — *See* millicurie.

MCi — *See* megacurie.

MCL — acronym for maximum contaminant level.

MCS — *See* multiple chemical sensitivity.

mean — (*statistics*) *See* arithmetic mean.

mean radiant temperature (MRT) — the temperature of a uniform black enclosure in which a solid body or person would exchange the same amount of radiant heat as in the existing nonuniform environment.

means of egress — an exit or way and method of passage to free and safe ground.

measurement error — (*statistics*) a statistical error made in a study because of poor selection of populations to be studied or the reliability and validity of the study instrument; basic sources of error are random and systematic.

meat — the part of the muscle of any cattle, sheep, swine, or goats which is skeletal or which is found in the tongue, diaphragm, heart, or esophagus, with or without the accompanying and overlying fat and the portions of bone, skin, sinew, nerve, and blood vessels which normally accompany the muscle tissue not separated in the process of dressing.

meat by-product — any part of cattle, sheep, swine, or goats capable of use as human food, other than meat.

mechanical aeration — the use of mechanical energy to inject air into water causing the waste stream to absorb oxygen from the air.

mechanical composting — a method of composting in which the compost is continuously, mechanically mixed and aerated.

mechanical ventilation — the air movement caused by a fan or other air moving device.

mechanical weathering — the disintegration of a rock by physical forces, such as freezing and thawing of water in the rock crevices, or the disruption of the rock by plant roots or burrowing animals reducing the rock to small fragments without changing its mineral composition.

medial tolerance limit (TL_{50}) — the concentration of a test material in a suitable diluent (maybe water) at which 50% of the test animals are able to survive for a specified period of exposure.

median lethal concentration (MLC) — the concentration of a test material that causes the death of 50% of a population within a given period.

median lethal dose (MLD) — the dose of radiation or other material ingested or injected that causes the death of a population of test organisms within a given period. Also known as LD_{50}.

mediation — the act or process of intervening between conflicting parties to promote reconciliation, settlement, or compromise.

medical benefits — benefits which are usually provided without dollar or time limits in the case of injury on the job, or with specific limitations for the normal maintenance of health.

Medical Literature Analysis and Retrieval System — *See* MEDLARS.

medical waste — any solid waste which is generated in the diagnosis, treatment, research, or in production or testing of biologicals in the immunization of human beings or animals.

Medical Waste Tracking Act (MWTA) — an act of Congress amending the Resource Conservation and Recovery Act by adding subtitle J, defining medical waste, and requiring the Environmental Protection Agency to establish a 2-year demonstration program on medical waste identification and disposal in several states.

Medicare — a nationwide health insurance program for people age 65 and over, or for persons under 65 but eligible for social security disability payments for over two years; a health insurance program supported by payroll taxes for people over 65 and disabled people.

medium — the environmental vehicle by which a pollutant is carried to the receptor.

medium texture — (*soils*) the texture exhibited by very fine sandy loams, loams, silt loams, and silts.

MEDLARS (Medical Literature Analysis and Retrieval System) — a computerized bibliographic system of the National Library of Medicine from which the Index Medicus is produced. *See also* MEDLINE.

MEDLINE — acronym for MEDLARS on-line, a computerized bibliographical computer system. *See also* MEDLARS.

MEFR — acronym for maximal expiratory flow rate.

mega- (M) — a prefix representing one million.

megacurie (MCi) — a unit of radiation equal to one million curies.

megaelectronvolt (MeV) — a unit of energy for expressing the kinetic energy of subatomic particles equal to 10^6 electron volts.

megavolt (MV) — a unit of electromotive force equal to one million electron volts.

megavolt per meter (MV/m) — one million volts per meter.

megawatt (MW) — a unit of electrical power equal to 1,000 kilowatts or one million watts.

megawatt days (MWD) — the amount of energy obtained from one megawatt power in one day.

meiosis — the type of cell division in which there is reduction of chromosomes to the haploid number during oogenesis and spermatogenesis and that produces four reproductive cells.

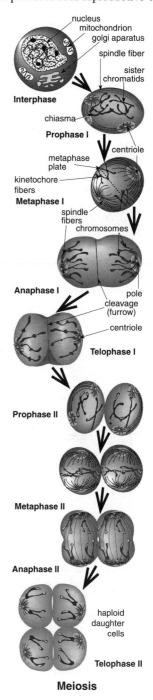

Meiosis

melanin — the pigment of the skin of animals or plants.

melanoma — a malignant tumor arising from the melanocytic system of the skin, in the eye, or rarely in the mucous membranes of the genitalia, anus, oral cavity, or other sites.

melting — the change of phase of a substance from solid to liquid. *See also* fusion.

melting point (MP) — the definite temperature at which a solid changes to a liquid.

membrane — a thin layer of tissue that covers a surface, lines a cavity, divides a space, or connects adjacent structures.

membrane filter — a filter medium made from various polymeric materials, such as cellulose, polyethylene, or tetrapolyethylene, used in collecting microscopic and submicroscopic particles.

membrane microfiltration system — a technique designed to remove solid particles from liquid waste forming filter cakes, typically ranging from 40–60% solids.

Meniere's disease — a disorder of the labyrinth of the inner ear with symptoms of tinnitus, heightened sensitivity to loud sounds, progressive loss of hearing, headache, and vertigo. It usually develops after a blow to the head or infection of the middle ear.

meninges — the three membranes covering the brain and spinal cord of vertebrates.

meningitis — inflammation of the meninges; the term does not refer to a specific disease entity, but rather to the pathological condition of inflammation of the tissues of the meninges.

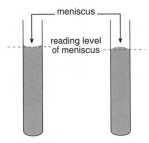

Meniscus

meniscus — the curved surface of a liquid in a cylinder which is caused by the attraction between a liquid and the walls of a cylinder.

menorrhagia — excessive menstruation. Also known as hypermenorrhea.

menstruation — periodic discharge of blood from the vagina from a nonpregnant uterus in women from puberty to menopause.

menu — (*computer science*) a set of options which can be used in the computer program.

meq — *See* milliequivalent.

mercurial barometer — a glass tube closed at the top and filled with mercury where the mercury column is supported by the pressure of the air; units are measured in inches, centimeters, or millibars.

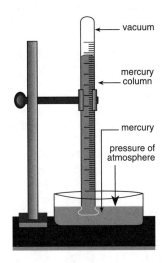

Mercury Barometer

mercury — Hg — a silver-white, heavy, odorless liquid; MW: 200.6, BP: 674°F, Sol: insoluble, sp gr: 13.6. It is used as a liquid cathode in electrolytic production of chlorine and caustic soda from brine; during the manufacture of industrial and medical apparatus; during the manufacture of inorganic and organic compounds for use as pesticides, antiseptics, germicides, and skin preparations; in the preparation of amalgams for use in tooth restorations, chemical processings, and molding operations; during the manufacture of mildew-proof paints and antifouling agents; in the manufacture of batteries, lamps, power tubes, tungsten-molybdenum wire and rods; in the manufacture of inorganic salts for use as catalysts in the production of urethanes, vinyl chloride monomers, anthraquinone derivatives, and other chemicals; as a chemical intermediate and in the manufacture of felt; as a flotation agent in the manufacture of bowling balls; as a laboratory reagent or as a working fluid in instruments; as a conductor during construction of military and nuclear power systems, and in air-rectifiers; in the manufacture of explosives; in the preparation of amalgams for use in artificial jewelry; in the manufacture of compounds for pulp and paper industry as controls for biological growths; during the mining and subsequent refining of ore containing cinnabar. It is hazardous to the skin, respiratory system, central nervous system, kidneys, and eyes; and is toxic by inhalation, absorption, and contact. Symptoms of exposure include cough, chest pain, dyspnea, bronchial pneumonitis, tremors, insomnia, irritability, indecision, headache, fatigue, weakness, stomatitis, salivation, gastrointestinal distress, anorexia, weight loss, proteinuria, irritation of the eyes and skin. Also known as quicksilver. OSHA exposure limit (TWA): 0.05 mg/m^3 [skin].

MESA — *See* Mining Enforcement and Safety Administration.

mesenchyma — the meshwork of embryonic connective tissues in the mesoderm from which the connective tissues of the body and the blood and lymphatic vessels are formed.

mesityl oxide — $(CH_3)_2C{=}CHCOCH_3$ — an oily, colorless to light-yellow liquid with a peppermint- or honey-like odor; MW: 98.2, BP: 266°F, Sol: 3%, Fl.P: 87°F, sp gr (59°F): 0.86. It is used as a solvent; as a paint and varnish remover and carburetor cleaner; in the preparation of roll-coating inks; during flotation processes for selective benefaction of ores. It is hazardous to the eyes, skin, respiratory system, and central nervous system; and is toxic by inhalation, ingestion, and contact. Symptoms of exposure include narcosis, coma, irritation of the eyes, skin, and mucous membrane; in animals: central nervous system effects. OSHA exposure limit (TWA): 15 ppm [air] or 60 mg/m^3.

meson — any unstable, elementary nuclear particle having a mass between that of an electron and a proton.

mesophiles — a microorganism having an optimal growth temperature between 20–45°C.

mesothelioma — a diffuse, rare form of cancer spreading over the surface of the lung and abdominal organs.

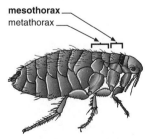

Mesothorax of Flea

mesothorax — the middle portion of the thorax of an insect bearing the second pair of legs and usually a pair of wings.

mesotrophic — having a nutrient load resulting in moderate productivity.

messenger RNA — the ribonucleic acid produced by the deoxyribonucleic acid in the nucleus and which functions as a template for protein building in the cytoplasm of a cell.

metabolism — the sum of all the physical and chemical processes by which a living organized substance is produced and maintained; the transformation by which energy is made available for the uses of the organism.

metabolite — any product of metabolism, especially a transformed chemical.

metal — a substance with the usual characteristic properties of luster, high strength, and good conductivity of heat and electricity and whose oxide combines with water to form a base. The bonding is provided by a large group of mobile electrons.

metal fume fever — an acute occupational condition caused by a brief high exposure to the freshly generated fumes and particles of metals. Also known as fume fever.

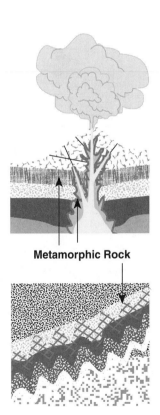

Metamorphic Rock

metamorphic rock — rock that has been changed from its original form which was igneous or sedimentary by great heat, pressure, or chemically active fluids; marble and slate are examples.

Earth's surface ☐
Earth's crust ◼

Metamorphism

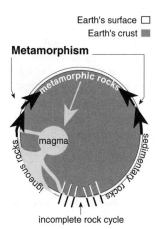

incomplete rock cycle

metamorphism — (*geology*) the process of change which rocks undergo when exposed to increasing temperatures and pressures causing their mineral components to become unstable.

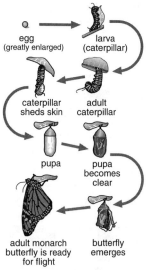

egg
(greatly enlarged)

larva
(caterpillar)

*caterpillar
sheds skin*

adult
caterpillar

pupa

pupa
becomes
clear

adult monarch
butterfly is ready
for flight

butterfly
emerges

Metamorphosis
of *Danaus plexippus*
(Monarch butterfly)

metamorphosis — a marked rapid change in structure of an animal during its growth, as it changes from larva to pupa and pupa to adult in the lifecycle of many invertebrates and amphibians.

metaphase — the stage of mitosis or meiosis during which the chromosomes are aligned on the equatorial plate.

metastasis — the transfer of disease from one organ or part of the body to another not directly connected with it; may be due either to the transfer of pathogenic microorganisms or to the transfer of cells, such as in malignant tumors.

mesothorax
metathorax

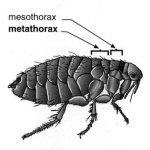

Metathorax of Flea

metathorax — the posterior portion of the thorax of an insect bearing the third pair of legs and usually a pair of wings.

meteoric water — the groundwater which originates as precipitation.

meter (m) — the international standard unit of length equal to 39.37 inches, 3.2808 feet, or 1.0936 yards.

metering pump — (*milk pasteurization*) a pump which is sealed at maximum speed and wired so it only operates if the flow diversion valve is fully operable; used to move milk under pressure to the holding tube and the rest of the high temperature short time pasteurization process.

meter-kilogram-second-ampere system — (*physics*) a system of electrical and mechanical units in which length, mass, time, and electrical current are the fundamental quantities.

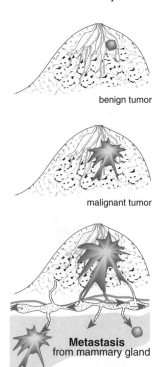

benign tumor

malignant tumor

Metastasis
from mammary gland

meter-kilogram-second system (mks system) — a metric system of units devised in 1901 which with some modifications forms the basis of the International System of Units.

methane — CH_4 — a colorless, odorless, simple, saturated hydrocarbon that comes from the decomposition of organic material and is the chief component of natural gas. Also known as marsh gas.

methanoic acid — *See* formic acid.

methanol — *See* methyl alcohol.

methanotrophic bioreactor system — an above-ground remedial technology for water contaminated with halogenated hydrocarbons, trichloroethylene, and related compounds.

methemoglobin — a compound formed from hemoglobin by oxidation of the iron atom from the ferrous to ferric state. Also known as ferrihemoglobin.

methemoglobinemia — methemoglobin in the blood usually due to toxic action of drugs or other agents, or to the hemolytic process; especially a problem of children up to age 5; nitrates are converted in the body to nitrites by bacterial action and the nitrites combine with hemoglobin thereby causing the hemoglobin to lose its ability to carry oxygen.

methicillin — an antibiotic which is a variation of penicillin.

methicillin resistant *Staphylococcus aureus* (MRSA) — a gram-positive bacterium growing in clusters not affected by the antibiotic methicillin and many other antibiotics.

methoprene — $C_{19}H_{34}O_3$ — (*insecticides*) an insect growth-regulator which is a true analog of the mosquito's own juvenile hormone; an amber liquid; MW: 310, Sol: 1–39 ppm in water, density: 8.56 lbs/gallon. It is used as a larvacide for mosquitos. No toxic effect is noted on non-target organisms. It is a biodegradable compound of very low persistence.

methoxychlor — $Cl_3CCH(C_6H_4OCH_3)_2$ — colorless to yellow crystals with a slight, fruity odor; MW: 345.7, BP: decomposes, Sol: insoluble, Fl.P: not known, sp gr ($77°F$): 1.41. It is used as an agricultural insecticide; an analogue of DDT. It is not known which organs are harmed by methoxychlor in humans; and is toxic by inhalation and ingestion in animals. No known symptoms of exposure in humans; symptoms of exposure in animals include fasciculation trembling, convulsions, kidney and liver damage; carcinogenic. OSHA exposure limit (TWA): 10 mg/m^3 [air].

Methyl Acetate

methyl acetate — CH_3COOCH_3 — a colorless liquid with a fragrant, fruity odor; MW: 74.1, BP: 135°F, Sol: 30%, Fl.P: 14°F, sp gr: 0.93. It is used as a solvent for nitrocellulose and cellulose acetate; as a chemical intermediate; in the preparation of artificial leather; during the formulation of lacquers, paints, perfumes, and any vinyl resin coatings. It is hazardous to the respiratory system, skin, and eyes; and is toxic by inhalation, ingestion, and contact. Symptoms of exposure include headache, drowsiness, optic atrophy, irritation of the nose and throat. OSHA exposure limit (TWA): 200 ppm [air] or 610 mg/m^3.

methyl acetylene — $CH_3C{=}CH$ — a colorless gas with a sweet odor; shipped as liquefied compressed gas; MW: 40.1, BP: –10°F, Sol: insoluble. It is used during the synthesis of pharmaceuticals and aromatics; liberated during high-temperature gas welding operations. It is hazardous to the central nervous system; and is toxic by inhalation. Symptoms of exposure include hyperexcitability, tremors, anesthesia, respiratory system irritation. OSHA exposure limit (TWA): 1,000 ppm [air] or 1,650 mg/m^3.

methyl acetylene-propadiene mixture — $CH_3C{=}CH/CH_2{=}C{=}CH_2$ — a colorless gas with a strong, characteristic, foul odor; shipped as a liquefied gas; MW: 40.1, BP: –36 to –4°F, Sol: insoluble. It is liberated from flame-hardening, metalizing, brazing, welding, and cutting. It is hazardous to the central nervous system, skin, and eyes; and is toxic by inhalation and contact. Symptoms of exposure include frostbite, disorientation, excitement. OSHA exposure limit (TWA): 1,000 ppm [air] or 1,800 mg/m^3.

methyl acrylate — $CH_2{:}CHCOOCH_3$ — a colorless liquid with an acrid odor; MW: 86.1, BP: 176°F, Sol: 6%, Fl.P: 25°F, sp gr: 0.96. It is used in the preparation of thermoplastic coatings; in the manufacture of acrylic fibers; in the synthesis of higher acrylates; during preparation of adhesives and sealants based on methyl acrylate; during polymerization in aqueous emulsions; during preparation of amphoteric surfactants for use in hair shampoos. It is hazardous to the respiratory system, eyes, and skin; and is toxic by inhalation, absorption, and contact. Symptoms of exposure include irritation of the eyes, skin, and upper respiratory system. OSHA exposure limit (TWA): 10 ppm [skin] or 35 mg/m^3.

methylal — $CH_3OCH_2OCH_3$ — a colorless liquid with a chloroform-like odor; MW: 32.1, BP: 111°F, Sol: 33%, Fl.P (open cup): –26°F, sp gr: 0.86. It is used as a solvent for adhesives, resins, gums, waxes, and protective coatings; as a solvent for the extraction of alkaloids, barbiturates, organic acids, and hydroxyacids; in the manufacture of artificial resins; as a gasoline and diesel fuel additive; as a special fuel for rocket and jet engines; in the manufacture of perfume; as a methylating agent or chemical intermediate. It is hazardous to the skin, respiratory system, and central nervous system; and is toxic by inhalation, ingestion, and contact. Symptoms of exposure include anesthesia, irritation of the skin, mild irritation of the eyes and upper respiratory system. OSHA exposure limit (TWA): 1,000 ppm [air] or 3,100 mg/m^3.

methyl alcohol — CH_3OH — a colorless liquid with a characteristic pungent odor; MW: 32.1, BP: 147°F, Sol: miscible, Fl.P: 52°F, sp gr: 0.79. It is used as a solvent for rotogravure inks, aniline dyes, and duplicator fluids; in the plastics industry to produce plasticizers, softening agents, and acrylic resins; as a solvent in the rubber industry; liberated during the application of surface coatings. It is hazardous to the eyes, skin, central nervous system, and gastrointestinal tract; and is toxic by inhalation, absorption, ingestion, and contact. Symptoms of exposure include headache, drowsiness, lightheadedness, nausea, vomiting, visual disturbances, blindness, eye irritation. Also known as methanol; wood alcohol. OSHA exposure limit (TWA): 200 ppm [skin] or 260 mg/m^3.

methylamine — CH_3NH_2 — a colorless gas with a fish- or ammonia-like odor; a liquid below 21°F; shipped as liquefied compressed gas; MW: 31.1, BP: 21°F, Sol: soluble, sp gr: 0.70. It is used as a chemical intermediate in the production of insecticides, herbicides, and fungicides, and in the production of surfactants; in the production of rocket fuels and explosives; in the production of pharmaceuticals and photographic chemicals; as an intermediate for dyes, textiles, dye assists, rubber, and anti-corrosive chemicals; as a polymerization inhibitor of hydrocarbons and in paint removers; to prevent webbing in natural and synthetic latex. It is hazardous to the respiratory system, skin, and eyes; and is toxic by inhalation, absorption, ingestion, and

contact. Symptoms of exposure include cough, dermatitis, conjunctivitis, skin and mucous membrane burns, irritation of the eyes and respiratory system. OSHA exposure limit (TWA): 10 ppm [air] or 12 mg/m^3.

methylation — a chemical process for the combination of a methyl radical with an organic compound.

methyl bromide — CH_3Br — a colorless gas with a chloroform-like odor at high concentrations; a liquid below 38°F; shipped as a liquefied compressed gas; MW: 95.0, BP: 38°F, Sol: 2%, Fl.P: not known, sp gr: 1.73. It is used as a space and soil fumigant in agriculture and industry; in food sterilization for pest control in fruits, vegetables, dairy products, nuts, and grains; in organic synthesis as a methylating agent; for preparation of quaternary ammonium compounds and organo-tin derivatives; as a selective solvent in aniline dyes; in laboratory procedures. It is hazardous to the central nervous system, skin, eyes, and respiratory system; and is toxic by inhalation, absorption, ingestion, and contact. Symptoms of exposure include headache, visual disturbances, vertigo, nausea, vomiting, malaise, hand tremors, convulsions, dyspnea, vesiculation, irritation of the skin and eyes; carcinogenic. OSHA exposure limit (TWA): 5 ppm [skin] or 20 mg/m^3.

methyl cellosolve® — $CH_3OCH_2CH_2OH$ — a colorless liquid with a mild, ether-like odor; MW: 76.1, BP: 256°F, Sol: miscible, Fl.P: 102°F, sp gr: 0.96. It is used in printing on plastic materials, rotogravure printing, and cellulose acetate pigment printing on textiles; in the manufacture of surface coatings and dyeing agents for stains, lacquers, dyes, and inks; as an anti-icing agent in military jet fuel and as an anti-stall additive in gasoline; liberated during the sealing of moisture-proof cellophane wrappers and packaging. It is hazardous to the central nervous system, blood, skin, eyes, and kidneys; and is toxic by inhalation, absorption, ingestion, and contact. Symptoms of exposure include headache, drowsiness, weakness, ataxia, tremors, somnolence, anemic pallor, eye irritation; carcinogenic. OSHA exposure limit (TWA): 25 ppm [skin] or 80 mg/m^3.

methyl cellosolve acetate® — $CH_3COOCH_2CH_2OCH_3$ — a colorless liquid with a mild, ether-like odor; MW: 118.1, BP: 293°F, Sol: miscible, Fl.P: 120°F, sp gr: 1.01. It is used as a solvent during the manufacture and spray or heat applications of surface coatings and adhesives; in the manufacture of photographic film; during dry cleaning operations. It is hazardous to the kidneys, brain, central nervous system, and peripheral nervous system; and is toxic by inhalation, absorption, ingestion, and contact. Symptoms of exposure include kidney and brain damage, eye irritation; in animals: narcosis, irritation of the eyes, nose, and throat. OSHA exposure limit (TWA): 25 ppm [skin] or 120 mg/m³.

methyl chloride — CH_3Cl — a colorless gas with a faint, sweet odor which is not noticeable at dangerous concentrations; shipped as a liquefied compressed gas; MW: 50.5, BP: –12°F, Sol: 0.5%. It is used in the manufacture of silicone resins and tetramethyl lead; as a methylating and chlorinating agent; as a dewaxing agent in petroleum refining; as a catalyst solvent in the production of butyl rubber; in the synthesis of a variety of other compounds; as an extractant for greases, oils, and resins; in the manufacture of pesticides, pharmaceuticals, and perfumes; as a propellant in aerosols; as a refrigerant. It is hazardous to the central nervous system, liver, kidneys, and skin; and is toxic by inhalation and contact. Symptoms of exposure include dizziness, nausea, vomiting, visual disturbances, staggering, slurred speech, convulsions, coma, liver and kidney damage, frostbite; carcinogenic. Also known as chloromethane. OSHA exposure limit (TWA): 50 ppm [air] or 105 mg/m³.

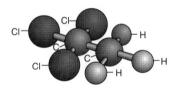

**Methyl Chloroform
(1,1,1–Trichloroethane)**

methyl chloroform — CH_3CCl_3 — a colorless liquid with a mild, chloroform-like odor; MW: 133.4, BP: 165°F, Sol: 0.4%, Fl.P: none, sp gr: 1.34. It is used as a solvent in the cold cleaning of metals and plastics; in vapor degreasing; in ultrasonic cleaning; in the dyeing and cleaning of fabrics and yarns; in organic synthesis in polymer manufacture; as a primary and carrier solvent in spot cleaners, adhesives, shoe polishes, stain repellents, hair sprays, Mace, insecticides, resins, inks, lubricants, protective coatings, asphalt extraction, and wastewater treatment; in aerosol manufacture; as a coolant; during printed circuit boards production; in liquid Drano production, and photographic film processing. It is hazardous to the skin, central nervous system, cardiovascular system, and eyes; and is toxic by

inhalation, ingestion, and contact. Symptoms of exposure include headache, lassitude, central nervous system depression, poor equilibrium, dermatitis, cardiac arrhythmias, eye irritation. Also known as trichloroethane. OSHA exposure limit (TWA): 350 ppm [air] or 1,900 mg/m³.

methylcyclohexane — $CH_3C_6H_{11}$ — a colorless liquid with a faint, benzene-like odor; MW: 98.2, BP: 214°F, Sol: insoluble, Fl.P: 25°F, sp gr: 0.77. It is used as a diluent solvent for cellulose during spray application of lacquers; during the manufacture of rotogravure inks; as a solvent for oils, fats, waxes, and rubber; in industrial organic synthesis in the hydro-reforming process; as a degreasing agent. It is hazardous to the respiratory system and skin; and is toxic by inhalation, ingestion, and contact. Symptoms of exposure include lightheadedness, drowsiness, irritation of the skin, nose, and throat. OSHA exposure limit (TWA): 400 ppm [air] or 1,600 mg/m³.

methylcyclohexanol — $CH_3C_6H_{10}OH$ — a straw-colored liquid with a weak odor like coconut oil; MW: 114.2, BP: 311–356°F, Sol: 4%, Fl.P: 154°F, sp gr: 0.92. It is used as a solvent in lacquers, oils, gums, waxes, and resins; as an auxiliary solvent for dry cleaning in soap manufacture; in the textile industry as a blending agent in textile soaps; as a degreaser; as an antioxidant in lubricants. It is hazardous to the respiratory system, skin, and eyes; in animals: central nervous system, liver, and kidneys; and is toxic by inhalation, absorption, ingestion, and contact. Symptoms of exposure include headache, irritation of the eyes and upper respiratory system; in animals: narcosis, liver and kidney damage. OSHA exposure limit (TWA): 50 ppm [air] or 235 mg/m³.

o-methylcyclohexanone — $CH_3C_6H_9O$ — a colorless liquid with a weak peppermint-like odor; MW: 112.2, BP: 325°F, Sol: insoluble, Fl.P: 118°F, sp gr: 0.93. It is used as a solvent in the plastics industry; in the manufacture of lacquers and varnishes; as a co-solvent with cyclohexanone; in the leather industry; as a rust remover. In animals, it is hazardous to the respiratory system, liver, kidneys, and skin; and is toxic by inhalation, absorption, ingestion, and contact. Symptoms of exposure in animals include narcosis, dermatitis, irritation of the eyes and mucous membrane. OSHA exposure limit (TWA): 50 ppm [skin] or 230 mg/m³.

methylene bisphenyl isocyanate — $CH_2(C_6H_4NCO)_2$ — white to light-yellow, odorless flakes; a liquid above 99°F; MW: 250.3, BP: 342°F, Sol: 0.2%, Fl.P (open cup): 396°F,

**M
N**

sp gr (122°F): 1.19. It is liberated during the manufacture of lacquer; during the production of component chemicals for foam systems; during the casting of high-density polyurethane elastomers; during the application of lacquer and sealant finishes; during flame lamination of fabrics. It is hazardous to the respiratory system and eyes; and is toxic by inhalation, ingestion, and contact. Symptoms of exposure include cough, pulmonary secretions, chest pain, dyspnea, asthma, irritation of the eyes, nose, and throat. OSHA exposure limit (TWA): ceiling 0.2 mg/m³ [air].

methylene blue — $C_{16}H_{18}N_3SCl\cdot3H_2O$ — (*chemistry*) — a dark green crystalline compound used as a dye, as a stain in bacteriology, in medicine as an antidote for cyanide poisoning, and as a reagent in oxidation-reduction processes.

methylene blue reduction test — a test designed to give a rough estimate of the bacterial load in food; in milk, decolorization in one half hour indicates a high bacterial load, whereas decolorization in 8 hours indicates a low bacterial load.

Methylene Chloride

methylene chloride — CH_2Cl_2 — a colorless liquid with a chloroform-like odor; a gas above 104°F; MW: 84.9, BP: 104°F, Sol: 2%, Fl.P: not known. It is used as a solvent in paint and varnish removers; in the manufacture of aerosols; as an extraction solvent for foods and furniture processing; as a cooling solvent in the manufacture of cellulose acetate; in organic synthesis; in plastics processing; as a solvent in vapor degreasing of thermal switches and thermometers; as a secondary refrigerant in air conditioning and scientific testing; as an extraction solvent for edible fats, coca, butter, beer flavoring, decaffeinated coffee, oleoresin manufacture, oils, waxes, perfumes, flavorings, and drugs; as a solvent for paints and lacquers, varnishes, enamels, adhesives, rubber cements; as a carrier for pharmaceutical tablet coatings; in the dyeing of synthetic fibers. It is hazardous to the skin, cardiovascular system, central nervous system, and eyes; and is toxic by inhalation, ingestion, and contact. Symptoms of exposure include fatigue, weakness, sleepiness, lightheadedness,

tingling and numb limbs, nausea, eye and skin irritation; carcinogenic. OSHA exposure limit (TWA): 500 ppm [air].

methyl formate — $HCOOCH_3$ — a colorless liquid with a pleasant odor; a gas above 89°F; MW: 60.1, BP: 89°F, Sol: 30%, Fl.P: –2°F, sp gr: 0.98. It is used as an insecticide and larvicide for fumigating dried fruits, nuts, tobacco, cereals, and infected clothing; as a general industrial solvent for greases, fatty acids, cellulose acetate, collodion, and celluloid; in organic synthesis. It is hazardous to the eyes, respiratory system, and central nervous system; and is toxic by inhalation, absorption, ingestion, and contact. Symptoms of exposure include chest depression, dyspnea, visual disturbances, central nervous system depression, irritation of the eyes and nose; in animals: pulmonary edema. OSHA exposure limit (TWA): 100 ppm [air] or 250 mg/m³.

5-methyl-3-heptanone — $CH_3CH_2CO(CH_2)_4CH_3$ — a colorless liquid with a pungent odor; MW: 128.2, BP: 315°F, Sol: insoluble, Fl.P: 138°F, sp gr: 0.82. It is used in the preparation of perfumes; as a specializing solvent in cleaning operations. It is hazardous to the eyes, skin, respiratory system, and central nervous system; and is toxic by inhalation, ingestion, and contact. Symptoms of exposure include headache, narcosis, coma, dermatitis, irritation of the eyes and mucous membrane. OSHA exposure limit (TWA): 25 ppm [air] or 130 mg/m³.

methyl hydrazine — CH_3NHNH_2 — a fuming, colorless liquid with an ammonia-like odor; MW: 46.1, BP: 190°F, Sol: miscible, Fl.P: 17°F, sp gr (77°F): 0.87. It is used in the preparation of liquid rocket propellants; as a chemical intermediate for the synthesis of pesticides; in the polymerization technology. It is hazardous to the central nervous system, respiratory system, liver, blood, eyes, cardiovascular system; and is toxic by inhalation, absorption, ingestion, and contact. Symptoms of exposure include vomiting, diarrhea, tremors, ataxia, anoxia, cyanosis, convulsions, irritation of the eyes and respiratory system; carcinogenic. OSHA exposure limit (TWA): ceiling 0.2 ppm [skin] or 0.35 mg/m³.

methyl iodide — CH_3I — a colorless liquid with a pungent, ether-like odor; turns yellow, red, or brown on exposure to light and moisture; MW: 141.9, BP: 109°F, Sol: 1%, sp gr: 2.28. It is used as a methylating agent in organic synthesis; as a laboratory reagent; as an insecticidal fumigant; in analytical chemistry. It is hazardous to the central nervous system, skin, and eyes; and is toxic by inhalation, absorption, ingestion, and contact. Symptoms of exposure include nausea, vomiting, vertigo, ataxia, slurred speech, drowsiness, dermatitis, eye irritation; carcinogenic. OSHA exposure limit (TWA): 2 ppm [skin] or 10 mg/m^3.

methyl isobutyl carbinol — $(CH_3)_2CHCH_2CH(OH)CH_3$ — a colorless liquid with a mild odor; MW: 102.2, BP: 271°F, Sol: 2%, Fl.P: 106°F, sp gr: 0.81. It is used as an extractant in dewaxing of mineral oils; as a frothier in froth flotation of various minerals; as an extractant in the manufacture of antibiotics; in the synthesis of surfactants; in the preparation of lubricating oil additives. It is hazardous to the eyes and skin; and is toxic by inhalation, absorption, ingestion, and contact. Symptoms of exposure include headache, drowsiness, dermatitis, irritation of the eyes. OSHA exposure limit (TWA): 25 ppm [skin] or 100 mg/m^3.

methyl isocyanate — CH_3NCO — a colorless liquid with a sharp, pungent odor; MW: 57.1, BP: 139°F, Sol (59°F): 10%, Fl.P: 19°F, sp gr: 0.96. It is used as a cross-linking agent; as an additive in polymer technology; in organic synthesis. It is hazardous to the respiratory system, eyes, and skin; and is toxic by inhalation, absorption, ingestion, and contact. Symptoms of exposure include cough, secretions, chest pain, dyspnea, asthma, eye and skin injury, irritation of the eyes, nose, and throat; in animals: pulmonary edema. OSHA exposure limit (TWA): 0.02 ppm [skin] or 0.05 mg/m^3.

methyl (n-amyl) ketone — $CH_3CO(CH_2)_4CH_3$ — a colorless to white liquid with a banana-like, fruity odor; MW: 114.2, BP: 305°F, Sol: 0.4%, Fl.P: 102°F, sp gr: 0.81. It is used in the preparation of synthetic resins, especially for metal coating; as a solvent for rubber and nitrocellulose. It is hazardous to the eyes, skin, respiratory system, central nervous system, and peripheral nervous system; and is toxic by inhalation, ingestion, and contact. Symptoms of exposure include headache, narcosis, coma, dermatitis, irritation of the eyes and mucous membrane. Also known as 2-heptanone. OSHA exposure limit (TWA): 100 ppm [air] or 465 mg/m^3.

methyl mercaptan — CH_3SH — a colorless gas with a disagreeable odor like garlic or rotten cabbage; a liquid below 43°F; shipped as a liquefied compressed gas; MW: 48.1, BP: 43°F, Sol: 2%, Fl.P (open cup): 0°F, sp gr: 0.90. It is used as a catalyst and activator; in wood processing; in the synthesis of chemical intermediates for the manufacture of resins, plastics, insecticides, and pressure-sensitive and oil-resistant adhesives; as an odorant and warning agent in

natural gas; in jet fuels. It is hazardous to the respiratory system and central nervous system; and is toxic by inhalation and contact. Symptoms of exposure include narcosis, cyanosis, convulsions, pulmonary irritation. OSHA exposure limit (TWA): 0.5 ppm [air] or 1 mg/m^3.

methyl methacrylate — CH_2=$C(CH_3)COOCH_3$ — a colorless liquid with an acrid, fruity odor; MW: 100.1, BP: 330°F, Sol: slight, Fl.P (open cup): 50°F, sp gr: 0.94. It is used during the casting of acrylic sheets; during polymerization to produce molding and extruding powders; during the manufacture of unsaturated polyester resins; during production of emulsion polymers for use in adhesives, sealants, fabrics, sizes, leather finishes, paper coatings, polishes. It is hazardous to the eyes, upper respiratory system, and skin; and is toxic by inhalation, ingestion, and contact. Symptoms of exposure include dermatitis, irritation of the eyes, nose, and throat. OSHA exposure limit (TWA): 100 ppm [air] or 410 mg/m^3.

methyl parathion — $C_8H_{10}NO_5PS$ — white solid crystals or a brownish liquid smelling like rotten eggs; MW: 263.23, BP: not known (decomposes above ambient temperature), decomposes violently at 120°C, Sol: at 25°C 50 ppm, Fl.P: not known, density at 20°C 1.358. It is used as a broad-spectrum, nonsystemic contact and stomach insecticide with some fumigant action; to control insects on a wide variety of crops. It is hazardous to the lungs, pulmonary system, liver, kidneys, and heart; and is toxic by inhalation and contact. Symptoms of exposure include pulmonary edema, bronchoconstriction and hypersecretion of bronchial glands, heart, liver, kidney, blood vessel lesions, cholinesterase, depression, lethargy, miosis, depressed plasma, and chromosomal abberations. OSHA exposure limit (TWA): 0.2mg/m^3 [air].

4-methylphenol — *See* p-cresol.

alpha-methyl styrene — $C_6H_5C(CH_3)$=CH_2 — a colorless liquid with a characteristic odor; MW: 118.2, BP: 330°F, Sol: insoluble, Fl.P: 129°F, sp gr: 0.91. It is used in the manufacture of specialized alpha-methyl-styrene-polyester and alkyd surface coating resins; in the manufacture of certain plasticizers in varnishes, adhesives, and plastics; in the manufacture of carbolic acid with cumene peroxidation process. It is hazardous to the eyes, respiratory system, and skin; and is toxic by inhalation, ingestion, and contact. Symptoms of exposure include drowsiness, dermatitis, irritation of the eyes, nose, and throat. OSHA exposure limit (TWA): 50 ppm [air] or 240 mg/m^3.

M
N

metric system — a system of scientific units based on the decimal system; its units of length, time, and mass are the meter, second, and kilogram, respectively.

Metropolitan Statistical Area (MSA) — except in New England as prescribed by federal rules and regulations, a county or group of contiguous counties which contains at least one city with a population of 50,000, or twin cities with a population of at least 50,000; they are essentially metropolitan in character and are socially and economically integrated with the central city.

MeV — *See* megaelectronvolt.

mg — *See* milligram.

mgd — notation for the amount equal to a million gallons per day.

mg/l — notation for milligram per liter equivalent to one part per million.

mg/m³ — *See* milligrams per cubic meter.

MHAPPS — acronym for Modified Hazardous Air Pollutant Prioritization System.

mho — a unit of electrical conductance equal to the conductance of a body through which one ampere of current flows when the potential difference is one volt; it is reciprocal to the ohm.

Mho (Ω) — *See* siemen.

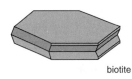

biotite

muscovite

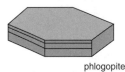

phlogopite

Mica

mica — colorless, odorless flakes or sheets of hydrous silicates containing less than 1% quartz; MW: 797, BP: not known, Sol: insoluble, sp gr: 2.6–3.2. It is used in the manufacture of asphalt shingles and roll roofing; in the manufacture of paint, wall paper, and bituminized cardboard; in the manufacture of molded rubber products, plastics, special greases; in the fabrication of windows and diaphragms; in the manufacture

of electrical insulation for low thermal conductivity and high dielectrical strength. It is hazardous to the lungs; and is toxic by inhalation. Symptoms of exposure include pneumoconiosis, cough, dyspnea, weakness, weight loss. OSHA exposure limit (TWA): 3 mg/m³ [resp].

micelle — an aggregation or cluster of molecules, ions, or minute submicroscopic particles.

micro (μ) — a prefix meaning one-millionth or 10^{-6}.

microanalysis — an analysis performed on a relatively small sample requiring the use of specialized instruments.

microbar — a unit of pressure used in acoustics equal to 1 dyne/cm². Also known as barye.

microbe — a microorganism, especially a bacterium of a pathogenic nature.

microbial insecticide — bacteria or other tiny plants or animals used to prevent damage by or destroy insects.

microbiology — the study of microorganisms including algae, bacteria, fungi, viruses, and protozoa.

microclimate — local climatic patterns caused by geographical, biological, and artificial factors.

microcurie (μCi) — a unit of measurement equal to one-millionth of a curie.

microelectronics — a branch of electronics concerned with the design, production, and application of electronic components, circuits, and devices of extremely small dimensions.

microenvironment — a well-defined place, such as a home, office, or car, in which a chemical or biological agent is present in a uniform manner.

microgram (μg) — a unit of mass equal to one-millionth of a gram.

microgram per cubic meter (μg/m³) — a measurement of weight per unit volume equal to one part per billion; micrograms of particulates per cubic meter of air sampled.

microgram per liter (μ/l) — a unit of weight per volume equal to one part per billion.

micrometer — *See* micron.

micromicro- — *See* pico.

micron — a metric unit of measurement equal to 1 one-thousandth part of a milliliter, 10^{-6} meters, and approximately 1/25,000 of an inch; now replaced by the micrometer.

micronutrient — a chemical element necessary in only small amounts for growth and development. Also known as trace element.

microorganism — an organism, including bacteria, protozoans, viruses, rickettsiae, and some algae and fungi, that can be observed only with the aid of a microscope.

microphone — an instrument that is actuated by electricity and transmits sound.

microphyte — (*botany*) a microscopic plant of plant-like organism, especially a fungus.

microplankton — plankton with a size range of 20–200 micrometers.

microtubule — (*genetics*) the abundant organella in eukaryotic cell cytoplasm consisting of aggregates of globular proteins arranged in strands of protofilaments.

microwave — a high-frequency electromagnetic wave usually 1 millimeter to 10 centimeters (sometimes 30 centimeters) in wavelength; causes structural or functional changes to the eyes, testicles, bone marrow, and cardiovascular and central nervous systems due to the heating caused by the microwave absorption.

microwave amplification by stimulated emission of radiation — *See* maser.

middle clouds — types of clouds, such as stratus or cumulus, with bases averaging 10,000 feet above the earth.

migrant — a seasonal farm worker who performs agricultural labor requiring travel.

migration — moving from place to place.

mildly alkaline detergent — a general-purpose cleaning material used in manual cleaning for floors, walls, ceilings, equipment, and utensils.

mile — a unit of measurement equal to 5,280 feet or 1.6093 kilometers.

mil gal — notation for the unit of measurement equal to one million gallons.

miliaria — *See* heat rash.

milk — fluid secretion of the mammary glands practically free of cholestrum, containing a minimum of 8.25% nonfat milk solids and 3.25% milk fat.

milk and food specialist — an environmental health practitioner concerned with the protection of the milk and food supply against contamination from biological, chemical, and physical hazards from source to consumer.

milk-fat — the fat of milk. Also called butter fat.

milk products — products made from milk, including cream, sour cream, half and half, whip cream, concentrated milk, flavored milk, and cottage cheese.

milli-(m) — a prefix meaning one-thousandth or 10^{-3}.

milliamperage — (*radiography*) the X-ray-tubed current during an exposure measured in milliamperes.

milliampere (mA) — a unit of current equal to one-thousandth of an ampere.

millibar (mb) — a unit of pressure used in meteorology and international weather observation equal to 1,000 dynes per square centimeter.

millicurie (mCi) — a unit of radiation equal to one-thousandth of a curie.

milliequivalent (meq) — one-thousandth of a chemical equivalent.

milligram (mg) — a unit of mass equal to one-thousandth of a gram.

milligrams per cubic meter (mg/m^3) — a unit of measurement of air concentrations of dusts, gases, mists, and fumes.

milliliter (ml) — a unit of volume equal to one-thousandth of a liter.

millimeter (mm) — a unit of length equal to one-thousandth of a meter.

millimeter of mercury (mm Hg) — the unit of pressure equal to the pressure exerted by a column of liquid mercury one millimeter high at a standard pressure.

millimicron (mμ) — a unit of length equal to one-thousandth of a micron or one-billionth of a meter.

millipore filter — trademark for a device used to filter solutions as they are administered.

millirad (mrad) — a unit of absorbed ionizing radiation equal to one-thousandth of a rad.

millirem (mrem) — a unit of radiation equal to one-thousandth of a rem.

milliroentgen (mr) — a unit of radioactive dose of electromagnetic radiation equal to one-thousandth of a roentgen.

millisecond (ms) — a unit of time equal to one-thousandth of a second.

millivolt (mV) — a unit of electromotive force equal to one-thousandth of a volt.

milliwatt (mW) — a unit of power equal to one-thousandth of a watt.

min — *See* minute.

Minamata disease — (*disease*) a severe neurological disorder due to alkyl mercury poisoning leading to permanent neurologic and mental disabilities or death; prevalent among those who eat seafood contaminated with mercury.

mine drainage — any drainage or water pumped or siphoned from an active mining area or a post-mining area.

mineral — a naturally occurring substance that has a characteristic chemical composition and, in general, a crystalline structure; identified by the properties of its crystal system, hardness, relative density, lustre, color, cleavage, and fracture.

M
N

mineralization — the conversion of an element from an organic form to an inorganic state as a result of microbial decomposition.

mineral soil — a soil consisting predominantly of or having its properties determined by mineral matter.

mineral spirits — flammable petroleum distillates that boil at temperatures lower than kerosene and used as solvents and thinners, especially in paints and varnishes.

Mine Safety and Health Administration (MSHA) — a federal administration which regulates coal mines, metal and non-metal mines, and constant exposure to asbestos and various carcinogens found in mines.

minimum risk level — an estimate of daily human exposure to a chemical that is likely to be without appreciable risk of deleterious or non-cancerous effects over a specified duration of exposure.

Mining Enforcement and Safety Administration (MESA) — a now defunct administration whose functions have been transferred to the Mine Safety and Health Administration.

mining wastes — residues which result from the extraction of raw materials from the earth.

minute (min) — a unit of time equal to 1/60th of an hour.

miosis — excessive contraction of the pupil of the eye.

misbranded food — a food whose labeling is false or misleading.

miscible — the capability of liquids to be soluble in each other.

misdemeanor — a crime punishable by a year or less in prison.

misfeasance — doing wrongfully and injuriously an act which might otherwise be done in a lawful manner.

mission — a stated self-imposed duty.

mist — liquid particles, up to one hundred micrometers in diameter, suspended in the air caused by the condensing of gases in the liquid state due to splashing, foaming, or atomizing; may cause irritation to the nose and throat, ulceration of the nasal passages, damage to the lungs, damage to other internal tissues depending on the chemical present, and may also cause dermatitis.

mite — any of numerous small to very minute arachnids that often infest animals, plants, or stored food.

miticide — a pesticide chemical or agent that prevents damage from or destroys mites and ticks. *See also* acaricide.

mitigation — alleviation, abatement, or diminution of a penalty or punishment imposed by law.

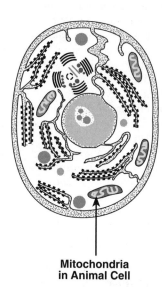

Mitochondria in Animal Cell

mitochondria — minute, rod-shaped bodies in the cytoplasm known to be centers of cellular respiration.

mitosis — cell division from which two identical cells result, each containing the same number and kind of chromosomes as the parent cell.

mix — (*ice cream*) the unfrozen combination of all ingredients of a frozen dessert with or without fruit, fruit juices, candy, nut meats, flavor, or harmless color.

mixed liquor suspended solids (MLSS) — (*wastewater*) suspended solids in a mixture of activated sludge and organic matter undergoing activated sludge treatment in the aeration tank.

mixing chamber — (*incineration*) a chamber usually placed between the primary and secondary combustion chambers of an incinerator; the products of combustion are thoroughly mixed there by turbulence created by increased velocity of gases or turns in the direction of the gas flow.

mixing depth — the atmospheric layer near the surface of a region in which warmer air rises, mixes with colder air, and creates turbulence that disperses pollutants.

mixture — an association of substances not chemically combined and able to be separated by mechanical means.

mks system — a system of measurement based on the meter, the kilogram, and the second.

mks units — *See* meter-kilogram-second system.

ml — *See* milliliter.

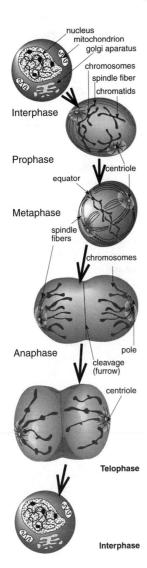

Mitosis

MLC — *See* median lethal concentration.

MLD — *See* median lethal dose.

MLSS — *See* mixed liquor suspended solids.

mm — *See* millimeter.

mm Hg — *See* millimeter of mercury.

MMWR — *See* Morbidity and Mortality Weekly Reports.

mobile environmental monitor — a field transportable analytical instrument designed to identify and measure organic pollutants in various environmental media by use of a spectrometer.

mobile food unit — a vehicle-mounted food establishment that is readily movable.

mobile home — a factory-assembled structure equipped with the necessary service connections and that is readily moveable as a unit; designed to be used as a dwelling without a permanent foundation.

mobile noise source — any noise source other than a fixed noise source.

mock lead — *See* sphalerite.

mock ore — *See* sphalerite.

model — a mathematical or physical system used to simulate real events and processes.

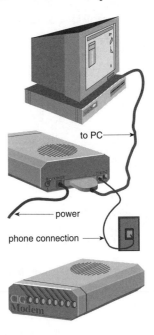

Modem

modem — (*computer science*) a device for the conversion of digital and analog signals to allow data transmission over telephone lines.

moderate limitations — (*soils*) limitations that can usually be overcome or modified with direct planning and careful design.

moderately well drained — (*soils*) soil in which the water is removed somewhat slowly so that the profile is wet for a small but significant part of the time.

moderator — a material used to slow neutrons in a reactor.

modulate — (*physics*) to vary the amplitude, frequency, or phase of electromagnetic waves, especially carrier waves.

modus operandi — manner of working.

moist soil — loose, very friable, friable, firm, very firm, or extremely firm soil.

moisture content — (*solid waste*) the weight loss expressed in percent in a sample of solid waste that is dried to a constant weight and a temperature of 100°C–105°C.

moisture penetration — the depth to which irrigation water or rain penetrates soil before the rate of downward movement becomes negligible.

mol — *See* mole.

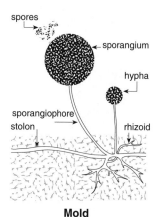

Mold

mold — a fungus of the order *Mucorales*.

mole (mol) — the unit of amount of substance, molecule, etc.; it replaces the term gram formula.

molecular absorption spectrophotometry — a technique using absorption of radiation by molecular species.

molecular epidemiology — measurement of the interaction of a toxic chemical or its derivative with a tissue constituent or a tissue alteration resulting from exposure to a chemical; provides indirect measure of individual exposure.

molecular fluorescence spectrophotometry — a technique used for special measurements of fluorescent radiation from luminescent compounds upon excitation by incident radiation.

molecular weight (MW) — the sum of the atomic weight of all the atoms in a molecule of a substance. Also known as relative molecular mass.

molecule — the smallest particle of a substance that can exist in a free state and retain all its chemical properties.

molluscicide — a pesticide chemical or agent that prevents damage by or kills slugs or snails.

mollusk — any of a large phylum of invertebrates Mollusca, such as snails or clams, with soft unsegmented bodies usually enclosed in a calciphoric shell. Mollusks are indicators of environmental contamination and environmental quality since they live in the mud or sand-bottoms of the aquatic ecosystems, feed by taking water through their bodies, and therefore reflect the chemical or microbiological conditions of the water.

molting — the shedding of the exoskeleton and secreting a new exoskeleton during periods of growth.

molybdenum and insoluble molybdenum — Mo — appearance, odor, and properties vary depending upon the specific soluble compound/insoluble metal: dark gray or black powder with a metallic luster; MW: 95.9, BP: 8717°F, Sol: insoluble, sp gr: 10.28. They are used as lubricants in greases, oil dispersions, resin-bonded films, and dry powders; as catalysts in petroleum refining and chemical processing; in pigment mixture for labeling glass bottles, for printing inks, for paints, and for plastics; as reagents in analytical chemistry laboratories; as catalysts in organic synthesis in medicinals, in decorative and protective coatings for metal; as feed additives; as a chemical intermediate and plating agent for mirrors; in the manufacture of cutting tools. Soluble Mo is hazardous to the respiratory system; in animals: kidneys, blood; there is no known harm to humans from metal Mo; they are toxic by inhalation, ingestion, and contact. Symptoms of exposure in animals include anorexia, incoordination, dyspnea, anemia, irritation of the eyes, nose, and throat, diarrhea, weight loss, listlessness, damage to the liver and kidneys. OSHA exposure limit (TWA): 5 mg/m³ and 10 mg/m³.

moniliasis — *See* candidiasis.

monition — (*law*) the summons to appear and answer issues listed on the complaint.

monitor — to observe, collect, check, and measure data to determine the capability of a given condition with predetermined parameters; a planned sequence of observations or measurements of critical limits designed to produce an accurate record and intended to ensure that the critical limit maintains product safety; a means of measuring concentrations and behavior of substances in environmental media or in human or other biological tissues.

monitoring well — a well used to obtain water samples for water quality analysis or to measure groundwater levels.

monochromatic light — light consisting of only one color.

monocyte — a mononuclear, phagocytic leukocyte, between 13 and 25 micrometers in diameter, having an ovoid- or kidney-shaped nucleus and azurophilic cytoplasmic granules.

monocytosis — an excess of monocytes in the blood.

monoenergetic radiation — radiation of a given type, such as alpha, beta, gamma, etc., in which all particles are photons originating with and having the same energy.

monomer — a compound of relatively low molecular weight which under certain conditions, either alone or with another monomer, forms various types and lengths of molecular chains called polymers or copolymers of high molecular weight. An example of this would be styrene, which when polymerized readily becomes the polymer polystyrene.

monomethyl aniline — $C_6H_5NHCH_3$ — a yellow to light-brown liquid with a weak, ammonia-like odor; MW: 107.2, BP: 384°F, Sol: insoluble, Fl.P: 175°F, sp gr: 0.99. It is used in organic synthesis and in dye intermediates. It is hazardous to the respiratory system, liver, kidneys, and blood; and is toxic by inhalation, absorption, ingestion, and contact. Symptoms of exposure include weakness, dizziness, headache, dyspnea, cyanosis, methemoglobinemia, pulmonary edema, liver and kidney damage. OSHA exposure limit (TWA): 0.5 ppm [skin] or 2 mg/m³.

monosaccharide — simple sugars, such as $C_6H_{12}O_6$, which cannot be hydrolyzed to simpler carbohydrates.

monotone — *See* monotonic.

monotonic — (*mathematics*) either never increasing or never decreasing as a function. Also known as monotone.

monthly catalog of United States government publications — a series of catalogs available from 1895 to the present of all United States government publications.

montmorillonite — an aluminosilicate clay mineral with an expanding structure, that is with two silicon tetrahedral layers enclosing an aluminum octahedral layer; considerable expansion may be caused by water moving between silica layers of contiguous units.

moor — *See* bog.

morbidity — the illness, injury, or disability in a defined population.

Morbidity and Mortality Weekly Reports (MMWR) — a Centers for Disease Control weekly publication that gives information on current trends in the nation's health.

morbidity rate — the rate of disease, injury, or disability from a specific cause.

morpholine — C_4H_8ONH — a colorless liquid with a weak, ammonia- or fish-like odor; a solid below 23°F; MW: 87.1, BP: 264°F, Sol: miscible, Fl.P (open cup): 98°F, sp gr: 1.00. It is used in the manufacture of rubber chemicals for rubber accelerators, catalysts, plasticizers, curing agents, stabilizers of halogenated butyl rubber, and emulsifying agents; as a corrosion inhibitor; in the manufacture of optical brightening agents in bleaches and detergents; in the compounding of waxes and polishes; as a chemical intermediate for the textile industry; in the pharmaceutical industry; as solvents for dyes, waxes, resins, and casein. It is hazardous to the respiratory system, eyes, and skin; and is toxic by inhalation, absorption, ingestion, and contact. Symptoms of exposure include visual disturbances, cough, liver and kidney damage, irritation of the nose, eyes, and skin, respiratory irritation. OSHA exposure limit (TWA): 20 ppm [skin] or 70 mg/m³.

morphology — the branch of biological science that deals with the study of the structure and form of living organisms.

mortality — death.

mortality rate — the ratio of the total number of deaths to the total population of a given community at a given time; usually expressed as deaths per 1,000, 10,000, or 100,000. Also known as death rate.

mortar — a hard mass made from calcium hydroxide, sand, and water used in masonry.

MOS — *See* margin of safety; (*physics*) abbreviation for metal oxide semiconductor.

M
N

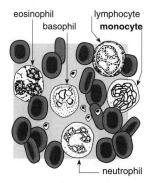

Monocyte

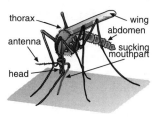

Mosquito

mosquito — small, long-legged, two-winged insect belonging to the order *Diptera* and family *Culicidae*; differs from other flies by having an elongated proboscis and scales on the winged veins and the winged margins.

most probable number — a statistical term inferring a particular microbial contamination level on the basis of a sample tested.

motile — capable of spontaneous movement.

motion — (*law*) an application to a court, either written or oral, for rule or order; (*mechanics*) a continuous change of position of a body.

motor nerve — a nerve containing only motor fibers.

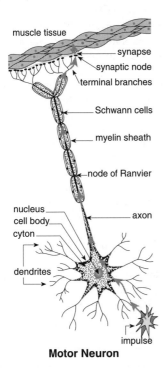

muscle tissue
synapse
synaptic node
terminal branches
Schwann cells
myelin sheath
node of Ranvier
nucleus
cell body
cyton
axon
dendrites
impulse

Motor Neuron

motor neuron — a neuron that carries impulses from the brain or spinal cord to a muscle or gland to initiate a physiological response.

motor unit — an anterior horn cell, its axon, and all of the muscle fibers innervated by the axon.

Motor Vehicle Air Pollution Control Act — a law passed in 1965 recognizing feasibility of setting auto emission standards, establishing national standards for auto emissions, and applying California state emission standards for hydrocarbons and carbon monoxide nationally.

mottled — (*soils*) a zone of chemical oxidation and reduction activity appearing as patches of red, brown, orange, or dull grays; spotty or uneven in appearance.

mouse — (*computer science*) a hand-steered device for controlling the position of the cursor on a video display screen.

moveable grate — a grate with moving parts designed to feed solid fuel or solid waste to a furnace.

MP — *See* melting point.

MPC — *See* maximum permissible concentration.

MPD — *See* maximum permissible dose.

MPE — *See* maximum permissible exposure.

MPL — *See* maximum permissible level.

Mppcf — acronym for million particles per cubic foot.

mr — *See* milliroentgen.

mrad — *See* millirad.

mrem — *See* millirem.

MRI — *See* magnetic resonance imaging.

MRIR — acronym for medium resolution infrared radiometer.

MRSA — *See* methicillin resistant *Staphylococcus aureus*.

MRT — *See* mean radiant temperature.

ms — *See* millisecond.

MS — *See* mass spectrograph.

MSA — *See* Metropolitan Statistical Area.

MSDS — *See* Material Safety Data Sheet.

MSHA — *See* Mine Safety and Health Administration.

MSS — *See* municipal sewage sludge; multispectral scanner system.

MSW — *See* municipal solid waste.

MSWI — *See* municipal solid waste incinerator.

MTD — acronym for maximum tolerated dose.

MTP — notation for maximum total trihalomethane potential.

μ — *See* micro-.

μCi — *See* microcurie.

μg — *See* microgram.

μg/m³ — *See* microgram per cubic meter.

μ/l — *See* microgram per liter.

muck soil — earth made from decaying plant materials.

mucociliary clearance — removal of materials from the respiratory tract via ciliary action.

mucous — of or pertaining to the cells of the mucous membrane; secreting mucus.

mucous membrane — the thin layer of protective mucus-secreting tissue which lines the digestive tract and other organs which have contact with air.

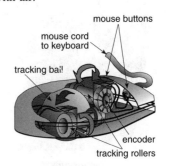

mouse buttons
mouse cord to keyboard
tracking ball
encoder
tracking rollers

Mouse

mucus — the fluid of the mucous membranes composed of the secretion of the glands, various salts, desquamated cells, and leukocytes.

mulch — a mixture of wood chips, dry leaves, straw, hay, or other material placed on the soil around plants to hold moisture in the ground, keep weeds from growing, soak up rain, or maintain soil temperature.

multiclone — *See* multicyclone.

multicyclone — a combination of cyclone collectors. Also known as multiplecyclone; multiclone.

multiphasic screening — the simultaneous use of multiple tests to detect severe pathological conditions.

multiple chamber incinerator — an incinerator consisting of two or more chambers arranged as in-line or retort types, interconnected by gas passage ports or ducts.

multiple chemical sensitivity — a condition of the modern age in which a person is considered to be sensitive to a number of everyday household, industrial, or agricultural chemicals at very low concentrations.

multiplecyclone — *See* multicyclone.

multiple dwelling — any dwelling containing more than two dwelling units.

multiple-family dwelling — any shelter containing two or more dwellings, units, or both.

multiple source — a source of pollution similar in character and wide-spread throughout the community, such as an automobile, home heating, or home incinerators, found in apartment houses.

multispectral remote sensing — the science of detection and analysis of an object at a distance.

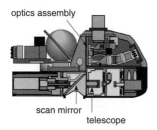

Multispectral Scanner

multispectral scanner system (MSS) — a device found in airplanes or satellites for recording radiation in several wave-bands at the same time.

municipal collection — the collection of solid waste by public employees and equipment under the supervision and direction of a municipal department or office.

municipal sewage sludge (MSS) — sludge produced during the operation of a municipal or public sewage plant operation.

municipal solid waste (MSW) — the normal, residential, and commercial solid waste generated within a community and usually collected at the structures.

municipal solid waste incinerator (MSWI) — a privately or publicly owned incinerator primarily designed and used to burn residential and commercial solid waste.

municipal waste — the combination of street litter, discarded auto bodies, power plant and incinerator ashes and residue, sludge from sewage, dead animals, and abandoned cars and trucks generated with a community.

municipal waste treatment plant (MWTP) — a municipal facility which treats sewage and releases the effluent into a body of water.

murine typhus fever — (*disease*) a mild form of typhus, producing symptoms of headache, fever, and some pain. Incubation time is 1–2 weeks, usually 12 days. It is caused by *Rickettsia typhi*; found worldwide. The reservoir of infection is rats. It is transmitted by fleas, usually *Xenopsylla cheopis*, whose feces contaminate the bite site of the person; fleas remain contaminated for life; not transmitted from person to person; general susceptibility. It is controlled by using pesticides where rats travel from area to area, and avoiding exposure to fleas which have been on rats.

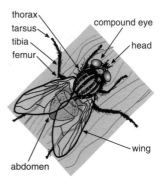

Musca domestica
(House Fly)

musca domestica — the house fly; a fly about 1/4″ long, gray or buff in color, with the male being a little smaller than the female; the body and legs are covered with hairs which hold microorganisms readily; the life cycle includes egg, larva, pupa, and adult.

muscarinism — a poisoning due to the ingestion of poisonous varieties of mushrooms.

M
N

muscle poison — a poison which causes a tingling of the neck and face, colic, vomiting, diarrhea, elevation of blood pressure, and death; includes barium salts, benzene, digitalis, and potassium salts.

MUSE — acronym for monitor of ultraviolet solar energy.

mussel poisoning — a poisoning due to the ingestion of mussels which have consumed contaminated plankton.

mustard gas — $C_4H_8Cl_2S$ — a general term referring to several chemicals, most commonly sulfur mustard. A colorless, odorless, oily liquid which becomes brown with a slight garlic smell when mixed with other chemicals. MW: 159.08, BP: 215–217°C, Sol: not available [%], Fl.P: 221°F, sp gr: not known. It is used as a vesicant chemical warfare agent. It is hazardous to the respiratory system, skin, and possibly the gastrointestinal system; and is toxic by inhalation and contact. Symptoms of exposure include painful coughing, vomiting, burning eyes, shock, chest tightness, sneezing, rhinorrhea, and sore throat. Delayed symptoms include chronic bronchitis, recurrent pneumonia, skin burns, nausea, anoxia, abdominal pain, diarrhea, headache, and lassitude. It is potentially carcinogenic. Also known as bis (2-chloroethyl) sulfide. Department of Defense exposure limit (TWA): 3mg/m^3 [air] or 0.1 mg/m^3 for the general population.

mutagen — a chemical or physical agent that induces a permanent change in the genetic material.

mutagenesis — the induction of changes in genetic material that are transmitted during cell division.

mutagent — an agent, such as a chemical or radiation, that damages or alters genetic material in cells.

mutant — an organism that contains a trait that is not inherited, but a result of a change in its genes.

mutation — permanent change in the genetic make-up of an organism resulting in a new characteristic which may be passed on to offspring.

mV — *See* millivolt.

MV — *See* megavolt.

MV/m — *See* megavolt per meter.

MVV — acronym for maximal voluntary ventilation.

mW — *See* milliwatt.

MW — *See* molecular weight; megawatt.

MWD — *See* megawatt days.

MWTA — *See* Medical Waste Tracking Act.

MWTP — *See* municipal waste treatment plant.

mycelium — the mass of interwoven filamentous hyphae that forms the vegetative portion or thallus of a fungus and is often submerged in the tissues of a host.

Mycobacterium — a genus of gram-positive, aerobic, acid-fast bacteria occurring as slightly curved or straight rods and containing many species, including the highly pathogenic organisms that cause tuberculosis and leprosy.

mycosis — any infection or disease caused by fungi.

mycotoxin — a toxic substance produced by a fungus, especially a mold.

myocardial — pertaining to the muscular tissue of the heart.

myocardial infarction — the local arrest or sudden insufficiency of arterial or venous blood supply of an area of the heart muscle.

myoclonic — shock-like, as with contractions of a muscle or a group of muscles.

N

n — *See* nano-.

N — notation for number of neutrons. *See* newton.

NAAQS — *See* National Ambient Air Quality Standards.

NADP — *See* National Atmospheric Deposition Process Network.

NAMS — acronym for National Air Monitoring Stations.

nano- (n) — a prefix in the metric system representing 10^{-9}.

nanocurie (nCi) — a unit of radiation equal to one-billionth of a curie.

nanogram (ng) — a unit of weight equal to one-billionth of a gram.

nanometer — a unit of length equal to one-billionth of a meter or 10^{-9} meter. Also known as nanon.

nanon — *See* nanometer.

nanosecond — a unit of time equal to one-billionth of a second, or 10^{-9} second.

nanotesla — a unit of magnetic flux density equal to 1×10^{-9} weber per square centimeter.

naphtha (coal tar) — a reddish-brown, mobile liquid with an aromatic odor; MW: 110 (approx), BP: 320–428°F, Sol: insoluble, Fl.P: 100–109°F, sp gr: 0.89–0.97. It is used in the preparation of coal-tar paints; in the preparation of coumarone and indene; as a solvent in the rubber industry in the manufacture of water-proof cloth, shoe adhesives, and rubber tires; as a solvent, diluent, or thinner in paint, varnish, and lacquer industries; in the formulation of nitrocellulose and ethylcellulose; as a solvent for polymerized styrol, phenolic varnishes, urea, resins, melamine, and other synthetic resins; as a solvent for DDT. It is hazardous to the respiratory system, eyes, and skin; and is toxic by inhalation, ingestion, and contact. Symptoms of exposure include lightheadedness, drowsiness, dermatitis, irritation of the eyes, nose, and skin. OSHA exposure limit (TWA): 100 ppm [air] or 400 mg/m^3.

naphthalene — $C_{10}H_8$ — a colorless to brown solid with an odor of mothballs; shipped as a molten solid; MW: 128.2, BP: 424°F, Sol: 0.003%, Fl.P: 174°F, sp gr: 1.15. It is used in the formulation of insecticides and moth repellent as flakes, powder, balls, or cakes; as a fumigant; in the manufacture of chemical intermediates for production of pharmaceuticals, resins, dyes, plasticizers, solvents, coatings, insecticides, pigments, rubber chemicals, tanning agents, surfactants, waxes, cable coatings, textile spinning lubricants, rodenticides, and in storage batteries. It is hazardous to the eyes, blood, liver, kidneys, skin, central nervous system, and red blood cells; and is toxic by inhalation, absorption, ingestion, and contact. Symptoms of exposure include confusion, headache, excitement, malaise, nausea, vomiting, abdominal pain, bladder irritation, profuse sweating, jaundice, hemoglobinuria, renal shutdown, and dermatitis. Also known as tar camphor. OSHA exposure limit (TWA): 10 ppm [air] or 50 mg/m^3.

alpha-naphthylamine — $C_{10}H_7NH_2$ — a colorless crystal with an ammonia-like odor; darkens in air to a reddish-purple color; MW: 143.2, BP: 583°F, Sol: 0.002%, Fl.P: 315°F, sp gr (77°F): 1.12. It is used in the manufacture of dyes, herbicides, and rubber antioxidants. It is hazardous to the bladder and skin; and is toxic by inhalation, absorption, ingestion, and contact. Symptoms of exposure include dermatitis, hemorrhagic cystitis, dyspnea, ataxia, methemoglobinemia, dysuria; carcinogenic. OSHA exposure limit (TWA): regulated use.

alpha-naphthyl thiourea (ANTU) — $C_{10}H_7NHC(NH_2)S$ — a white crystalline or gray odorless powder; MW: 202.3, BP: decomposes, Sol: 0.06%, Fl.P: N/A. It is used as a rodenticide. It is hazardous to the respiratory system; and is toxic by inhalation and ingestion. Symptoms of exposure after ingestion of large doses include vomiting, dyspnea, cyanosis, coarse pulmonary râles, mild liver damage. OSHA exposure limit (TWA): 0.3 mg/m^3 [air].

Naphthalene

narcosis — a state of stupor, unconsciousness, or arrested activity produced by the influence of narcotics or other chemicals.

narcotic — a chemical agent that diminishes awareness of sensory impulses and in large doses can cause coma.

NAS — *See* National Academy of Sciences.

NASA — *See* National Aeronautics and Space Administration.

nascent — pertaining to something which is just forming.

natal — the ratio of the total number of births to the total population of a given community at a given time. Also known as birth rate.

natality — birth rate.

National Academy of Sciences (NAS) — a private, honorary organization of scholars in scientific and engineering research chartered by an act of Congress in 1863 to serve as an advisory agency to the federal government on questions of science and technology.

National Aeronautics and Space Administration (NASA) — a federal government agency whose function is to conduct research for the solution of problems of flight within and without the earth's atmosphere, and also to develop, construct, test, and operate aeronautical and space vehicles.

National Ambient Air Quality Standards (NAAQS) — standards established by the federal government for regulated pollutants in ambient air.

National Atmospheric Deposition Process Network (NADP) — a major monitoring network involving cooperative effort among numerous federal, state, and private research agencies in the area of acid rain.

National Cancer Institute (NCI) — one of the National Institutes of Health whose function is to expand existing scientific knowledge on cancer causes and prevention as well as on the diagnosis, treatment, and rehabilitation of cancer patients.

National Center for Health Statistics (NCHS) — a part of the Centers for Disease Control whose function is to collect, maintain, analyze, and disseminate data on health status and services.

National Center for Toxicological Research (NCTR) — a program within the Department of Health and Human Services whose function is to study the biological effects of potentially toxic chemical substances found in the environment, emphasizing the determination of the health effects resulting from long-term, low-level exposure to chemical toxicants.

National Coal Association (NCA) — an organization of coal companies representing the coal industry in all matters except labor relations.

National Committee on Radiation Protection (NCRP) — an advisory group of scientists and professionals that make recommendations for radiation protection in the United States.

National Cooperative Soil Survey — a system for classifying soil material based on textural classes and the relative proportions of sand, silt, and clay in the soil fraction.

National Disaster Medical System (NDMS) — an emergency medical system housed at the United States Public Health Service; based on the concept of the Civilian Military Contingency Hospital System.

National Emission Standards for Hazardous Air Pollutants (NESHAPS) — standards set by the Environmental Protection Agency in compliance with the Clean Air Act for pollutants that could contribute to an increase in mortality or serious illness.

National Environmental Health Association (NEHA) — a professional association of environmental health and protection specialists which provides certification, training, continuing education, consulting, and professional liaisons; all members are in the environmental health field.

National Environmental Policy Act (NEPA) — an act of Congress passed in 1969 and setting forth the responsibilities of federal agencies declaring a national policy which will encourage productive and enjoyable harmony between people and the environment, to promote efforts which will prevent or eliminate damage to the environment and biosphere, stimulate the health and welfare of people to enrich the understanding of ecological systems and natural resources important to the nation, and to establish a council on environmental quality.

National Environmental Training Association (NETA) — a professional association of environmental trainers in air pollution, noise pollution, hazardous waste, water pollution, and wastewater treatment.

National Fire Academy (NFA) — a component of the Federal Environmental Management Agency's National Emergency Training Center. It provides fire protection and control training for fire services and allied services.

National Fire Codes — sixteen volumes of codes, standards, recommended practices, and manuals developed by the National Fire Protection Association technical committees.

National Fire Protection Association (NFPA) — an international organization which promotes improved fire protection and prevention and establishes safeguards against loss of life and property by fire.

National Hazardous Materials Information Exchange (NHMIE) — an organization providing information on hazardous materials (HAZMAT) training courses, planning techniques, events and conferences, and emergency response experiences for the handling of hazardous materials.

National Health Services Corps (NHSC) — a unit within the Public Health Service that is responsible for providing physicians for areas that are underserved.

National Health Survey — an on-going health survey by the National Center for Health Statistics which includes studies to determine the extent of illness and disability in the population of the United States; describes the use of health services by Americans and gathers related information.

National Highway Traffic Safety Administration (NHTSA) — an agency of the federal government whose function is to carry out programs related to the safety performance of motor vehicles and related equipment, motor vehicle drivers, pedestrians, and to promote a uniform nationwide speed limit.

National Institute for Occupational Safety and Health (NIOSH) — a federal agency whose function is to identify substances that pose potential health problems and recommend exposure levels to the Occupational Safety and Health Administration. NIOSH does in-depth research on potential hazards in industry and carries out a variety of field studies.

National Institute of Environmental Health Sciences (NIEHS) — an institute of the National Institute of Health that conducts and supports fundamental research concerned with defining, measuring, and understanding the effects of chemical, biological, and physical factors in the environment on the health and well-being of people.

National Institute of Health (NIH) — an important component of the Department of Health and Human Services composed of several specialized institutes in various critical health areas; their objective is to improve the general health by conducting and supporting biomedical research into the causes, prevention, and cure of disease; research, training, and the development of research sources; and the provision of biomedical information.

National Library of Medicine (NLM) — the nation's chief medical information source, authorized to provide medical library services and on-line bibliographic searching capabilities, such as MEDLINE, TOXLINE, and others, to public and private agencies, organizations, and individuals.

National Occupational Exposure Survey (NOES) — a survey conducted in a sample of nearly 5,000 establishments from 1981 to 1983 to compile data on the types of potential exposure agents found at the workplace and the kinds of safety and health programs implemented at the plant level.

National Oceanic and Atmospheric Administration (NOAA) — a major organization within the Department of Commerce whose objective is to explore, map, and chart the global ocean and its living resources; to manage, use, and conserve these resources; to describe, monitor, and predict conditions in the atmosphere, ocean, sun, and space environment.

National Ocean Pollution Planning Act — a law passed in 1978 and updated through 1988; states that the activities of people in marine environment can have profound short- and long-term impact on the environment and affect ocean and coastal resources; calls for a federal plan for ocean pollution research, monitoring of materials that have been dumped, fate of material, and effects of pollutants on marine environment.

National Pollution Discharge Elimination System (NPDES) — the national program for issuing, modifying, revoking, reissuing, terminating, monitoring and enforcing permits, and imposing and enforcing pretreatment requirements under the Clean Water Act by the Environmental Protection Agency; effluent limitation discharge standards for industrial categories and sewage treatment plants; the standards are based on the best available technology economically feasible.

National Primary Drinking Water Regulations (NPDWR) — drinking water regulations establishing maximum levels of chemical, physical, and biological agents permitted in drinking water.

National Priorities List (NPL) — the Environmental Protection Agency's official list of hazardous waste sites which need to be corrected under Superfund.

National Research Council (NRC) — a council established by the National Academy of Sciences in 1916 to serve as the operating arm of

M
N

the Academy and the National Academy of Engineering by providing scientific and technical advice to the government, the public, and the scientific and engineering communities.

National Response Center (NRC) — a communication center for activities related to response actions; located at Coast Guard headquarters.

National Response Team (NRT) — a group consisting and representative of 14 government agencies prepared for a response in the event of an emergency.

National Safety Council (NSC) — an independent, nonprofit organization with the goal of reducing the number and severity of accidents and industrial illnesses by collecting and distributing information about the causes and means of prevention of accidents and illnesses.

National Sanitation Foundation (NSF) — a unique organization consisting and representative of industry and government to determine standards for a variety of equipment used in the environmental health science field; it runs elaborate research and testing laboratories, and conducts training and education programs.

National Swimming Pool Foundation (NSPF) — an educational research organization of the swimming pool, spa, and hot tub industry.

National Technical Information Service (NTIS) — a federal agency involved in the development and publishing of advanced information, products, and services for the achievement of productivity and innovative goals.

National Toxicology Program (NTP) — a program in operation since 1961 which has developed testing techniques for determining the carcinogenicity of chemicals. The program runs long-term animal studies and other tests determining carcinogenic activity.

natural draft — the gas flow created by the height of a stack or chimney and the difference in temperature between flue gases and the atmosphere.

natural gas — a combustible gas composed largely of methane and other hydrocarbons with variable amounts of nitrogen and noncombustible gases; obtained from natural earth fissures or wells; used as a fuel in the manufacture of carbon black and in chemical synthesis.

natural history of disease — the history of a disease from when it is acquired, manifests, and terminates; includes information on susceptibility, presymptomatic stage, clinical disease, and stage of disability.

natural immunity — an immunity present in an individual at birth and not artificially acquired.

natural radioactivity — the property of radioactivity exhibited by more than 50 naturally occurring radionuclides.

natural selection — the process of the survival of the fittest by which organisms that successfully adapt to their environments survive and those that do not disappear; first propounded by Charles Darwin.

natural soil drainage — the condition of soil in which the removal of water occurs by natural means.

natural ventilation — the air movement caused by wind or temperature differences.

nausea — an unpleasant sensation usually felt when nerve endings in the stomach and other parts of the body are irritated and accompanied by a tendency to vomit.

NCA — *See* Noise Control Act; National Coal Association.

NCAB — acronym for National Cancer Advisory Board.

NCHS — *See* National Center for Health Statistics.

NCI — *See* National Cancer Institute.

NCP — acronym for National Contingency Plan.

NCRP — *See* National Committee on Radiation Protection.

NCTR — *See* National Center for Toxicological Research.

NDIR — notation for nondispersed infrared analyzer.

NDMS — *See* National Disaster Medical System.

Necrobia rufipes — *See* Red-legged Ham Beetle.

necropsy — *See* autopsy.

necrosis — the morphological changes indicative of cell death as a result of injury, disease, or other pathogenic state.

NEDS — acronym for National Emission Data System.

needs assessment — the identification of the difference between current or projected future levels of concern and the desired levels.

negative pressure — a condition that exists when less air is supplied to a space than is exhausted from the space, so the air pressure within the space is less than that in the surrounding area; a vacuum.

negligence — the act of not taking prudent care.

NEHA — *See* National Environmental Health Association.

neighborhood — an area comprising all the public facilities and conditions required by the average family for their comfort and existence.

Neisseria — a genus of gram-negative aerobic or facultatively anaerobic cocci which are part of the normal flora of the oropharynx, nasopharynx, and genitourinary tract.

NEL — the no-effect level in high-to-low dose extrapolation models.

nematocide — a pesticide chemical or agent that prevents injury from or destroys nematodes.

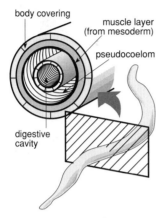

Nematoda

Nematoda — the class of unsegmented round worms with elongated bodies pointed at both ends and to which hookworms, trichina, and *Ascaris* belong, many of which are parasites.

neonatal — pertaining to the period within 28 days after birth.

neonatal death — the death of a live-born infant under 28 days of age.

neoplasia — a condition characterized by the presence of new growths or tumors.

neoplasm — any new and abnormal growth, specifically one in which cell multiplication is uncontrolled and progressive; a benign or malignant tumor.

neoprene — synthetic rubber closely paralleling natural rubber in flexibility and strength; used in paints, putties, adhesives, the soles of shoes, and tank linings.

NEPA — *See* National Environmental Policy Act.

neper (Np) — division of a logarithmic scale.

nephelometer — (*optics*) an instrument for measuring the concentration of suspended matter in a liquid dispersion by measuring the amount of light scattered by the dispersion.

nephelometric turbidity unit (NTU) — unit of measurement for water clarity.

nephometry — the measurement of clouds.

nephritis — inflammation of the kidney.

nephrologist — licensed physician who specializes in the treatment of kidney disease.

nephron — the structural and functional unit of the kidney; each nephron is capable of forming urine by itself.

nephrosis — any kidney disease, especially disease marked by purely degenerative lesions of the renal tubules.

nephrotoxin — a substance that causes injury to the kidneys, including halogenated hydrocarbons and uranium.

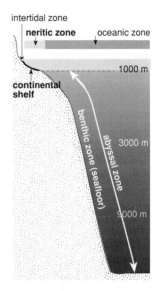

Neritic Zone

neritic — of or pertaining to the part of the ocean that is associated with the continental shelf from the seacoast to approximately 660 feet.

nerve — a bundle of nerve fibers bound together by connective tissue leading to or from the central nervous system.

nervous system — the entire network of nerve cells that enable an organism to be aware of its environment and react to it; in vertebrates, the system consists of the ganglia, cranial and peripheral nerves, spinal cord, brainstem, and brain.

NESHAPS — *See* National Emission Standards for Hazardous Air Pollutants.

nested design — (*epidemiology*) the combination of two study designs of a problem used to gather more data, inexpensively and quickly.

NETA — *See* National Environmental Training Association.

net heating value — gross heating value minus the latent heat of vaporization of the water formed by the combustion of the hydrogen in the fuel. Also known as low heat value.

M
N

Token ring topology

Star topology

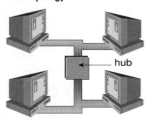

hub

Bus Topology

BUS

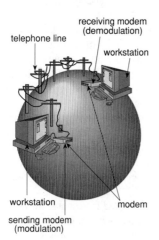

telephone line

receiving modem
(demodulation)

workstation

workstation

modem

sending modem
(modulation)

Network

network — two or more interconnected computer systems for the implementation of specific functions.

neuralgia — a pain in a nerve or along the course of one or more nerves.

neural loss — hearing loss.

neuraxon — *See* axon.

neurite — *See* axon.

neuritis — noninflammatory lesions of the peripheral nervous system.

neurology — the branch of medical science dealing with the nervous system.

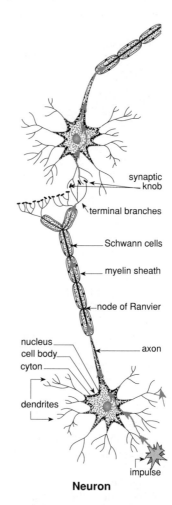

synaptic knob

terminal branches

Schwann cells

myelin sheath

node of Ranvier

nucleus
cell body
cyton

axon

dendrites

impulse

Neuron

neuron — the nerve cell and its branches along which nerve impulses travel.

neurosis — an emotional disorder that can interfere with normal health and daily living; anxiety is the chief symptom.

neurotoxicity — the occurrence of adverse effects on a nervous system following exposure to a chemical.

neurotoxin — a poisonous chemical which produces primary toxic effects on the nervous system with symptoms of narcosis, behavioral changes, and decrease in motor functions; chemicals include mercury and carbon disulfide.

neurotrauma — mechanical injury to a nerve.

neutral — neither basic nor acidic; referring to the absence of a net electrical charge.

neutralization — (*chemistry*) a chemical reaction between an acid and a base.

neutralize — (*chemistry*) to make a solution neutral by adding a base to an acidic solution or an acid to a basic solution.

neutrino — a neutral particle having zero rest mass, originally postulated to account for the continuous distribution of energy among particles in the beta-decay process.

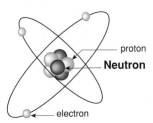

neutron — an electrically neutral subatomic particle with approximately the same mass as a proton and that is within the nucleus of an atom, having a mass of approximately greater than 1.

neutrophil — a polymorphonuclear granular leukocyte that stains easily with neutral dyes.

newborn — a human offspring from the time of birth through the 28th day of life.

newton (N) — the SI unit of force equal to the force required to give a mass of 1 kilogram an acceleration of 1 meter per second squared; formerly known as large dyne.

newton-meter of energy — See joule.

NEXUS — See numerical examination of urban smog.

NFA — See National Fire Academy.

NFPA — See National Fire Protection Association.

ng — See nanogram.

NHMIE — See National Hazardous Materials Information Exchange.

NHSC — See National Health Services Corps.

NHTSA — See National Highway Traffic Safety Administration.

nickel-cadmium cell — a rechargeable, dry cell containing nickel oxide, cadmium, and an electrolyte of potassium hydroxide.

nickel carbonyl — $Ni(CO)_4$ — a colorless to yellow liquid with a musty odor; a gas above 110°F; MW: 170.7, BP: 110°F, Sol: 0.05%, Fl.P: <–4°F, sp gr (63°F): 1.32. It is used in the Mond process for nickel refining; in plating operations on foundry patterns, steel, and electronics manufacture; as a reagent in synthesis of acrylic esters; as a catalyst or reagent in organic synthesis; in petroleum and petrochemical processing. It is hazardous to the nasal cavities, lungs, and skin; and is toxic by inhalation, ingestion, and contact. Symptoms of exposure include sensitization dermatitis, allergic asthma, pneumonitis; carcinogenic. OSHA exposure limit (TWA): 0.001 ppm [air] or 0.007 mg/m³.

nickel metal and other compounds — Ni — a lustrous, silvery metal; MW: 58.7, BP: 5139°F, Sol: insoluble, sp gr: 8.90. It is used in the manufacture of more than 3,000 alloys; in electronic tube parts, coins, heavy machinery, tools, instrument parts, magnets, food and chemical processing equipment, flatware, jet engines, automotive parts, zippers, surgical and dental instruments, and cooking utensils; in chemical synthesis; in the textile industry. It is hazardous to the lungs, paranasal sinus, and central nervous system; and is toxic by inhalation, ingestion, and contact. Symptoms of exposure include headache, vertigo, nausea, vomiting, epigastric pain, substernal pain, cough, hyperpnea, cyanosis, weakness, leukocytosis, pneumonitis, delirium, convulsions; carcinogenic. OSHA exposure limit (TWA): 0.1 mg/m³ for soluble compounds, 1 mg/m³ for metal and insoluble compounds.

nicotine — $C_5H_4NC_4H_7NCH_3$ — a pale yellow to dark brown liquid with a fish-like odor when warm; MW: 162.2, BP: 482°F (decomposes), Sol: miscible, Fl.P: 203°F, sp gr: 1.01. It is used as a pesticide and fumigant on vegetable crops, fruit, grasses and turf, and greenhouse plants. It is hazardous to the central nervous system, cardiovascular system, lungs, and gastrointestinal tract; and is toxic by inhalation, absorption, ingestion, and contact. Symptoms of exposure include nausea, salivation, abdominal pain, vomiting, diarrhea, headache, dizziness, hearing and vision disturbance, confusion, weakness, incoordination, paroxysmal atrial fibrillation, convulsions, dyspnea. OSHA exposure limit (TWA): 0.5 mg/m³ [skin].

NIEHS — See National Institute of Environmental Health Sciences.

NIH — See National Institute of Health.

nimbostratus — true rain clouds, darker than ordinary stratus with a base; height from near the earth to 6,500 feet up; diffuse from continuous streaks of rain often extending to the ground.

nimbus — a family of clouds usually producing precipitation; the term is usually used in combination, such as in cumulonimbus.

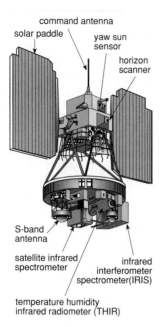

NIMBUS 4 Satellite

NIMBUS — a series of research-oriented meteorological satellites used to develop sensor technology for determining vertical atmospheric temperature and moisture profile; uses microwave spectrometer and electrically scanning microwave radiometer for vertical temperature profiles.

NIMBUS I — a satellite which photographs nighttime cloud cover from space.

NIOSH — *See* National Institute for Occupational Safety and Health.

Nissl bodies (*cytology*) — large, granular bodies that stain with basic dyes, forming the substance of the reticulum of the cytoplasm of a nerve cell.

nitrate — a salt or ester of nitric acid.

nitrate-forming bacteria — bacteria that converts nitrites into compounds that can be used by green plants to build proteins.

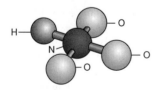

Nitric Acid

nitric acid — HNO_3 — a colorless, yellow, or red fuming liquid with an acrid, suffocating odor; MW: 63.0, BP: 181°F, Sol: miscible, sp gr (77°F): 1.50. It is used in metallurgy as a pickling agent; in metal refining, ore recovery, metal etching, and photoengraving; in acidulation of phosphate rock and in the manufacture of nitrogen solutions for use in the fertilizer industry; in the wood pulping industry; during organic synthesis in the manufacture of fertilizers, explosives, herbicides, antibiotics, meat-curing, pickling, ceramics, and pharmaceuticals; during organic synthesis in the manufacture of nitrating and oxidizing agents, nylons, foams, lubricants, insecticides, dyes, explosives, photographic films, lacquers, and celluloids. It is hazardous to the eyes, respiratory system, skin, and teeth; and is toxic by inhalation, ingestion, and contact. Symptoms of exposure include delayed pulmonary edema, pneumonitis, bronchitis, dental erosion, irritation of the eyes, mucous membrane, and skin. OSHA exposure limit (TWA): 2 ppm [air] or 5 mg/m^3.

nitric oxide — NO — a colorless gas; MW: 30.0, BP: –241°F, Sol: 5%. It is used as an intermediate in the synthesis of nitric acid, acrylonitrile, hydroxylamine, nitrosyl halides, nitrosyl hydrogen sulfate, and nitrogen dioxide; as a decomposition agent in certain gaseous products; as an additive to rocket propellants; in the bleaching of rayon. It is hazardous to the respiratory system; and is toxic by inhalation and contact. Symptoms of exposure include drowsiness, unconsciousness, irritation of the eyes, nose, and throat. OSHA exposure limit (TWA): 25 ppm [air] or 30 mg/m^3.

nitrification — the biochemical oxidation of ammonium to nitrates by certain bacteria.

nitrilotriacetate — the salt of nitrilotriacetic acid; has the ability of complex metal ions and is proposed as a builder for detergents.

nitrilotriacetic acid (NTA) — $N(CH_2COOH)_3$ — a white powder; MP: 240°C; used as a chelating agent in the laboratory. It is hazardous to a developing fetus; and is toxic by ingestion. Also known as TGA.

nitrite — any salt or ester of nitrous acid.

nitrite-forming bacteria — bacteria that combines ammonia with oxygen to form nitrites.

m-Nitroaniline

m-nitroaniline — $C_6H_6N_2O_2$ — yellow needle-like crystals with a burning sweet odor; MW:

138.12, BP: 581–584.6°F (decomposes), Sol: 90 mg/100 ml, Fl.P: no data, sp gr: 1.43. It is used as a dye-stuff intermediate. Other information is not available. OSHA exposure limit (TLV-TWA): not established.

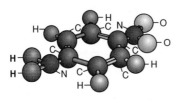

p-Nitroaniline
(4 Nitroaniline)

p-nitroaniline — $NO_2C_6H_4NH_2$ — a bright yellow, crystalline powder with a slight ammonia-like odor; MW: 138.1, BP: 630°F, Sol: 0.08%, Fl.P: 390°F, sp gr: 1.42. It is used as a gasoline-gum inhibitor; as an intermediate in the pharmaceutical industry; in the production of intermediates for dyestuff manufacture; during antioxidation and antiozonation operations in rubber manufacture. It is hazardous to the blood, heart, lungs, and liver; and is toxic by inhalation, absorption, ingestion, and contact. Symptoms of exposure include cyanosis, ataxia, tachycardia, tachypnea, dyspnea, irritability, vomiting, diarrhea, convulsions, respiratory arrest, anemia, methemoglobinemia. OSHA exposure limit (TWA): 3 mg/m³ [skin].

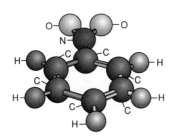

Nitrobenzene

nitrobenzene — $C_6H_5NO_2$ — a yellow, oily liquid with a pungent odor like paste shoe polish; a solid below 42°F; MW: 123.1, BP: 210.8°C, Sol: 0.2%, Fl.P: 190°F, sp gr: 1.20. It is used in the production of chemical intermediates; in solvent refining of lubricating oils; in the production of intermediates in the synthesis of rubber chemicals, photographic chemicals, explosives, liquid propellants, and pharmaceuticals; as a solvent in specialized surface coatings; as a solvent in organic synthesis; as perfume in the manufacture of toilet and household soaps; in the synthesis of insecticides and germicides. It is hazardous to the blood, liver, kidneys, cardiovascular system, and skin; and is toxic by inhalation, absorption, ingestion, and contact. Symptoms of exposure include anoxia, dermatitis, anemia, eye irritation; in animals: liver and kidney damage. OSHA exposure limit (TWA): 1 ppm [skin] or 5 mg/m³.

4-nitrobiphenyl — $C_6H_5C_6H_4NO_2$ — a white to yellow, needle-like crystalline solid with a sweetish odor; MW: 199.2, BP: 644°F, Sol: insoluble, Fl.P: 290°F. It is used in the production of rocket fuel; as an industrial solvent; as an oxidant; as an additive in lubricants. It is hazardous to the bladder and blood; and is toxic by inhalation, absorption, ingestion, and contact. Symptoms of exposure include headache, lethargy, dizziness, dyspnea, ataxia, weakness, methemoglobinemia, urinary burning, acute hemorrhagic cystitis; carcinogenic. OSHA exposure limit (TWA): regulated use.

p-nitrochlorobenzene — $ClC_6H_4NO_2$ — yellow crystalline solid with a sweet odor; MW: 157.6, BP: 468°F, Sol: insoluble, Fl.P: 261°F, sp gr: 1.52. It is used in the production of chemical intermediates for the manufacture of pesticides, fungicides, and preservatives; in the manufacture of pharmaceuticals, rubber chemicals, antioxidants, gasoline gum inhibitors, corrosion inhibitors, and photographic chemicals. It is hazardous to the blood, liver, kidneys, and cardiovascular system; and is toxic by inhalation, absorption, ingestion, and contact. Symptoms of exposure include anoxia, unpleasant taste, mild anemia; carcinogenic; in animals: hemoglobinuria. OSHA exposure limit (TWA): 1 mg/m³ [skin].

nitroethane — $CH_3CH_2NO_2$ — a colorless, oily liquid with a mild, fruity odor; MW: 75.1, BP: 237°F, Sol: 5%, Fl.P: 82°F, sp gr: 1.05. It is used as a solvent in coatings and adhesives on cellulose esters and synthetic resins; as an intermediate in the synthesis of organic dyes, insecticides, pesticides, nitroplasticizers, pharmaceuticals, and other organic chemicals; as a propellant; as a reaction-media fluid; as an extraction solvent in petroleum fractionation; as a recrystallization solvent. It is hazardous to the skin; and is toxic by inhalation, ingestion, and contact. Symptoms of exposure include dermatitis; in animals: lacrimation, dyspnea, pulmonary râles, edema, narcosis, liver and kidney injury. OSHA exposure limit (TWA): 100 ppm [air] or 310 mg/m³.

M
N

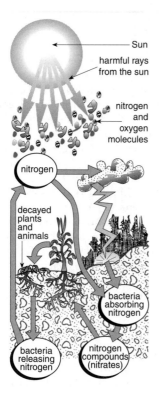

Nitrogen Cycle

nitrogen cycle — the natural sequence through which nitrogen circulates in the biosphere; nitrogen compounds in the soil are taken in and stored by plants; when the plants are eaten by animals, these compounds are broken down into a variety of new compounds and a portion is passed on as waste; along with nitrogen in the atmosphere, the nitrogen passed into the soil or by the decay of plants or animals is then converted by nitrogen-fixing bacteria into nitrogen compounds used by plants.

nitrogen dioxide — NO_2 — a yellowish-brown liquid or reddish-brown gas (above 70°F) with a pungent acrid odor; in solid form (below 15°F) it is found structurally as N_2O_4; MW: 46.0, BP: 70°F, Sol: reacts, sp gr: 1.44 (a liquid at 68°F). It is used during metal surface treatment with nitric acid; in production of intermediates in manufacture of sulfuric acid, nitric acid, and fertilizers; in chemical synthesis; in the synthesis of dyes; in the manufacture of nitrocellulose paints and lacquers; in the production and handling of rocket propellants. It is hazardous to the respiratory system and cardiovascular system; and is toxic by inhalation, absorption, ingestion, and contact. Symptoms of exposure include cough, mucoid frothy sputum, dyspnea, chest pain, pulmonary edema, cyanosis, tachypnea,

tachycardia, eye irritation. Nitrogen dioxide pollution reacts with water vapor in the air affecting visibility and contributing to acid rain. OSHA exposure limit (TWA): short term 1 ppm [air] or 1.8 mg/m³.

nitrogen fixation — the utilization of atmospheric nitrogen in synthesizing amino acids by some bacteria and blue-green algae.

nitrogen-fixing bacteria — bacteria that convert nitrogen from the atmosphere or soil into ammonia that can then be converted to plant nutrients by nitrite- and nitrate-forming bacteria.

nitrogenous waste — animal or plant residue that contains large amounts of nitrogen.

nitrogen oxides — NO_x — gases formed from nitrogen and oxygen during combustion in engines or in industries during conditions of high temperature and pressure.

nitrogen trifluoride — NF_3 — a colorless gas with a moldy odor; shipped as a liquefied compressed gas; MW: 71.0, BP: –200°F, Sol: insoluble. It is used in the manufacture of flash cubes; in welding, brazing, and cutting of metals using hydrogen-nitrogen trifluoride torch; in research as an oxidizer for rocket and other high-energy fuels. In animals, it is hazardous to the blood; and is toxic by inhalation. Symptoms of exposure in animals include methemoglobin, anoxia, cyanosis, weakness, dizziness, headache, injury of the liver and kidneys. OSHA exposure limit (TWA): 10 ppm [air] or 29 mg/m³.

nitroglycerine — $CH_2NO_3CHNO_3CH_2NO_3$ — a colorless to pale yellow, viscous liquid or solid (below 56°F); MW: 227.1, BP: begins to decompose at 122–140°F, Sol: 0.1%, Fl.P: explodes, sp gr: 1.60. It is used during formulation and filling operations in the manufacture of industrial explosives and propellants; in the preparation of dosage forms including tablets and solutions in the manufacture of pharmaceuticals. It is hazardous to the cardiovascular system, blood, and skin; and is toxic by inhalation, absorption, ingestion, and contact. Symptoms of exposure include throbbing headache, dizziness, nausea, vomiting, abdominal pain, hypotension, flush, palpitations, methemoglobin, delirium, depressed central nervous system, angina, skin irritation. OSHA exposure limit (TWA): short term 0.1 mg/m³ [air].

nitromethane — CH_3NO_2 — a colorless, oily liquid with a disagreeable odor; MW: 61.0, BP: 214°F, Sol: 10%, Fl.P: 95°F, sp gr: 1.14. It is used as a solvent in coatings and adhesives on cellulose esters and synthetic resins; as a propellant or fuel additive; as an intermediate in the synthesis of organic dyes, textiles, surfactants, insecticides, pharmaceuticals, and explosives; as a reaction-media fluid; as a recrystallization solvent; as a stabilizer for halogenated alkanes, aerosol formulations, and paste formulations for inks. It is hazardous to the skin; and is toxic by inhalation, ingestion, and contact. Symptoms of exposure include dermatitis. OSHA exposure limit (TWA): 100 ppm [air] or 250 mg/m^3.

1-nitropropane — $CH_3CH_2CH_2NO_2$ — a colorless liquid with a somewhat disagreeable odor; MW: 89.1, BP: 269°F, Sol: 1%, Fl.P: 96°F, sp gr: 1.00. It is used as a thinner and solvent for cellulose compounds, lacquers, and dopes; in vinyl resins for industrial coatings and printing inks; in synthetic finish removers; as an extraction solvent for purification, separation, recrystallization, and recovery for natural and synthetic resins, tars, coating materials, fats, and oils; as a reaction medium in polymer technology; as a catalyst, initiator, and solvent; in organic chemical synthesis for preparation of amines, nitrated alcohols, acids, and chloronitroparaffins; in the manufacture of explosives. It is hazardous to the eyes and central nervous system; and is toxic by inhalation, ingestion, and contact. Symptoms of exposure include headache, nausea, vomiting, diarrhea, eye irritation. OSHA exposure limit (TWA): 25 ppm [air] or 90 mg/m^3.

2-nitropropane — $CH_3CH(NO_2)CH_3$ — a colorless liquid with a pleasant, fruity odor; MW: 89.1, BP: 249°F, Sol: 2%, Fl.P: 75°F, sp gr: 0.99. It is used in the manufacture of explosives; as a thinner and solvent; in organic chemical synthesis; as a propellant in rocket motors. It is hazardous to the respiratory system and central nervous system; and is toxic by inhalation, ingestion, and contact. Symptoms of exposure include headache, anorexia, nausea, vomiting, diarrhea, irritation of the respiratory system; carcinogenic. OSHA exposure limit (TWA): 10 ppm [air] or 35 mg/m^3.

nitrosamine — any of various neutral compounds characterized by the grouping NNO, some of which are powerful carcinogens.

nitrosation — the reaction between nitrous acid and secondary or tertiary amines.

N-nitrosodimethylamine — $(CH_33)_2N_2O$ — a yellow, oily liquid with a faint, characteristic odor; MW: 74.1, BP: 306°F, Sol: soluble, Fl.P: not known, sp gr: 1.00. It is used in the production of rocket fuel; as a solvent in the fibers and plasticizers industries; as an oxidant; as an additive in lubricants. It is hazardous to the liver, kidneys, and lungs; and is toxic by inhalation, absorption, ingestion, and contact. Symptoms

of exposure include nausea, vomiting, diarrhea, abdominal cramps, headache, fever, enlarged liver, jaundice, reduced function of the liver, kidneys, and lungs; carcinogenic. OSHA exposure limit (TWA): regulated use.

nitrotoluene — $NO_2C_6H_4CH_3$ — yellow liquids (o-, m-isomers) or crystalline solid (p-isomers) with a weak, aromatic odor; MW: 137.1, BP: 432/450/460°F, Sol: 0.07/0.05/0.04%, Fl.P: 223/232/223°F, sp gr: 1.16/1.16/1.12. It is used for the production of dyes, intermediates for vulcanization accelerators, gasoline inhibitors, and flotation agents; as an analytical reagent; in the production of explosives; in organic synthesis. It is hazardous to the blood, central nervous system, cardiovascular system, skin, and gastrointestinal tract; and is toxic by inhalation, absorption, ingestion, and contact. Symptoms of exposure include anoxia, cyanosis, headache, weakness, dizziness, ataxia, dyspnea, tachycardia, nausea, vomiting. OSHA exposure limit (TWA): 2 ppm [skin] or 11 mg/m^3.

nitrous — pertaining to or containing nitrogen in its lowest valence.

nitrous oxide — N_2O — a colorless gas with a sweet odor and taste; used as an anesthetic. Also known as laughing gas.

NLM — *See* National Library of Medicine.

NMR — acronym for nuclear magnetic resonance; now known as magnetic resonance imaging.

NOAA — *See* National Oceanic and Atmospheric Administration.

NOAEL — acronym for no observed adverse effect level.

noble gas — *See* inert gas.

NOC — *See* not otherwise classified.

nocturnal — referring to activity during the night.

node — a small mass of tissue in the form of a swelling knot or protuberance, either normal or pathological.

nodule — a small mass of rounded or irregular shaped cells or tissues; a small node.

NOEL — acronym for no observed effects level.

NOES — *See* National Occupational Exposure Survey.

NOHS — acronym for National Occupational Hazard Survey.

noise — unwanted sounds.

Noise Control Act — a law passed in 1972 and amended in 1978 that states inadequately controlled noise presents growing danger to health and welfare of the population of the United States; that major sources of noise include transportation vehicles, equipment, machinery, appliances, and other products used in commerce;

M

N

the primary responsibility for noise belongs to state and local governments; calls for federal action to deal with major noise sources in commerce.

noise control specialist — an environmental health and safety practitioner who prevents loss of hearing, minimizes the psychological effects of noise, and reduces the nuisance factors associated with noise.

noise exposure — a combination of amount of noise, kind of noise, and duration of noise.

noise-induced hearing loss — the slow, progressive inner ear loss that results from exposure to continuous noise over a period of time as contrasted to loss from a trauma or injury.

noise level — physical quantity of unwanted sound measured, usually expressed in decibels.

nolle prosequi — not willing to prosecute either as the sum of accounts, sum of defendants, or all.

nolo contendere — literally "I will not contest it"; an implied confession of guilt in a criminal case having the same legal effect as a guilty plea.

nomenclature — a classified system of technical terms.

nonaggressive sampling — an environmental sampling method performed in a quiescent atmosphere.

nonauditory effects of noise — the stress, fatigue, health, work efficiency, and performance effects due to continuous loud noise.

noncommunicable disease — a disease the causative agents of which cannot be passed or carried from one person to another directly or indirectly.

noncommunity water system — a public water system that serves less than 15 connections or 25 individuals on a daily basis.

noncompliance — failure or refusal to conform to or follow rules, regulations, or the advice or wishes of another person.

nonconductor — a substance that does not readily transmit electricity, light, or heat.

non-critical soils — those undisturbed soil materials that can support a conventional private sewage disposal system where at least the lower portion of the soil absorption part of the system can be installed in original, uncompacted, or undisturbed soil.

nondirectional — *See* omnidirectional.

nonelectrolyte — a compound which when dissolved in water does not separate into charged particles and is incapable of conducting an electrical current.

nonexpanding clay — clay particles that do not change appreciably in volume when dried by weather.

non-fat — skim milk which contains not more than 0.1% of milk-fat. Also known as fat free.

nonfeasance — the omission to carry out a task which should be done as dictated by law.

nonferrous metal — any metal that contains no iron or iron alloys.

nonflammable — incapable of being easily ignited or rapidly burning.

non-food-contact surfaces — exposed surfaces other than food-contact surfaces.

nonfouling — a property of cooling water that allows it to flow over steam condenser surfaces without accumulation of impediments.

nonintact skin — skin that is chapped, abraded, weeping, or that has rashes or eruptions.

nonionic — not dependent on a surface-active anion for effect.

nonionic detergent — a synthetic detergent that produces electrically neutral colloidal particles in solution.

nonionizing radiation — electromagnetic radiation, including ultraviolet, infrared, laser, microwave, and radio wave, which does not produce ionization.

nonpersistent — lasting only for a few weeks or less.

nonpoint source — a scattered, diffuse source of pollution.

nonpoint-source water pollution — the polluted water coming typically from rural areas that enters a receiving body from many small, scattered sources.

● hydrogen electron

○ oxygen electron

Nonpolar Molecule

nonpolar — any molecule or liquid that has a reasonable degree of electrical symmetry such that there is low or no separation of charge, such as benzene and carbon tetrachloride.

nonpotable water — any water not defined as potable.

nonreactive liquid — a simple solvent which dissolves the vapor or releases it with no chemical reaction involved, such as water or heavy carbon oil.

nonregenerative absorbent — an absorbent that is irreversibly converted and must be discharged.

nonselective herbicide — a pesticide chemical or agent that will prevent growth or kill most of the plants it contacts.

nonventilated air sampler — an instrument constantly in contact with the air sample directly without the aid of an air mover, but with the aid of natural ventilation.

nonvolatile — a pesticide chemical that does not evaporate by turning into a gas or vapor.

norepinephrine — $C_8H_{11}O_3N$ — a hormone which acts as a neurotransmitter of most sympathetic postganglionic neurons and also of certain tracts in the central nervous system; produces vasoconstriction, an increase in heart rate, and elevation of blood pressure.

norm — a numerical or statistical measure of average observed performance.

normal — (*math*) perpendicular to; a mean or average.

normal solution — a solution that contains one equivalent of the active reagent in grams in 1 liter of solution.

normal temperature and pressure — *See* standard temperature and pressure.

Norwalk type disease — diarrhea caused by Norwalk virus, a 27 nanometer particle, which is a parvovirus-like agent, is a limited mild disease with symptoms of nausea, vomiting, abdominal pain, malaise, and low grade fever. The occurrence of the disease is worldwide, with people being the only reservoir of infection. The mode of transmission is probably the oral fecal route through food and water. The incubation period is 24–48 hours, with communicability occurring during the acute state of disease and shortly thereafter. The reservoir of infection is probably people; general susceptibility. Also known as epidemic viral gastroenteritis.

nosocomial — pertaining to originating in a hospital; of a disease or infection caused or aggravated by hospitalization.

notice — a legal warning from an administrative agency advising an individual or individuals concerning a violation of rules and regulations.

not otherwise classified (NOC) — risks that do not have a classification definition which precisely describes their operation.

NOVI — notation for notice of violation.

NO$_x$ — scientific notation for nitrogen oxide.

noy — a unit of perceived noise level.

Np — *See* neper.

NPDES — *See* National Pollution Discharge Elimination System.

NPDWR — *See* National Primary Drinking Water Regulations.

NPIRS — acronym for National Pesticide Information Retrieval System.

NPL — *See* National Priorities List.

NRC — *See* National Research Council; Nuclear Regulatory Commission; National Response Center.

NRDC — acronym for National Resources Defense Council.

N$_{Re}$ — *See* Reynolds number.

NRT — *See* National Response Team.

NSC — *See* National Safety Council.

NSF — *See* National Sanitation Foundation.

NSPF — *See* National Swimming Pool Foundation.

NSPI — acronym for National Swimming Pool Institute.

NSPS — acronym for new source performance standards.

NTA — *See* nitrilotriacetate.

NTIS — *See* National Technical Information Service.

NTOF — acronym for National Traumatic Occupational Fatality Database.

NTP — *See* National Toxicology Program.

NTU — *See* nephelometric turbidity unit.

nuclear battery — a device in which the energy emitted by the decay of a radioisotope is converted first to heat and directly to electricity.

nuclear bombardment — the rapid movement of atomic particles toward nuclei to split the atom or to form a new element.

nuclear change — the process by which an element is converted into another element through a change in the number of protons in its atoms, such as uranium 238, which is radioactive and decays through a series of nuclear reactions to form lead, atomic weight: 206, a stable, nonradioactive isotope.

nuclear disintegration — a spontaneous nuclear transformation of radioactivity characterized by the emission of energy and/or mass in the nucleus.

nuclear energy — the energy released by the spontaneous or artificially produced fission, fusion, or disintegration of the nuclei of atoms. Also known as atomic energy.

nuclear fission — a nuclear transformation in which a nucleus is divided into at least two nuclei with a release of energy, such as when

M
N

uranium 235 is bombarded with neutrons and split into barium 144 and krypton 90.

nuclear fuel — a substance that is consumed in a reactor during nuclear fission or fusion.

nuclear fusion — a nuclear transformation in which two smaller atomic nuclei fuse into one larger nucleus and release energy.

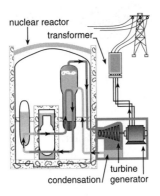

Nuclear Power Plant

nuclear power plant — an energy plant that utilizes uranium 235 in fuel rods to produce heat which is converted into energy for populations. Also known as atomic power.

nuclear reaction — any reaction involving a change in nuclear structure, such as fission, fusion, neutron capture, or radioactive decay.

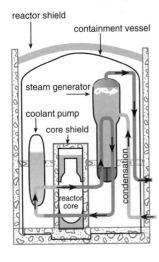

Nuclear Reactor

nuclear reactor — a device in which a controlled chain reaction of fissionable material can be produced and sustained.

Nuclear Regulatory Commission (NRC) — a federal agency which regulates the civilian uses of nuclear energy to protect the public health, safety, and the environment.

nuclear transformation — the process by which a nuclide is transformed into a different nuclide by absorbing or emitting a particle. Also known as transmutation.

nucleic acid — the large organic molecule made of nucleotides that functions in the transmission of hereditary traits in protein synthesis and in the control of cellular activities.

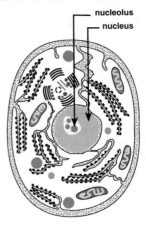

Nucleolus

nucleolus — a rounded refractile body in the nucleus of most cells which is the site of synthesis of ribosomal RNA, becoming enlarged during periods of synthesis and smaller during quiescent periods.

nucleon — the common name for a proton or neutron, the main constituent particles of the nucleus.

nucleonics — the science that deals with the constituents and changes in the atomic nucleus.

nucleon number — *See* mass number.

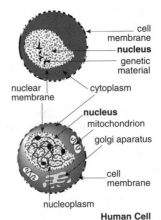

Human Cell

Nucleus

nucleus — (*cells*) the dense mass of living material enclosed within a membrane in the central part of a cell. It dictates how protein synthesis occurs and directs the other cell activities, such as active transport, metabolism, growth, and heredity. The nucleus stores, transcribes, and transmits genetic information stored in the deoxyribonucleic acid molecules present in the nucleus.

nuclide — a species of atom characterized by the constituents of its nucleus or by its atomic number, mass number, and atomic mass.

nuisance — an action or situation that is annoying, unpleasant, or potentially hazardous, such as the presence of rodents or sewage on top of the ground.

nuisance dust — dust that has little or no adverse effect on the lungs or health of individuals.

nuisance law — common law which states the use of private property is unrestricted only as long as it does not injure another person or his/her property; an injury or potential injury would constitute a nuisance.

null hypothesis — (*statistics*) a hypothesis of a research study involving two or more groups where the assumption is that there is no difference between the groups; if a difference is shown, the null hypothesis is rejected, and the results are said to be statistically significant.

numerical examination of urban smog (NEXUS) — mathematical model of air pollution which uses a computer for simulation of transport and diffusion of pollutants in the atmosphere.

nunc pro tunc — "now for then"; a phrase applied to acts allowed to be done after the time that they should have been done with a retroactive effect.

N-unit — a measure of radiation due to fast neutrons; a unit of index of refraction.

nutrient — an organic and inorganic chemical necessary for the growth or reproduction of an organism.

nutrition — the science of food, the nutrients and other substances therein, their action, interaction and balance in relation to health and disease, and the processes by which an organism ingests, digests, absorbs, transports, utilizes, and excretes food substances.

NWQSS — acronym for National Water Quality Surveillance System.

NWS — acronym for National Weather Service.

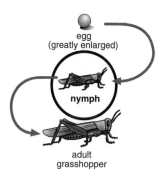

egg
(greatly enlarged)

nymph

adult
grasshopper

Grasshopper Nymph

nymph — the immature larval stage of various insects, such as grasshoppers and roaches.

nystagmus — an involuntary movement of the eyeball; occurs among workers who subject their eyes to abnormal and unaccustomed movements for extended time periods.

M
N

O

O$_2$ — scientific notation for oxygen.

O and MP — *See* operation and maintenance program.

OAQPS — acronym for Office of Air Quality Planning Standards.

objective — a quantified statement of a desired future state or condition to be achieved within a specified time period.

oblatespheroid — a flattened sphere in which the diameter of its equatorial circle exceeds the length of the axis of revolution.

obligate aerobe — an organism that cannot live without atmospheric oxygen.

obligate anaerobe — an organism that cannot grow in the presence of atmospheric oxygen.

oblique — having a slanted direction or position.

observational study — research in which the investigator can only observe the occurrence of disease (or other characteristics) in persons already categorized on the basis of some experience or exposure, the degree to which is not under the investigator's control.

oc — *See* open cup.

occlude — to close, shut, or stop up a passage.

occupant — any individual over one year of age living, sleeping, cooking, eating in, or having possession of a dwelling, dwelling unit, or rooming unit.

occupational health nurse — a registered nurse who provides acute and rehabilitative care, and participates in preventive and health maintenance programs for workers.

occupational health physician — a physician who conducts medical practices, diagnosis and treatment activities, directs medical departments in the workplace, and collaborates with other occupational safety and health personnel to identify, investigate, and prevent the recurrence of occupational safety and health problems.

occupational injury — an injury occurring in the course of employment and caused by inherent or related factors arising from the operation of materials of one's occupation.

Occupational Safety and Health — a database from 1972 to present covering virtually all aspects of occupational health and safety including hazardous agents, unsafe workplace, environment, and toxicology; coverage of over 150 journals plus government publications.

Occupational Safety and Health Act — a 1970 law establishing the Occupational Safety and Health Administration (OSHA) and the National Institute for Occupational Safety and Health (NIOSH); the Congressional purpose of the Act was to assure as far as possible that every working person in the nation will have safe and healthful working conditions, and thereby preserve our human resources.

Occupational Safety and Health Administration (OSHA) — an organization formed in 1971 as a result of the Occupational Safety and Health Act of 1970; it establishes and enforces regulations and standards to control occupational health and safety hazards, such as exposure to carcinogens, diseases, or disorders related to exposure to hazardous chemicals or agents.

occupational safety specialist — an environmental health practitioner who plans, develops, implements, and administers occupational health and injury control programs in industry to identify and correct hazardous conditions that may cause disease and injury to employees, damage to equipment and facilities, or loss of materials. Also known as safety management specialist; environmental safety specialist.

occupational skin diseases or disorders — a series of diseases or disorders including contact dermatitis, eczema, or rash caused by primary irritants, poisonous plants, oil acne, chrome ulcers, chemical burns, or inflammations.

occurrence — an accident, including continuous or repeated exposure to conditions which result in bodily injury or property damage, which the owner or operator neither expected or intended.

ocean — any portion of the high sea beyond the contiguous zone occupying the depressions in the earth's surface; banded by continents and imaginary lines.

ocean disposal — the deposition of waste into an ocean or estuary.

oceanic — pertaining to marine waters seaward of the continental shelf margin.

Oceanic Abstracts — a database from 1964 to the present, indexing technical literature, published worldwide on marine-related subjects, including marine pollution, from journals, books, technical reports, conference proceedings, and government publications.

ochronosis — a blue or brownish-blue discoloration of body tissues caused by deposits of alkapton bodies as a result of a metabolic disorder.

octachloronaphthalene — $C_{10}Cl_8$ — a waxy, pale yellow solid with an aromatic odor; MW: 403.7, BP: 824°F, Sol: insoluble, sp gr: 2.00. It is used as an insulating material; as an inert compound of resins or polymers for coatings or impregnating textiles, wood, and paper to impart flame resistance, waterproofness, and fungicidal and insecticidal properties; as an additive for cutting oil in various operations preformed on metal; as an additive to special lubricants. It is hazardous to the skin and liver; and is toxic by inhalation, absorption, ingestion, and contact. Symptoms of exposure include acne-forming dermatitis, liver damage, jaundice. OSHA exposure limit (TWA): 0.1 mg/m³ [skin].

octane — $CH_3(CH_2)_6CH_3$ — a colorless liquid with a gasoline-like odor; MW: 114.2, BP: 258°F, Sol: insoluble, Fl.P: 56°F, sp gr: 0.70. It is used in the preparation of gasoline and rocket fuels; as an industrial solvent as a lacquer diluent, phosphate manufacture, and preparation of liquid soaps and detergents; in printing ink manufacture; as an additive and solvent in polymer manufacture; in the manufacture of benzene, toluene, and xylene aromatics; as a blowing agent for foam rubber used in rocket propellants. It is hazardous to the skin, eyes, and respiratory system; and is toxic by inhalation, ingestion, and contact. Symptoms of exposure include drowsiness, dermatitis, chemical pneumonia, irritation of the eyes and nose. OSHA exposure limit (TWA): 300 ppm [air] or 1,800 mg/m³.

octane number — a conventional rating for gasoline based upon its behavior under standard conditions in a standard internal combustion engine as compared with isooctane, whose octane number is 100.

octanol-water partition coefficient (Kow) — the equilibrium ratio of the concentration of a chemical in octanol and water in dilute solution.

octave — the interval between any two frequencies having the ratio 2:1.

octave band — a band in frequency where the lower frequency is related to the upper frequency by the ratio of 2:1.

octave band analyzer — an instrument used to determine where noise energy lies in the frequency spectrum.

ocular — pertaining to the eye.

ocular route — a route of entry into the body through the eye.

OD — notation for outside diameter. *See* optical density.

odds ratio — the ratio of the probability of a specific event occurring relative to the probability of the event not occurring.

odor — the property of a substance which affects the sense of smell.

odorant — material added to odorless gases to impart an odor for safety purposes.

odor threshold — the lowest concentration of an airborne odor that a human can detect.

OFAP — *See* Office of Federal Agency Programs.

offal — intestines and discarded parts including manure remaining after animals have been slaughtered.

Office of Federal Agency Programs (OFAP) — the organizational unit of the Occupational Safety and Health Administration which provides federal agencies with guidance to develop and implement occupational safety and health programs for federal employees.

Office of Management and Budget (OMB) — a federal government agency within the Executive Branch whose function is to evaluate, formulate, and coordinate management procedures and program objectives within and among federal departments and agencies.

Office of Pesticides and Toxic Substances (OTS) — a branch of the Environmental Protection Agency whose function is to develop national strategies for control of toxic substances, direct enforcement activities, develop criteria, and assess the impact of existing chemicals and new chemicals.

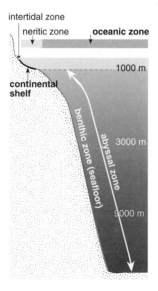

Oceanic Zone

O
P

Office of Research and Development (ORD) — a branch of the Environmental Protection Agency whose function is national research in pursuit of technological controls of all forms of pollution.

Office of Solid Waste (OSW) — a branch of the Environmental Protection Agency whose function is to provide policy guidance and direction for the agency's hazardous waste and emergency response programs.

Office of Solid Waste and Emergency Response (OSWER) — a federal government agency within the Environmental Protection Agency whose function is to provide policy, guidance, and direction for the agency's hazardous waste and emergency response program.

Office of Technology Assessment (OTA) — an agency of Congress whose function is to provide objective analysis of major public policy issues related to scientific and technological change.

Office of Workers Compensation Program (OWCP) — a federal program within the Department of Labor whose function is to administer federal laws which compensate workers in a variety of special areas.

ohm (Ω) — the mks and SI unit of electrical resistance equal to the resistance through which a current of 1 ampere will flow when there is a potential difference of 1 volt across it.

OHM's law — the law that states that electric current flowing through an electric circuit is equal to the voltage divided by the resistance.

OHMTADS — acronym for Oil and Hazardous Material Technical Assistance Data System.

oil mist (*mineral*) — a colorless, oily liquid aerosol dispersed in air; has an odor like burned lubricating oil; MW: varies, BP: 680°F, Sol: insoluble, Fl.P (open cup): 275–500°F, sp gr: 0.90. It is used as a coolant or quenching agent; in steel rolling operations; in pressroom operations in printing; as an insecticide. It is hazardous to the respiratory system and skin; and is toxic by inhalation. No reported symptoms of exposure. OSHA exposure limit (TWA): 5 mg/m³ [air].

Oil Pollution Act — a 1924 law preventing oily discharges into coastal waters.

oil shale — a generic name for the sedimentary rock that contains substantial organic materials known as kerogen. The kerogen content of oil shale may vary between 5–80 gallons of oil per equivalent ton. The extraction of the oil from the oil shale occurs when the shale is heated to 350–550°C under an inert atmosphere. The products include oil vapor, hydrocarbon gases, and a carbonaceous residue. This retorting shale process produces one ton of spent shale per barrel of oil. The volume of the shale increases by more than 50% during the crushing process, and a considerable amount of alkaline minerals contained in the oil shale may leach into the groundwater supply. A large amount of water is needed for controlling dust and reducing the alkalinity of the spent shale.

oil solution — a pesticide chemical dissolved in oil.

oil spill — an accidental discharge of oil into bodies of water which may be controlled by chemical dispersion, combustion, mechanical containment, and absorption.

olefiant gas — *See* ethylene.

olefins — C_nH_{2n} — a class of unsaturated hydrocarbons characterized by relatively great chemical activity; butene, ethylene, and propylene are examples. *See also* alkene.

oleum — an oily, corrosive liquid consisting of a solution of sulfur trioxide and anhydrous sulfuric acid that turns to fume in moist air. Also known as sulfuric acid.

olfactory — pertaining to the sense of smell.

oligotrophic lake — a deep, clear lake with low nutrient supply containing little organic matter and a high dissolved oxygen level.

oliguria — diminished urine secretion in relation to fluid intake.

OMB — *See* Office of Management and Budget.

omnidirectional — the sending or receiving of signals in all directions. Also known as nondirectional.

oncogene — a gene found in the chromosomes of tumor cells whose activation is associated with the initial and continuing conversion of normal cells into cancer cells.

oncogenic — causing tumor formation.

oncology — the study of the causes, development, and treatment of neoplasms.

on-site disposal — any method or process to eliminate or reduce the volume or weight of solid waste on the property of the generator.

on-site incinerator — an incinerator that burns solid waste on the property utilized by the generating person or person.

on-site sewage — the gray and black water containing solids coming from a structure which receives primary settlement in a septic tank, and whose effluent is disposed on-site in the ground.

onus — some task, duty, or responsibility that involves or necessitates considerable effort or that is a burden.

oocyst — the encysted or encapsulated zygote of some sporozoa.

ookinete — the fertilized form of the malarial parasite in a mosquito's body formed by fertilization by a microgamete and which develops into an oocyst.

opacity — the degree of obstruction of light.

opacity reading — the apparent obscuration of an observer's vision that equals the apparent obscuration of smoke of a given rating on the Ringelmann chart.

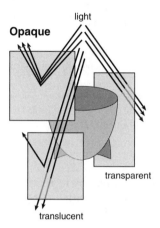

opaque — unable to transmit light; totally absorbent of rays of a specified wavelength and not transparent to the human eye.

open burning — uncontrolled burning of waste in an open area, such as a dump.

open cup (oc) — a method for determining the flashpoint of a chemical.

open dump — any facility or site where solid waste is disposed of that is not a sanitary landfill.

open hearth furnace — a long, wide, shallow furnace used to produce steel from cast or pig iron.

open-loop system — a system of heat exchange where well water at 55°F is used as the condenser; the water is pumped into a house and through a refrigeration process where the heat from the water can be removed and used to heat the property; when cooling the property, the hot air can be passed across the 55°F water to raise the temperature of the water and to decrease the temperature of the air within the property.

open-pit incinerator — a burning device that has an open top and a system of closely spaced nozzles that places a stream of high-velocity air over the burning zone.

operating flow rate — the flow rate at which a facility actually operates as opposed to the design flow rate.

operating humidity range — the range of ambient relative humidity over which a machine or instrument will meet all performance specifications.

operating room (OR) — a specially equipped room in which an operation on a patient is conducted in a sterile manner.

operating temperature range — the range of ambient temperatures over which a machine or instrument will meet all performance specifications.

operation and maintenance program (O and MP) — a program of training, work practices, and periodic surveillance to maintain friable asbestos-containing building material in good condition, clean-up of asbestos fibers previously released, and prevent further release by minimizing and controlling friable asbestos containing building material disturbance or damage.

operator — an environmental health or safety practitioner qualified by education, training, and practice to manage plant and systems processes necessary for environmental health or safety control and prevention activities; certification is usually required.

Opthalmoscope

ophthalmoscope — an instrument used for examining the interior of the eye through the pupil.

opisthotonos — a form of spasm of the back muscles in which the head and heels are bent backward and the body bowed forward.

opportunistic infection — an infection that is usually warded off by a healthy immune system, but which can affect an individual whose immune system has been suppressed.

opportunistic organisms — usually saprophytic or commensal organisms but which also can be pathogenic under conditions of impaired host resistance; such organisms include systemic fungi, and *Pneumocystis carinii*; important agents of nosocomial infections.

opsonin — an antibody in blood serum which renders bacteria and other cells susceptible to phagocytosis.

optical density (OD) — a logarithmic expression of the degree of opacity of a translucent medium.

optical microscopy — a basic system of magnification using light rays reflected from the object which bend as they pass through one or more lenses; used to determine particle size and to identify solid pollutants.

optical pyrometer — a temperature-measuring instrument that matches the intensity of radiation at a single wavelength from a tungsten filament with the intensity of the radiation at the same wavelength emitted by a heat source.

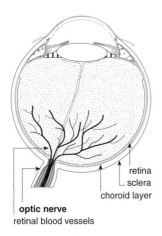

retina
sclera
choroid layer
optic nerve
retinal blood vessels

Optic Nerve

optic nerve — the sensory nerve leading from the retina of the eye to the optic lobe of the brain.

OR — *See* operating room.

oral route — a route of entry into the body by ingestion.

oral toxicity — the ability of a chemical to sicken or kill an organism if eaten or swallowed, usually measured in LD_{50}.

orbivirus — *See* reovirus.

ORD — *See* Office of Research and Development.

order — (*law*) a formal written statement from a court requiring action or stating a ruling.

order of magnitude — a range of magnitudes of a quantity extending from some value of the quantity to some small multiple of the quantity (usually 10).

ordinance — a statute enacted by the legislative body of a local, generally municipal government.

ordinary summer conditions — a temperature of 10°F below the highest recorded temperature in a locality during the most recent ten year period.

ordinary winter conditions — a temperature of 15°F above the lowest recorded temperature in a locality during the most recent ten year period.

ore — a naturally occurring mineral substance from which economically valuable elements may be obtained.

organ — an organized collection of tissues that have a specific and recognized function.

organic — referring, relating, or deriving from living organisms.

organic acid — a chemical compound containing one or more carboxyl group (–COOH); examples are formic, acidic, propionic, chloracetic, and trifluoracetic; causes severe damage to skin and mucosal surfaces, eye injury, and may cause damage to the enzyme systems of the body.

organic chemistry — the scientific study of compounds containing carbon.

organic compound — compounds containing carbon which refer, relate, or derive from living organisms; can also be produced by chemical synthesis.

organic gas — a gas containing carbon, usually hydrogen and possibly other elements bound to the carbon; examples include hydrocarbons such as methane, butane, octane, benzene, acetylene, ethylene, and butadiene; aldehydes and ketones such as formaldehyde and acetone; other organics such as chlorinated hydrocarbons, benzopyrene, and alcohols.

organic material in soils — the organic matter in soils consisting of decaying tissue, residues, decaying matter, and organic by-products created by microorganisms which break down plant and animal tissue.

organic matter — the carbonaceous waste contained in plant or animal matter and originating from domestic or industrial sources.

organic nitrogen — nitrogen combined in organic molecules, such as proteins and amino acids.

organic peroxide — an organic compound that contains the bivalent –O–O structure and may be considered a structural derivative of hydrogen peroxide, as one or both of the hydrogen atoms has been replaced by an organic radical.

organic pesticide — any chemical containing carbon, which alone in chemical combination, or in formulation with one or more substances, is an "economic poison" within the meaning of the Federal Insecticide, Fungicide, and Rodenticide Act.

organic soil — a soil which contains a high percentage of organic matter throughout the solum.

organic solid — (*sewage*) solid matter of animal or vegetable origin including waste products, dead animal matter, plant tissues, organisms, and synthetic organic compounds.

organic solvent — a liquid organic material, including diluents and thinners, used to dissolve solids, reduce viscosity, or clean substances.

organism — an animal, plant, fungus, or microorganism able to carry out all life functions.

organogenesis — the secretion of tissues in the different organs during embryonic development.

organoleptic — pertaining to or perceived by a sensory organ.

organophosphate — one of a group of pesticide chemicals containing phosphorous; it interferes with the normal working of the nervous systems of animals and humans.

orifice — (*pesticides*) the opening or hole in a nozzle through which liquid material is forced out and broken up into a spray; an opening through which a fluid can pass.

orifice meter — a flow meter used to measure differential pressures between upstream and downstream on a restriction within a pipe or duct.

original container — the package, can, bag, or bottle in which a company sells a chemical.

orographic precipitation — a precipitation caused by topographic barriers, such as mountains that force moisture-laden air to rise and cool.

ORP — acronym for oxidation-reduction potential. *See* redox.

Orsat analyzer — an apparatus used to analyze flue gases volumetrically by measuring the amount of carbon dioxide, oxygen, and carbon monoxide present.

orthotolidine — an organic test agent which turns yellowish-green in the presence of chlorine, bromine, or iodine.

orthotolidine colorimetric comparison — a test in which a colorless solution of orthotolidine is added to a chlorine solution, producing an orange-brown colored compound proportional to its concentration which is then compared to a standardized color.

Oryzaephilus surinamensis — *See* Saw-Toothed Grain Beetle.

oscillate — to swing backward and forward like a pendulum; to move or travel periodically back and forth between two points.

oscillation — variation with time of the magnitude of a quantity with respect to a designated reference when the magnitude is alternately greater and smaller than the reference.

oscillator — a radiant tube or circuit that produces steady alternating current, usually at high frequency.

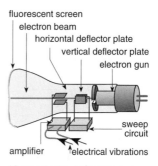

(Cathode Ray) **Oscilloscope**

oscilloscope — a type of cathode-ray tube in which sound waves are changed to electrical impulses and are displayed on a screen; used to make visible the pattern of electrical vibrations, sound waves. Also known as cathode ray oscilloscope.

OSHA — *See* Occupational Safety and Health Administration.

osmium tetroxide — OsO_4 — a colorless, crystalline solid or pale yellow mass with an unpleasant, acrid, chlorine-like odor; a liquid above 105°F; MW: 254.2, BP: 266°F, Sol (77°F): 6%, sp gr: 5.10. It is used as an intermediate for separation of metals and ores of osmium from platinum; as a catalyst in organic synthesis; as a staining and fixing agent in pathological and histological analysis; as a laboratory reagent. It is hazardous to the eyes, respiratory system, and skin; and is toxic by inhalation, ingestion, and contact. Symptoms of exposure include lacrimation, visual disturbances, conjunctivitis, headache, cough, dyspnea, dermatitis. OSHA exposure limit (TWA): 0.0002 ppm [air] or 0.002 mg/m³.

osmosis — the diffusion of a solvent through a semipermeable membrane from a region of greater concentration to a region of lesser concentration.

O
P

ossification — the process of the formation of bone.

OSW — *See* Office of Solid Waste.

OSWER — *See* Office of Solid Waste and Emergency Response.

OTA — *See* Office of Technology Assessment.

OTC — notation for over-the-counter drug.

other waste — substances, not including sewage or industrial waste, which may cause or tend to cause pollution such as garbage, refuse, wood, residues, sand, lime, cinders, ashes, offal, oil, tar, dye stuffs, acids, and chemicals.

otitis media — inflammation of the middle ear.

otosclerosis — formation of spongy bone just inside the inner ear, causing progressive deafness.

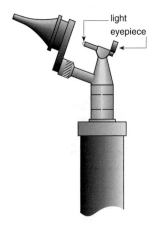

Otoscope

O
P

otoscope — an instrument for inspecting or auscultating the ear and rendering the tympanic membrane visible.

OTS — *See* Office of Pesticides and Toxic Substances.

ounce (oz) — a unit of mass in avoirdupois measure equal to 1/16 of a pound or approximately 28.3495 grams.

outcome — a final consequence, result, or effect.

outfall — the point at which an effluent is discharged into receiving waters.

OVA — acronym for organic vapor analyzer.

ovary — the female sex gland in which ova are formed in vertebrates.

overburden — the overlying layer of sediment that must be removed to reach a mineral or coal deposit; (*mineral engineering*) to change in a furnace too much ore and flux in proportion to the amount of fuel.

overexposure — exposure beyond the specified limits.

overfire air — air introduced above or beyond a fuel bed by a natural, induced, or forced draft. Also known as secondary combustion air.

overfire air fan — a fan used to provide air above a fuel bed.

overflow system — (*swimming pools*) a series of devices or channels used to remove overflow water from a pool, including surface skimmers and surface water collection systems of various design and manufacture.

overhead — the general costs of operating an entity which are distributed to all the revenue-producing operations, but which are not directly attributable to a single activity.

overlay — (*mapping*) the process of stacking digital representations of various spatial data on top of one another so each position in the area covered can be analyzed in terms of these data.

ovicide — a pesticide chemical or agent used to destroy insect, mite, or nematode eggs.

ovoid — shaped like an egg.

ovulation — the process in which an ovum is discharged from an ovary.

ovum — an egg; a female gamete.

OWCP — *See* Office of Workers Compensation Program.

OX — *See* photochemical oxidant.

oxalic acid — HOOCCOOH–$2H_2O$ — a colorless, odorless powder or granular solid; the anhydrous form is an odorless, white solid; MW: 126.1, BP: sublimes, Sol: 14%, Fl.P: not known, sp gr: 1.65. It is used in metal cleaning and polishing operations to remove carbonaceous deposits from steel; in textile cleaning and bleaching operations for rust, ink, and stain; bleaching of cotton and other fabrics; as a sour in treatment of woolen and other pieces of goods; in bleaching and tanning of hides in leather manufacture; in the manufacture of household and industrial cleaning and bleaching agents; in

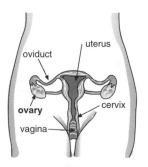

Ovary

metallurgical processing of tin; in the synthesis of other chemicals for the manufacture of dyes and blue-print photography; in extraction and purification operations; in lithography and photoengraving operations; as an analytical reagent; in the manufacture of inks and lacquers; in the synthesis of pharmaceuticals. It is hazardous to the respiratory system, skin, kidneys, and eyes; and is toxic by inhalation, absorption, ingestion, and contact. Symptoms of exposure include eye burns, local pain, cyanosis, shock, collapse, convulsions, irritation of the eyes, mucous membrane, and skin. OSHA exposure limit (TWA): 1 mg/m^3 [air].

oxidant — *See* oxidizing agent.

oxidation — a chemical union of oxygen with any substance; a chemical reaction in which an element loses electrons and thus increases in oxidation number.

oxidation number — the charge which an ion has, or which an atom appears to have, when its electrons are counted.

oxidation pond — a shallow lake or other body of water in which wastes are consumed by bacteria. Also known as a sewage lagoon.

oxidation potential — (*physics*) a measure in volts of an element's tendency to oxidize or lose electrons; used to predict how the element will react with another substance.

oxidation-reduction — (*chemistry*) a chemical reaction in which one or more electrons are transferred from one atom or molecule to another.

oxidation-reduction potential — (*chemistry*) the electromotive force resulting from oxidation-reduction, usually measured relative to a standard hydrogen electrode.

oxidation state — (*chemistry*) the number of electrons lost or gained by an element or atom in a compound; expressed as a positive or negative number indicating the ionic charge and the number of electrons lost or gained is equal to the valence.

oxidative phosphorylation — the transfer of phosphate ion onto an acceptor molecule, usually ADP, to establish a more energy-rich molecule driven by the coupling of hydrogen transport down an electron transport chain.

oxide — a compound of two elements, one of which is oxygen.

oxidize — to combine with oxygen.

oxidizer — a chemical, other than a blasting agent or explosive, that initiates or supports combustion in another material, causing fire either by itself or through the release of oxygen or other gases.

oxidizing agent — a substance that gives up its oxygen readily, removes hydrogen from a compound, or attracts electrons from elements. Also known as an oxidant.

oxirane — *See* ethylene oxide (ETO).

oxyacetylene welding — an oxyfuel gas welding process that produces coalescence of metals by heating them with a gas flame obtained from the combustion of acetylene with oxygen. Also known as acetylene welding.

oxyfuel gas welding — the process that produces coalescence by heating materials with an oxyfuel gas flame, with or without the application of pressure, and with or without the use of filler metal.

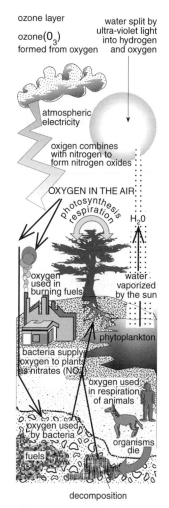

Oxygen Cycle

oxygen cycle — the circulation and reutilization of oxygen in the biosphere.

oxygen deficient — pertaining to an atmosphere with a partial pressure of oxygen (PO_2) less than 132 millimeters of mercury; normal air at sea-level contains approximately 21% oxygen at a PO_2 of 160 millimeters of mercury.

oxygen difluoride — OF_2 — a colorless gas with a peculiar, foul odor; MW: 54.0, BP: –230°F, Sol: 0.02%. It is used as an oxidizer in rocket propellants. It is hazardous to the lungs and eyes; and is toxic by inhalation and contact. Symptoms of exposure include intractable headache, respiratory system irritation, pulmonary edema, eye and skin burns. OSHA exposure limit (TWA): ceiling 0.05 ppm [air] or 0.1 mg/m³.

oxygen enriched atmosphere — any oxygen concentration greater than 25%.

oxygen sag — the declined oxygen concentration downstream from a pollution source when there is a high biological oxygen demand.

oxygenation — the process whereby the blood is supplied with oxygen from the lungs; the process of treating, infusing, or combining with oxygen.

oxyhemoglobin — $HHbO_2$ — the oxygen in the blood combined with hemoglobin in the red blood cells.

oz — *See* ounce.

ozone — O_3 — a colorless to blue gas with a very pungent odor; MW: 48.0, BP: –169°F, Sol: insoluble. It is liberated during welding operations; during the oxidizing process of fine organic chemicals; during operations involving high-intensity ultra-violet light; during operations involving high-voltage electrical equipment; during bleaching operations; during food preserving operations for mold and bacteria control. It is hazardous to the eyes and respiratory system; and is toxic by inhalation. Symptoms of exposure include pulmonary edema, chronic respiratory disease, irritation of the eyes and mucous membrane, reduced resistance to infection, and possibly a premature aging of the lung tissue. OSHA exposure limit (TWA): ceiling 0.1 ppm [air] or 0.2 mg/m³.

ozone layer — *See* ozonosphere.

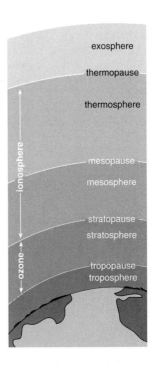

Ozonosphere

ozonosphere — the region in the upper atmosphere where most atmospheric ozone is concentrated from about 8 miles to 30 miles above the earth with maximum ozone occurring about 12 miles above the earth. Also known as ozone layer.

P

p — *See* pico.

Pa — *See* pascal.

PA — notation for permanent abeyance.

pacemaker of the heart — *See* sinoatrial node.

package — (*computer science*) a set of computer programs that can be used for various operations.

packaged foods — candy bars and other foods stored in specific containers.

package treatment plant — a small wastewater treatment facility serving from a few homes to large institutions.

packed tower — a control device in which a gaseous pollutant is removed from a rising air stream by absorption in a liquid flowing downward through a closely packed, chemically inert substance.

PAH — *See* polycyclic aromatic hydrocarbon; polynuclear aromatic hydrocarbon.

pain — a feeling of distress, suffering, or agony caused by stimulation of specialized nerve endings.

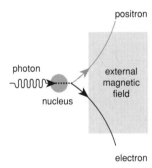

Pair Production

pair production — an absorption process for x- and gamma-radiation in which the incident photon is annihilated in the vicinity of the nucleus of the absorbing atom with subsequent production of an electron and positron pair.

PAIS Bulletin — *See* Public Affairs Information Services Bulletin.

pallor — paleness of the skin.

palpitation — an abnormally rapid heart beat of which a person is acutely aware; usually over 120 beats per minute.

PAN — *See* peroxyacylnitrates.

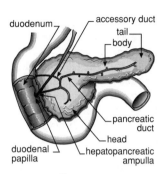

Pancreas

pancreas — a digestive and endocrine gland located between the stomach and intestine in most vertebrates.

pandemic — disease widespread throughout a geographic area.

paper — a thin, felted matrix of interlacing cellulose fibers.

paperboard — a paper product which is heavier, more rigid, and thicker than paper.

paperstock — paper waste that is recovered and reused.

papilloma — a small growth or tumor of the skin or mucous membrane.

papovavirus — a group of relatively small ether-resistant deoxyribonucleic acid viruses, including papilloma and polyoma viruses, many of which are oncogenic or potentially oncogenic.

PAPR — acronym for powered air-purifying respirator.

papule — a small, circumscribed, solid, elevated lesion of the skin.

paraffin series — any of the methane series of hydrocarbons.

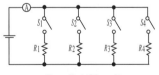

Parallel Circuit
S=Switch   A=Ammeter
R_x =Resistor

parallel circuit — an arrangement of circuit elements such that each has the same voltage.

paralysis — loss or impairment of motor or sensory function in a part due to a lesion of the neural or muscular mechanism.

paramedic — a health professional who has completed a course in advance life support care and has been certified by the Emergency Medical Services Commission.

parameter — the property of a system that can be measured numerically; (*mathematics/statistics*) an arbitrary constant, such as a population mean or standard deviation.

paraquat — $(CH_3(C_5H_4N)CH_3)$–2Cl — a yellow solid with a faint ammonia-like odor; may also be found commercially as a methyl sulfate salt; MW: 257.2, BP: 347–356°F decomposes, Sol: miscible, sp gr: 1.24. It is used as an agricultural herbicide, desiccant, defoliation agent; to control aquatic weeds. It is hazardous to the eyes, respiratory system, heart, liver, gastrointestinal tract, and kidneys; and is toxic by inhalation, absorption, ingestion, and contact. Symptoms of exposure include dermatitis, fingernail damage, damage to the heart, liver, and kidneys, acute pulmonary inflammation, irritation of the gastrointestinal tract, irritation of the eyes, nose, and epistaxis. OSHA exposure limit (TWA): 0.1 mg/m^3 [resp and skin].

parasite — an organism living in or on another organism from whose body the parasite takes nutrients.

parasympathetic nervous system — a division of the autonomic nervous system.

parathion — $(C_2H_5O)_2P(S)OC_6H_4NO_2$ — a pale yellow to dark brown liquid with a garlic-like odor; a solid below 43°F; MW: 291.3, BP: 707°F, Sol: 0.001%, Fl.P: not known, sp gr: 1.27. It is used as a pesticide on agricultural crops, vegetables, and ornamentals. It is hazardous to the respiratory system, central nervous system, cardiovascular system, eyes, skin, and blood cholinesterase; and is toxic by inhalation, absorption, ingestion, and contact. Symptoms of exposure include miosis, rhinorrhea, headache, tight chest, wheezing, laryngeal spasm, salivation, cyanosis, anorexia, nausea, vomiting, abdominal cramps, diarrhea, sweat, muscle fasciculation, weakness, paralysis, giddiness, confusion, ataxia, convulsions, coma, low blood pressure, cardiac irregularities, irritation of the eyes and skin. OSHA exposure limit (TWA): 0.1 mg/m^3 [skin].

paratyphoid fever — (*disease*) a bacterial enteric infection in humans with abrupt onset of fever, malaise, headache, enlargement of the spleen, diarrhea, and involvement of lymphoid tissues. Incubation time is 1–3 weeks for enteric fever, 1–10 days for gastroenteritis; caused by *Salmonella paratyphi A*, *S. paratyphi B*, and *S. paratyphi C*; found in the United States and Canada. The reservoir of infection is people. It is transmitted by direct or indirect contact with feces or carrier; spread by food, milk, milk products, shellfish, flies, water, and the contaminated individual's hands. It is communicable as long as the infectious agent is in the excreta,

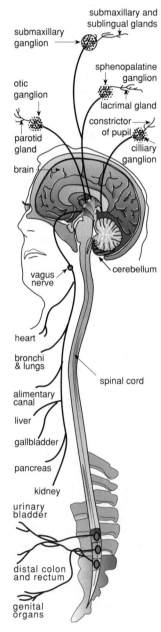

**Parasympathetic
Nervous System**

which may range from weeks to months, or permanently; general susceptibility. It is controlled by treatment of the disease, good health habits, elimination of flies and feces, and protection of food and water from the microorganisms.

parent — a radionuclide that upon disintegration yields a specified nuclide, the daughter, either directly or as a later member of a radioactive series. Also known as precursor.

parenteral route — a route of entry into the body through injection.

paresis — a slight paralysis.

paresthesia — a burning, prickling, or tingling sensation of the skin.

Parshall flume — a calibrated device for measuring the flow of liquid in an open conduit; consists of a contracting length, a throat, and an expanding length. At the throat, the flow passes a sill opening at critical depths, and the upper and lower heads are each measured at a definite distance from the sill; the lower head need not be measured unless the sill is submerged more than 67%.

particle — a small discrete mass of solid or liquid matter.

particle accelerator — (*nuclear physics*) a device for speeding up charged subatomic particles to high enough energies to smash the nuclei of the target atoms; used to produce radioisotopes.

particle board — board product composed of distinct particles of wood or other lignocellulosic materials not reduced to fibers but bonded together with an organic or inorganic resin binder and then cured.

particle conditioning — the increase in the effective size of small particles so they can be collected more efficiently by a variety of mechanisms.

particle size — the effective diameter of a particle usually measured by sedimentation or sieving.

particle-size distribution — the amounts of the various soil separated in a soil sample and usually expressed as weight percentage.

particulate — unburned carbon usually present when coal or residual oil is used as fuel; appears in the form of black smoke.

particulate matter (PM) — a suspension of fine solid or liquid particles in air; dust, fog, fume, mist, smoke, or spray are examples. Also known as aerosol.

parts per billion (ppb) — the number of parts of a chemical or substance found in one billion parts of a particular gas, liquid, or solid.

parts per million (ppm) — the number of parts of a chemical or substance found in one million parts of a particular gas, liquid, or solid.

parvovirus — (*disease*) a virus belonging to the family *Parvoviridae* containing linear single-stranded DNA surrounded by a protein capsid of about 22 nanometers in diameter which may cause viral gastroenteritis in all age groups with symptoms of diarrhea, nausea, vomiting, malaise, abdominal pain, and fever. The mode of transmission is the oral-fecal route, from person to person, ingestion of contaminated food (especially shellfish), and contaminated water. The incubation period is 10–70 hours. The disease has been reported in England, Australia, and the United States. The reservoir of infection is people; general susceptibility. Also known as acute nonbacterial infectious gastroenteritis; viral gastroenteritis.

pascal (Pa) — a unit of pressure equal to the force of 1 newton per square meter.

passive immunity — immunity acquired by the injection or transfer of antibodies in another organism.

passive smoking — an involuntary second-hand inhalation of tobacco smoke by nonsmokers.

Pasteurella — a genus of gram-negative nonmotile facultatively anaerobic ovoid to rod-shaped bacteria sometimes causing infection in humans.

pasteurization — the process of heating milk or other liquids for a specified time to destroy microorganisms that would cause spoilage and/or disease; the process of heating every particle of milk or milk products to at least 145°F for 30 minutes, or 161°F for 15 seconds and holding it at such temperature continuously.

pathogen — an organism that produces disease in a host and may alter the response of the host to other diseases processes.

pathogenesis — the cellular events and reactions and other pathologic mechanisms occurring in the development of disease.

pathogenic — disease-causing.

pathological — abnormal or diseased.

pathologist — a licensed physician whose practice of medicine is limited to or specializes in the study of the origin, nature, and course of disease.

pathology — the study of the essential nature, cause, and effects of diseases and their abnormalities.

pathway — a history of the flow of a pollutant from source to receptor, including qualitative descriptions of emission-type transport medium and exposure route.

O
P

patient — the individual receiving diagnosis and/ or treatment for disease or injury.

Pb — *See* lead.

PBB — *See* polybrominated biphenyl.

PbS — *See* galena.

PBZ — *See* personal breathing zone.

PCB — *See* polychlorinated biphenyl.

PCDD — notation for polychlorinated dibenzo-p-dioxin. *See* 2,3,7,8 tetrachlorodibenzo-p-di-oxin.

PCDF — notation for polychlorinated dibenzo-furan.

PCE — notation for pentachloroethane.

pCi — *See* picocurie.

PCM — *See* phase contrast microscope.

PCP — *See* pentachlorophenol.

pct — *See* percent.

PCV valve — a valve that regulates the flow rate of unburned crank case vapors back to the combustion chamber in order to control these from leaving the automobile and becoming pollutants.

PD — *See* potential difference.

PDP-CVS — acronym for positive displacement pump-constant volume sampler.

PDS — acronym for personal decontamination station.

peak optical response — the wavelength of maximum sensitivity of an instrument.

peat — partially decomposed dark brown or black plant material found in marshes and other wet places.

Pea Weevil (*Vruchus pisorum*) — similar to the Bean Weevil but 1/5″ long, brownish flecked with white, and with black to gray patches or scales; a stored food product insect.

Ped — (*soils*) a natural soil aggregate.

pediculosis — (*disease*) an infestation of the head and hairy parts of the body with lice causing severe itching. Incubation time is 15–17 days; caused by *Pediculus capitis*, the head louse; *P. humanus*, the body louse; and *Phthirus pubis*, the crab louse; found worldwide with head lice being especially common in schools and institutions among children. The reservoir of infection is infested people; transmitted by direct contact or contact with clothing, combs, or towels. It is communicable as long as lice or eggs remain alive on the person or clothing; general susceptibility. It is controlled by avoidance of physical contact with infested people, their clothing, and bedding; good health habits; inspection of children for head lice, and appropriate treatment.

pedology — (*soils*) the scientific studies of soils. Also known as soil science; (*medicine*) the study of the physiological and psychological aspects of childhood.

pedon — (*soils*) the smallest volume in a soil body which displays the normal of variations in properties of a soil profile.

pedosphere — (*soils*) the soil layers of the earth's crust.

peer review — the evaluation by practitioners or other professionals of the effectiveness and efficiency of services ordered, performed, or professional studies completed and prepared for publication.

pel — *See* pixel.

PEL — *See* permissible exposure limit.

pelagic — of, relating to, living, or occurring in the open sea.

pellagra — a dietary deficiency disease due to a lack of niacin characterized by skin lesions, inflammation of the soft tissues of the mouth, diarrhea, and central nervous system disorders.

pendemte lite-pending — (*law*) the litigation continues (or is pending).

penetration — (*protective clothing*) the flow of a chemical through closures, porous materials, seams, pinholes, or other imperfections in a protective clothing material on a non-molecular level; the action of a liquid entering into porous materials from cracks and pinholes.

penicillin — the collective name for an antibiotic produced by the mold *Penicillium* which is toxic to a number of bacteria, both pathogenic and nonpathogenic.

pen plotter — (*mapping*) a device for drawing maps and figures using a computer-guided pen.

pentaborane — B_5H_9 — a colorless liquid with a pungent odor like sour milk; MW: 63.1, BP: 140°F, Sol: reacts, Fl.P: 86°F, sp gr: 0.62. It is used in chemical research as jet and rocket fuels, catalysts, corrosion inhibitor-fluxing agents, and oxygen scavengers; as a chemical intermediate. It is hazardous to the central nervous system, eyes, and skin; and is toxic by inhalation, absorption, ingestion, and contact. Symptoms of exposure include dizziness, headache, drowsiness, lightheadedness, incoordination, tremors, spasms of the face, neck, and limbs, convulsions, irritation of the eyes and skin. OSHA exposure limit (TWA): 0.005 ppm [air] or 0.01 mg/m³.

pentachloronaphthalene — $C_{10}H_3Cl_5$ — a pale yellow or white solid or powder with an aromatic odor; MW: 300.4, BP: 636°F, Sol: insoluble, Fl.P: 180°F, sp gr: 1.73. It is used as an insulating material; as an inert compound of resins or polymers for coating or impregnating textiles, wood, and paper to impart flame resistance, waterproofness, and insecticidal properties; as an additive for cutting oil in various operations preformed on metals; as an additive

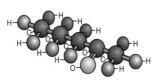

2–Pentanone

to special lubricants. It is hazardous to the skin, liver, and central nervous system; and is toxic by inhalation, absorption, ingestion, and contact. Symptoms of exposure include headache, fatigue, vertigo, anorexia, pruritus, acne-form skin eruptions, jaundice, liver necrosis. OSHA exposure limit (TWA): 0.5 mg/m³ [skin].

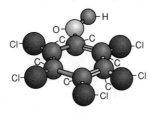

Pentachlorophenol

pentachlorophenol (PCP) — C_6Cl_5OH — a colorless to white, crystalline solid with a benzene-like odor; MW: 266.4, BP: 588°F, Sol: 0.001%, sp gr: 1.98. It is used as a preservative for wood, starch, paint, adhesives, leather, latex, and oils; in slime-algae control; as a pesticide, herbicide, and snail control agent. It is hazardous to the cardiovascular system, respiratory system, eyes, liver, kidneys, skin, and central nervous system; and is toxic by inhalation, absorption, ingestion, and contact. Symptoms of exposure include sneezing, cough, weakness, anorexia, weight loss, sweating, headache, dizziness, nausea, vomiting, dyspnea, chest pain, high fever, dermatitis, irritation of the eyes, nose, and throat. OSHA exposure limit (TWA): 0.5 mg/m³ [skin].

n-pentane — a colorless liquid with a gasoline-like odor; a gas above 97°F; MW: 72.2, BP: 97°F, Sol: 0.04%, Fl.P: –57°F, sp gr: 0.63. It is used as a gasoline additive; in the synthesis of amyl chlorides for intermediates; in the synthesis of polychlorocylopentanes as intermediates; as a solvent; as a component of lighter fluids and blow-torch fuel; as a heat-exchange medium in the manufacture of artificial ice. It is hazardous to the skin, respiratory system, and eyes; and is toxic by inhalation, ingestion, and contact. Symptoms of exposure include drowsiness, dermatitis, chemical pneumonia, irritation of the eyes and nose. OSHA exposure limit (TWA): 600 ppm [air] or 1,800 mg/m³.

2-pentanone — $CH_3COCH_2CH_2CH_3$ — a colorless to water-white liquid with a characteristic acetone-like odor; MW: 86.1, BP: 215°F, Sol: 6%, Fl.P: 45°F, sp gr: 0.81. It is used in the manufacture of pharmaceuticals and some flavorings; as an extractant in dewaxing petroleum products. It is hazardous to the respiratory system, eyes, skin, and central nervous system; and is toxic by inhalation, ingestion, and contact. Symptoms of exposure include headache, dermatitis, narcosis, coma, irritation of the eyes and mucous membrane. OSHA exposure limit (TWA): 200 ppm [air] or 700 mg/m³.

pepsin — (*gastric juices*) a digestive enzyme which changes proteins to peptones and proteoses in mammals, birds, reptiles, and fish.

peptization — the breaking down of materials in the particles of colloidal size by chemical action.

peptone — a stage in protein digestion prior to the formation of amino acids.

percent (pct; %) — 1 part per 100 parts.

percent by weight — a measurement of the amount of a substance in a mixture based on its weight compared to the total weight of the mixture.

percent volatile — a measurement of the percentage of a liquid or solid by volume that will evaporate at an ambient temperature of 70°F.

perched water table — the water table of a discontinuous saturated zone in a soil.

perchloromethyl mercaptan — Cl_3CSCl — a pale yellow, oily liquid with an unbearable acrid odor; MW: 185.9, BP: 297°F (decomposes), Sol: insoluble, sp gr: 1.69. It is used in chemical synthesis for agricultural and dye chemicals; as an organic intermediate. It is hazardous to the eyes, respiratory system, liver, kidneys, and skin; and is toxic by inhalation, absorption, ingestion, and contact. Symptoms of exposure include lacrimation, eye inflammation, cough, dyspnea, deep breathing pain, coarse râles, vomiting, pallor, tachycardia, acidosis, anuria, irritation of the nose and throat. OSHA exposure limit (TWA): 0.1 ppm [air] or 0.8 mg/m³.

perchloryl fluoride — $ClFO_3$ — a colorless gas with a characteristic, sweet odor; shipped as a liquefied compressed gas; MW: 102.5, BP: –52°F, Sol: 0.06%. It is used as a liquid oxidizer in rocket fuels; in chemical synthesis; in flame photometry for analysis of different elements. It is hazardous to the respiratory system, skin, and blood; and is toxic by inhalation and contact. Symptoms of exposure include respiratory irritation, skin burns; in animals: methemoglobin, anoxia, cyanosis, weakness, dizziness, headache,

O P

pulmonary edema, pneumonitis. OSHA exposure limit (TWA): 3 ppm [air] or 14 mg/m^3.

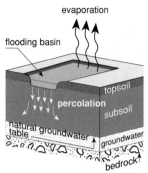

Percolation

percolation — the flow or trickling of a liquid downward into contact of a filtering medium; (*soil*) the flow of water down through the soil.

percutaneous absorption — the absorption of a substance into the skin.

percutaneously — pertaining to entry through the skin.

perennial stream — a stream that flows throughout the year.

perfect gas — *See* ideal gas.

perforate — to puncture or pierce.

pericardial cavity — the area of the body in which the heart lies.

pericardial fluid — the clear fluid contained in the pericardial cavity.

pericardium — the membranous sack surrounding the heart.

perinatal — pertaining to an event happening just before, during, or immediately after birth.

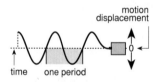

Period

period — the time for one complete cycle, vibration, or oscillation; the time required for a single wave to pass a given point.

periodic table — the systematic classification of the elements according to atomic numbers or atomic weights and by physical and chemical properties.

peripheral nervous system (PNS) — nerve cells lying outside the skull and the vertebral column, including the somatic nervous system and the autonomic nervous system.

perishable food — any food of such type or in such condition as it may spoil unless hermetically sealed in containers processed by heat to

prevent spoilage; properly packed, dehydrated; or dried to so low a moisture content as to preclude the development of microorganisms.

peristalsis — a series of circular contractions by which the digestive tract propels its contents.

peristaltic — of or pertaining to peristalsis.

peritoneal fluid — fluid contained in the membrane lining of the abdominal cavity.

peritoneum — the membrane lining the abdominal cavity.

peritonitis — an inflammation of the peritoneum.

permanent injunction — a decree that when signed by the court perpetually restrains the defendant from engaging in specified violative practices and remains in force until final termination.

permanent partial disability — a condition under Worker's Compensation Insurance which actually or presumptively results in a partial loss of earning power.

permanent total disability — a condition under Worker's Compensation Insurance which actually or presumptively is considered the equivalent of a complete and permanent loss of earning power.

permeability — the ease at which liquids move through a soil allowing substances to pass through certain membranes; the capacity of a porous medium to conduct or transmit fluids.

permeation — the process by which a chemical moves through protective clothing on a molecular level; involves sorption of a chemical into the contacted material, diffusion of the chemical molecules within the material, and desorption from the opposite surface of the material; (*chemistry*) the movement of atoms, molecules, or ions into or through a porous or permeable substance.

permethrin — (*insecticides*) a synthetic pyrethroid used for control of adult mosquitos in residential and recreational areas.

permissible dose — the amount of radiation which may be safely received by an individual within a specified period; formerly known as tolerance level.

permissible exposure limit (PEL) — an Occupational Safety and Health Administration standard designating the maximum occupational exposure permitted as an 8-hour time-weighted average.

permit — a means of registering a source of air, water, or land pollution.

permittivity ($\in$) — the ability of a dielectric to store electrical energy.

peroxyacylnitrates (PAN) — any of a group of organic compounds that can come from several sources and are formed by the chemical reactions

O
P

of other pollutants; they proceed more rapidly in intense sunlight and are extremely reactive, oxidizing agents; causes eye irritation and respiratory distress.

persist — to stay or to remain.

persistent — the length of time a pesticide remains in the soil or on the crops after it has been applied and degraded only slowly by the environment.

personal breathing zone (PBZ) — the region of air inhaled by a worker in the normal execution of duties; samples are collected by a device secured to the worker's lapel.

personal noise dosimeter — that portion of the personal noise dosimeter set which must actually be worn by the worker, including the microphone.

personal protection equipment — devices worn to protect against environmental hazards.

personnel monitoring — the monitoring of radioactive contamination based on measurements from any part of an individual, including breath, excretions, or clothing. The monitoring of hazardous chemicals in the breathing zone of a person.

person-rem — (*radiation*) the product of the average individual dose in a population times the number of individuals in the population.

person-year — the activity or exposure of an individual during the course of an entire year; used in epidemiological study design.

perspiration — the excretion of moisture through the pores of the skin. Also known as sweat.

PERT — *See* program evaluation and review technique.

perturbation — a disturbance of motion, course, arrangement, or state of equilibrium.

perturbed electric field — modification of electric field by the presence of a conducting object.

pest — an unwanted plant or animal.

pesticide — any agent that destroys pests.

pesticide formulation — the substance or mixture of substances comprised of all active and inert ingredients (if any) of a pesticide product.

pesticide solid waste — the residue resulting from the manufacturing, handling, or use of chemical pesticides.

pesticide tolerance — the amount of a pesticide chemical that can safely remain on or in food for human consumption or for consumption by livestock.

pet — a domesticated dog, cat, fish, bird, or snake.

petri dish — a flat-bottom, transparent dish and cover used in culturing microorganisms and tissues.

petrochemicals — chemicals isolated or derived from petroleum or natural gas.

petroleum — a complex naturally-occurring green to black mixture of gaseous, liquid, and solid hydrocarbons; commercial petroleum products are obtained from a crude petroleum distillation cracking chemical treatment.

petroleum distillate (naphtha) — a colorless liquid with a gasoline- or kerosene-like odor; a mixture of paraffins that may contain a small amount of aromatic hydrocarbons; MW: 99 (approx), BP: 86–460°F, Sol: insoluble, Fl.P: –40 to –86°F, sp gr: 0.63–0.66. It is used as solvents in the rubber industry during the manufacture of water-proof cloth, shoe adhesives, and rubber tires; as extractants; in the preparation of paint, varnish, and lacquer as solvents, diluents, and thinners; as solvents in pesticides; during dry cleaning operations. It is hazardous to the skin, eyes, respiratory system, and central nervous system; and is toxic by inhalation, ingestion, and contact. Symptoms of exposure include dizziness, drowsiness, headache, nausea, dry and cracked skin, irritation of the eyes, nose, and throat. OSHA exposure limit (TWA): 400 ppm [air] or 1,600 mg/m^3.

PF — *See* protection factor.

PFE — acronym for pneumatic fracturing extraction.

pg — *See* picogram.

phagocyte — a cell that engulfs foreign material and consumes debris and foreign bodies.

phagocytic cell — a cell that engulfs bacteria and digests them with enzymes, including lysozymes.

O
P

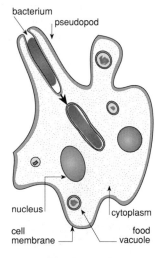

Phagocytosis

phagocytosis — the engulfing of microorganisms or other cells and foreign particles by phagocytes.

Phalanges

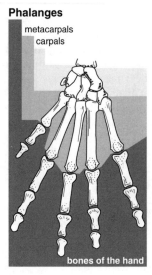

bones of the hand

phalanges — the bones of the fingers and toes in most vertebrates.

Pharmaceutical Manufacturer's Association (PMA) — a trade association made up of producers of ethical pharmaceuticals and biological products sold under their own labels.

pharmaceuticals — the drugs and related chemicals produced by companies for medicinal purposes.

pharmacokinetics — the science of the quantitative analysis between an organism and a drug; the exposure to a drug or chemical which can lead to a variety of chemical and biological reactions in the body. It includes the absorption into the body, metabolism into other substances, distribution to other organs and tissues, or removal from the body. It also includes the various biochemical pathways in the body that activate, metabolize, detoxify, transport, and excrete chemicals, and the relationship between the administered doses and the effective doses.

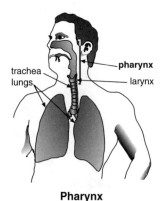

Pharynx

pharynx — the muscular throat cavity extending up over the soft palate and to the nasal cavity in vertebrates.

phase — the measure of the progression of a cyclic wave or motion form in time or space from a chosen instant or position.

phase contrast microscope (PCM) — a magnifying instrument which uses the differences in phases of light passing through or reflected by the object being examined.

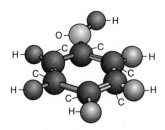

Phenanthrene

phenanthrene — a polycyclic aromatic hydrocarbon formed during the incomplete combustion of coal, oil, gas, garbage or other organic substances and found especially at hazardous waste sites. See polycyclic aromatic hydrocarbons.

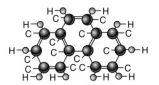

Phenol

phenol — C_6H_5OH — a colorless to light pink, crystalline solid with a sweet, acrid odor; liquefies by mixing with about 8% water; MW: 94.1, BP: 359°F, Sol (77°F): 9%, Fl.P: 175°F, sp gr: 1.06. It is used in plywood manufacture; in the manufacture of molded articles; in industrial coatings; in the synthesis of thermosetting phenolic resins, epoxy, polycarbonate, phenoxy, and polysulfone; in the synthesis of aprolactam; in the synthesis of agricultural chemicals and intermediates; in the synthesis of pharmaceuticals; in the synthesis of stabilizers and preservatives for dyes, perfumes, and fungicides; in the synthesis of intermediates in polyester production; in the synthesis of surface-active agents and detergent intermediates; in the synthesis of explosives; in the manufacture of disinfectant agents and products; in the synthesis of synthetic cresols and xyenols. It is hazardous to the liver, kidneys, and skin; and is toxic by inhalation, absorption, ingestion, and contact. Symptoms of exposure include anorexia, weight loss, weakness, muscle aches, pain, dark urine, cyanosis, liver and kidney damage, skin burns, dermatitis, ochronosis, tremors, convulsions, twitches, irritation of the eyes, nose, and throat. OSHA exposure limit (TWA): 5 ppm [skin] or 19 mg/m^3.

phenolic resins — a class of resins produced as the condensation product of phenol, formaldehyde, or other aldehydes.

phenolphthalein — $(C_6H_4OH)_2COC_6H_4CO$ — pale yellow crystals used as an organic indicator in volumetric analysis; colorless in acid solutions below pH 8 and red in the presence of ^-OH ions above pH 9.6.

phenol-red — an organic dye used in swimming pool testing which is yellow at a pH of 6.8 and turns progressively deeper red as the pH increases to 8.4.

phenotype — the visible properties of an organism that are produced by the interaction of the genotype and the environment.

p-phenylene diamine — $C_6H_4(NH_2)_2$ — a white to slightly red crystalline solid; MW: 108.2, BP: 513°F, Sol: 5%, Fl.P: 312°F. It is used in dye and dyestuff intermediates for fur, hair, leather, cotton, and synthetics; in photographic application as a fine-grain developing agent, for preparation of color developers, and for photochemical measurements; in polymer technology; as a catalyst in the preparation of epoxy resins, synthetic fibers, heat-resistant polymers, and coatings for leather, paper, and textiles; in rubber technology for natural and synthetic rubber for vulcanization; in organic synthesis; as an analytical reagent, and as a special solvent. It is hazardous to the respiratory system and skin; and is toxic by inhalation, absorption, ingestion, and contact. Symptoms of exposure include sensitization dermatitis, bronchial asthma, irritation of the pharynx and larynx. OSHA exposure limit (TWA): 0.1 mg/m^3 [skin].

phenyl ether (vapor) — $C_6H_5OC_6H_5$ — a colorless, crystalline, solid or liquid (above 82°F) with a geranium-like odor; MW: 170.2, BP: 498°F, Sol: insoluble, Fl.P: 234°F, sp gr: 1.08. It is used as a high-temperature, heat-transfer medium; as an intermediate in organic synthesis; as an odorant in the manufacture of soaps and perfumes; as a flavoring agent in the manufacture of food products. It is hazardous to the eyes, skin, and respiratory system; and is toxic by inhalation, ingestion, and contact. Symptoms of exposure include nausea, irritation of the eyes, nose, and skin. OSHA exposure limit (TWA): 1 ppm [air] or 7 mg/m^3.

phenyl ether-biphenyl mixture (vapor) — $C_6H_5OC_6H_5/C_6H_5C_6H_5$ — a colorless to straw-colored liquid or solid (below 54°) with a disagreeable, aromatic odor; a mixture typically contains 75% phenyl ether and 25% biphenyl; MW: 166 (approx), BP: 495°F, Sol: insoluble, Fl.P (open cup): 255°F, sp gr (77°F): 1.06. It is used in the chemical, petroleum, and nuclear industries as a high-transfer medium used in heating and cooling; as a dye carrier for printing and dyeing polyester textiles. It is hazardous to the eyes, skin, and respiratory system; and is toxic by inhalation, ingestion, and contact. Symptoms of exposure include nausea, irritation of the eyes, nose, and skin. OSHA exposure limit (TWA): 1 ppm [air] or 7 mg/m^3.

phenyl glycidyl ether — $C_9H_{10}O_2$ — a colorless liquid; a solid below 38°F; MW: 150.1, BP: 473°F, Sol: 0.2%, Fl.P: 248°F, sp gr: 1.11. It is used as a coupling agent, catalyst, or reactive diluent in curing rubber, thermostable epoxy resins, tire cord, and electric insulating material; as a stabilizer of halogenated compounds; in the treatment of fabric; as a copolymer in the production of epoxy polymers; as a chemical intermediate. It is hazardous to the skin, eyes, and central nervous system; and is toxic by inhalation, ingestion, and contact. Symptoms of exposure include narcosis, irritation of the skin, eyes, and upper respiratory system; carcinogenic. OSHA exposure limit (TWA): 1 ppm [air] or 6 mg/m^3.

phenylhydrazine — $C_6H_5NHNH_2$ — a colorless to pale yellow liquid or solid (below 67°F) with a faint, aromatic odor; MW: 108.1, BP: 470°F (decomposes), Sol: slight, Fl.P: 190°F, sp gr: 1.10. It is used in the manufacture of pharmaceuticals, photographic developers, and polymethene dyes; as an analytical reagent, intermediate in organic synthesis, or reducing agent. It is hazardous to the blood, respiratory system, liver, kidneys, and skin; and is toxic by inhalation, absorption, ingestion, and contact. Symptoms of exposure include skin sensitization, hemolytic anemia, dyspnea, cyanosis, jaundice, kidney damage, vascular thrombosis; carcinogenic. OSHA exposure limit (TWA): 5 ppm [skin] or 20 mg/m^3.

pheromone — a chemical substance produced by a body and acting as a stimulus to individuals of the same species to elicit a behavioral response.

phon — unit of loudness level.

phosdrin — $C_7H_{13}O_4PO_2$ — a pale yellow to orange liquid with a weak odor; MW: 224.2, BP: decomposes, Sol: miscible, Fl.P (open cup): 347°F, sp gr: 1.25. It is used as an insecticide on agricultural crops, fruits, and vegetables, and in sewage treatment plants. It is hazardous to the respiratory system, central nervous system, cardiovascular system, skin, and blood cholinesterase;

O
P

and is toxic by inhalation, absorption, ingestion, and contact. Symptoms of exposure include miosis, rhinorrhea, headache, tight chest, wheezing, laryngeal spasms, salivation, cyanosis, anorexia, nausea, vomiting, abdominal cramps, diarrhea, paralysis, ataxia, convulsions, low blood pressure, cardiac irregularities, irritation of the skin and eyes. OSHA exposure limit (TWA): 0.01 ppm [skin] or 0.1 mg/m^3.

phosgene — $COCl_2$ — a colorless gas with a suffocating odor like musty hay; a fuming liquid below 47°F; shipped as a liquefied compressed gas; MW: 98.9, BP: 47°F, Sol: slight, sp gr: 1.43. It is used in sand bleaching operations in glass manufacture; in the synthesis of isocyanates in the manufacture of polyurethane plastics; liberated during fire-extinguishing with agents containing chlorinated hydrocarbons. It is hazardous to the respiratory system, skin, and eyes; and is toxic by inhalation, absorption, ingestion, and contact. Symptoms of exposure include eye irritation, dry and burning throat, vomiting, cough, foamy sputum, dyspnea, chest pain, cyanosis, skin burns. OSHA exposure limit (TWA): 0.1 ppm [air] or 0.4 mg/m^3.

phosphatase — any of a group of enzymes capable of catalyzing the hydrolysis of esterified phosphoric acid with liberation of inorganic phosphate; found in practically all tissues, body fluids, and cells, including erythrocytes and leukocytes.

phosphatase test — a test used to determine the efficiency of pasteurization and quality of milk.

phosphate — PO_4^{-3} — a generic term for any compound containing a phosphate group.

phosphine — PH_3 — a colorless gas with a fish- or garlic-like odor; shipped as a liquefied compressed gas; pure compound is odorless; MW: 34.0, BP: –126°F, Sol: slight. It is used as a doping agent in the manufacture of solid-state components for electronic circuits and in the manufacture of lasers; in preparations for textile treatments; in organic intermediates; as a fumigant for stored grain; liberated during the use of acetylene; in the manufacture of safety matches and pyrotechnics. It is hazardous to the respiratory system; and is toxic by inhalation. Symptoms of exposure include nausea, vomiting, abdominal pain, diarrhea, thirst, chest pressure, dyspnea, muscle pain, chills, stupor, or syncope. OSHA exposure limit (TWA): 0.3 ppm [air] or 0.4 mg/m^3.

phosphorescence — the capacity of a substance to glow for more than 0.1 nanoseconds after removal of a light source.

phosphoric acid — H_3PO_4 — a thick, colorless, odorless, crystalline solid often used in an aqueous solution; MW: 98.0, BP: 415°F, Sol: miscible, sp gr (77°F): 1.87. It is used in the manufacture of aluminum products; in the synthesis of intermediates in the manufacture of soil fertilizers; in the manufacture of poultry and livestock feed; during synthesis of detergent and soap builders and water-treatment chemicals; as an acidulant and flavor agent in the manufacture of carbonated beverages and jellies and preserves; in the manufacture of cleaning preparations and disinfectants; as a bonding agent in the manufacture of refractory bricks; during lithography and photoengraving operations; as a catalyst in the synthesis of other chemicals; in the synthesis of textile and leather processing chemicals, clays, ceramics, cements, and clay-thinning agents; in the synthesis of pharmaceuticals; as a laboratory reagent; during the manufacture of opal glass; in the manufacture of dental cements and dentrifice adhesives; in the manufacture of electric lights. It is hazardous to the respiratory system, skin, and eyes; and is toxic by inhalation, ingestion, and contact. Symptoms of exposure include dermatitis, irritation of the upper respiratory tract, eyes, and skin, and burns of the skin and eyes. OSHA exposure limit (TWA): 1 mg/m^3 [air].

phosphorus (yellow) — P_4 — a white to yellow, soft, waxy allotrope with acrid fumes in air; usually shipped or stored in water; MW: 124.0, BP: 536°F, Sol: 0.0003%, Fl.P: not known, sp gr: 1.82. It is used in the synthesis of high purity phosphoric acid salts for use as fertilizers, water treatment chemicals, food products, beverages, and dentrifices; in the synthesis of inorganic phosphorous compounds for use as pesticides, flame retardants, and gasoline and lube oil additives; in the synthesis of inorganic and organic compounds; in the manufacture of phosphorus alloys; in the manufacture of explosives, munitions, and pyrotechnics, for military use; as an ingredient of rat poison and roach powders. It is hazardous to the respiratory system, liver, kidneys, jaw, teeth, blood, eyes, and skin; and is toxic by inhalation, ingestion, and contact. Symptoms of exposure include abdominal pain, nausea, jaundice, anemia, cachexia, dental pain, excess salivation, jaw pain, swelling, skin and eye burns, irritation of the eyes and respiratory tract. OSHA exposure limit (TWA): 0.1 mg/m^3 [air].

phosphorus cycle — the circulation of phosphorus in the biosphere; it is found in sediment and bones and is taken up by plants, is consumed by animals, is used to produce and fortify teeth and bones, and returns to the soil upon the animals' deaths.

phosphorus pentachloride — PCl_5 — a white to pale yellow, crystalline solid with a pungent, unpleasant odor; MW: 208.3, BP: sublimes, Sol: reacts. It is used as a chlorinating agent in organic synthesis or chemical reactant; as a catalyst in organic synthesis in the production of polyethylene from ethylene. It is hazardous to the respiratory system, skin, and eyes; and is toxic by inhalation, ingestion, and contact. Symptoms of exposure include bronchitis, dermatitis, irritation of the eyes and respiratory system. OSHA exposure limit (TWA): 1 mg/m^3 [air].

phosphorus pentasulfide — P_2S_5/P_4S_{10} — a greenish-gray to yellow, crystalline solid with an odor of rotten eggs; MW: 222.3/444.6, BP: 957°F, Sol: reacts, Fl.P: not known, sp gr: 2.09. It is used in the preparation of lubrication oil additives for low-lead gas and reduction of air pollution. It is hazardous to the respiratory system, central nervous system, skin, and eyes; and is toxic by inhalation, ingestion, and contact. Symptoms of exposure include apnea, coma, convulsions, conjunctivitis pain, lacrimation, photosensitivity, kerato-conjunctivitis, corneal vesiculation, irritation of the respiratory system, dizziness, headache, fatigue, irritability, insomnia, gastrointestinal distress. OSHA exposure limit (TWA): 1 mg/m^3 [air].

phosphorus trichloride — PCl_3 — a colorless to yellow, fuming liquid with an odor like hydrochloric acid; MW: 137.4, BP: 169°F, Sol: reacts, sp gr: 1.58. It is used during the synthesis of plasticizers and intermediates; in the produc-

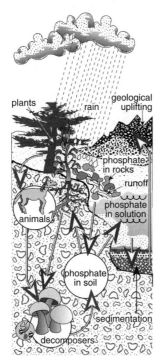

Phosphorus Cycle

tion of intermediates and during chemical synthesis of dyes, pharmaceuticals, other chlorinating agents, and other organic chemicals; in the treatment of polypropylene before drying in the manufacture of knitted fabrics. It is hazardous to the respiratory system, skin, and eyes; and is toxic by inhalation, ingestion, and contact. Symptoms of exposure include pulmonary edema, eye and nose burns, irritation of the eyes, nose, and throat. OSHA exposure limit (TWA): 0.2 ppm [air] or 1.5 mg/m^3.

photobiologic effect — a photosensitive effect associated with specific chlorinated hydrocarbons, particularly the chlorobenzols, the diphenyls, and the triphenyls.

photochemical — influenced or initiated by light.

photochemical oxidant (OX) — any of the products of secondary atmospheric reactions which occur during smog.

photochemical process — chemical changes caused by the radiant energy of the sun acting upon various polluting substances to produce photochemical smog.

photochemical reaction — a chemical reaction between different air pollutants involving absorption or emission of radiation, particularly ultra-violet light.

photochemical smog — a mixture of secondary air pollutants including ozone, organic nitrates, and others produced by photochemical reactions from primary pollutants, such as nitrogen oxides and hydrocarbons.

photoconductive cell — a photoelectric cell in which the electrical resistance varies inversely with the intensity of light that strikes its active material.

photodegradable — a chemically-degradable action due to exposure to light.

photodisintegration — a nuclear reaction involving the interaction of a gamma ray with a nucleus of an atom.

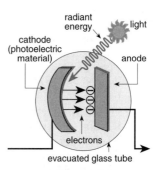

Photoelectric Effect
in a Photocell

photoelectric effect — the ejection of electrons in a substance by incident electromagnetic radiation.

O
P

photographic dosimeter — a photographic film or badge used in personal radiation monitoring to measure the cumulative dosage of radiation to which an individual is exposed.

photolysis — the degradation of a chemical caused by exposure to light or ultraviolet radiation.

photometer — an instrument for measuring luminous intensity, luminous flux, illumination, or brightness; a direct-reading physical instrument using the relative radiant power of a beam of radiant energy measured as it passes through a gas-air measure, or the suspension of solid or liquid particles.

photometry — the branch of physics dealing with measurements of the intensity of light, light distribution, elimination, and luminous flux.

photon — an invisible unit of electromagnetic radiation having zero rest mass, yet traveling at the speed of light; the energy (E) of each photon in radiation of frequency (v) is given by formula E = hv, where h is Planck's constant. Also known as quantum.

photophobia — the abnormal sensitivity to and discomfort from light.

photosensitivity — the capacity of an organ, organism, or certain chemicals in plants to be stimulated to activity by light or to react to light.

photosensitize — to make sensitive to the influence of radiant energy, especially light.

photosensitizer — a substance that increases the response of a material to electromagnetic irradiation.

photosynthesis — the biochemical process by which green plants and some bacteria in the presence of chlorophyll use light energy to produce chemical bonds resulting in the consumption of carbon dioxide and water, and the production of oxygen and simple sugars; symbolized by the equation $CO_2 + 2H_2O \rightarrow [CH_2O] + H_2O + O_2$. Virtually all atmospheric oxygen originates from oxygen released during photosynthesis.

photovoltaic cell — an energy-conservation device that captures solar energy and converts it directly into electrical energy. Also known as photovoltaic solar cell.

photovoltaic solar cell — *See* photovoltaic cell.

phthalic anhydride — $C_6H_4(CO)_2O$ — a white solid with a characteristic, acrid odor; MW: 148.1, BP: 563°F, Sol: 0.6%, Fl.P: 305°F, sp gr: 1.53. It is used in the manufacture of plasticizers; in the manufacture of unsaturated polyester resins for use in structural building parts, swimming pools, automotive parts, and luggage; in the synthesis of dyes; in the manufacture of chemicals and chemical intermediates; in the manufacture of pharmaceuticals and pharmaceutical intermediates; in the manufacture of metallic and acid salts; in the manufacture of epoxy resins; in the manufacture of fire retardants. It is hazardous to the respiratory system, skin, eyes, liver, and kidneys; and is toxic by inhalation, ingestion, and contact. Symptoms of exposure include conjunctivitis, nasal ulcer bleeding, upper respiratory irritation, bronchitis, bronchial asthma, dermatitis. OSHA exposure limit (TWA): 1 ppm [air] or 6 mg/m³.

pH value — measure of the acidity or alkalinity of a solution on a scale from 0–14 with 7 being neutral, less than 7 being acid, and greater than 7 being alkaline.

physical change — a change that does not produce a new substance.

physical geology — the overall study of the earth, including its composition and the physical changes occurring in it.

physical hazard — (*hazardous chemicals*) a chemical classified or described as a combustible liquid, compressed gas, explosive, flammable, organic peroxide, oxidizer, pyrophoric, unstable reactive, or water-reactive.

physics — the study of the laws and phenomena of nature, especially of forces and general properties of matter and energy.

physiological half-life — the rate at which pollutants are eliminated from the body.

physiology — the basic, biomedical science dealing with the vital functions of living organisms such as nutrition, respiration, reproduction, and excretion.

phytoplankton — microscopic, free-floating, autotrophic organisms that function as producers in aquatic ecosystems.

phytotoxic — of or pertaining to a substance or state that is injurious to a plant.

PIC — acronym for product of incomplete combustion.

pica — the consumption of nonfood items; an abnormal appetite as seen in some children and pregnant women.

pickling — (*air pollution/industry*) using various industrial baths for cleaning or processing; commonly concentrated with sulfuric acid. Also known as acid pickling.

pico (p) — a prefix meaning 10^{-12}. Also known as micromicro-.

picocurie (pCi) — a measurement equal to one-millionth of a microcurie.

picocurie per liter — pCi/l — a measurement equal to 2.2 radioactive disintegrations per minute in one liter of air.

picogram (pg) — a measurement equal to 10^{-12} grams.

O
P

picomole (pmol) — a unit of quantity equal to 10^{-12} mole.

picornavirus — an extremely small, ether-resistant RNA virus.

picric acid — $C_6H_2OH(NO_2)_3$ — a yellow, odorless solid; usually used as an aqueous solution; MW: 229.1, BP: explodes above 572°F, Sol: 1%, Fl.P: 302°F, sp gr: 1.76. It is used in munitions and explosives manufacture; in the synthesis of dye and dye intermediates in the textile industry; in the manufacture of medicinals; in the manufacture of pyrotechnics and compounds for pyrotechnics as color intensifiers and as oxidizers in matches; as a chemical reagent; in the manufacture of colored glass; in the manufacture of electric batteries. It is hazardous to the kidneys, liver, blood, skin, and eyes; and is toxic by inhalation, absorption, and contact. Symptoms of exposure include sensitization dermatitis, eye irritation, yellow-stained hair and skin, weakness, myalgia, anuria, polyuria, bitter taste, gastrointestinal disturbance, hepatitis, hematuria, albuminuria, nephritis. OSHA exposure limit (TWA): 0.1 mg/m³ [skin].

PID — acronym for photo ionization detector.

piezoelectric — the property of certain crystals having the ability to produce a voltage when subjected to a mechanical stress, such as bending, or to produce a mechanical force when voltage is applied.

piezoelectricity — the electricity of certain asymmetric crystalline materials that relates the electric field to the mechanical strain.

piezometer — an instrument used to measure pressure or compressibility of a fluid; an instrument for measuring the change of pressure of a material subjected to hydrostatic pressure.

pig — a lead container used to ship radioactive material.

pigment — a chemical that has color because it reflects light of only certain wavelengths; any coloring matter in plant or animal cells.

pile — (*nuclear physics*) a nuclear reactor; (*solid waste*) any non-containerized accumulation of solid, nonflowing hazardous waste that is used for treatment or storage.

pilot program — a program that is initiated on a limited basis for the purpose of facilitating a future full-scale program.

pindone — $C_9H_5O_2C(O)C(CH_3)_3$ — a bright yellow powder with almost no odor; MW: 230.3, BP: decomposes, Sol (77°F): 0.002%, Fl.P: not known, sp gr: 1.06. It is used in pesticide formulations; as an intermediate in pharmaceutical synthesis. It is hazardous to the blood prothrombin; and is toxic by inhalation. Symptoms of exposure include epistaxis, excess bleeding from minor cuts and bruises, smoky urine, black tarry stools, pain of the abdomen and back. OSHA exposure limit (TWA): 0.1 mg/m³ [air].

pint (pt) — a unit of volume equivalent to 2 cups or 1/8 gallon.

pinworm — the nematode that causes enterobiasis. Also known as human threadworm; seatworm.

pipe lagging — asbestos containing material used to insulate pipes carrying heated or refrigerated liquids or vapors.

PIR — *See* proportional incidence ratio.

piscicide — a pesticide specifically designed to kill unwanted fish.

pitch — a characteristic of auditory sensation in which noises are ordered on a scale which extends from low to high, depending on the frequency of the sound's stimulus, sound pressure, and wave form of the stimulant.

pitless installation device — an assembly of parts which permits water to pass through the casing or extension thereof, provides access to the well and

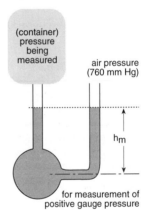

for measurement of
positive gauge pressure

h_m=variation in distance
between center of
pressure vessel and
liquid height

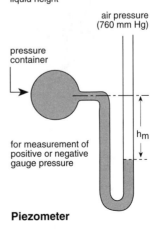

Piezometer

to the parts of the water system within the well, and provides for the transportation of the water and the protection and water therein from surface or near-surface contaminants.

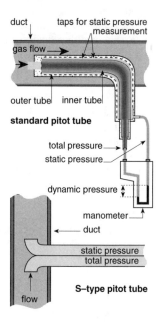

Pitot Tube

pitot tube — an instrument for measuring the total static and dynamic pressure of a fluid stream consisting of a tube of small bore connected at one end to a manometer, the other end being open and pointing upstream; used to determine velocity pressure. Also known as impact tube.

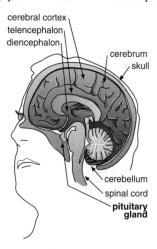

Pituitary Gland

pituitary gland — an endocrine gland in the head of higher vertebrates that helps coordinate the work of other endocrine glands. Also known as hypophysis.

pival — $C_{14}H_{14}O_3$ — a yellow powder with a mild, moldy, acrid odor suggesting marigolds; used as an anticoagulant; causes internal bleeding in rats or mice and therefore prevents developing an avoidance of the chemical. Also known as tertiary butyl valone.

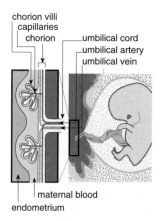

Pixel

pixel (pel) — (*mapping*) a contraction for picture element; the smallest unit of information in a grid, cell, map, or scanner image.

placard — a legal notice posted in a public place.

placebo — a drug or treatment that has no known pharmacological effect on disease but works because the patient believes in its efficacy.

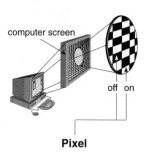

Placenta

placenta — a large, thin membrane in the uterus of most mammals which transports substances between the mother and developing fetus by means of the umbilical cord.

plague — (*disease*) an infectious bacterial disease of rodents and humans caused by *Yersinia pestis* or *Pasterella pestis*. It is transmitted to humans by the bite of an infected flea (*Xenopsylla cheopis*) or by inhalation. Symptoms of exposure include inflammation and tenderness of the lymph nodes, fever, septicemia, and pneumonia. Incubation time is 2–6 days for plague, and 1–6 days for plague pneumonia.

Found in various parts of the world, including the wild rodent population of the Western United States. The reservoir of infection is wild rodents. It is not usually transmitted from person to person, but pneumonic plague is highly communicable; general susceptibility. It is controlled by checking the rodent population for fleas, rat control on ships and docks, and immunization where necessary. Also known as black death; bubonic plague.

plaintiff — the party initiating a suit or action in a civil case; the opposite of the defendant.

plan — a detailed formulation of a program of action; a design of desired future states.

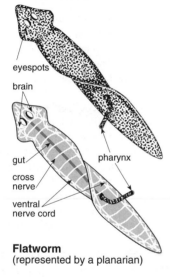

Flatworm
(represented by a planarian)

planaria — non-parasitic flatworm.

Planck's constant (h) — a universal constant having the value 6.626076×10^{-34} Js. *See also* Planck's law.

Planck's law — the law that forms the basis of quantum theory. The energy of electromagnetic radiation is confined to small indivisible packets or photons, each of which has an energy hf where h is the Planck constant and f is the frequency of radiation.

Xenopsylla cheopis
(Rat Flea)
a vector for bubonic plague

Plague

plankton — microscopic plants and animals that drift in water, usually because of currents and tides.

plant regulator — any substance or mixture of substances which through physiological action accelerates or retards the rate of growth or rate of maturation, or otherwise alters the behavior of plants; does not include plant nutrients, trace elements, nutritional chemicals, plant inoculants, and soil amendments.

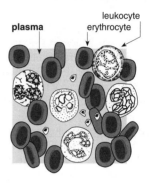

Plasma

plasma — the fluid portion of the lymph or blood containing approximately 92% water, 7% proteins, and less than 1% inorganic salts, organic substances other than proteins, dissolved gases, hormones, antibodies, and enzymes.

plasma arc cutting — a process that severs metal by melting a localized area with a constricted arc and removes the molten material with a high-velocity jet of high-temperature ionized gas issuing from the orifice.

plasma arc welding — a process that produces coalescence of metals by heating them with a constricted arc between an electrode and the work piece, or an electrode and the constricting nozzle.

plasma state — a phase of matter consisting of nuclei and free electrons in contrast to liquid, solid, and gas states.

plastic — a generic name for any one of a large and various group of materials which consists of or contains as an essential ingredient an organic substance of large molecular weight, and which, while solid in the finished state at some stage in its manufacture, has been or can be cast, molded, etc., into various shapes by application of heat and/or pressure. Each plastic has its own individual physical, chemical, and electrical properties. Adaptability, uniformity of composition, lightness, and good electrical properties make plastic substances of wide application, though relatively low resistance to heat, strain, and weather are limiting factors in use.

O
P

plasticizers — high-boiling, organic, liquid chemicals used in modifying plastics, synthetic rubber, and other materials to give them special properties such as elongation, flexibility, and toughness.

plastic limit — (*soils*) a soil moisture content at the point of transition between plastic and semi-solid states.

plastic soil — (*soils*) a soil capable of being molded or deformed continuously and permanently by relatively moderate pressure into various shapes.

plate — (*geology*) any of the huge moveable segments into which the earth's crust is divided and which float on or travel on the mantle.

plateau — (*dose-response assessment*) the point in the dose response curve where, once reached, higher levels of the drug will not produce any greater response.

plate count — a bacteriological technique involving the taking of a sample, diluting it, transferring it to nutrient agar in a petri plate, incubating it, and then counting the colonies.

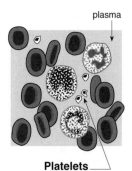

plasma

Platelets

platelet — the smallest of the solid components of the blood releasing thromboplastin in clotting. Also known as thrombocyte.

plate tower — a vertical shell in which are mounted a large number of equally spaced, circular, perforated, sieve plates; the gases or vapors bubble upward through the liquid seal above each plate.

platinum (Pt) {soluble salts} — appearance, odor, and properties vary depending upon the specific soluble salt. It is used as a catalyst in the production of high octane gasoline, nitric and sulfuric acids, vinyl esters, petrochemicals, and pharmaceuticals; in photographic and related industries as a photographic paper sensitizer. It is hazardous to the respiratory system, skin, and eyes; and is toxic by inhalation, ingestion, and contact. Symptoms of exposure include cough, dyspnea, wheezing, cyanosis, skin sensitization,

lymphocytosis, irritation of the eyes and nose. OSHA exposure limit (TWA): 0.002 mg/m^3 [air].

Platy Soil

platy soil — (*soils*) a soil structure in which the aggregates are flat or plate-like with horizontal dimensions greater than vertical; plates overlap causing slow permeability, usually found in subsurface or the lower A horizon of timber and claypan soil.

pleading — (*law*) process performed by the parties to a suit or action, each presenting written statements of their contention and each serving to narrow the field of controversy until there revolves a single point affirmed on one side and denied on the other called the "issue," upon which the parties go to trial.

plenum — a receiving enclosure for gases in which the static pressure at all points is relatively uniform.

Plesiomonas shigelloides — (*disease*) gastroenteritis caused by *Plesiomonas shigelloides*, a gram-negative, rod-shaped bacterium which has been isolated from freshwater, freshwater fish, shellfish, and many animals; is usually a mild self-limiting disease with fever, chills, abdominal pain, nausea, diarrhea, or vomiting. In severe cases, diarrhea may be greenish-yellow, foamy, and blood-tinged. Although this condition cannot yet be considered a definite cause of human disease, the organism has been isolated in the stools of patients with diarrhea, but also sometimes isolated from healthy individuals. The disease occurs primarily in tropical or subtropical areas with rare infections reported in the United States or Europe. The mode of transmission is probably from contaminated water, drinking and recreational, or water used to rinse foods which are uncooked or unheated. The mode of transmission is probably through the oral-fecal route. The incubation period is 20–24 hours. The reservoir of infection is not clear; possibly people and/or animals. There is general susceptibility with children under 15 and people who are immunocompromised.

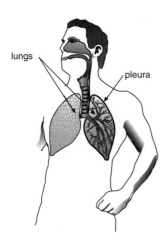

Pleura

pleura — the serous membrane covering the lungs and lining the walls of the thoracic cavity.

pleural fluid — a fluid contained in the membrane covering the lung and lining the chest cavity.

pleurisy — inflammation of the pleura caused by infection, injury, or tumor; a condition where the visceral and parietal pleura lose their lubricating properties resulting in friction and causing pain and irritation.

PLM — acronym for polarized light microscopy.

Plodia interpunctella — *See* Indian Meal Moth.

plumbago — *See* graphite.

plumbing — the practice, materials, and fixtures used in installing, maintaining, and altering all pipes, fixtures, appliances, and appurtenances that connect with sanitary or storm drainage, a venting system, and a public or individual water supply system.

plumb line — a line of cord that has at one end a weight; usually used to determine vertical distance.

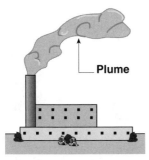

Plume

plume — the column of smoke emitted into the atmosphere through a smoke stack or chimney.

plutonium — Pu — found in five different states; plutonium dioxide {PuO_2}, plutonium nitride

{PuN}, plutonium hexafluoride {PuF_6}, plutonium oxalate {$Pu(C_2O_4)_2 \cdot 6H_2O$}, and as plutonium tetrafluoride {PuF_4}; found as a variety of isotopes, plutonium-236, plutonium-237, plutonium-238, plutonium-239, plutonium-240, plutonium-241, plutonium-242, plutonium-243; plutonium, a silvery-white radioactive metal which exists as a solid under normal conditions; produced when uranium absorbs an atomic particle; small amounts occur naturally though larger amounts have been produced in nuclear reactors; MW: 242.00, BP: 3,232°C, Sol: not known, Fl.P: not known, density at 20°C, 19.84; plutonium dioxide, yellowish-green, MW: 274.00, BP: not known, Sol: not known, Fl.P: not known, density at 20°C, 11.46; plutonium nitride, black in color, MW: 256.01, BP: not known, Sol: in water at 20°C hydrolyzed in cold water, Fl.P: not known, density at 20°C, 14.25; plutonium hexafluoride, reddish brown, MW: 355.99, BP: 62.3°C, Sol: in water at 20°C decomposes in cold water, Fl.P: not known, density at 20°C, no data; plutonium oxalate, yellowish-green solid, MW: 526.13, BP: not known, Sol: in water at 20°C insoluble, Fl.P: not known, density not known; plutonium tetrafluoride, pale brown, MW: 317.99, BP: not known, Sol. in water at 20°C insoluble in water, Fl.P: not known, density at 20°C, 7.0. It is used primarily in the form of plutonium-239 as an ingredient in nuclear weapons; plutonium-238 is a heat source in thermo-electric power devices, such as in satellites and for powering artificial hearts. Adverse health effects and death may occur in occupational settings; in treated animals, in rats, mice, hamsters, dogs, and baboons from a single acute inhalation; mortality in people is uncertain since the people working are extremely healthy and therefore are not a typical population to be compared with; radiation pneumonitis with alveolar edema, fibrosis pulmonary hyperplasia, and metaplasia have been observed in dogs, mice, rats, hamsters, and baboons; biological effects occur in a hematopoietic system of animals; increases in liver enzymes occur in dogs; no conclusive information concerning genotoxic effects although chromosomal aberrations have been observed in rhesus monkeys and Chinese hamsters; causal link between plutonium exposure and cancer has not been demonstrated in workers; in areas surrounding production plants, a somewhat higher incidence than normal for all cancers is observed; radiation induced cancer occurs in experimentally exposed dogs in the lungs; also osteosarcomas are found; gastrointestinal effects are observed in neonatal rats; liver

O P

tumors have been observed in dogs. OSHA exposure limits do not exist; Nuclear Regulator Commission limits in air for accumulative annual dose for the general population from nuclear power plant operations 0.5 rem/yr.

PM — *See* preventive maintenance; particulate matter.

PM10 — the particulate matter ten corrected by a sampler that indicates the particles under ten micrometers in diameter which have the greatest adverse effect on human health.

PMA — *See* Pharmaceutical Manufacturer's Association.

pmol — *See* picomole.

pneumoconiosis — a pneumonia-like disease characterized by the development of fibrotic tissue formation in the alveoli of the lungs caused by dust inhalation. Also known as black lung disease; dust diseases of the lungs.

pneumonia — inflammation of the lung with solidification of a fluid with a high content of protein and cellular debris that has escaped from blood vessels and has been deposited in tissues or on tissue surfaces (called exudate).

pneumonitis — inflammation of lung tissue.

PNS — *See* peripheral nervous system.

pocket dosimeter — a direct-reading portable dosimeter shaped like a pen with a pocket clip used to measure x- and gamma-radiation. It consists basically of a quartz fiber, a scale, a lens to observe the movement of the fiber across the scale, and an ionization chamber; the fiber is charged electrostatically until it reaches zero on the scale; as the fiber is exposed to radiation, some of the air atoms in the chamber become ionized; this allows the static electricity charge to leak from the quartz fiber in direct relationship to the amount of radiation present; the fiber then moves to a new position on the scale indicating the amount of radiation exposure.

podzol — (*soils*) any of a group of zonal soils that develop in a moist climate under coniferous or mixed forest and have an organic mat and a thin organic-mineral layer.

POHC — acronym for principle organic hazardous constituent.

point of contact measurement — a measurement of the contact of a chemical and person over time while the exposure is taking place.

point of supply — location where water is obtained from a specific source.

point of use — the location where water is actually used in a process or incorporated into a product.

point source — a stationary source that emits a given pollutant.

point-source water pollution — the water pollutants from urban areas or from industries and entering a receiving body from a single pipe.

poison — any highly toxic chemical or agent that through ingestion, injection, absorption, or inhalation can cause injury, harm, or destruction to organs, tissue, or life.

poisoning — the morbid condition produced by a poison which may be swallowed, inhaled, injected, or absorbed through the skin.

● hydrogen electron
○ oxygen electron

Polar Molecule

polar — a molecule that has a strong overall charge or polarity.

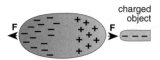

Polarization

polarization — (*electromagnetic waves*) the direction of the electric field vector.

polarized light — light which vibrates in only one plane.

polarography — a method of quantitative or qualitative analysis based on current-voltage curves obtained during electrolysis of a solution with a steadily increasing electromotive force.

polar solvents — solvents, such as alcohols and ketones, that contain oxygen.

policy — a statement of the principles which guide and govern the activities, procedures, and operations of a program.

pollen — the male microspores of seed plants produced by the anther of a flower, or by the male cone of a conifer.

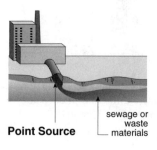

Point Source

sewage or waste materials

pollutant — a polluting agent or medium; a contaminant.

pollution — the presence of a foreign substance — organic, inorganic, radiological, or biological — that tends to degrade the quality of the environment so as to create a health hazard.

Pollution Abstracts — a database covering articles on air, noise and water pollution, environmental quality, pesticides, and solid waste from the years 1970 to the present.

pollution audit — a thorough assessment of a company's product and processes which may be contributing to environmental degradation.

polybrominated biphenyl (PBB) — a chemical substance the composition of which, without regard to impurities, consists of brominated biphenyl molecules having the molecular formula $C_{12}H_xBr_y$.

polychlorinated biphenyl (PCB) — a class of 209 colorless, synthetic liquid compounds with the trade name aroclor. It is used as an insulating fluid in electrical equipment; as a plasticizer; in surface coatings, inks, adhesives, pesticide extenders, and in carbonless duplicating paper. It is hazardous to the upper respiratory tract, the digestive system, liver, blood, eyes, and skin; and is toxic by inhalation, ingestion, and skin absorption. Symptoms of exposure include eye and upper respiratory tract irritation; cough; tightness of the chest; chest pain; loss of appetite; anorexia; weight loss; nausea; vomiting; abdominal pain after eating; epigastric distress and pain; liver enlargement; increased cholesterol; chloracne and skin rashes; increased urinary excretion; headache, dizziness, depression, fatigue, nervousness; possibly carcinogenic. OSHA exposure limit (TWA): chlorodiphenyl (42% chlorine) 1 mg/m^3 [air]; (54% chlorine) 0.5 mg/m^3.

polychromatic light — light composed of several colors.

polycyclic aromatic hydrocarbon (PAH) — a class of 15 colorless, white, or pale yellow-green solids, including acenaphthene, acenaphthylene, anthracene, benz (a) anthracene, benzo (a) pyrene, benzo (b) fluoranthene, benzo (g,h,i) perylene, benzo (k) fluoranthene, chrysene, didenzo (a,h) anthracene, fluoranthene, fluorine, indeno (1, 2, 3-cd) pyrene, phenanthrene, and pyrene; formed during the incomplete combustion of coal, oil, gas, garbage, or other organic substances and found especially at hazardous waste sites. Only antracene is used for more than research — as an intermediate in dye production, in smoke screens, scintillation canter crystals; inorganic semi-conductor research; as an insecticide and fungicide; in the manufacture of plastics. Exposure to emissions containing polycyclic aromatic hydrocarbons may increase mortality due to lung cancers. Animal research indicates exposure may cause respiratory tract tumors induced in the nasal cavity, pharynx, larynx, and trachea of hamsters; decreased survival time due to benzo (a) pyrene in mice; preneoplastic hepatocytes in animals correlated with cancer promotion; renal effects in rats, developmental effects in mice; produce skin disorders such as regressive verruca, local villus eruptions, and skin lesions. Also known as polynuclear aromatic hydrocarbon. OSHA exposure limits (PEL): 0.2 mg/m^3 [air].

polyelectrolyte — a natural or synthetic chemical used to speed the removal of solids from sewage; it causes solids to flocculate or clump together more rapidly than chemicals like alum or lime.

polyethylene — a thermoplastic material with good flexibility, tensile strength, and resistance to solvents.

polyethylene sheeting structure — a microscopic bundle, cluster, fiber, or matrix of polyethylene which may contain asbestos.

polymer — material of high molecular weight formed by the joining together of many simple molecules.

polymerization — a chemical reaction in which two or more similar molecules combine to form larger, very stable compounds with the same chemical proportions but greater molecular weight; used extensively in production of plastics.

polynominal — (*mathematics*) an expression having a finite number of terms of the form ax + bx^2... nx^n; a mathematical expression of one or more algebraic terms each of which consists of a constant multiplied by one or more variables raised to a positive integral power.

polynuclear aromatic hydrocarbon (PAH) — *See* polycyclic aromatic hydrocarbon.

polypeptide — a compound formed by the peptide bonds of two or more amino acids.

polypropylene — $(C_3H_6)_x$ — a thermoplastic material, which is a polymer of propylene, prone to ozone and ultraviolet attack but tolerant to chemicals and extremes of temperatures, with good tensile strength and low permeability to water and solvents.

polysaccharide — a group of large, complex carbohydrates, such as starch and cellulose, with the general formula $(C_6H_{10}O_5)_n$.

polystyrene resins — synthetic resins formed by polymerization of styrene.

O
P

polythemia — a condition of excess numbers of red corpuscles in the blood.

polyvinyl chloride (PVC) — $(H_2CCHCl)_x$ — a white thermoplastic substance with good mechanical and electrical properties and a high resistance to chemicals; the inhalation of the vinyl chloride may be a health hazard; produced by the polymerization of vinyl chloride in the presence of initiators, such as benzoyl peroxide.

POM — acronym for polycyclic organic matter.

ponding — (*sewage*) a condition where the spaces between a stone become clogged with solids and air cannot pass through the filter due to excessive growths on the filter, plugging up the filter with primary tank solids, or the gradual breaking down and breaking up of filter stones.

poorly drained soil — soil in which the water is removed so slowly that the soil remains wet for a large part of the time.

poor soil suitability — a condition in which one or more of the soil properties are somewhat unfavorable for cover material.

population attributal risk — (*epidemiology*) the attributal risk multiplied by the prevalence of exposure to a factor in a total population; the number of new cases of the diseases in a total population that can be attributed to the factor.

population equivalent — an average waste loading equivalent to that produced by one person, equal to 100 gallons per day in volume, or any lesser amount containing a bio-chemical oxygen demand of 0.17 pounds per day.

population risk — (*epidemiology*) the excess number of cases of disease in an exposed population above the expected number of cases.

pork tapeworm disease — (*disease*) *See* taeniasis.

porosity — the ratio of the volume in any porous material that is not filled with solid matter to the total volume occupied.

porphyria — a genetic, hereditary disorder characterized by a disturbance in porphyrin metabolism with resultant increase in the formation and excretion of porphyrins or their precursors; characterized by photosensitivity and porpyhrinuria.

porphyrin — any of a group of iron- or magnesium-free cyclic tetrapyrrole derivatives, occurring universally in protoplasm and forming the basis of respiratory red pigments of animal and plants; in combination with iron it forms hemes.

portal — place of entrance.

Port and Tanker Safety Act — a 1978 law updated through 1986 in which navigation, vessel safety, and protection of marine environment are matters of major national importance; states that the handling of dangerous articles and substances immediately adjacent to navigable waters be subjected to appropriate care and advanced planning for critical areas.

portland cement — a gray, odorless powder. It is used in the manufacture of mortar for building blocks, bricks, stone, and pre-cast items; as a moisture sealant for exterior concrete blocks; in concrete on highway paving, domestic and commercial building construction. It is hazardous to the respiratory system, eyes, and skin; and is toxic by inhalation, ingestion, and contact. Symptoms of exposure include eye and nose irritation, cough, expectoration, exertional dyspnea, wheezing, chronic bronchitis, dermatitis. OSHA exposure limit (TWA): 10 mg/m^3 [resp].

positive-displacement pump — a pump which forces or displaces a liquid under pressure through a pumping mechanism.

positive pressure — a condition when more air is supplied to a space than is exhausted, so the air pressure within that space is greater than that in surrounding areas.

positive-pressure fabric filter — a fabric filter with the fans on the upstream side of the filter bags.

positron — a positively-charged particle of electricity with about the same weight but opposite charge as the electron.

posterior — referring to the tail or back end of an organism.

postmortem — *See* autopsy.

postnatal — occurring after birth.

postpartum — the period following childbirth during which generative organs are returning to their normal, nonpregnant state.

post-processor — (*computer science*) a computer program that is used to convert the results of another operation into a standard format ready for further analysis.

potable water — fresh water of a quality for drinking, culinary, and domestic purposes not less than that prescribed in international standards for drinking water, especially as concerns bacteriological requirements and chemical and physical requirements.

potassium alum — $KAl(SO_4)_2$ — white, odorless crystals used as a flocculate in sand filters; in medicines and baking powder; in dyeing, papermaking, and tanning. Also known as alum; aluminum potassium sulfate; potassium aluminum sulfate.

potassium aluminum sulfate—*See* potassium alum.

potency — the amount of a chemical that will produce an effect.

potential difference (PD) — a measure of force produced between charged objects that move free electrons; the unit is called the volt. Also known as voltage or electromotive force.

potential dose — the amount of a chemical contained in material ingested, inhaled, or absorbed by the skin.

potential energy — the energy due to the position of a body or to the configuration of its particles. Also known as stored energy.

potential ionization — the ionization necessary to separate one electron from an atom resulting in the formation of an ion pair.

potentially hazardous food — any natural or synthetic perishable food or ingredient, such as milk, eggs, meat, poultry, or shellfish, in a form capable of supporting the rapid and progressive growth of infectious or toxigenic microorganisms, or the slower growth of *C. botulinum*.

potential occupational carcinogen — any substance, combination, or mixture of substances which causes an increased incidence of benign and/or malignant neoplasm, or a substantial decrease in the latency period between exposure and onset of neoplasms in human or in one or more experimental species in a million; results from any oral, respiratory, dermal, or other exposure which causes induction of tumors at a site other than the site of administration.

potentiation — the process of a non-toxic or relatively non-toxic substance increasing the effect of another toxic substance.

potentiometry — a technique of analysis of a gas which causes a change in the hydrogen-ion concentration (pH) when the gas reacts with a reagent, which is sensed by a galvanic cell; the change in pH is a measure of concentration of the gas.

POTW — acronym for publicly owned treatment works (municipal treatment plant).

poultry — domesticated birds, such as chickens, turkeys, ducks, geese, and pigeons, that are raised for food consumption.

poultry processing plant — any food establishment or portion thereof in which poultry are killed, dressed, processed, stored for sale, or offered for live sale.

pound (lb) — a unit of mass equal to 0.4536 kilograms or 16 ounces.

pound foot (lb-ft) — a unit of energy; the force needed to move a substance weighing 1 pound 1 foot. *See also* foot-pound of torque.

power — the rate at which work is done.

power density — the intensity of electromagnetic radiation per unit area expressed in watts per centimeter squared.

ppb — *See* parts per billion.

PPE — notation for personal protection clothing and equipment.

ppm — *See* parts per million.

precancerous — pertaining to a pathologic process that tends to become cancer.

precious metal — any relatively scarce, valuable metal, such as gold, silver, or platinum group metals, and the principal alloys of those metals.

precipitate — an insoluble compound formed in the chemical reaction between two or more substances in solution; to form a precipitate.

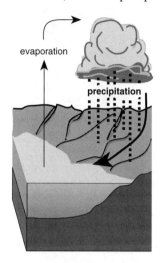

Precipitation

precipitation — the physical settling of particles; any form of water particles falling from the atmosphere to the ground.

Poultry Processing Plant

O
P

precipitator — a device using mechanical, electrical, or chemical means to collect particulates.

precipitin — an antibody that specifically aggregates the macromolecular antigen to give a visible precipitate.

precision — the degree of accuracy of agreement of repeated measurements of the same property; (*mathematics*) the number of significant digits to the right of the decimal point; (*statistics*) the degree of variation about the mean.

precision of measurement — (*statistics*) *See* random error.

precoat — a layer of diatomaceous earth deposited on the filter at the start of the filter run.

precoat feeder — a chemical feeder designed to inject diatomaceous earth into a filter in sufficient quantity to coat the filter septa at the start of a filter run.

precordia — the region over the heart and lower thorax.

precursor — a pollutant that takes part in a chemical reaction resulting in the formation of one or more new pollutants. *See also* parent.

predator — an insect or other animal that attacks, feeds on, or destroys other insects or animals.

preliminary injunction — (*law*) the intermediate segment of the injunction process, which is initiated through the filing of a motion of preliminary injunction. If granted and signed by the court, the defendant is preliminarily restrained from committing future violations of the act. It is not a final process and is an unadjudicated civil case as far as the court is concerned. *See also* temporary restraining order.

premises — a lot, parcel, or plot of land either occupied or unoccupied by any dwelling or non-dwelling structure and includes any such buildings, accessory structures, or other structures thereon.

premium — an amount of money paid to an insurance company in return for insurance protection.

prenatal — preceding birth.

presbycusis — a condition of hearing loss due primarily to aging.

prescription (Rx) — a written directive for the compounding, dispensing, or administration of drugs or other services to a particular patient.

presentence investigation — an investigation of parties convicted in criminal cases prior to the imposition of sentence conducted by a probation officer.

preservative — any substance added to a product to destroy or inhibit multiplication of microorganisms in a foodstuff.

pressed wood product — a group of materials used in building and furniture construction that are made from wood veneers, particles, or fibers bonded together with an adhesive under heat and pressure; may lead to indoor air quality problems.

pressure — the force per unit area.

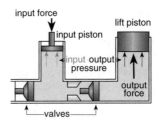

Pressure Differential
in a hydraulic lift

pressure differential — the difference of pressure between two points in a hydraulic system.

pressure drop — the difference in static pressure measured at two locations in a ventilation system due to friction or turbulence.

pressure head — the height of a column of fluid required to produce a given pressure at its base.

pressure loss — energy lost from a pipe or duct through friction or turbulence.

pressure sewer — a small diameter pipe line shallowly buried and following the profile of the ground.

pressure sore — *See* decubitus ulcer.

pretreatment — the reduction of the amount of pollutants, the elimination of pollutants, or the alteration of the nature of pollutant properties in wastewater to a less harmful state prior to treatment facilities.

pretrial conference — a conference of the parties in a suit or case called by the court to consider various actions in order to limit the various issues for trial; the discussion between the agency attorney and witnesses before a trial.

prevalence — (*epidemiology*) *See* prevalence rate.

prevalence rate — the number of cases of disease that exist in a population at some point in time; expressed per unit of population.

prevention — a means of control where proper planning, maintenance of equipment, location of sources of pollutants, and an interruption in the spread of the pollutant occurs; a means of avoidance of disease.

preventive maintenance (PM) — a management-initiated process of inspection and corrective action undertaken prior to any actual failure of the facility assets, including the physical structure and related equipment.

PRF laser — *See* pulsed recurrence frequency laser.

prickly heat — *See* heat rash.

primacy — the state of being first in importance.

prima facie — (*law*) a fact presumed to be true, unless it is proved by evidence to the contrary.

primary air pollutants — those pollutants emitted to the atmosphere as the result of some heating or industrial process.

primary burner — a burner that dries out and ignites material in the primary combustion chamber.

primary combustion air — air admitted to a combustion system at the point where the fuel is first oxidized.

primary combustion chamber — the chamber of an incinerator where waste is ignited and burned.

primary drinking water standard — a regulation which applies to public systems and specifies contaminants which may have adverse effects on human health.

primary health care — the level of routine outpatient care provided by a supervised primary health care physician.

primary ionization — (*collision theory*) the ionization produced by the primary particle.

primary irritant — a chemical which causes dermatitis by direct action on the normal skin at the site of contact if permitted to act in sufficient intensity or quantity for a sufficient length of time.

primary irritation dermatitis — an inflammation and irritation of the skin due to brief contact with a concentrated chemical such as an acid, alkali, or irritant gas; may also be caused by extended exposure to a dilute chemical, friction, cold, or heat.

primary metals — ferrous or non-ferrous metals produced by the smelting of ore in industry; the smelting process may cause metallic oxide pollution as well as the emission of carbon monoxide, smoke, dust, sulfur dioxide, lead, and other contaminants.

primary pollutant — a pollutant directly emitted from a polluting source.

primary sedimentation — the first major process in wastewater treatment plants in which solids settle by gravity.

primary septicemia — *See Vibrio vulnificus.*

primary sewage treatment — a system in which preliminary treatment consisting of the use of screens and grit chambers, comminuting devices, preaeration tanks, sedimentation tanks, chemical treatment, and discharge of the effluent to the receiving stream may occur; the effluent may go on to secondary treatment; the sludge is removed to a sludge digester.

primary sludge — that portion of the raw wastewater solids contained in the effluent which is directly captured and removed in the primary sedimentation process.

primary standard — a substance with a known property that can be defined, calculated, or measured and is readily reproducible.

primary treatment — the first stage in wastewater treatment where substantially all floating or setteable solids are removed by flotation and/or sedimentation.

primary voltaic cell — a pair of plates in an electrolyte from which electricity is derived by chemical action. Also known as voltaic cell.

principle component analysis — (*geographic information system*) a technique of analyzing multivariate data expressing their variation in small numbers of principle components or linear combinations of the original data.

priority — a rational assignment of a preferential rating to something meriting attention among competing alternatives.

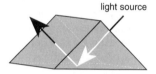

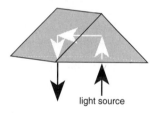

Prism
showing internal reflection

prism — a three-, four-, six-, eight-, or twelve-sided piece of glass that separates light into its colors and is used in a number of optical instruments.

Prismatic Soil

prismatic soil — (*soils*) a soil structure which is without rounded caps, prism-like with the vertical axis greater than the horizontal, usually found in subsoil or B horizon.

privacy — the existence of conditions that permit an individual or individuals to be without interruption or interference, either by sight or sound of unwanted individuals.

private sector — a division of people as related to business to produce products or services using private capital and the profit motive.

private water system — any water system for the provision of water for human consumption if the system has fewer than 15 service-connections and does not readily serve an average of at least 25 people per day at least 60 days out of the year.

privy — any sanitary, waterless device for the collection and storage of human excreta, also including chemical commodes or other portable receptacles.

probability — (*statistics*) the relative frequency with which a specific event occurs expressed as the number of times the event occurs divided by the total number of trials.

probe — a tube used for sampling or measuring pressure or temperature at a distance from the actual collection or measuring apparatus; (*soils*) a tube used for taking a soil sample.

proboscis — a tubular mouth part in certain insects and mammals.

procarcinogen — a chemical substance that becomes carcinogenic only after it is altered by metabolic processes.

procedure — a sequence of actions which collectively produce a desired outcome.

procedure in admiralty — (*law*) seizure actions taken pursuant to the Food, Drug, and Cosmetic Act stating that the government may initiate confiscation of articles it deems to be contraband without first having to show proof of the allegation.

proceeding — (*law*) any rulemaking, adjudication, or licensing conducted by a governmental agency under applicable statutes or under regulations used to implement them.

process — a particular method of doing something generally involving a number of steps or operations.

processed meat — meat plus additives, spices, and other ingredients receiving special treatment such as curing, smoking, or canning.

processed water — water that comes in contact with an end-product or with materials incorporated in an end-product.

process emission — the particulate matter which is collected by a capture system.

processing — (*solid waste*) any method, system, or other treatment designed to change the physical form or chemical content of solid waste.

processor — any person who processes a chemical substance or mixture.

pro confesso — (*law*) as if conceded as confessed.

prodromal stage — a premonitory symptom; a symptom indicating the onset of a disease; (*radiation*) the initial period for acute radiation syndrome.

producer — (*food chain*) an autotrophic member of the food chain able to make its food from inorganic substances and upon whom the other plants and animals in the chain depend.

product of combustion — gases, vapors, and solids that result from the combustion of a fuel.

pro forma — (*law*) as a matter of form.

progeny — the decay products resulting after a series of radioactive decay; descendants.

program — an organized activity with a definable purpose; (*computer science*) a precise, sequential set of instructions directing the computer to perform a task.

program evaluation and review technique (PERT) — an objective analysis using an evaluation of project study and periodic reviews to enhance the effectiveness of management, supervision, and control.

project — a task, that upon completion, will require concerted efforts and will yield data, material, and products.

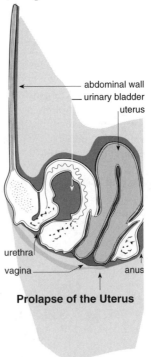

Prolapse of the Uterus

prolapse — downward displacement of a part or organ.

prolate — an elongated sphere; a cigar-shaped object.

proliferation — the reproduction or multiplication of cells.

prolonged exposure — the continuous exposure of workers to a chemical agent or chemical hazard over a long period of time; may lead to adverse health effects.

promoter — a chemical that when administered after an initiator has been given, promotes the change of an initiated cell into a cancer.

promotion — (*cancer*) an event that causes a cell to grow rapidly and to acquire the capacity to metastasize.

promulgate — to publish or make known officially.

propagation of flame — the spread of flame through the entire volume of the flammable vapor-air mixture from a single source of ignition.

propane — $CH_3CH_2CH_3$ — a colorless, odorless gas; a foul-smelling odorant is often added when used for fuel purposes; shipped as a liquefied compressed gas; MW: 44.1, BP: –44°F, Sol: 0.01%. It is used as a basic material in chemical synthesis for oxidation, alkylation, nitration, and chlorination; as a solvent and extractant in deasphalting and degreasing of crude oils; as feedstock in cracking process in the production of ethylene and propylene and motor gasoline; as a refrigerant, in petroleum refining, and gas processing operations, low-temperature crystallizers, and helium recovery from natural gas; as a fuel in welding and cutting operations; in the desalination of water. It is hazardous to the central nervous system; and is toxic by inhalation and contact. Symptoms of exposure include dizziness, disorientation, excitation, frostbite. OSHA exposure limit (TWA): 1,000 ppm [air] or 1,800 mg/m^3.

propellant — a chemical or other agent used in aerosol bombs to force or push the pesticide chemical out through the nozzle.

properly connected — the condition of a connection or installation being in accordance with all applicable codes and ordinances.

property — a parcel of land for which legal title has been recorded; a characteristic by which a substance may be identified; physical properties include state of matter, color, odor, and density; chemical properties include behavior in reaction with other materials.

property right — (*law*) a legal right in or against specific property.

property tax — a tax levied on real or personal property.

property transfer assessment — a determination of those areas of potential legal and economic liability that a buyer may assume by purchasing a business or industry that has actual or potential environmental problems.

prophase — the first stage of nuclear division by mitosis or meiosis.

prophylaxis — any substance or steps taken to prevent an event.

proportional counter — a gas-filled radiation-detection tube in which the pulse produced is proportional to the number of ions formed in the gas by the primary ionizing particle.

proportional incidence ratio (PIR) — a measure of risk that compares the observed numbers of events with the expected number of events.

proportional mortality ratio — the ratio of deaths for a specific disease among all deaths in an exposed group to the ratio of deaths for that disease among all deaths in an unexposed group.

proprietary — of or pertaining to organizations that possess, own, or hold exclusive rights to something.

n-propyl acetate — $CH_3COOCH_2CH_2CH_3$ — a colorless liquid with a mild, fruity odor; MW: 102.2, BP: 215°F, Sol: 2%, Fl.P: 55°F, sp gr: 0.84. It is used in the manufacture of lacquers and adhesives; as a solvent for rubber; in the preparation of flavoring agents and perfumes. It is hazardous to the respiratory system, eyes, skin, and central nervous system; and is toxic by inhalation, ingestion, and contact. Symptoms of exposure include narcosis, dermatitis, irritation of the eyes, nose, and throat. OSHA exposure limit (TWA): 200 ppm [air] or 840 mg/m^3.

n-propyl alcohol — $CH_3CH_2CH_2OH$ — a colorless liquid with a mild, alcohol-like odor; MW: 60.1, BP: 207°F, Sol: miscible, Fl.P: 72°F, sp gr: 0.81. It is used during the printing on plastic film and sheeting with polyamide-based inks; during textile leather processing; in the manufacture of surface coatings as a solvent; in the manufacture of cleaning preparations and polishing agents; in the extraction of vegetable oils, castor oil, and pharmaceuticals; during cellulose processing; in the manufacture of brake fluids. It is hazardous to the skin, eyes, respiratory system, and gastrointestinal tract; and toxic by inhalation, absorption, ingestion, and contact. Symptoms of exposure include dry cracking skin, drowsiness, headache, ataxia, gastrointestinal pain, abdominal cramps, nausea, vomiting, diarrhea, mild irritation of the eyes, nose, and throat. OSHA exposure limit (TWA): 200 ppm [air] or 500 mg/m^3.

O
P

**Propylene Dichloride
(1,2–Dichloropropane)**

propylene dichloride — $CH_3CHClCH_2Cl$ — a colorless liquid with a chloroform-like odor; MW: 113.0, BP: 206°F, Sol: 0.3%, Fl.P: 60°F, sp gr: 1.16. It is used as a soil fumigant for protection of fruit and nut crops, field crops, beets, and tobacco against nematodes; in cleaning, degreasing, and spot removal operations including paint and varnish removal; during rubber compounding and vulcanizing operations; during extraction processing of fats, oils, lactic acid, and petroleum waxes; in the manufacture of tetrachloroethylene and propylene oxide; as an additive and lead scavenger in antiknock fluids. It is hazardous to the skin, eyes, respiratory system, liver, and kidneys; and is toxic by inhalation, ingestion, and contact. Symptoms of exposure include drowsiness, eye irritation, lightheadedness, skin irritation; carcinogenic; in animals: liver and kidney disease. OSHA exposure limit (TWA): 75 ppm [air] or 350 mg/m³.

propylene imine — C_3H_7N — a colorless, oily liquid with an ammonia-like odor; MW: 57.1, BP: 152°F, Sol: miscible, Fl.P: 25°F, sp gr: 0.80. It is used as a polymer modifier; in the manufacture of polypropylene imines and polymers for coating materials, adhesives, chelating agents, emulsifying agents, and fire-proofing agents in textile, rubber, paint, ink, and paper industries; as a chemical intermediate in the pharmaceutical and chemical industries. It is hazardous to the eyes and skin; and is toxic by inhalation, absorption, ingestion, and contact. Symptoms of exposure include eye and skin burns; carcinogenic. OSHA exposure limit (TWA): 2 ppm [skin] or 5 mg/m³.

propylene oxide — C_3H_6O — a colorless liquid with a benzene-like odor; a gas above 94°F; MW: 58.1, BP: 94°F, Sol: 41%, Fl.P: –35°F, sp gr: 0.83. It is used in the production of polyurethane foam; in the manufacture of propylene glycol for use as a solvent, emulsifier, and mold inhibitor; in the manufacture of dipropylene glycol for use as a solvent, ink formulations; as a chemical intermediate in the production of lubricants, surfactants, and miscellaneous chemicals for pharmaceutical, petroleum, textile, rubber, and soap industries; as a low-boiling solvent for cellulose derivatives, hydrocarbons, commercial gums, and various resins; as a fumigant, herbicide, germicide, insecticide, and stabilizer of vinyl resin lacquers and discoloration preventer of hydrocarbons; as a food preservative; during hydropropylating of wheat flour. It is hazardous to the eyes, skin, and respiratory system; and is toxic by inhalation, ingestion, and contact. Symptoms of exposure include irritation, blistering, and burning of the skin, irritation of the eyes, upper respiratory system, and lungs; carcinogenic. OSHA exposure limit (TWA): 20 ppm [air] or 50 mg/m³.

n-propyl nitrate — $CH_3CH_2CH_2NO_3$ — a colorless to straw-colored liquid with an ether-like odor; MW: 105.1, BP: 231°F, Sol: insoluble, Fl.P: 68°F, sp gr: 1.07. It is used in liquid rocket propellants. No known hazards to humans; toxic by inhalation, ingestion, and contact to animals. Symptoms of exposure in animals include methemoglobin, anoxia, cyanosis, dyspnea, weakness, dizziness, headache, irritation of the eyes and skin. OSHA exposure limit (TWA): 25 ppm [air] or 105 mg/m³.

prosecution — the party beginning action in a criminal case; the institution of a criminal proceeding against an individual.

prospective study — (*epidemiology*) a study of susceptible exposed and unexposed individuals. Also known as longitudinal study.

protection factor (PF) — the ratio of the ambient, airborne concentration of a contaminant to the concentration inside the facepiece of respiratory protective equipment.

protective barrier — a barrier of radiation-absorbing material such as lead, concrete, plaster, and plastic used to reduce or eliminate radiation exposure.

protective clothing — clothing that protects a person against injury, exposure, or death while handling toxic chemicals.

protective equipment — equipment that protects a person against injury, exposure, or death while handling toxic chemicals.

protein — any of a class of complex nitrogen-containing compounds made by living organisms from amino acids which may consist of one or more polypeptide chains.

proteinuria — an excess of serum proteins in the urine.

proteolytic enzyme — an enzyme that helps break down protein; either rennet or casease.

proteose — a secondary protein derivative formed by hydrolytic cleavage of the protein molecule.

Proteus — a genus of gram-negative, facultatively anaerobic, motile, rod-shaped bacteria found in fecal material, especially in patients treated with oral antibiotics.

protocol — the plan and procedures to be followed in conducting a test, performing a service, or setting up a program.

protofilament — the filament extending along the axoneme of a cilium and flagellum.

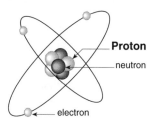

Proton

proton — a positively charged elementary particle that together with the neutron forms atomic nuclei.

protoplasm — an organized colloidal complex system of substances that constitutes the living material of the cell.

protoplasmic poisons — poisons which cause damage or death to cells resulting in an inflammatory change altering cell membranes and inhibiting enzymes; they include ammonia, formaldehyde, dimethyl sulfate, arsenic, and cyanide.

protozoan — single-celled microorganisms belonging to the phylum protozoa.

protraction dose — a measure of administering radiation by delivering it continuously over a relatively long period at a low dose rate.

proximate analysis — analysis of a solid fuel to determine on a percentage basis how much moisture, volatile matter, fixed carbon, and ash the sample contains; this also helps establish the fuel's heat value.

pruritus — a subjective, unpleasant itching sensation due to irritation of sensory nerve endings.

PSD — acronym for prevention of significant deterioration.

pseudopodium — a temporary protrusion or retractile process of the protoplasm of a cell that serves as a locomotor or food-gathering function.

PSI — acronym for pollution standards index; notation for pressure in pounds of force per square inch.

psig — notation for pressure per square inch gauge. *See* gauge pressure.

psittacosis — (*disease*) an acute generalized chlamydial disease of humans and birds with human symptoms of fever, headache, myalgia, chills, and upper or lower respiratory tract disease. Incubation time is 4–15 days, usually 10 days. It is caused by *Chlamydia psittaci*; found worldwide especially where there are various types of birds. Reservoirs of infection include parakeets, parrots, pigeons, turkeys, ducks, and other birds. It is transmitted by inhaling the chlamydia from dry droppings and secretions of infected birds in an enclosed space. Communicability may last for weeks or months from diseased or healthy birds; rarely transmitted from person to person; general susceptibility. It is controlled by regulating the importing, raising, and selling of birds of the parrot family, surveillance of pet shops and bird areas, and educating the public of the potential for the disease.

psoriasis — a usually chronic, recurrent skin disease of unknown origin marked by discrete bright red macules, papules, or patches covered with lamellated silvery scales.

PSS-TLVs — notation for particulate size-selected TLVs.

psychogenic deafness — a loss of hearing originating in or produced by the mental reaction of an individual to the physical or social environment.

psychoneuroses — a mental disorder of psychogenic origin but presenting the symptoms of a functional nervous disease.

psychosocial factors — psychological, organizational, and personal stressors that could produce symptoms similar to poor indoor air quality, such as shortness of breath.

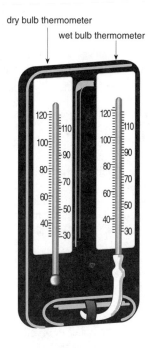

Psychrometer

psychrometer — a hygrometer consisting of two similar thermometers with the bulb of one being kept wet so that the cooling that results from evaporation makes it register a lower temperature than the dry one, with the difference between the readings constituting a measure of the dryness of the atmosphere; used to measure relative humidity.

psychrometric chart — a graphical representation of the thermodynamic properties of moist air.

psychrophile — an organism that thrives in temperatures of 0–5°C and that causes spoilage in refrigerated foods; found on uncultivated soil in lakes and streams, on meats, and in ice creams.

psychrophilic — of or pertaining to an organism that thrives at relatively low temperatures, usually at or below 15°C.

pt — *See* pint.

ptosis — prolapse of an organ or part.

Public Affairs Information Services Bulletin (PAIS) — a subject index published in quarterly paperback issues, and accumulated into annual hardcover volumes; covers areas of economics, political science, public administration, international law, sociology, and demography.

public health law — a body of statutes, regulations, and precedence that protects and promotes individual and community health; founded on the preamble of the Constitution and Section 8, Article 1 of the Constitution for the promotion of the general welfare.

public hearing — a public session in which witnesses are heard and testimony is taken.

public market — the space, stall, or enclosure designated for the vending of farm products.

public sewage disposal — a community sewage disposal system in which sewage from assorted homes, businesses, and possibly industry is collected and transported through a piping system to a sewage treatment plant, where it may be treated in a conventional way through settling and biological oxidation, or through advanced wastewater treatment methods.

public water supply — water distributed from a public water system.

public water system — a system for the provision of piped water for public human consumption with at least 15 service connections and 25 people serviced.

puck — a hand-held device for entering data from a digitizer.

pulmonary — pertaining to the lung.

pulmonary edema — fluid in the lungs.

pulmonary emphysema — a pathological accumulation of air in tissues or organs in which the terminal bronchials become plugged with mucus and the aveoli lose their elasticity, making breathing difficult.

pulmonary function — the performance of the respiratory system in supplying oxygen to and removing carbon dioxide from the body by way

of circulating blood and moving air in and out of the alveoli.

pulp — fiber material produced by chemical or mechanical means from such raw materials as virgin wood, secondary fibers, and rags that are used in the manufacture of paper and paperboard.

pulsed laser — a class of laser where emission occurs in one or more flashes of short duration.

pulsed recurrence frequency laser (PRF laser) — a laser with properties similar to a continuous wave laser.

pulse-purge — that part of the sterilizer cycle after evacuation/exhaust consisting of repeated cycles of operation of the vacuum pump followed by vacuum relief.

pulsus alternans — alternating pulse; a regular alternation of weak and strong beats without changes in cycle length.

pulverization — the crushing or grinding of material into small pieces. Also known as comminution.

pumice — a natural silicate of volcanic ash or lava used as an abrasive.

pump curve — a graph of performance characteristics of a given pump under varying power flow resistance factors.

pumping station — a machine installed on sewers to pull sewage uphill.

pump strainer — a device containing a removable strainer basket designed to protect a pump from debris in the water flow when installed in the pump action line.

puncture — the act of piercing or penetrating with a pointed object or instrument.

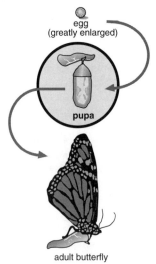

egg
(greatly enlarged)

pupa

adult butterfly

Pupa of Monarch Butterfly

pupa — the resting stage of an insect having completed metamorphosis that follows the larva stage and precedes the adult stage.

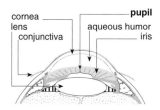

Pupil

pupil — the opening in front of the vertebrate eyeball the size of which is controlled by the iris.

pure tone — sound wave in which the instantaneous sound pressure is a simple sinusoidal function of time; a sound sensation that is characterized or signaled by a single pitch.

purging — the method by which gases, vapors, or other airborne impurities are displaced from a confined space.

purine — the parent substance of adenine, chignon, and other naturally occurring purine "bases" not known to exist as such in the body.

purpura — extensive hemorrhage into the skin, joints, or mucous membrane.

purulent — consisting of or containing pus.

pus — a collection of dead bacteria or white blood corpuscles at the site of an infection.

pustule — an elevation on the skin containing pus.

putrefaction — decomposition of organic matter by microorganisms producing disagreeable odors.

putrescible — of or pertaining to organic matter capable of being decomposed by microorganisms.

PVC — *See* polyvinyl chloride.

pycnocline — a layer of water that exhibits rapid change in density with depth.

pyogenic — pus-producing.

pyorrhea — a discharge of pus.

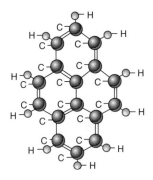

Pyrene

pyrene — a polycyclic aromatic hydrocarbon formed during the incomplete combustion of coal, oil, gas, garbage or other organic substances and found especially at hazardous waste sites. See polycyclic aromatic hydrocarbons.

pyrethrin — one of two oily liquid esters $C_{21}H_{28}O_3$ and $C_{22}H_{28}O_5$ that have insecticidal properties and that occur naturally in old-world ehrysanthemums; an excellent knock-down agent.

pyrethrum — $C_{20}H_{28}O_3/C_{21}H_{28}O_3/C_{22}H_{28}O_5$ — a brown, viscous oil or solid; MW: 316–372, BP: not known, Sol: insoluble, Fl.P: 180 to 190°F. It is used as an insecticide on pre- and post-harvest agricultural crops, cattle, and poultry, food cartons, and confined areas. It is hazardous to the respiratory system, skin, and central nervous system; and is toxic by inhalation, ingestion, and contact. Symptoms of exposure include erythema, dermatitis, papules, pruritus, rhinorrhea, sneezing, asthma. OSHA exposure limit (TWA): 5 mg/m³ [air].

Pyridine

pyridine — C_5H_5N — a colorless to yellow liquid with a nauseating, fish-like odor; MW: 79.1, BP: 240°F, Sol: miscible, Fl.P: 68°F, sp gr: 0.98. It is used in the manufacture of pharmaceuticals; as a solvent in the manufacture of polycarbonate resins used in hand tools, small appliances, camera parts, safety helmets, and electrical connectors; as a starting material used in the manufacture of chemical intermediates and products; in the manufacture of rubber accelerators, epoxy resins, and pharmaceuticals; as a solvent reaction medium or catalyst in paint manufacture, drug manufacture, and rubber manufacture; as a reagent in chemical analysis; in textile treatment as a water-proofing agent; as a denaturant for ethyl alcohol; as a coupling assistant in azo dye manufacture; in purification of mercury fulminate in explosives manufacture; during thermal decomposition of flexible polyurethane foams. It is hazardous to the central nervous system, liver, kidneys, skin, and gastrointestinal tract; and is toxic by inhalation, absorption, ingestion, and contact. Symptoms of exposure include headache, nervousness, dizziness, insomnia, nausea, anorexia, frequent urination, eye irritation, dermatitis, liver and kidney damage. OSHA exposure limit (TWA): 5 ppm [air] or 15 mg/m³.

pyrimidine — $C_4H_2N_2$ — an organic base that is a constituent of deoxyribonucleic acid and ribonucleic acid.

O
P

pyrite — FeS_2 — a mineral compound of sulfur and a metal, usually iron, often found in coal. Also known as fool's gold.

pyrogen — (*bacteria*) a fever-producing substance of bacterial origin, believed to be complex polysaccharides, attached to another radical containing nitrogen or phosphorous.

pyroligneous acid — a combination of acetic acid, methyl alcohol, and acetone which comes from the destructive distillation of wood.

pyrolysis — an endothermic reaction causing a destructive distillation of solid material in the presence of heat and the absence of air or oxygen.

pyrometer — an instrument used for measuring or recording temperatures.

pyrophoric — pertaining to a chemical that ignites spontaneously in air at a temperature of 13°F or below.

pyroxene — (*mineralogy*) any of a group of silicate materials, usually calcium, magnesium, and iron silicate, often found in igneous rocks.

O
P

q_1^* — the upper-bound estimate of the low-dose slope of the dose-response curve as determined by the multistage procedure; can be used to calculate an estimate of carcinogenic potency, the incremental excess cancer risk per unit of exposure (usually $\mu g/L$ for water, $mg/Kg/day$ for food, and $\mu g/m^3$ for air).

QA — *See* quality assurance.

QC — *See* quality control.

QCP — acronym for quiet communities program.

QF — *See* quality factor.

Q fever — (*disease*) an acute, febrile, rickettsial disease with sudden chills, headache, weakness, malaise, and severe sweats. Pneumonitis may occur along with chest pains, acute pericarditis, acute hepatitis, and generalized infections. Incubation time is 2–3 weeks; caused by *Coxiella burnetii* (*Rickettsia burneti*); found in all parts of the world. Reservoirs of infection include cattle, sheep, goats, ticks, and some wild animals; transmitted by airborne dissemination of rickettsiae in dust. Transmission from person to person is rare; general susceptibility. It is controlled by understanding how the disease occurs, pasteurization of milk, and immunization.

Q-switched laser — a pulsed laser capable of extremely high peak power for a very short duration.

qt — *See* quart.

quad — a unit of heat energy equal to one quadrillion or a million billion British thermal units.

quadrant — (*mathematics*) a quarter of a circle measured in units of 90 degrees.

quadratic polynominal — (*mathematics*) one in which the highest degree of terms is 2.

quagmire — *See* bog.

qualitative analysis — determination of the components of a mixture of substances.

quality — the degree to which techniques contribute to the improvement or maintenance of a product, service, or situation.

quality assessment — the process of evaluating and measuring the effectiveness of an improvement or maintenance of a product, service, or situation.

quality assurance (QA) — a system of actions designated to promote confidence that a product or service will satisfy needs.

quality control (QC) — a group of activities designed to ensure adequate quality, especially in manufactured products; consists of random sampling and inspectional techniques.

quality factor (QF) — the linear-energy-transfer dependent factor by which absorbed doses are multiplied to obtain, for radiation protection purposes, a quantity that expresses the effectiveness of the absorbed dose on a common scale for all ionizing radiation.

quantitative analysis — the determination of relative amounts of significant components present in a substance or mixture of substances.

quantum — the smallest indivisible quantity of radiant energy; a photon.

quantum mechanics — the branch of physics dealing with matter and electromagnetic radiation, and the behavior of particles whose specific properties are given by quantum numbers.

quantum number — a number used to describe a specific property of a subatomic particle in mathematical terms.

quarry sanitary landfill method — a variation of the area landfill method in which the waste is spread and compacted in a depression and cover material is brought in from elsewhere.

quart (qt) — a unit of liquid volume equal to 32 fluid ounces or 2 pints.

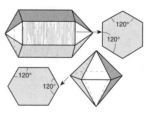

Quartz

quartz (SiO_2) — a colorless, transparent, very hard mineral composed of silica and found in many different types of rocks, such as sandstone and granite; the most abundant and widespread mineral.

quash — to vacate, annul, or make void.

quaternary ammonium compounds — any of numerous strong bases and their salts derived from ammonium by replacement of the hydrogen atoms with organic radicals; important as surface-active agents and disinfectants.

quench trough — a water-filled trough into which burning residue drops from an incinerator furnace.

quicksilver — *See* mercury.

quinone — $C_6H_4O_2$ — a pale yellow solid with an acrid, chlorine-like odor; MW: 108.1, BP: sublimes, Sol: slight, Fl.P: 100 to 200°F, sp gr: 1.32. It is used as an oxidizing agent in the synthesis of organic chemicals and intermediates; in photography as developer hyperquinone; in agriculture in the production of insecticides and fungicides; in the pharmaceutical industry for production of cortisone and addition compound with barbiturates; in the polymer and resin industry as an inhibitor and retarder-antioxidant, curing agent, and catalyst; as a toner and intensifier in the photographic industry; as a tanning agent for the leather industry; in the manufacture of quinhydrone electrodes for use for pH determinations. It is hazardous to the eyes and skin; and is toxic by inhalation, ingestion, and contact. Symptoms of exposure include conjunctivitis, keratitis, eye and skin irritation. OSHA exposure limit (TWA): 0.4 mg/m^3 [air].

Q-value — the energy liberated or absorbed in a nuclear reaction.

R

R — an abbreviation or symbol of hydrocarbon radical; radius; Rankine; resistance; roentgen; Rhydberg constant.

R&D — *See* research and development.

rabbit fever — (*disease*) *See* tularemia.

rabies — (*disease*) a viral encephalomyelitis with a headache, fever, malaise, sense of apprehension, paralysis, muscle spasm, delirium, and convulsions after the bite of an infected animal. Incubation time is 2–8 weeks or as short as 10 days; caused by the rabies virus which is a rhabdovirus; found worldwide. Reservoirs of infection include wild and domestic dogs, foxes, coyotes, wolves, cats, raccoons, and other biting mammals; also found in bats. It is transmitted in the saliva from the bite of a rabid animal; communicable in dogs and cats for 3–5 days before the onset of clinical signs and during the course of the disease; general susceptibility. It is controlled by vaccinations of all pets for rabies, 10-day detention for observation of dogs and cats who have bitten people, immediate evaluation of heads of animals for virus isolation in the brain when suspected of being rabid, and immediate immunization of an individual bitten by a rabid animal.

RACT — *See* reasonable available control technology.

rad — *See* radiation absorbed dose.

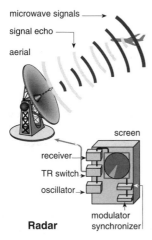

Radar

radar (radio detection and ranging) — a system for locating natural and artificial objects, navigating, measuring distance or altitude, homing, or bombing by means of radio signals.

radian — an arc of a circle equal in length to the radius.

radiant energy — *See* radiation.

radiant heat transfer — an energy transfer occurring when there is a large difference between the temperatures of two surfaces exposed to each other but not touching.

radiant temperature — the temperature of the body resulting from absorbed radiant energy.

Radiation

radiation — the emission and propagation of energy in space or through a material medium in the form of waves; the emission and propagation of the electromagnetic waves of sound and elastic waves; the transfer of heat or light by waves of energy; a stream of alpha- or beta-particles from a radioactive source or neutrons from a nuclear reactor; usually refers to electromagnetic radiation, but can also be applied to sound waves and emitted particles. Also known as radiant energy.

radiation absorbed dose (rad) — a measure of the energy imparted to matter by ionizing radiation per unit mass of irradiated material; equal to 100 ERGs per gram; equal to one gray (Gy).

radiation equilibrium — the condition in the radiation field where the energy of the radiations entering a volume equals the energy of the radiations leaving that volume.

radiation gas flow counter — a counter in which an appropriate atmosphere is maintained in the

317

counter tube by allowing a suitable gas to flow slowly through the sensitive volume.

radiation illness — an illness caused by the harmful effect of radiation which predisposes people to the development of symptoms of general malaise, nausea, and vomiting.

radiation pyrometer — a device that determines temperature by measuring the intensity of radiation at all wavelengths emitted by material having a high temperature.

radiation sickness — the self-limited syndrome of radiation characterized by nausea, vomiting, diarrhea, and psychic depression following exposure to appreciable doses of ionizing radiation, particularly to the abdominal region; its mechanism is unknown and there is no satisfactory treatment; it usually appears a few hours after irradiation and may subside within a day, or may be more intense, depending on exposure, and cause death.

radiation source — an apparatus or material emitting or capable of emitting ionizing radiation.

radiation standards — regulations that govern exposure to permissible concentrations of radioactive materials as well as their transportation.

radical — a group of atoms of different elements acting as a single unit in a chemical reaction. They are normally incapable of existing separately. A radical may be negatively charged, positively charged, or without a charge. Also known as group.

radioactive — referring to an unstable isotope that decays spontaneously and releases subatomic particles or energy.

radioactive carbon dating — *See* carbon dating.

radioactive contamination — deposition of radioactive material in any place where it is not desired, particularly where its presence may be harmful.

radioactive decay — the change in the nuclei of radioactive isotopes that spontaneously emits high-energy electromagnetic radiation and/or subatomic particles, while gradually changing into a different isotope or element.

radioactive equilibrium — in a radioactive series, the state which prevails when the rates of decay of two or more successive members of a series remains constant.

radioactive half-life — the time required for a radioactive substance to lose 50% of its activity by decay.

radioactive waste — waste generated by the emission of particulate or electromagnetic radiation resulting from the decay of the nuclei of unstable elements.

radioactivity — the quality of emitting or the emission of particulate or electromagnetic radiation as a consequence of the decay of the nuclei of unstable elements.

radiobiology — the study of the principles, mechanisms, and effect of radiation on living organisms.

radiocarbon dating — *See* carbon dating.

radiochemical — any compound or mixture containing a sufficient portion of radioactive elements to be detected by a Geiger counter.

radiochemistry — the phase of chemistry concerned with the properties and behavior of radioactive materials.

radio detection and ranging — *See* radar.

radiodiagnosis — a means of diagnosis that uses X-ray examination.

radioecology — the study of the effects of radiation on plants and animals in natural communities.

radio-frequency (rf) — electromagnetic radiations produced by very rapid reverses of current in a conductor; suitable for radio transmission above approximately 10^4 Hz and below 3×10^{12} Hz.

radiography — the making of film records of internal structures of the body by exposure of film specially sensitized to X-rays or gamma rays.

radioisotope — a radioactive atomic species of an element with the same atomic number and usually identical chemical properties where the substances are both unstable and decaying; used as biological tracers, in industrial control operations, and in the diagnosis and treatment of disease.

radiological health specialist — an environmental health practitioner responsible for protection of workers and the general public from ionizing and nonionizing radiation hazards.

radiological monitoring — the periodic or continuous determination of the amount of ionizing radiation or radioactive contamination present in an occupied region as a safety measure for health protection.

radiological survey — an evaluation of the radiation hazards in an area that may occur as a result of production, use, or existence of radioactive materials or other sources of radiation under certain conditions.

radiometer — an instrument for measuring the intensity of radiant energy by twisting on its own axis of suspended waves that are blackened on one side and exposed to a source of radiant energy; an instrument for measuring electromagnetic or acoustical energy.

radionuclide — a radioactive nuclide which has the capability of spontaneously emitting radiation; an isotope of an element that is unstable and exhibits radioactive decay; a cancer-causing, hazardous air pollutant that may be naturally-occurring or produced by accelerators, reactors, and fission by-products.

radiophotoluminescence — the luminescence exhibited by certain materials after exposure to ionizing radiation.

radiopoison — a radioactive poison, such as strontium-90.

radiosensitive — relative susceptibility of cells, tissues, organs, organisms, or any living substance to the injurious action of radiation.

radiosonde — a balloon-borne instrument used to measure and transmit temperature, pressure, and humidity at various heights in the atmosphere.

radiotherapy — treatment of disease with the application of relatively high roentgen doses.

radiotoxicity — having the potential of an isotope to cause damage to living tissue by absorption of energy from the disintegration of the radioactive material introduced into the body.

radium — Ra — a naturally-occurring, silvery-white, radioactive metal formed by the decay of uranium and isotopic forms, the most abundant being radium 226. It is used in medicine, industrial radiography, and as a source of neutrons and radon. Previously used in luminescent paints for watches and clock faces, instrument panels in airplanes, and military instruments and compasses. It is hazardous to the blood, bones, liver, eyes, hematopoietic tissues, and immunological system; and is toxic by inhalation, absorption, ingestion, and contact. Symptoms of exposure include anemia, necrosis of the jaw, osteogenic sarcoma, bone marrow failure, chronic liver disease, cataracts, breast and liver cancer, and dental disease. OSHA does not regulate radium. EPA limit for radium 228 and radium 226: 5(2 $\times 10^{-1}$) picocuries per liter of water.

radius (R) — any line segment going straight from the center to the outside of a circle or sphere.

radon — Rn — a naturally occurring, colorless, odorless, tasteless, radioactive gas formed by the decay of radium which may be found in rocks, soils, and uranium ore. Radon produces a series of daughter products through rapid radioactive decay which are characterized by relatively short half-lives and are easily adsorbed on the dust particles present in the air. The particles in the air containing the adsorbed gases are deposited on the surface of the respiratory tract, emitting alpha radiation doses to the epithelium and diffusing from the lungs into the blood. It was used from 1930–1950 in the medical treatment of malignancies, obliterative arthritis, and atherosclerosis of the lower extremities. It is hazardous to the respiratory system, kidneys, and blood; and is toxic by inhalation. Symptoms of exposure include increased mortality due to cancer and nonneoplastic diseases, silicosis, chronic obstructive pulmonary disease, fibrosis, and emphysema; chronic nephritis and renal sclerosis, hematological effects, lung cancer, and genotoxic effects with chromosomal aberrations. Radon is released from soils and can enter buildings through cracks in foundations and other openings. EPA's annual atmospheric release rate environmental and indoor air from residual radioactive material from inactive uranium processing sites: 20 picocuries per m² per second [air].

Radon Gas and Indoor Air Quality Research Act of 1986 — a federal law enacted because of the potential health effects of high levels of radon gas in structures in certain areas of the country, and because existing federal radon and indoor air pollution research programs were fragmented, under-funded, and did not provide adequate information concerning exposure to radon and indoor air pollutants.

ragweed — any of a variety of weeds that produce highly allergenic pollen and potential allergic reactions.

rain — water falling in drops condensed in vapor in the atmosphere.

rain forest — a forest with high humidity, constant temperature, and a large rain fall of approximately 150 inches per year.

râle — a sound or noise in the lungs accompanying normal sounds of respiration.

ramp sanitary landfill method — variation of the area landfill method in which a cover material is obtained by excavation in front of the working face.

random allocation — the experimental procedure in which allocation of subjects to treatment groups is controlled by a table of random numbers; an essential part of valid clinical trials.

random error — (*statistics*) the measure of random variations around a specific value, such as taking 10 samples of a gaseous air pollutant in a confined area and the samples range around the specific value; an error that can only be predicted on a statistical basis.

random incident field — a sound field in which the angle of arrival of sound at a given point in space is random in time.

Q
R

random noise — oscillation whose instantaneous magnitude cannot be specified for a given point in time.

random sampling — (*air*) a process in which grab or integrated samples are collected at random intervals; (*statistics*) a sample so drawn from the total group that every item in the group has an equal chance of being chosen.

Rankine (R) — *See* Rankine temperature scale.

Rankine cycle — an ideal steam-engine cycle since it is theoretically reversible. Also known as a steam cycle.

Rankine temperative scale (R) — an absolute temperature scale on which the unit of measurement equals a fahrenheit degree and on which the freezing point of water is 491.69° and the boiling point is 671.69°.

rapid sand filter — a closed, pressurized tank containing filter media made up of filter sand, and free of carbonates or other foreign material; used to filter swimming pool water and/or drinking water.

rash — a general term for any abnormal reddish coloring or blotching on some part of the skin.

rasper — a grinding machine in the form of a large vertical drum containing heavy, hinged arms that rotate horizontally over a rasp-and-sieve floor.

raster — (*mapping*) a regular grid of cells covering an area.

raster database — (*mapping*) a database containing all mapped, spatial information in the form of regular grid cells.

raster display — (*mapping*) a device for displaying information in pixels on a visual display unit.

raster map — a map encoded in the form of a regular array of cells.

raster-to-vector — the process of converting an image made up of cells into one described by lines and polygons.

rat-bite fever — (*disease*) either of two diseases, streptobacillus fever and spirillary fever, transmitted by the bite of a rat. *See* streptobacillary fever; spirillary fever.

rate — (*pesticides*) the actual amount of a pesticide chemical put in or on a plant, animal, or part of a building per unit of time; the cost of a unit of insurance through a specified period of time; the measure of probability of an event in a population.

rate coding — the process of controlling muscular force by regulating the firing rate of motor neurons.

rated capacity — the number of tons of solid waste that can be processed at an incinerator in a 24-hour period.

rate of flow — the quantity of water flowing past a given point in a unit of time; usually measured in gallons per minute.

rate of flow indicator — a flow meter which measures differentials across a calibrated orifice and indicates the rate of flow at that point, usually in gallons per minute.

rate of refrigeration — a measure of temperature control influenced by heat transfer properties of food, volume of food to be refrigerated, kind of containers used, heat conductivity of containers, agitation of food, and temperature difference between food and refrigerator unit.

rationale — a discussion of a concern, some issues involved, and some methods in dealing with the concern.

Raynaud's disease — a primary or idiopathic vasospastic disorder characterized by bilateral and symmetrical pallor and cyanosis of the fingers with or without local gangrene; caused by exposure to cold or emotional upset.

RBC — red blood cell. *See* red blood count; erythrocyte.

RBE — *See* relative biologic effectiveness.

RCRA — *See* Resource Conservation and Recovery Act.

RDF — *See* refuse-derived fuel.

reaction — (*chemistry*) a change involving the rearrangement of the atoms and molecules of one or more additional substances that often have different properties.

reactivation — the treatment of activated carbon to remove adsorbed organic material and restore its adsorption capabilities. Also known as regeneration.

reactive bond — (*chemistry*) readily given to or involving a reaction; (*physiology*) readily responding to a stimulus.

reactive liquid — any liquid which removes the gaseous pollutant from a chemical reaction transforming the pollutant to a less offensive form.

reactivity — (*chemical*) a substance's susceptibility to undergo a chemical reaction or change that may result in explosion or burning and release of corrosive or toxic emissions.

reactor — a nuclear reactor.

reagent — any substance used in a chemical reaction to produce, measure, examine, or detect another substance.

real conductivity — a property of material which when multiplied by the electric field gives the free current density.

real number — a number that has both an integer and a decimal component.

real time — tasks or functions executed so rapidly that the user gets an impression of continuous visual feedback.

**Q
R**

reasonable available control technology (RACT) — the level of air pollution emissions control required to be imposed on all existing sources in nonattainment areas, pursuant to the Clean Air Act.

rebuttal — an answer or response to a statement; to refute, oppose, or contradict that which has been stated.

recall — removal of goods from the market through voluntary action by the producer in lieu of seizure.

recarbonation — the diffusion of carbon dioxide gas through water to restore the carbon dioxide removed by adding lime to water in water softeners.

receiving waters — rivers, lakes, oceans, or other watercourses that receive treated or untreated wastewaters.

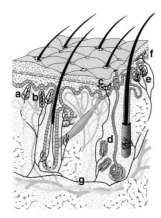

(a) heat (Ruffini's end organ)
(b) touch, pressure
       (Meissner's corpuscle)
(c) touch, pressure (Merkel's discs)
(d) deep pressure
       (Pacinian corpuscle)
(e) cold (end bulb of Krause)
(f) pain (naked nerve endings)
(g) cutaneous nerve

Receptors

receptor — a special molecule that recognizes and binds to a foreign chemical and is involved in the initial steps in a toxic response; a sense organ.

receptor antagonism — the process that occurs when a chemical blocks the effect of a second chemical.

recharge — to add water to the zone of saturation, as in recharging of an aquifer.

recharge zone — the area through which water is added to an aquifer. Also known as an intake area.

recidivism — a tendency to relapse into a previous condition or mode of behavior.

recirculating system — the entire system of pipes, pumps, and filters which allows water to be taken from the pool, filter-treated, and returned to the pool.

reclamation — the restoration to a better or more useful state, such as land reclamation by sanitary land filling or the obtaining of useful materials from solid waste.

recombinant DNA — deoxyribonucleic acid artificially introduced into a cell that alters the genotype and phenotype of the cell and is replicated along with the natural DNA.

recommended action — a statement of activity which, if implemented, is expected to contribute to the achievement of an objective.

recommended exposure limit (REL) — the National Institute of Occupational Safety and Health recommendation for maximum occupational exposure to a hazardous chemical.

recondition — (*law*) the process of reworking a lot of goods under seizure pursuant to a consent decree of condemnation and the attempt to bring the goods into compliance with the act.

reconstitute — to recombine dehydrated food products with water or other liquids.

reconstituted milk — a product resulting from the recombining of milk constituents with water and which complies with the standards for milkfat and solids-not-fat of milk.

record — (*computer science*) a set of related contiguous data in a computer file.

recorder controller — (*milk*) equipment that records pasteurization temperatures and time, and automatically controls the position of flow diversion valve.

recoverable resource — the capacity and likelihood of a valuable substance, such as metal, being recovered from solid waste for a commercial or industrial use.

recreational vehicle — a vehicle which may or may not be motor propelled and is used primarily for recreational purposes; includes motor homes, travel trailers, vans, etc.

recreational vehicle park — any tract of land used for parking five or more self-contained recreational vehicles including any roadway, building, structure, or enclosure used or intended for use as part of the park facilities, and any tract of land subdivided or leased for individual lots.

recruitment — the successive activation of the same and additional motor units with increasing strength of voluntary muscle contraction.

recycle — to separate waste material from the waste stream and process it so that it can be used again.

red blood count (RBC) — the number of erythrocytes in a cubic millimeter of blood.

red corpuscles — cells in blood containing hemoglobin.

Q
R

redevelopment — (*housing*) the elimination of underutilized areas and the development of land for other purposes, such as businesses, industries, highways, or new housing.

Red Flower Beetle (*Tribolium castaneum*) — an elongate reddish-brown flying beetle about 1/7″ long with a distinct joint between the thorax and abdomen; the last three segments abruptly enlarged; a stored food product insect.

Red-legged Ham Beetle (*Necrobia rufipes*) — a shiny blue to green beetle, 1/7″–1/4″ long with reddish legs; a stored food product insect especially problematic in the mid-Atlantic states.

redox — (*chemistry*) producing or containing the processes of reduction, where there is a gain of electrons in a chemical reaction, and oxidation, where there is a loss of electrons in a chemical reaction.

red tide — a proliferation of reddish ocean plankton that may kill large numbers of fish; a natural phenomenon which may be stimulated by the addition of nutrients. Also known as red water.

reduced hemoglobin — hemoglobin which has given up its oxygen.

reduced oxygen packaging — (*food*) the reduction of the amount of oxygen in a package by mechanical evacuation, displacement with another gas or gases, or controlling the oxygen level to keep it under 21%.

reducing agent — a substance that loses it valence electrons to another element and is readily oxidized.

reduction — a chemical reaction in which an element gains electrons and thereby decreases in valence; the addition of hydrogen or removal of oxygen in a substance; (*geology*) the extraction of a metal from its ore.

reduction control system — an emission control system which reduces emissions from sulfur-recovering plants by converting these emissions to hydrogen sulfide.

red water — *See* red tide.

re-entrainment — the process of air being exhausted from a building and immediately brought back into the system through the air intake and other openings in the building envelope so that contaminants are reintroduced into the building air.

reentry interval — (*pesticides*) the period of time immediately following the application of a pesticide to an area during which unprotected workers should not enter the area.

reference dose (RfD) — an estimate of the daily exposure of the human population to a potential hazard that is likely to be without risk of deleterious effects during a lifetime.

reference method — a method of sampling and analyzing ambient air for an air pollutant that is specified in the Code of Federal Regulations.

refinery — a plant at which gasoline or other petroleum-derived products are produced.

refinery gas — any form or mixture of gases gathered from equipment in a petroleum refinery.

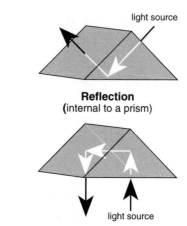

Reflection
(internal to a prism)

reflection — the return of a wave or ray after striking and bouncing off of a surface.

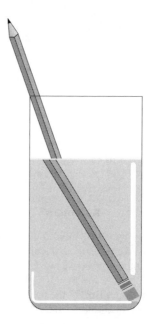

Refraction

refraction — the bending of light waves or rays as they pass at an angle from one material to another.

refractive index — the ratio of the velocity of light rays in a vacuum to that of a specified medium; the concentration and measuring of the bending of light rays by a refractometer.

refractometer — an instrument which measures the refraction of light rays.

refractory — a substance with a high melting point and good physical resistance to the effect of high temperature; resisting ordinary treatment.

refractory erosion — the wearing away of refractory surfaces by the washing action of moving liquids, such as molten slags, metals, or the action of moving gases.

refrigeration — the reduction of temperature of substances to a level below the ambient temperature.

refuse — all putrescible and nonputrescible solid wastes, except sewage and body waste; it includes garbage, rubbish, ashes, dead animals, discarded appliances and vehicles, trash, junk, and similar materials.

refuse burner — a device for either central or on-site volume reduction of solid waste by burning.

refuse chute — a pipe, duct, or trough through which solid waste is conveyed pneumatically or by gravity to a central storage area.

refuse container — a water-tight container constructed of metal or other durable material, impervious to rodents, and capable of being serviced without creating unsanitary conditions. All openings to the container, including covers and doors, must be tight-fitting.

refuse-derived fuel (RDF) — energy produced by burning the remainder of processed solid wastes to remove metal, glass, and other unburned materials.

regeneration — *See* reactivation.

regenerative absorbent — a liquid which releases the gaseous pollutant by application of heat or steam.

regenerative process — the replacement of damaged cells by new cells.

regenerator — a piece of equipment used in milk processing in which heat exchange occurs while the pasteurized milk is always under positive pressure on one side of the plate, and the unpasteurized milk is always under negative pressure on the other side of the plate.

regional response center (RRC) — the physical housing of the regional response team.

regional response team (RRT) — a group of official federal, regional, and state representatives who review the plans of the Local Emergency Planning Committee and offer necessary advice; also responds to large emergencies.

registered environmental health specialist (REHS) — a person who meets the educational and experience criteria, and passes appropriate examinations to be qualified to do the work of an environmental health specialist.

registered pesticide — a pesticide chemical accepted by the United States Department of Agriculture for use as stated on the label of the container.

registered trademark — the name, symbol, or mark used by a business or person to separate products.

registrant — (*pesticides*) any manufacturer or formulator who obtains registration for a pesticide-active ingredient or product.

registration — the process by which a qualified individual is listed on an official roster maintained by a governmental or nongovernmental agency or board; registration may also require minimum practice standards.

regulation — the legal mechanism for ensuring that a statute, law, or act's directives are to be carried out; the process of holding constant a property or characteristic.

regulatory letter — a letter to the management of a firm warning of violation of a specific law or regulation and the statement that unless immediate remedy is initiated, regulatory action will be taken by the governmental agency.

rehabilitation — (*housing*) the elimination of those properties which cannot be brought up to a standard and the rebuilding of properties that can be brought up to standard; a less expensive alternative to redevelopment; the process by which physical sensory and mental functional capacities are restored or developed after damage.

reheat — use of a heat exchanger or reintroduction of some flue gas downstream from a gas scrubber to raise the temperature of the gas to prevent condensation of water vapor in the stack.

REHS — *See* registered environmental health specialist.

Reid vapor pressure (RVP) — the absolute vapor pressure of volatile crude oil and volatile nonviscous petroleum liquids, except liquified petroleum gases.

reinfestation — the return of insects or other pests after extermination or migration.

REL — *See* recommended exposure limit.

relabeling — the application of different labels to a product.

relapsing fever — (*disease*) a systemic spirochetal disease in which periods of fever, lasting 2–9 days, alternate with no fever for 2–4 days, and the relapses vary from 1–10 times or more; the louse-borne disease averages 13–16 days; the tick-borne disease lasts longer. The incubation period 5–15 days, usually 8 days. It is caused by *Borrelia recurrentis*, a spirochete, for the louse-borne disease, and for the tick-borne disease, a variety of different strains of the

Q
R

spirochete. It is found in epidemic state when spread by lice and endemic state when spread by ticks. The tick-borne disease may be found in the United States. Reservoir of infection for the louse-borne disease is people. Reservoirs of infection for the tick-borne disease include rodents and ticks in the transovarian transmission; not transmitted from person to person. The epidemic relapsing fever occurs through the crushing of an infective louse with the organisms getting into the wound; communicability for the louse is 4–5 days after ingestion of blood from an infected person, and remains so for life for infected ticks; general susceptibility. It is controlled through the control of lice and ticks.

relational database — (*computer science*) a method of structuring data in the form of sets of records so relations between different entities and attributes can be used for data access and transformation.

relative atomic mass — *See* atomic weight.

relative biologic effectiveness (RBE) — a term used in radiobiology referring to the relative effectiveness of a given kind of ionizing radiation and producing a particular biological response, as compared with 250 keV x- or gamma rays; for any living organism or part of the organism, RBE is the ratio of the absorbed dose of a reference radiation that produces a specified biological effect to the absorbed dose of the radiation of the substance under evaluation that produces the same biological effect.

relative density — *See* specific gravity.

relative humidity — a measure of the water vapor content of the atmosphere calculated as the ratio of the vapor pressure in the atmosphere to the saturation vapor pressure at the existing temperature.

relative molecular mass — *See* molecular weight.

relative risk — a measure of the association between exposure to a factor and occurrence of a disease expressed as a ratio of the incidence rate in exposed persons to the incidence rate in unexposed persons.

release — any spilling, leaking, pumping, pouring, emitting, emptying, discharging, injecting, escaping, leaching, dumping, or disposing of a pollutant into the environment; excludes release into the workplace, emissions from engine exhaust, and release from a nuclear incident.

release mechanism — the means by which a hazardous material enters a transporting medium.

release source — any hazardous material, or substance contaminated with hazardous material, that can release the contaminant in any medium.

reliability — a measure of the consistency of a method in producing the same results with each trial.

relief valve — a device to vent the sterilizer in the event that pressure exceeds the chamber designed pressure.

rem — a unit of absorbed energy that takes into account the different relative biological effectiveness of different types of radiation; a unit of dose equivalent numerically equal to the absorbed dose in rads multiplied by the quality factor, distribution factor, and any other necessary modifying factors. *See also* sievert.

remediation — (*law*) the act or process of recovering a right or preventing or readdressing a wrong.

remote optical sensing of emissions (ROSE) — a mobile spectrometer system using long path, high resolution infrared wavelengths to analyze gaseous air pollutants.

remote sensing — the quantitative or qualitative determination of air pollutants or of meteorological parameters by means of instruments not in physical contact with the sample being examined.

remote sensing air sampler — a sampler used to measure pollution concentration by means of a sensor which detects the absorbed or scattered electro-optical beam as affected by pollutants at a distance.

renal — referring to the kidney.

rendering — the process of recovering fatty substances from animal parts by heat treatment, extraction, and distillation.

reovirus — a group of animal viruses spread by mosquitos and ticks and cause fevers. Also known as orbivirus.

rep — *See* roentgen equivalent physical.

repair — specific and limited intervention for corrective maintenance after a failure.

repellent — a pesticide chemical or agent that drives or keeps insects or other pests away from a plant, animal, product, or treated area.

replication — the act or process of duplicating or reproducing.

reportable quantity (RQ) — quantity of a hazardous substance that is considered reportable under the Comprehensive Environmental Response, Compensation, and Liability Act.

report of disease — an official report, usually by a doctor, of the occurrence of a communicable or other disease of humans or animals to departments of health, agriculture, or both as locally required; reports include those diseases requiring epidemiological investigation or initiation of special control measures.

representative sample — a large sample of a universe or whole which can be expected to exhibit the average properties of the universe or whole.

reprocessing — the action of changing the condition of a scrap material, as in aluminum into chairs and other products.

reproductive toxicant — an agent that affects post-pubertal reproductive or sexual function.

reproductive toxicity — the occurrence of adverse effects on the reproductive system that may result from exposure to a chemical.

reproductive toxin — a substance that affects the reproductive system and may impair fertility or cause chromosomal damage to a fetus.

reregistration — (*pesticides*) the reevaluation and relicensing of existing pesticidal active ingredients originally registered prior to current scientific and regulatory standards.

res — "things"; includes real and personal property.

resazurin test — a test in which the dye resazurin is added to fresh milk; reduction of the dye is indicated by a gradual change through various shades of purple and mauve to fading pink; the longer the time required to reduce the dye, the better the quality of the milk; similar to methylene blue test.

rescission — enacted legislation cancelling budget authority previously provided by the legislative authority.

research and development (R&D) — studious inquiry, examination, investigation, and experimentation aimed at the discovery of new information, products, and their practical implementation.

reservoir — any human being, animal, arthropod, plant, or inanimate matter in which an infectious agent usually lives. Also known as reservoir of infection.

reservoir of infection — *See* reservoir.

residential burner — a device used to burn the solid waste generated in an individual dwelling.

residential drain — the horizontal piping in a house drainage system which receives the discharge from soil waste and drainage pipes inside the walls of the house and conveys it to the residential sewer.

residential sewage disposal system — all equipment and devices necessary for proper conduction, collection, storage, treatment, and on-site disposal of sewage from a one- or two-family dwelling.

residential sewer — the horizontal piping beginning two feet outside of the house which carries discharges from the residential drain to the sanitary sewerage system or a residential sewage disposal system.

residential solid waste — all solid waste that normally originates in a residential environment.

residential waste — a combination of garbage, refuse, ashes, and bulky materials which come from residences.

residual chlorine — the chlorine remaining in water, sewage, or industrial waste at the end of a specified contact period, and which will react chemically and biologically with materials present.

residual oil — the heavy oil left after refining processes have separated out most of the lower-molecular weight compounds in crude oil.

residual pesticide — the amount of a pesticide chemical still effective for more than a few hours after application.

residual volume (RV) — the amount of air remaining in the lungs after maximum expiratory effort.

residual waste — the solid, liquid, or sludge waste from human activities in urban, agricultural, mining, and industrial environments remaining after collection and necessary treatment.

residue — material that remains after gases, liquids, or solids have been removed; (*pesticides*) the amount of a pesticide chemical that remains after it has been applied to a plant, animal, product, or area.

residuum — the residue from the weathering of sedimentary rock.

resin — a solid or semi-solid amorphous organic compound or mixture of such compounds only soluble in organic solvents, with no definite melting point, and no tendency to crystallize; may be of vegetable, animal, or synthetic origin; having distinctive physical and chemical properties and may be molded, cast, or extruded; used as adhesives or protective coatings.

resistance (R) — (*medicine*) the sum total of body mechanisms placing barriers to the progress or invasion of pathogenic organisms; (*electricity*) the ratio of the voltage across an object to the current flowing through it.

resistance welding — a process that produces coalescence of metals by the application of pressure and with the heat obtained from resistance of the work to electric current by circuit that includes the work.

resistant — able to survive in conditions which are harmful to other organisms through existing immunity or built-in immunity.

resistant species or strain — a type of plant, animal, or microorganism that has developed a tolerance for a pesticide chemical.

res judicata — (*law*) matters that have been decided and are not to be decided again; in food, the product has been seized for misbranding and adjudicated by order of the court. Therefore, the same article with the same labeling contained in the same misbranding may be seen subsequently and some rejudgment obtained without a lawsuit under this rule.

Q
R

resolution — the smallest spacing between two display elements.

resonance — oscillation of molecules between two or more structures, each possessing identical atoms but different arrangements of electrons; a marked increase occurring in the amplitude of oscillation of a system when a system is subjected to an oscillating force whose frequency is the same as or very close to the natural frequency of the system.

resource conservation — reduction of the amounts of solid waste generated by reduction of overall resource consumption and utilization of recovered resources.

Resource Conservation and Recovery Act (RCRA) — a federal act passed in 1976 and updated through 1988 authorizing the United States Environmental Protection Agency to identify hazardous wastes and regulate their generation, transportation, treatment, storage, and disposal.

resource recovery — the recovery of materials or energy from waste.

resp — *See* respiratory route.

respirable — of a size small enough to be inhaled deeply into the lungs.

respirable dust monitor — a device using beta-attenuation by respirable dust where the aerosols are drawn through a cyclone to remove non-respirable dust and impacted on a surface positioned between a beta source and a counter; the attenuation of the beta radiation is directly related to the amount of collector particulates.

respirable particulate mass threshold limit values (RPM-TLVs) — those materials which are hazardous when deposited in the gas-exchange region.

respiration — the exchange of oxygen and carbon dioxide between the atmosphere and the body cells, including inspiration and expiration, diffusion of oxygen from the pulmonary alveoli to the blood and of carbon dioxide from the blood to the alveoli, and the transport of oxygen to and carbon dioxide from the cells; to breathe.

respirator — a face mask used to filter out poisonous gases and dust particles from the air so that a person can breath and work safely.

respiratory protection — any of a group of devices that protect the respiratory system from exposure to airborne contaminants; usually a mask with a fitting to cover the nose and mouth.

respiratory route (resp) — a route of entry into the body through the breathing process.

respiratory system — a system consisting of the nose, mouth, nasal passages, nasal pharynx, pharynx, larynx, trachea, bronchi, bronchioles, and the muscles of respiration involved with the intake, exchange, and exhalation of oxygen in vertebrates.

respiratory toxicity — ability of a chemical to sicken or kill an animal or human by entering the lungs or other breathing organs as dust particles, gases, fine droplets, smoke, or vapors.

respiratory toxin — a chemical, such as silica or asbestos, that irritates or damages the pulmonary tissue causing symptoms of coughing, tightness in the chest, and shortness of breath.

respondent — (*law*) one who answers in notice of a hearing; responses given by the party to whom notice has been given, he or she is then the respondent.

response — the efforts to minimize the risks created in an emergency by protecting people, the environment, property, and returning to normal pre-emergency conditions.

response time — (*computer science*) the time that elapses between sending a command to a computer and receiving the results at the work station.

resting spore — a cell in a period of reorganization and temporary dormancy; often initiated by adverse or changing environmental conditions.

restricted use pesticide — a pesticide that is classified for restricted use under the provisions of the Federal Insecticide, Fungicide, and Rodenticide Act, such as fluoroacetamide (1081).

retail food processing establishment — any food establishment or portion thereof undertaking any of the activities of such establishment for sale or delivery of retail including grocery stores, meat or poultry markets, fish markets, fruit or vegetable markets, bakeries, confectionery stores, restaurants, delicatessens, and soft drink stands.

retail food store — any establishment or section of an establishment where food and food products are offered to the consumer and intended for off-premises consumption.

reticuloendothelial system — *See* hematopoietic system.

reticulum — a small network, especially a protoplasmic network, in cells.

retina — the inner, photoreceptive layer of the vertebrate eye formed from the expanded end of the optic nerve.

retort — a vessel within which substances are subjected to distillation or undergo decomposition.

retort-type incinerator — a multiple-chamber incinerator in which the gases travel from the end of the ignition chamber and pass through the mixing and combustion chamber.

Q
R

retrofit — a modification to equipment or a facility added to existing equipment of a facility.

return date — (*law*) the expiration date of the interval between seizure by the United States Marshall when a claimant may appear before the court. If no one appears as a claimant before expiration of the return date and proceedings, the United States Attorney will secure a default decree of condemnation forfeiting and disposing of the article seized.

return of service — (*law*) proof of service to the court of a person serving the process.

reuse — the reintroduction of a waste material or product into the economic stream without any chemical or physical change.

revenue bond — a bond issued by a public agency authorized to build, acquire, or improve a property and payable out of revenue from that property.

reverberation — the continuing of a sound in an enclosed area resulting from multiple reflections after the sound source has ended.

reverberation time — the time that would be required for the mean-square sound pressure level originally in a steady state to fall 60 decibels after the source is stopped.

reversal — (*law*) the act of reversing a court order or verdict; an Appellate Court may reverse a decision rendered by a District Court.

reverse osmosis — a membrane process in which water is forced to flow from a solution of high salt concentration to one of lower concentration.

reversible reaction — any reaction that reaches an equilibrium or which can be made to proceed in either direction.

revocation of probation — the return to the original penalty imposed on a person convicted of a criminal charge because the person violated the law during the period of probation.

revolution — (*geology*) the earth's motion around the sun in a 600 million mile orbit; the motion of a body around a closed orbit.

revolutions per minute (rpm) — the number of revolutions of an object per minute of time.

Reynolds number (N_{Re}) — a dimensionless parameter computed by dividing the product of pipe diameter, average velocity, and fluid density by the fluid viscosity.

rf — *See* radio-frequency.

RfD — *See* reference dose.

RF energy — notation for radiofrequency energy.

normal configuration of hand

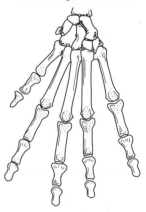

Rheumatoid Arthritis

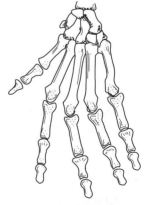

some results of deformity and
disintegration of joints and bones

rheumatoid arthritis — a chronic systemic disease with inflammatory changes occurring throughout the body's connective tissues.

rhinitis — inflammation of the mucous membrane of the nose.

rhinorrhea — the free discharge of a thick nasal mucus.

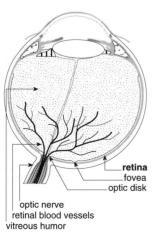

retina
fovea
optic disk
optic nerve
retinal blood vessels
vitreous humor

Retina

Q
R

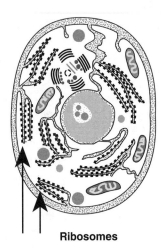

Ribosomes

rhodium (metal fume and insoluble compounds) — Rh — a white, hard, ductile, malleable metal with a bluish-gray luster; MW: 102.9, BP: 6741°F, Sol: insoluble, sp gr: 12.41. It is used in the manufacture of platinum alloys for use in thermocouples, windings for resistance furnaces, laboratory crucible and catalysts for chemical processing, and preparation of dental castings; in applying thin reflective coatings by sublimation to scientific instruments, radio and radar equipment, searchlight reflectors, cinema projectors, and headlight reflectors. No known hazards to humans; toxic by inhalation. No known symptoms of exposure. OSHA exposure limit (TWA): 0.1 mg/m^3 [air].

rhodium (soluble compounds) — Rh — appearance and odor vary depending upon the specific soluble compound; properties vary depending upon the specific soluble compound. It is used during refining and extraction of metal from platinum; in electroplating baths for finishing scientific instruments, camera fittings, radio equipment, and jewelry. It is hazardous to the eyes; and is toxic by inhalation, ingestion, and contact. Symptoms of exposure in animals include mild eye irritation, central nervous system damage. OSHA exposure limit (TWA): 0.001 mg/m^3 [air].

R horizon — (*soils*) the bedrock such as limestone, sandstone, or shale.

Rhydberg constant (R) — (*physics*) an atomic constant which appears in the formulas of wave numbers in all atomic spectra.

Rhyzopertha dominica — *See* Lesser Grain Borer.

ribonucleic acid (RNA) — a nucleic acid used for transportation and translation of genetic code found on DNA molecules.

ribose — $C_5H_{10}O_5$ — a 5-carbo sugar present in ribonucleic acid.

ribosomes — tiny, dense granules attached to the endoplasmic reticulum and lying between its folds; contain RNA and protein-synthesizing enzymes.

Rice Weevil (*Sitophilus oryza*) — a small reddish-brown to brown snout beetle 1/8″–1/6″ in length with small, brown pits on the thorax and two reddish or yellowish spots on each wing cover; a stored food product insect.

Richter scale — a logarithmic scale for expressing the magnitude of a seismic disturbance in terms of the energy dissipated; indicates the smallest earthquake; 4.5 on the scale is an earthquake causing slight damage, and 8.5 on the scale is a severe earthquake with considerable damage.

ricin — a white, toxic powder derived from castor beans.

rickettsia — a group of organisms intermediate in size between viruses and bacteria and which cause diseases.

rickettsial pox — (*disease*) a rickettsial infection with a skin lesion occurring at the site of a bite of a mite followed by fever and skin rash; caused by *Rickettsia akari*.

ringbuoy — a ring-shaped floating buoy capable of supporting a drowning person.

Ringelmann chart — a series of charts numbered from 0–5 that simulate various smoke densities ranging from white to black; a measure of 1 is equal to 20% black while a measure of 5 is equal to 100% black.

Ringelmann number — a rating from 0–5 on the Ringelmann chart.

riparian rights — rights of a land owner to water on or bordering his or her property including the right to prevent upstream water from being diverted or misused.

risk — a measure of the probability of an adverse or untoward outcome and the severity of the result and harm to the health of individuals in a defined population associated with a specific type of situation; (*emergencies*) the probability that damage to life, property, and/or the environment will occur if a hazard manifests itself.

risk assessment — the evaluation of short and long term risks by hazard identification, dose-response assessment, exposure assessment, and risk characterization; the process of integrating available information in the inherent toxicity of the chemical with information on how much of the chemical may come in contact with the individual.

risk characterization — the exposure assessment and the dose-response assessment combined to

Q
R

estimate some measure of the risk of toxicity and any uncertainties in the assessment of risk; the summary of the strengths and weaknesses of each component of the assessment is presented along with major assumptions, scientific judgment, and, to an extent, possible estimates of the uncertainties.

risk estimate — a description of the probability that organisms exposed to a specific dose of a chemical will develop an adverse response.

risk factor — characteristics, such as race, sex, age, weight, or variables, such as smoking or occupational exposure level, associated with increased probability of a toxic effect.

risk management — the development of regulatory options; evaluation of public health; and economic, social, and political consequences of the regulatory options leading to agency decisions and actions.

risk perception — the magnitude of the risk as perceived by an individual or population.

Risk Reduction Engineering Laboratory (RREL) — a division of the Environmental Protection Agency responsible for planning, implementing and managing research, development and demonstration programs to provide an authoritative, defensible engineering basis in support of the policies, programs, and regulations of the EPA with respect to drinking water, wastewater, pesticides, toxic substances, solid waste, hazardous waste, and Superfund-related activities.

river basin — the land area drained by a river and its tributaries.

Rivers and Harbors Act — a law passed in 1899 prohibiting the illegal discharge or deposit of refuse or sewage into navigable United States' waters to keep them free for boat traffic; industrial polluters can be subject to prosecution under this act.

RMCL — acronym for recommended maximum contaminant level.

rms value — *See* root mean square value.

RNA — *See* ribonucleic acid.

RO — symbol representing a hydrocarbon radical generated from an organic emission, such as formaldehyde or an aromatic derivative.

roach — a broad, flattened, dark or light brown or black insect; usually active only at night to seek food. Also called a cockroach.

rocky mountain spotted fever — (*disease*) a rickettsial disease with sudden onset of moderate to high fever lasting for 2–3 weeks, malaise, deep muscle pain, severe headache, chills, and rash. Incubation time is 3–14 days; caused by

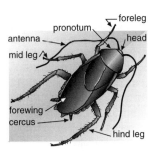

Roach

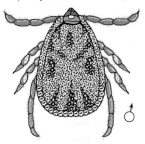

Dermacentor andersoni
(Rocky Mountain wood tick)

Dermacentor variabilis
(American dog tick)

Amblyomma americanum
(Lone Star tick)

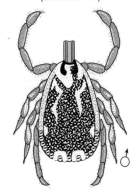

**Rocky Mountain
Spotted Fever**
(most common vectors)

Q
R

Rickettsia rickettsii; found throughout the United States. Reservoirs of infection include nature and transmission of the organism by transovarial and transstadial passage. Rickettsia can be transmitted to dogs, rodents, and other animals by the bite of an infected tick; the tick remains infected for life; not directly transmitted from person to person; general susceptibility. It is controlled by avoidance of tick infested areas.

rodent harborage — any location where rodents can live, nest, or seek shelter.

rodenticide — a pesticide chemical or agent that prevents damage by or destroys rodents.

rodentproof — a type of construction which prevents the ingress or egress of rodents from a given space or building, or their access to food, water, or harborage.

roentgen (R) — the amount of ionization made by a beam of gamma radiation or X-ray radiation in air that produces 2.082×10^9 ion pairs per cm^3 of standard air; 1 roentgen of exposure will usually produce about 1 rad of absorbed energy in soft tissue. *See also* exposure unit.

roentgen equivalent physical (rep) — a measurement of ionizing radiation absorbed in human tissue.

roentgen ray — *See* X-ray.

ronnel — $(CH_3O)_2P(S)OC_6H_2Cl_3$ — a white to light tan, crystalline solid; a liquid above 95°F; MW: 321.6, BP: decomposes, Sol (77°F): 0.004%, sp gr (77°F): 1.49. It is used as an insecticide and pest control agent for agricultural and livestock operations. It is hazardous to the skin, liver, kidneys, and blood plasma; and is toxic by inhalation, ingestion, and contact. Symptoms of exposure in animals include cholinesterase inhibition, eye irritation, liver and kidney damage. OSHA exposure limit (TWA): 10 mg/m³ [air].

rooming house — any dwelling or part thereof in which a specified portion may be used separately as a single habitable rooming unit.

rooming unit — any room or group of rooms forming a single habitable unit used or intended to be used for living and sleeping, but not for dining and/or cooking.

root mean square value (rms value) — the square root of the arithmetic mean of the squares of a set of related values.

ROSE — *See* remote optical sensing of emissions.

rosin — any of a large class of synthetic products that has some of the physical properties of natural resins but is different chemically and is used primarily in plastics; any of various products made from a natural resin or a natural polymer.

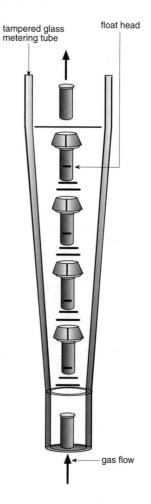

tampered glass metering tube

float head

gas flow

Rotameter

rotameter — a flow-measuring device that operates at a constant pressure and consists of a weight or float in a tapered tube; as the flow rate increases the float moves to a region of larger area and seeks a new position of equilibrium in the tapered tube related to flow rate.

rotary kiln stoker — a cylindrical, slightly inclined kiln that rotates causing the solid waste to move in a slow cascading motion forward.

rotary lime kiln — a kiln with an inclined, rotating drum used to produce a lime product from limestone by calcination.

rotary screen — an inclined, meshed cylinder that rotates on its axis and screens material placed at its upper end.

rotation — (*geology*) a whirling motion of the earth on its own axis once in about every 24 hours at a speed of approximately 1,000 miles per hour at the equator; motion about a fixed point.

Roundworm
(Ascaris lumbricoides)

rotavirus — (*disease*) a virus belonging to the family *Reoviridae* containing spherical RNA 65–75 nanometers in diameter with a wheel-like appearance, which may cause gastroenteritis with vomiting followed by diarrhea, malaise, fever, abdominal pain, and dehydration. Death may occur from dehydration or aspiration of vomitus. The mode of transmission is person to person, or possibly through contaminated food. Identical or similar viruses are found in various animals. The disease has been reported in the United States, and is the most common cause of diarrhea in infants and young children. The reservoir of infection is humans and possibly animals; general susceptibility. Also known as infantile gastroenteritis.

rotenone — $C_{23}H_{22}O_6$ — a colorless to red, odorless, crystalline solid; MW: 394.4, BP: decomposes, Sol: insoluble, Fl.P: not known, sp gr: 1.27. It is used as a pesticide, ascaricide, and insecticide on agricultural crops and livestock. It is hazardous to the central nervous system, eyes, and respiratory system; and is toxic by inhalation, ingestion, and contact. Symptoms of exposure include eye irritation, numbness of the mucous membrane, nausea, vomiting, abdominal pain, muscle tremors, incoordination, clonic convulsions, stupor, pulmonary irritation, and skin irritation. OSHA exposure limit (TWA): 5 mg/m³ [air].

rough fish — those species not prized for the purpose of fishing sports or for human consumption; more tolerant of changing environmental conditions than game fish.

round off — pertaining to a number shortened to a specific decimal place. Also known as truncating.

roundworm — a phylum of round, unsegmented nematode, including trichina and hookworms.

route of entry — *See* route of exposure.

route of exposure — the avenue by which a microorganism or chemical comes in contact with an organism, such as by inhalation, ingestion, dermal contact, or injection. Also known as route of entry.

rpm — *See* revolutions per minute.

RPM-TLVs — *See* respirable particulate mass threshold limit values.

RQ — *See* reportable quantity.

RRC — *See* regional response center.

RREL — *See* Risk Reduction Engineering Laboratory.

RRT — *See* regional response team.

RSD — acronym for relative standard deviation.

RTECS — Registry of Toxic Effects of Chemical Substances produced by the National Institute of Occupational Safety and Health.

rubbish — a general term for solid waste, excluding food waste and ashes taken from residences, commercial establishments, and institutions; nonputrescible solid waste, excluding ashes consisting of either combustible waste including paper, cardboard, plastic containers, vehicle tires, yard clippings, wood and similar materials, or noncombustible waste including tin cans, glass crockery, and similar materials.

rubble — broken pieces of masonry and concrete.

RUD — acronym for reflectance unit density.

run — the net period during which an emission sample is collected.

runoff — the portion of precipitation or irrigation water that drains from an area as surface flow.

rust — a type of corrosion in which iron is converted to hydrated Fe_2O_3 by the combined action of atmospheric oxygen and water.

RV — *See* residual volume.

RVP — *See* Reid vapor pressure.

Rx — *See* prescription.

Q
R

S

s — *See* second.

S — *See* siemen.

S & A — *See* surveillance and analysis.

SA — notation for specific activity; acronym for *Staphylococcus aureus*.

SAAQS — *See* Standard Ambient Air Quality Standards.

SAB — acronym for the Scientific Advisory Board to the Environmental Protection Agency.

SADT — *See* structure analysis and design technique.

SAED — acronym for selected area electron defraction.

safe — a condition of being reasonably free from danger and hazards that may cause unintentional injury or disease.

Safe Drinking Water Act (SDWA) — a law passed in 1974 and updated through 1988 to ensure a safe drinking water supply by regulating water provided by public water systems and hazardous substances, including carcinogens, in drinking water; directed the Environmental Protection Agency to prescribe national drinking water standards; allows states to enforce requirements for water purification, establishes a system for emergency allocation of water purification, and provides protection for underground sources of drinking water.

safe temperature — (*food*) a temperature of 41°F or below and 140°F or above as applied to potentially hazardous foods.

safety belt — a device usually worn around the waist or as a harness which secures a person into a vehicle or a worker to a structure.

safety cabinet — enclosure used for processing of biological materials.

safety can — an approved, closed container of not more than 5 gallons having a flash-arresting screen, spring-closing lid, and spot cover; designed to safely relieve internal pressure when subjected to fire exposure.

safety factor — a factor that is used to provide a margin of error when extrapolating from animal experimentation to estimate human risk.

safety glass — a transparent material that is prepared by laminating a sheet of transparent plastic between sheets of glass so that it resists shattering; especially used for car window shields.

safety guard — an enclosure designed to restrain the pieces of the grinding wheel and furnish all possible protection in the event that the wheel is broken in operation.

safety management specialist — *See* occupational safety specialist.

safety shoe — a steel-toed shoe or boot which is waterproof and impervious to chemicals.

safety solvent — a solvent which has relatively low toxicity and low flammability such as inhibited 1, 1, 1-trichloroethane.

sagittal — the longitudinal median plane of the body or any plane parallel to it.

salinity — the degree of dissolved salts in water measured by weight in parts per thousand.

salinization — the accumulation of salt in soil.

saliva — a fluid secreted into the mouth by the salivary glands to lubricate the passage of food and sometimes to carry out part of its digestion.

Salmonella — a complex genus of gram-negative, facultatively aerobic, usually motile, rod-shaped, pathogenic bacteria for people and animals.

salmonellosis — (*disease*) a bacterial disease with acute enterocolitis, sudden onset of headache, abdominal pain, diarrhea, nausea and sometimes vomiting, severe dehydration, and fever in infants; septicemia may occur. Incubation time is 6–72 hours, usually 12–36 hours; caused by varying serotypes of *Salmonella*, which are pathogenic to animals and people. It is found worldwide in small or large outbreaks, related to institutions, food, or water. Reservoirs of infection include domestic and wild animals, poultry, pigs, cattle, rodents, turtles, chickens, dogs and cats, as well as people who may be carriers. It is transmitted by ingestion of the microorganisms in food contaminated by feces of infected animals or people; also may be transmitted through water which is contaminated. It is communicable throughout the time of the infection, which can be several days to several weeks and the carrier state may last for more than a year; general susceptibility. It is controlled by proper cooking of foods, avoidance of raw foods, prevention of contamination of the foods by sick food handlers, and good personal hygiene.

salt — a usually crystalline compound containing positive ions from a base and negative ions from an acid in which the hydrogen of the acid has been replaced by metal or other positive ions; the minerals that water picks up as it passes through the air, over and under the ground, and from household and industrial use.

saltmarsh — a maritime community of salt-tolerant plants growing on intertidal mud and brackish conditions in sheltered estuaries and bays where excess sodium chloride is the primary environmental feature.

salt water intrusion — the invasion or displacement of fresh water or groundwater by salt water because of its greater density.

salvage — the recovery and utilization of waste materials for reuse or refabrication.

sample — a representative portion or specimen of an entity presented for inspection or analysis.

sampler — a device used with or without flow measurement to obtain an adequate portion of water, air, or waste for analytical purposes.

sand — small rock or mineral fragments ranging from 0.05 millimeter to 2.0 millimeter in diameter, distinguishable by the naked eye.

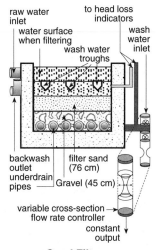

Sand Filter

sand filter — a water filter using sand or sand and gravel as a filter medium; a system which removes some suspended solids from sewage.

sandstone — a sedimentary rock consisting of sand-size particles, usually a quartz.

sandstone aquifers — geological formations found in many areas along with shale formations, supplying a free flowing spring and well water; shale serves as an aquiclude, in effect forming a conduit through which the water flows.

sandy loam — a soil which consists largely of sand but has sufficient quantities of silt and clay to impart stability.

sanitarian — *See* environmental health practitioner.

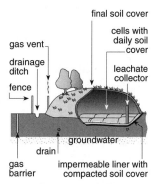

Sanitary Landfill

sanitary landfill — a site where solid waste is disposed of using sanitary landfilling techniques.

sanitary landfilling — an engineered method of disposing of solid waste on land in a manner that protects the environment by spreading the waste in thin layers, compacting it to the smallest compactible volume, and covering it with soil by the end of each working day.

sanitary landfill liner — an impermeable barrier, manufactured, constructed, or existing in a natural condition, utilized to collect leachate.

sanitary sewer — a sewer that carries sewage and to which storm, surface, and groundwaters are not intentionally admitted.

sanitary sewerage system — a sewer or a system of sewers which conveys sewage away from its origin to a wastewater treatment facility owned and operated by an incorporated city or town, conservation district, regional sewer district, or private utility.

sanitary survey — an on-site review of facilities, equipment, operation, and maintenance of water sources and sewage system problems.

sanitation — the control of physical factors in the human environment that can harm development, health, or survival.

sanitization — the act of reducing microbial organisms on cleaned food-contact surfaces to a safe level.

sanitizer — any product used for the effective bacteriocidal treatment of clean surfaces of equipment and utensils by a process which provides sufficient accumulated heat or concentration of chemicals for a proper amount of time to reduce the bacterial count, including pathogens, to an acceptable level by public health standards.

saponification — the hydrolysis of esters into acids and alcohols by the action of alkalis or acids, by boiling with water, or by the action of superheated steam.

saponin — any of numerous plants producing toxins that hemolyze red blood cells.

saprophyte — an organism living on dead organic matter.

SAR — *See* structure-activity relationship; supplied air respirator.

SARA — *See* Superfund Amendments and Reauthorization Act.

sarcoma — a malignant tumor of mesenchymal derivation.

satellite vehicle — a small collection vehicle that transfers its load into a larger vehicle operating in conjunction with it.

saturated organic compound — an organic compound that cannot take additional hydrogen atoms into its structure; compounds that are less reactive than unsaturated organic compounds.

saturated solution — a solution containing the maximum equilibrium amount of solute at a given temperature; the dissolved substance is in equilibrium with the undissolved substance.

saturated vapor pressure — *See* vapor pressure.

saturated zone — *See* zone of saturation.

saturation point — the maximum concentration of water vapor the air can hold at a given temperature.

sausage — a highly seasoned mixture of ground meat enclosed in a tube casing of animal intestine.

Saw-Toothed Grain Beetle (*Oryzaephilus surinamensis*) — a small, active, brown beetle, 1/10″ long with a flattened body and 6 sawtoothed projections on each side of the thorax; a major stored food product insect found throughout the world.

SBS — *See* sick building syndrome.

scabies — (*disease*) an infectious disease of the skin caused by a mite which penetrates the skin and lays eggs. Incubation time is 2–6 weeks before onset of intense itching; if previously infested, 1–4 days after re-exposure; caused by a mite, *Sarcoptes scabiei*. It is widespread, especially in areas with poor hygiene, during war, and economic disturbance. The reservoir of infection is people; transmitted by direct skin-to-skin contact, clothes, or infected individuals; communicable until mites and eggs are destroyed, usually 1–2 weeks; some individuals are resistant. It is controlled by isolating and treating infested individuals, their underclothing, clothing, and bed clothing.

scaffold — any temporary elevated platform and structure used for supporting workers, materials, or both.

scale — (*mapping*) the relationship between the size of an object on a map and its real size.

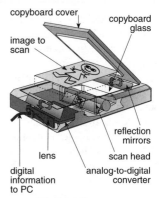

Flatbed Scanner

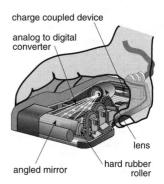

Hand-held Scanner

scanner — a device for automatically converting images into digital form; (*geographic information systems*) used to convert images from maps, photographs, or views of the real world into digital form.

scattered radiation — radiation which during its passage through a substance has been deviated in direction and may also have been modified by a decrease in energy.

scavenger — (*solid waste*) a person who participates in the uncontrolled removal of materials at any point in the solid waste stream; (*chemistry*) a substance added to a system to remove impurities.

SCBA — *See* self-contained breathing apparatus.

Schick test — a skin test to determine immunity to diphtheria.

schistosome — any of a genus of elongated trematode worms having separate sexes and parasitizing the blood vessels of birds and mammals. Also known as blood fluke.

schistosomiasis — (*disease*) a trematode infection of humans during which parasitic adult worms inhabit mesenteric or vesical veins of the host, resulting in symptoms of diarrhea, abdominal pain, urinary problems, liver fibrosis, hypertension, bacterial infection, and possibly bladder cancer. Incubation time is 2–6 weeks after exposure. It is caused by three

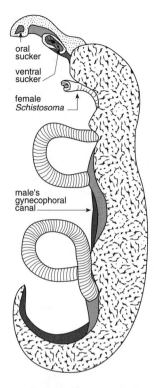

Schistosoma mansoni
(Blood Fluke)
carrier of schistosomiasis

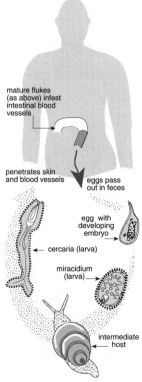

Life cycle of *S. mansoni*
Schistosomiasis

major species of blood flukes: *Schistosoma mansoni, S. haematobium,* and *S. japonicum* and is found in Africa, South America, the Caribbean, the Middle East, and Asia. The reservoir of infection for *S. haematobium* and *S. mansoni* is humans. The reservoirs of infection for *S. japonicum* are humans, dogs, cats, pigs, cattle, and horses. It is transmitted through water when free-swimming larval forms develop in freshwater snails, and then leave the snails to penetrate the skin of mammals. It is not communicable from person to person, but an infected individual can excrete the organisms for many years; general susceptibility. It is controlled by the proper disposal of excreta, the destruction of snails, and the wearing of proper clothing to prevent the organism from penetrating the skin.

scientific method — an orderly method used in scientific research generally consisting of identifying the problem, gathering data, formulating hypothesis, performing experiments, interpreting results, and reaching a conclusion.

scientific name — the accepted name used throughout the world by scientists for each animal and plant; made up of a genus and species.

scientific notation — a short form of mathematical notation used by scientists in which a number is expressed as a decimal number between 1 and 10 multiplied by a power of 10. Also known as standard notation.

scintillation counter — a counter in which light flashes produced on a scintillator by ionizing radiation are converted into electrical pulses by a photomultiplier tube.

scintillation detector — a detector operating on the principle of energy being transferred from radiation to a substance which in turn produces visible or near-visible light that may be picked up on a photosensitive vacuum tube and developed into an electrical pulse.

S
T

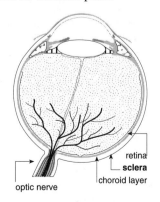

Sclera

sclera — the hard, white, outer coat of the eye.
sclerosis — hardening of tissue.

Scombroid **poisoning** — (*disease*) poisoning caused by eating *scombroid* fish or fish of the family Mahi-mahi; poisoning due to a histamine-like substance produced by several species of *Proteus* bacteria or other bacteria.

scooter — a three-wheeled satellite vehicle equipped with a flat bed, dump box, or packer body used to gather solid waste.

scotomo — an area of lost or depressed vision within the visual field, surrounded by an area of less depressed or normal vision.

scrap — any discarded or rejected material or parts of material that result from manufacturing or fabricating operations and may be recycled.

screen — (*solid waste*) a device for separating material according to size by passing it through variously sized mesh and retaining sorted materials on the surfaces.

scrubber — a device using a liquid spray to remove aerosols and gaseous pollutants from an air stream by making the particles heavier than the surrounding gas, aiding in separation of the pollutants. Also known as a wet collector.

scrubbing — the washing of any impurities from a process gas stream.

scrub typhus — (*disease*) a rickettsial disease causing a skin ulcer from an infected mite. Symptoms include fever, headache, profuse sweating, red eruption on the body, cough, and pneumonitis. Incubation time is 10–12 days, varying from 6–21 days. It is caused by *Rickettsia tsutsugamushi*, found in Asia and Japan. The reservoir of infection is the infected larval stages of mites transmitted by the infected larval mite; not communicable from person to person; general susceptibility. It is controlled by preventing contact with infected mites and not using contaminated bedding and clothing. Also known as Tsutsugamushi disease.

SCS — *See* Soil Conservation Service.

SD — *See* standard deviation.

SDR — acronym for standard dimension ratio.

SDWA — *See* Safe Drinking Water Act.

seafood — marine and fresh-water animal food products.

sealed — the condition of being free of cracks or other openings that permit the passage or entry of unwanted elements.

sealing — (*wells*) the complete filling of a well (with grout) in order to protect the aquifer from contamination.

seatworm — *See* pinworm.

sebaceous — oily or greasy; relating to a hair follicle which produces sebum.

sebum — the greasy secretion of the sebaceous glands constituting fat, cellular debris, and it lubricates the skin.

sec — *See* second.

SEC — *See* Securities and Exchange Commission.

secator — a separating device that throws mixed material onto a rotating shaft where heavy, resilient materials are propelled to one side and light, elastic materials are propelled to the other side.

second (s, sec) — the 60th part of a minute of time; (*mathematics*) the 60th part of a minute of angular measure.

secondary air pollutant — any pollutant produced by the reaction of two or more primary pollutants in the presence of catalysts and sunlight.

secondary attack rate — the number of cases of a disease developing in a stated period of time among members of a closed group at risk.

secondary burner — a burner installed in a secondary combustion chamber of an incinerator to maintain a minimum temperature and to complete the combustion of incompletely burned gases. Also known as an afterburner.

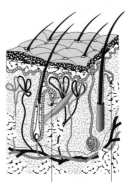

Sebaceous Oil Gland

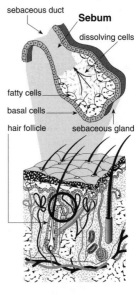

sebaceous duct

Sebum

dissolving cells

fatty cells

basal cells

hair follicle

sebaceous gland

Sebum

secondary combustion air — *See* overfire air.

secondary combustion chamber — the chamber of an incinerator where combustible solids, vapors, and gases from the primary chamber are burned and where fly ash settles.

secondary infection — infection acquired by person-to-person transfer from a primary case or from subsequent secondary cases.

secondary material — material utilized in place of a primary or raw material in manufacturing.

secondary maximum contaminant level (SMCL) — (*EPA*) maximum permissible level of contaminant allowed in free-flowing public water supply where the water can affect public welfare.

secondary pollutant — a pollutant formed in the atmosphere by chemical changes taking place between primary pollutants and sometimes other substances found in the air, such as acid rain.

secondary radiation — radiation which results from an absorption of other radiation in matter; may be either electromagnetic or particulate.

secondary sewage treatment — a system which uses primary sewage treatment followed by some type of biological oxidation, such as a trickling filter, activated sludge, and stabilization ponds, to reduce the solids present and biochemical oxygen demand of the effluent before entering a receiving stream.

secondary standard — a unit having a property that is calibrated against a primary standard to a known accuracy.

secondary treatment — the second step in most waste treatment systems in which bacteria consume the organic parts of the waste, which is accomplished by bringing the sewage and bacteria together in trickling filters or in the activated sludge process.

second-hand smoke — tobacco smoke inhaled by nonsmokers.

second law of thermodynamics — the law that states that heat will flow of its own accord from a hot body to a cooler body.

Securities and Exchange Commission (SEC) — a quasi-judicial commission formed by an act of Congress to provide the fullest possible disclosure to the investment public, to regulate investments, and to protect the public and investors against malpractice in the securities and financial markets.

sediment — (*soils*) the fine particles of soil produced by weathering which become suspended in water, air, or ice and finally settle.

sedimentary rock — a rock formed as the result of the weathering of preexisting rocks, erosion, and deposition.

sedimentation — the deposition of solids by gravity during wastewater treatment.

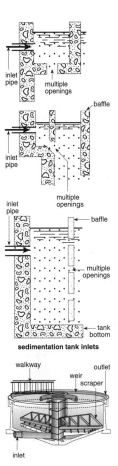

sedimentation tank inlets

Sedimentation Tank
(Circular)

sedimentation tanks — tanks in which sewage flows slowly and solids settle to the bottom or float on the top as scum; the scum is skimmed off the top and solids on the bottom are pumped to digestors with the effluent moving on for further treatment.

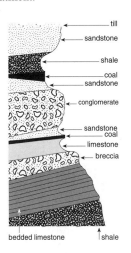

Sedimentary Rock

S
T

seepage — the slow movement of water or gas through soil or other medium without forming definite channels.

segregation — *See* recondition; separation.

seizure — confiscation by the United States Marshall of goods through a court order pending determination of the case.

selective absorption — the trapping of radiant energy from the sun by the carbon dioxide layer and the conversion of that energy into heat.

selenium compounds — Se — an amorphous or crystalline, red to gray element; occurs as an impurity in most sulfide ores; MW: 79.0, BP: 1265°F, Sol: insoluble, Fl.P: not known, sp gr: 4.28. It is used in the glassware industry for decolorization of fiberglass, scientific glassware, vehicular tail lights, traffic and other signal lenses, and infrared equipment; in the manufacture of electrical components; in the manufacture of photography and photocopy devices; in the manufacture of dyes, pigments, and colored glazes; in the manufacture of lubricating oils; in the rubber industry; in the manufacture of pharmaceuticals, fungicides, and dermatitis control; as a catalyst for hardening fats; in the manufacture of insecticides, parasiticides, bactericides, and herbicides; in the manufacture of flame-proofing agents on textiles and electric cables; in the manufacture of delayed action blasting caps; as solvents in paint and varnish removers; in the refining of copper, silver, gold, or nickel ores; in miscellaneous operations in the manufacture of insect repellents, activators, hardeners, special ceramic materials, plasticizers, and mercury vapor detectors; for preparation of feed additives for poultry and swine. It is hazardous to the upper respiratory system, eyes, skin, liver, kidneys, and blood; and is toxic by inhalation, absorption, ingestion, and contact. Symptoms of exposure include eye, nose, and throat irritation, visual disturbance, headache, chills, fever, dyspnea, bronchitis, metallic taste, garlic breath, gastrointestinal disturbances, dermatitis, eye and skin burns; in animals: anemia, liver and kidney damage. OSHA exposure limit (TWA): 0.2 mg/m³ [air].

selenium hexafluoride — SeF$_6$ — a colorless gas; MW: 193.0, BP: –30°F, Sol: insoluble. It is used as a gaseous electric insulator. No known hazards to humans; and is toxic by inhalation and contact. Symptoms of exposure in animals include pulmonary irritation and edema. OSHA exposure limit (TWA): 0.05 ppm [air] or 0.4 mg/m³.

self-contained breathing apparatus (SCBA) — a portable respiratory protection device that consists of a supply of air, oxygen, or oxygen-generating material, including a mask and hood, carried by the wearer; a device which provides complete respiratory protection against toxic gases and oxygen deficiency.

self-contained recreational vehicle — any recreational vehicle which can operate independent of connections to sewer and water and has plumbing fixtures or appliances which are connected to sewage holding tanks located within the vehicle.

self-insure — to assume liability for workers' compensation and avoid administrative costs associated with insurance policies.

self-purification — a process in which the stream proceeds by physical, chemical, and biological means to dispose of materials which may include suspended solids or other organic material.

SEM — acronym for scanning electron microscope; scanning electron microscopy.

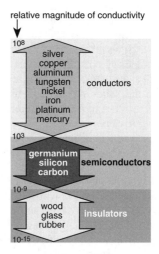

Semiconductor

semiconductor — a material, such as silicon, whose electrical resistance is somewhere between conductors and insulators.

semipermeable membrane — a membrane through which solvent molecules diffuse easily but through which dissolved substances diffuse slightly or not at all.

sensation — an impression produced by impulses conveyed by an afferent nerve to the sensorium.

sensation of sound — sensation created when sound waves pass through the outer ear, which is shaped to collect the waves and channel them down the ear canal to the ear drum.

sense organ — a body organ sensitive to a particular kind of sensitivity such as vision, hearing, smell, taste, and touch.

sensible — capable of being perceived by the sense organs.

sensible heat — the heat generated or absorbed by a substance during a change in temperature which is not accompanied by a change of state.

sensitive ingredient — any ingredient historically associated with a known microbiological hazard.

sensitivity — the ratio of the number of persons testing positive for the disease divided by the total number of persons tested; the minimum amount of contamination that can be repeatedly detected by an instrument.

sensitization — the process of making an individual sensitive to the action of a chemical.

sensitization dermatitis — an allergic reaction of the skin to certain chemicals after repeated exposure with production of antibodies that produce skin reactions in sensitized people.

sensitizer — a chemical that causes a substantial proportion of exposed people or animals to develop an allergic reaction in normal tissue after repeated exposure.

sensor — a device that measures a physical quantity or the change in a physical quantity, such as temperature, pressure, flow rate, pH, or liquid level. *See also* detector.

sensorium — the part of the cerebral cortex that receives and coordinates all the impulses sent to individual nerve centers.

sensory nerve — a nerve composed only of the fibers of sensory neurons.

sensory neurons — those neurons that carry impulses from a receptor to the central nervous system.

sentence — the formal judgment pronounced by the court or judge upon the defendant after conviction in a criminal prosecution.

separation — the segregation of solid waste into designated categories as general as paper, met-als, and glass, or as specific as colors of glass. Also known as segregation.

sepsis — the bacteriological process of decay.

septage — the liquid content, including sludge and scum, from a septic tank periodically pumped out and transported to another site for disposal.

septicemia — a disease condition due to the presence of pathogenic bacteria and their toxins in the blood. Also known as blood poisoning.

septic filter field use rating — (*soils*) a determination of the limitations of a soil to the absorption of water.

septic system — an individual sewer system employing a septic tank and soil treatment seepage trenches that are partially or totally in original soil material. Also known as conventional septic system.

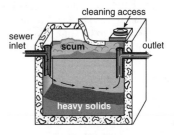

Septic Tank

septic tank — a water-tight underground receptacle which receives sewage and is designed and constructed to provide for sludge storage.

septum — a membrane between bodily spaces or masses of soft tissue.

sequela — a harmful consequence persisting after recovery from a disease.

sequential sampling — a process in which a series of individual grab or integrated samples are collected one after the other at regular, predetermined intervals accumulating results until a determination is made.

sequestering agent — a substance that removes a metal ion from a solution system by forming a complex ion that does not have the chemical reactions of the ion that is removed.

sequestrants — chelates used to deactivate undesirable properties of metal ions without the necessity for removing these ions from solutions.

sequestration — the process of surrounding or tieing up metal atoms in large, ring-shaped molecules.

SERC — *See* State Emergency Response Commission.

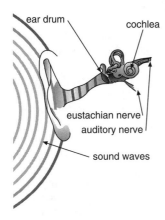

Sensation Of Sound

S
T

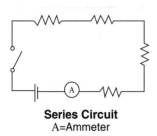

Series Circuit
A=Ammeter

series circuit — an arrangement of circuit elements such that a single path is formed for the current.

serious injury — any injury requiring an emergency service response, medical treatment as determined by the emergency medical response personnel, and/or resulting in medical attention at a hospital emergency room or admittance to a hospital.

serious level — the level of single pollutant combinations likely to lead to chronic disease or significant alteration of physiological function.

seroconversion — the process by which a person previously known to be antibody-negative converts to testing positive for human immunodeficiency virus antibodies.

serological — pertaining to serums and their study.

serologic test — any of a group of tests performed on blood.

seropositive — a condition in which antibodies to a disease-causing agent are found in the blood.

serum — a substance, usually an extract of blood, containing antibodies taken from an animal inoculated with bacteria or their toxins that is used to immunize people or animals; the clear liquid of the blood.

serum albumin — the principal blood protein necessary for absorption.

serum globulin — a globular blood protein containing antibodies.

serum hepatitis — (*disease*) *See* viral hepatitis B.

service — (*law*) the delivery of a legal document by an authorized person to a person who is thereby officially notified of some action or proceeding in which he or she is concerned, and is thereby advised or warned of some action which he or she is commanded to take or forebear.

service connection — (*water*) that point at which the water system enters any structure, building, or dwelling.

sesquicarbonate — a salt whose composition is between a carbonate and a bicarbonate.

sesquioxide — a compound of oxygen and a metal element in the proportion of three atoms of oxygen to two of the other; it is a semi-obsolete term.

seta — the erect, aerial part of the spore-producing structure of masses or liver warts; a bristle-like structure in Annelid worms.

settable solids — the matter in wastewater which will not stay in suspension during a preselected settling period but either settles to the bottom or floats to the top.

settlement — the sinking of the surface of a sanitary landfill or other engineered structure because of decomposition, consolidation, drainage, or underground failure.

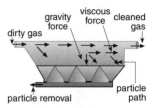

Settling Chamber

settling chamber — a dry collector air pollution control device consisting of an enclosed compartment in which the velocity of the carrier gas is reduced sufficiently to allow the particles to settle by gravity.

settling tank — a container that gravimetrically separates oil, grease, and dirt from petroleum solvent; a holding area for wastewater where heavier particles sink to the bottom and are siphoned off.

settling velocity — (*air pollution*) a velocity at which a given dust will fall out of dust-laden gas under the influence of gravity only.

seven-day average — the arithmetic mean of pollutant parameter values for samples collected in a series of seven consecutive days.

severe limitations — limitations that are difficult and usually costly to overcome or modify for subsurface seepage fields.

Sevin® — *See* carbaryl.

sewage — the water-carried human or animal wastes from residences, buildings, industrial establishments, or other places together with such groundwater infiltration and surface water as may be present; when fresh, gray in color with a musty odor; when old, black in color with a foul odor.

sewage disposal system — *See* sewage treatment system.

sewage lagoon — *See* oxidation pond.

sewage sludge — a semiliquid substance consisting of settled sewage solids combined with varying amounts of water and dissolved materials; it may either be raw or fresh, digested, elutriated, dewatered, or dry.

sewage treatment residue — coarse screening grit or sludge resulting from wastewater treatment units.

sewage treatment system — any sewage handling or treatment facility receiving domestic sewage and having a ground surface discharge, or any sewage handling or treatment facility receiving domestic sewage and having no ground surface discharge. Also known as sewage disposal system.

sewage treatment works — municipal or domestic waste treatment facilities of any type which are publicly owned or regulated.

sewer — an underground pipe or conduit used for carrying sewage.

sewerage — the entire system of sewage collection, treatment, and disposal.

sewerage system — any community or individual system publicly or privately owned for the collection and disposal of sewage or industrial waste of a liquid nature, including treatment facilities for sewage or industrial waste.

shale — a soft, sedimentary rock formed by consolidated clay or silt.

shale oil — any oil derived by retorting crushed oil-bearing rock and characterized by a large proportion of unsaturated hydrocarbons, alkenes, and dialkenes.

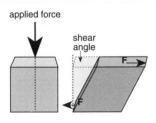

Shear

shear — a strain resulting in a change in shape.

shear shredder — a size reduction machine that cuts material between two large blades or between a blade and a stationary ledge used in solid waste disposal.

sheen — an iridescent appearance on the surface of water caused by the addition of a dye used as a tracking agent to a potential source of contamination to determine if the water is contaminated from that source.

shell — a pattern of electrons surrounding a nucleus.

shellfish — a group of mollusks, including oysters, mussels, and clams, usually enclosed in a self-secreted shell or exoskeleton.

shellstock — raw, in-shell molluscan shellfish.

shielded metal arc welding — a process that produces coalescence of metals by heating them

with an arc between a covered metal electrode and the work.

shielding — a control technique in which equipment or substances are physically kept away from people by barriers.

shield volcano — *See* basaltic domes.

shigellosis — (*disease*) an acute bacterial disease involving the large and small intestine with diarrhea, fever, nausea, sometimes toxemia, vomiting, and cramps; may cause convulsions in young children; also blood, mucus, and pus are present due to an enterotoxin produced by the bacteria. Incubation time is 1–7 days, usually 1–3 days. It is caused by four species of *Shigella*: *Shigella dysenteriae*, *S. flexneri*, *S. boydii,* and *S. sonnei.* It is found worldwide with most cases occurring in individuals under ten years of age, usually in overcrowded or poor environmental conditions, institutions for children, day-care centers; jails, mental hospitals, and on ships. The reservoir of infection is people. It is transmitted directly or indirectly through the oral-fecal route from person to person; the carrier state may exist; also transmitted through water, food, flies, or the person's hands. It is communicable for at least four weeks and in the carrier state, for many months; general susceptibility. It is controlled through proper supervision of institutions, especially children's centers, food, water, personal hygiene practices, control of feces and contaminated articles, as well as proper sewage disposal.

shipper — the party whose name and address appear on a bill of lading or freight bill as the one who introduced an article into interstate commerce.

shock — a condition of acute peripheral circulatory failure due to derangement of circulatory control or loss of circulating fluid.

Shore Protection Act — a 1988 law stating that a vessel may not transport municipal or commercial waste in coastal waters without a permit from the Secretary of the Department of Transportation and without displaying a number or marking on the vessel as prescribed by the secretary; the secretary has the right to enforce regulations concerning loading, securing, offloading, and clean-up.

short circuit — reduction to zero resistance or impedance across a voltage source usually resulting in damage if the circuit is not opened elsewhere; the condition occurring when the supply air flows to exhaust registers before entering the breathing zone.

short-term concentration (STC) — concentration measurements derived from a combination

S
T

of site monitoring and modeling information averaged over a relatively short period of time, usually 10–90 days; the information is used to evaluate potential subchronic effects.

short-term exposure limit (STEL) — the maximum concentration to which workers can be exposed continually, up to 15 minutes; no more than four periods are allowed per day and at least 60 minutes must expire between exposure periods.

short-term test — any test that can be completed in a short time that is used to examine genetic changes in laboratory cultures of cells in humans, animals, or lower organisms.

short ton — a measure of weight of a substance equal to 2,000 pounds.

shredder — a machine that reduces discarded automobiles and other low-grade sheet and coated metal to small pieces in a continuous operation.

shrink-swell potential — the amount of shrinkage or swelling which will occur in a soil when wet or dry.

shucked shellfish — shellfish which have been removed from their shells.

shute-fed incinerator — an incinerator that is charged through a shute that extends two or more floors above it. Also spelled chute fed incinerator.

SI — *See* surface impoundment.

SIC — *See* Standard Industrial Classification Code.

sick building syndrome (SBS) — a generalized term used to describe situations in which building occupants experience acute health and/or comfort effects that appear to be linked to time spent in a particular building, but where no specific illness or cause can be identified. Symptoms include acute discomfort, headaches, eye, nose, or throat irritation, dry cough, dry or itchy skin, dizziness, nausea, difficulty in concentrating, fatigue, and sensitivity to odors. Typically, the cause of the symptoms are not known, but the complainants report relief of the symptoms when they leave the building.

SIDS — *See* sudden infant death syndrome.

siemen (S) — in the meter, kilogram, second, and ampere system of measurement, it is a unit of conductance. Formerly known as Mho (Ω).

sievert (sv) — a unit of radiation dose equal to approximately 8.38 roentgens; 1 sv = 1 joule/kg = 100 rems. Formerly known as rem.

siftings — the fine materials that fall from a fuel bed through its grate openings during incineration.

signal power — signal voltage divided by the source impedance.

significant deterioration — the generation of pollution from a new source in previously unpolluted areas.

significant discharge — (*EPA*) any point source discharge for which timely management action must be taken to meet water quality objectives within the period of the operative water quality management plan.

significant hazard to public health — any level of contaminant which causes or may cause potable water to exceed the maximum contaminant level set forth in any promulgated National Primary Drinking Water Standard, or which may otherwise adversely affect health, or which may require a public water system to install additional treatment to prevent such adverse effects.

significant impairment — (*EPA*) impairment of visibility which interferes with the management, protection, preservation, or enjoyment of a visual experience by visitors at a national park or forest.

significant risk — pertaining to a situation posing a moderate likelihood of causing an unacceptable health risk.

SIL — *See* speech interference level.

silica (amorphous) — SiO_2 — a transparent to gray, odorless powder; MW: 60.1, BP: 4046°F, Sol: insoluble, sp gr: 2.20. It is used in the manufacture of insulating materials; as a filter medium in food and beverage manufacture; in the manufacture of construction bonding materials; in the manufacture of abrasive cleaning and polishing agents; in the manufacture of surface coatings; as a filler agent in paints, lacquers, and varnishes; in the manufacture of pharmaceuticals and as a constituent of pill masses, dentrifices, and salves; in pottery manufacture; in water treatment; as a carrier for nickel catalysts in petroleum and the petrochemical industries. It is hazardous to the respiratory system; and is toxic by inhalation. Symptoms of exposure include pneumoconiosis. OSHA exposure limit (TWA): 6 mg/m³ [air].

silica (crystalline — as respirable dust) — SiO_2 — a colorless, odorless solid; a component of many mineral dusts; MW: 60.1, BP: 4046°F, Sol: insoluble, sp gr: 2.66. It is used in the metallurgy industry for foundry molds, iron and steel casting, flux in smelting basic ores; in the manufacture of fiberglass; in the ceramic industry; as an abrasive in scouring and polishing soaps and powders, flint sandpaper, metal polishes, and sand blast work; in the processing of synthetic quartz; in the manufacture of refractories and building products; in grading and classification of electronic and optical grade quartz; in the manufacture of optical equipment in prisms, wedges, and lenses; in a variety of processes as in dental composition, in rocket engines and spacecraft, as a paint extender, and in graining lithographic plates. It is hazardous to the respiratory system; and is toxic by inhalation. Symptoms of exposure include cough, dyspnea, wheezing, impaired pulmonary function, progressive symptoms; carcinogenic. OSHA exposure limit (TWA): 0.1 to 0.05 mg/m³ [air].

silica gel — a regenerative absorbent consisting of the amorphous silica manufactured by the action of hydrochloric acid on sodium silicate used as a dehumidifying and dehydrating agent.

silicate — the largest group of minerals whose basic structure is a product of the joining of silicon, oxygen, and one or more metals and may contain hydrogen; the dust causes nonspecific dust reaction.

silicon — Si — a nonmetallic element being one of the primary constituents of the earth's crust.

silicon carbide — SiC — a bluish-black refractory material that is very dense, resists abrasion, and has a high melting point.

silicone — a group of compounds (SiR_2O) made by molecular combination of the element silicon or certain of its compounds with organic chemicals; produced in a variety of forms including silicone fluids, resins, and rubber with special properties such as water repellency, wide temperature resistance, and durability.

silicosis — a pneumoconiosis caused by the prolonged inhalation of silica dust, usually ten years or more, with symptoms of progressive shortness of breath and steady, dry, unproductive coughing in the early stages, later followed by mucous tinged with blood, loss of appetite, chest pain, and general weakness; the silica produces a nodular fibrotic reaction that scars the lungs and makes them receptive to the further complications of bronchitis, emphysema, and increased susceptibility to tuberculosis.

sill — (*hydraulics*) the top level of a weir or the lowest level of a notch; (*construction*) a beam or threshold.

silt — (*soils*) soil particles in an intermediate size range between clay particles and sand grains, ranging from 0.002 millimeter to 0.05 millimeter in diameter; sediment carried or deposited by water.

silt loam — (*soils*) a soil in which a moderate amount (0–50%) of fine grades of sand, a small amount (0–27%) of clay, and a large amount (50–88%) of silt particles occur.

silver (metal dust and soluble compounds) — Ag — a white, lustrous metal; MW: 107.9, BP: 3632°F, Sol: insoluble, sp gr: 10.49. It is used in the manufacture of silver nitrate for use in photography, mirrors, plating, inks, dyes, and porcelain; as germicides, antiseptics, caustics, and analytical reagents; in the manufacture of silver salts as catalysts; in chemical synthesis; in the manufacture of glass; in silver plating; in photography; in medicine. It is hazardous to the nasal septum, skin, and eyes; and is toxic by inhalation, ingestion, and contact. Symptoms of exposure include skin irritation, ocular burns, ulceration, and gastrointestinal distress. OSHA exposure limit (TWA): 0.01 mg/m³ [air].

silvicide — a pesticide chemical used to destroy woody shrubs and trees.

silviculture — management of forest land for timber; may sometimes contribute to water pollution.

simple asphyxiant — a physiologically inert gas that dilutes or displaces atmospheric oxygen below amounts needed to maintain blood levels adequate for normal tissue respiration; examples include carbon dioxide, ethane, helium, hydrogen, methane, and nitrogen.

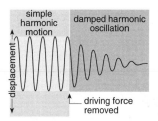

Simple Harmonic Motion

simple harmonic motion — periodic motion from a fixed point in which a particle goes equal distances and opposite directions in the equilibrium position. The particle completes oscillatory motion about a point in the same amount of time. The force exerted on the particle and the resulting acceleration are directly proportional to the displacement from the equilibrium position.

simulated workplace protection factor (SWPF) — a surrogate measure of the workplace protection provided by a respirator.

simulation — a mock accident or release designed to test emergency response methods or for use as a training tool; the determination of possible outcomes of situations expressed in the form of mathematical models.

single-blind design — an experiment in which participants are unaware of their allocation to treatments.

single-family dwelling — an unattached dwelling unit inhabited by an adult person plus one or more related persons.

Single-Grain Soil

single-grain soil — (*soils*) a soil in which structure is lacking and individual soil particles exist separately and do not form aggregates, usually found in sub-stratum or C horizon.

single-service article — tableware, flatware, hollowware, carry-out utensils, and other items such as bags, containers, stirrers, straws, toothpicks, and wrappers which are designed, fabricated, and intended by the manufacturer for one-time use.

single sources — industrial and/or municipal sources which release contaminants into the air.

sinkhole — (*geology*) a closed, funnel-shaped cavity formed in limestone regions by the removal of rock through the action of water.

sinks — (*physics*) any natural or artificial means of absorbing or removing a substance or a form of energy from a system.

sinoatrial node — a collection of atypical muscle fibers in the wall of the right atrium where the rhythm of cardia contraction is usually established. Also known as pacemaker of the heart.

sintering — a heat treatment that causes adjacent particles of a material to cohere below a temperature that would cause them to melt; used in powder metallurgy.

sinus — the air cavity in the bones of the skull; a cavity, recess, or depression in a part of an animal body.

sinusitis — inflammation of a sinus.

SIP — acronym for state implementation plans.

siphon — a closed conduit, a portion of which lies above the hydraulic grade line resulting in a pressure less than atmospheric and acquiring a vacuum within the conduit to start a flow; a siphon utilizes atmospheric pressure to affect or increase the flow of water through the conduit.

SIR — *See* standardized incidence ratio.

site — the land or water area before activity is physically located or conducted, including adjacent land use in connection with the facility of activity; any location where acutely toxic chemicals are manufactured, processed, stored, handled, used, or disposed.

SITE — *See* Superfund Innovative Technology Evaluation.

Sitophilus granarius — *See* Granary Weevil.

Sitophilus oryza — *See* Rice Weevil.

Sitotroga cerealella — *See* Angoumois Grain Moth.

SI units — Système International d'Unités. *See* International System of Units.

SIVE — *See* steam injection and vacuum extraction.

skeletal muscle — a striated, voluntary muscle attached to a bone.

skimmer — a device other than an overflow trough for continuous removal of surface water and floating debris from the pool.

skimmer weir — the part of the skimmer which adjusts automatically to small changes of water level to ensure a continuous flow of water to the skimmer.

skim milk — milk from which a sufficient portion of milk-fat has been removed to reduce its milk-fat content to less than 3-1/4%; milk which contains 1% butterfat.

skin absorption — the process by which some hazardous chemicals pass directly through the skin and enter the blood stream.

slag — the top-layer, nonmetallic waste product formed when the flux reacts with the impurities of an ore in a metallurgical process, such as smelting and refining. Also known as dross.

slaking — the addition of water to quick lime (CaO) to produce hydrated lime $(CaOH)_2$.

SLAMS — acronym for state or local air monitoring stations.

slaughter — to kill an animal for food.

slaughter house — any food establishment or portion thereof in which cattle, sheep, swine, goats, or horses are slaughtered for transportation, sale, or processing as food for human consumption.

sleet — transparent or translucent beads of ice occurring when rain dropping from upper warm air falls through a layer of freezing air; the rain drops first become freezing rain and turn into sleet or ice pellets.

slight limitations — (*soils*) limitations which are easy to overcome in soils in relationship to water permeability.

slope — (*dose-response assessment*) the linear portion of the curve which defines the potency of the agent; (*mathematics*) the deviation of a surface from the horizontal expressed as a percentage by a ratio or in degrees.

slot velocity — the linear flowrate of air through a slot.

slough — a wet or marshy area.

sludge — the semi-solid mixture of organic and inorganic materials that settle out of wastewater at a sewage treatment plant; the digested or partially digested solid material accumulated in a septic tank.

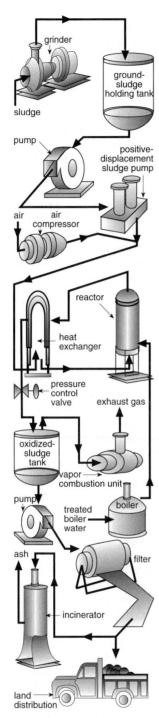

**Thermal Sludge
Conditioning Process
Schematic**

sludge conditioning — a pretreatment of a sludge to facilitate removal of water in a thickening or dewatering process by means of chemical, with inorganic and organic methods, elutriation, or heat treatment.

sludge reduction — any of the processes including incineration, wet air oxidation, and pyrolysis which primarily yield a major reduction in the volatile sludge solids.

sludge stabilization — a technique for converting raw, untreated sludge into a less offensive form with regard to odor, putrescibility, weight, and pathogenic organism content through anaerobic digestion, aerobic digestion, lime treatment, chlorine oxidation, heat treatment, and composting.

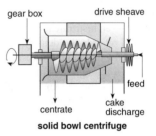

solid bowl centrifuge

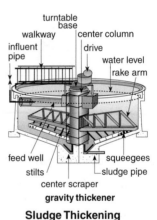

gravity thickener

Sludge Thickening

sludge thickening — a technique for increasing solid concentration by gravity, flotation, or centrifugation.

slurry — a mixture of liquid and insoluble material, such as portland cement.

slurry feed — a feed of diatomaceous earth introduced as a liquid slurry onto a piece of equipment to aid in filtration.

small quantity generator (SQG) — a generator that produces less than 1,000 kilograms of hazardous waste per month, accumulates less than 1,000 kilograms at any one time, produces less than 1 kilogram of acutely hazardous waste per month, or accumulates less than 1 kilogram of acutely hazardous waste at one time.

SMCL — *See* secondary maximum contaminant level.

smelting — the treatment of an ore by heat to separate out the desired metal.

S
T

s/mn² — notation for structures per square millimeter.

smog — a hazy mixture of smoke and fog resulting from the sun's action on certain pollutants in the air, especially those from automobile exhaust and factories; may also be any air pollution problem which reduces visibility.

smoke — the solid or liquid particles under one micron in diameter dispersed in a gaseous medium; (*incineration*) an aerosol consisting of all the dispersible particulates produced by incomplete combustion of carbonaceous materials entrained in flue gas.

smoke alarm — an instrument that continuously measures and records the density of smoke by determining how much light is obscured when a beam is shown through the smoke.

smoke density — the amount of solid matter contained in smoke, often measured by a system of grayness of the smoke to an established standard.

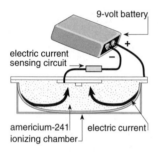

Smoke Detector

smoke detector — a device that can detect the presence of products of combustion in the air even in minute concentrations.

smoke number (SN) — the dimensionless term quantifying smoke emissions.

smokestack — a vertical pipe or flue designed to exhaust gases and any particulate matter suspended in the gases.

smooth muscle — the involuntary muscle found lining the walls of the intestines, stomach, and arteries.

SMR — *See* standardized mortality ratio.

SMS — acronym for synchronous meteorological satellite.

SMSA — *See* standard metropolitan statistical area.

SN — *See* smoke number.

Snellen chart — a vision-measuring chart consisting of block letters in diminishing sizes; the most common industrial test for vision acuity.

snow — precipitation in the form of small white ice crystals formed directly from the water vapor of the air at a temperature less than 32°F.

SO₂ — *See* sulfur dioxide.

SO₃ — *See* sulfur trioxide.

soap — a type of detergent that can be cast into bars; ordinarily a metal salt of a fatty acid, usually sodium stearate, sodium oleate, sodium palmitate, or some combination of those materials.

soap-bubble meter — a device used to calibrate high-flow and low-flow air-sampling pumps.

soapstone (containing less than 1% quartz) — an odorless, white-gray powder; MW: 379.3, BP: not known, Sol: insoluble, sp gr: 2.7–2.8. It is used in the manufacture of acid-proof coverings in floors, tables, sinks; in switchboard panels for high-electrical resistance; in the Kraft process of pulp manufacture; in fume cupboards and fireless cookers; in crayons for marking cloth, metal, and glass. It is hazardous to the lungs and cardiovascular system; and is toxic by inhalation and contact. Symptoms of exposure include cough, dyspnea, digital clubbing, cyanosis, basal crackles. OSHA exposure limit (TWA): 6 mg/m³ [resp].

SOC — acronym for synthetic organic chemical.

Social Security Administration (SSA) — a major division of the Department of Health and Human Services; its function is to operate a national program of contributory social insurance, to provide funds for retirement, death, and disability benefits, and for health benefits under Medicare.

soda ash — Na₂CO₃ — sodium carbonate powder used in glass manufacturing and petroleum refining, and for soaps and detergents.

sodium bicarbonate — NaHCO₃ — a white, crystalline chemical used to raise total alkalinity of the pool with no change in pH. Also known as baking soda.

sodium bisulfate — NaHSO — a dry white powder which produces an acid solution when dissolved in water; used as a laundry-rinse neutralizer, preservative, antiseptic, in glass etching, and tin plating.

sodium fluoroacetate — FCH₂C(O)ONa — a fluffy, colorless to white (sometimes dyed black), odorless powder; a liquid above 95°F; MW: 100.0, BP: 332°F, Sol: miscible. It is used in the formulation of pesticides. It is hazardous to the cardiovascular system, lungs, kidneys, and central nervous system; and is toxic by inhalation, absorption, ingestion, and contact. Symptoms of exposure include vomiting, apprehension, auditory hallucinations, facial paresthesia, twitching of the facial muscles, pulsus altenans, ectopic heartbeat, tachycardia, ventricle fibrillation, pulmonary edema, nystagmus, convulsions. OSHA exposure limit (TWA): 0.05 mg/ m³ [skin]. Also known as 1080.

sodium hydroxide — NaOH — a colorless to white, odorless solid (flakes, beads, granular form); MW: 40.0, BP: 2534°F, Sol: 11%, sp gr: 2.13. It is used in chemical manufacture; in the explosives industry; in boiler water and as a laboratory reagent; in pH control in the textiles, paper, and chemical industries; in the manufacture of pulp and paper; in pulping kraft process; in the manufacture of insulating board; in metal processing and refining; in petroleum refining; as a floatation reagent; in the manufacture of soaps and detergents; in food processing; in glass manufacture. It is hazardous to the eyes, skin, and respiratory system; and is toxic by inhalation, ingestion, and contact. Symptoms of exposure include nose irritation, pneumonitis, eye and skin burns, temporary loss of hair. OSHA exposure limit (TWA): ceiling 2 mg/mg³ [air].

sodium hypochlorite — NaOCL — pale green crystals with a sweet aroma and containing 12–15% available chlorine used as a bleaching agent for paper pulp and textiles, as a chemical intermediate, and in disinfection.

sodium paratoluene sulfonchloramide — *See* chloramine-T.

sodium thiosulfate — $Na_2S_2O_3$ — a chemical solution used to remove all chlorine from a test sample to avoid false pH test readings; in photography since it dissolves silver halides; in tanning, dyeing, and manufacture of chemicals.

soft swell — a can bulged at both ends, but not so tightly that the ends cannot be pushed in somewhat with thumb pressure.

soft water — water with very few minerals or other dissolved chemicals.

soil — unconsolidated particles from weathered rock, water, air, and humus over bedrock from which plants obtain essential materials.

soil absorption — a process which utilizes the soil to treat and dispose of effluent from a septic tank.

soil absorption system — pipes laid in a system of trenches or elevated beds into which the effluent from the septic tank is discharged for soil absorption.

soil aggregate — a group of soil particles cohering so as to behave mechanically as a unit.

soil air — air and other gases found in the voids between soil particles; similar to atmospheric air, but not mobile.

soil application — the process of applying a pesticide chemical on the soil.

soil boring — a soil core taken intact and undisturbed by a probe.

soil bulk density — the mass of dry soil per unit of bulk volume determined before drying to constant weight at 105°C.

soil characteristic limitation — those limits which preclude installation of an on-site sewage system because of certain soil characteristics, such as a high water table, high clay, or high silt levels.

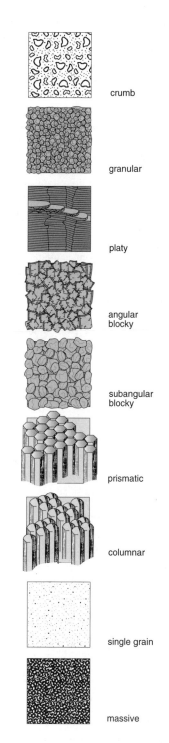

crumb

granular

platy

angular blocky

subangular blocky

prismatic

columnar

single grain

massive

Soil Structures

soil cohesion — the mutual attraction exerted on soil particles by molecular forces and moisture films.

soil conditioner — a synthetic chemical or natural material added in small quantities to improve the structure of soil.

Soil Conservation Service (SCS) — an agency of the Department of Agriculture whose function is to develop and carry out national soil and water conservation programs in cooperation with landowners, land-users, developers, and governmental agencies.

soil consistence — the combined properties of soil material that causes the aggregates to hold together or fall apart, resisting deformation.

soil deposition — the movement of material removed from soil and laid down as sediments.

soil drainage — the ability of the soil pores to avoid saturation and therefore drain freely.

soil erosion — the wearing away or removal of soil by the action of water or wind.

soil fumigant — a pesticide chemical used to kill pests in the soil; evaporates quickly.

soil gas — any gas such as radon, volatile organics, and pesticides entering a building from the surrounding ground.

soil horizon — a layer of soil or soil material in a soil profile differing from adjacent related layers in physical, chemical, and biological properties or characteristics such as color, structure, texture, consistency, kind, and number of organisms present and degree of acidity or alkalinity.

soil injection — the placing of a pesticide chemical below the soil surface with little or no soil mixing.

soil interpretation — a description of the soil horizon organized and presented to provide an understanding for soil use and management.

soil limitations — *See* soil characteristic limitation.

soil map — a map showing the distribution of soil, soil types, or other soil mapping units in relation to the prominent physical and cultural features of the earth's surface.

soil mineralogy — a study of the kinds and proportions of minerals present in soil.

soil morphology — the physical constitution, particularly the structural properties or soil profile, as exhibited by the kinds, thickness, and arrangement of the horizons in the profile, and by the texture, structure, consistency, and porosity of each horizon.

soil organic matter — plant and animal residue that decomposes and becomes a part of the soil.

soil permeability — the quality of a soil that enables water and air to move through it; a measure of this quality is the rate at which soil will transmit water under saturated conditions.

soil pipe — a pipe that conveys sewage containing fecal matter to the building drain or building sewer.

soil plasticity — the property of a soil that allows it to be deformed or molded in a moist condition without cracking or falling apart.

soil profile — a vertical cross-section of soil horizons to a depth of 2–5 feet.

soil profile analysis — the observation and evaluation of the physical characteristics of the soil horizon or layers to a depth of at least five feet, or to a shallower layer which cannot be readily penetrated.

soil saturation — the state when all the pores in a soil are filled with water.

soil science — the study of the principles of soil classification and mapping. *See also* pedology.

soil sealant — a chemical or physical agent used to plug porous soils to prevent leaching or percolation.

soil separate — any of a group of mineral particles separated on the basis of arrangement and sizes; the principle separates are sand, silt, and clay.

soil series — the basic unit of soil classification consisting of soils which are essentially alike in all major profile characteristics.

soil slope — the incline of the surface of soil area.

soil solution — the aqueous liquid phase of a soil and its solutes consisting of ions associated from the surfaces of the soil particles and other soluble materials.

soil sterilant — a pesticide chemical or agent that destroys all plants and animals in the treated soil for extended periods of time.

soil structure — the combination or arrangement of individual soil particles into definable aggregates which are characterized and classified on the basis of size, shape, and degree of distinctiveness.

soil suction — a measure of the force of water retention in unsaturated soil; equal to a force per unit area that must be exceeded by an externally applied suction to initiate water flow from the soil.

soil suitability — the determination of soil properties to assess their various uses.

soil survey — the systematic examination, description, classification, and mapping of soils in an area.

soil textural class — soil grouped on the basis of twelve specified ranges and textural classes.

soil textural classification — soil particles, sizes, or textures which correspond to soil classifications in the Department of Agriculture's Soil Survey Manual.

soil texture — the relative proportions of size groups in sand, silt, and clay of individual soil particles.

soil type — a subdivision of a soil series based on differences in the texture of the A horizon.

soil washing — a technique used to remove contaminants concentrated in a fine-size fraction of soil including silt, clay, and soil organic matter.

soil water — the moisture trapped between and around soil particles that contains nutrients available to plants and may contain pesticides, fertilizers, and other contaminants.

soil water tension — the expression in positive terms of the negative hydraulic pressure of soil water.

Sol — *See* solubility.

solar energy — energy produced by the sun and absorbed by the earth or trapped in various mechanical pieces of equipment producing thermal, electrical, mechanical, or chemical energy which is eventually reradiated into space as heat.

solder — a general term for a low-melting metal or alloy used to join to adjacent surfaces of less fusible metals or alloys; the principle types are soft solder and brazing solder.

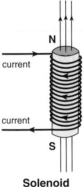

Solenoid

solenoid — a coil of wire, commonly in the form of a long cylinder, that when carrying a current resembles a bar magnet so that a movable core is drawn into the coil when a current flows.

solid flammable — any solid, other than a blasting agent or explosive, that is liable to cause fire, friction, absorption of moisture, spontaneous chemical change, retain heat from manufacturing or processing, or that can be readily ignited and burns so vigorously and persistently as to create a serious hazard.

solid-gas extraction — a chemical technique used to remove a desired gas from a mixture of gases.

solid-liquid extraction — a chemical technique used to remove impurities from a mixture by dissolving the substances from the desired precipitate, and the reaction of a solvent with a component to make it either soluble or insoluble.

solids-not-fat — the solids in milk such as protein, lactose, and minerals which are not fat.

solid-sorbent tube — a device used to capture insoluble or nonreactive gas and vapor contaminants from air through tubes filled with granular sorbents, such as activated charcoal.

solids retention time (SRT) — the average residence time of suspended solids in a biological waste treatment system equal to the total weight of suspended solids in the system divided by the total weight of suspended solids leaving the system per unit time.

solid waste — any garbage, refuse, or sludge from a waste treatment plant, water supply treatment plant, or air pollution control facility; discarded material including solid, liquid, semi-solid, or contained gaseous material resulting from industrial, commercial, mining and agriculture operations, and from domestic activities.

solid waste desorption process — a technique using superheated steam (up to 900°F) as a continuous conveying and stripping gas in a pneumatic system to treat contaminated solids.

solid waste disposal — the final disposal of refuse from which it cannot be salvaged or recycled.

solid waste management — the purposeful, systematic control, generation, storage, collection, transport, separation, processing, recycling, recovery, and disposal of solid waste.

solid waste manager — an environmental health practitioner responsible for the planning, administration, and regulation of storage, transportation, and disposal of solid waste.

solid waste storage container — a receptacle used for the temporary storage of solid waste before collection.

solubility (Sol) — the extent to which one substance will dissolve in another; usually expressed as the weight in grams of a substance needed to saturate 100 grams of a solvent at a given temperature.

soluble powder — a dust that will dissolve in water or other liquid.

solum — (*soils*) the upper part of a soil profile comprising the A and B horizons.

solute — the substance dissolved in a solvent to produce a solution.

solution — a uniform mixture of a solute in a solvent; a preparation made by dissolving a solid, liquid, or gaseous substance into another substance without a chemical change taking place.

solvent — (*chemistry*) a substance, usually a liquid, that can dissolve other substances.

SOM — acronym for sensitivity of methods.

somatic cell — any nonreproductive body cell which has two sets of chromosomes and is not a germ cell.

S
T

somewhat poorly drained — (*soil classification*) a classification describing soil in which the water is removed slowly enough to keep it wet for significant, but not constant periods.

sonar (sound navigation and ranging) — an electronic method of ranging using ultrasonic waves for underwater surveying, detection, navigation, and communication. A narrow pulsed beam of ultrasonic waves is sent through water and reflected back to the receiver. The length of time it takes for the reflected waves to return is the measure of the distance of the reflecting surface.

sonic boom — the thunderous noise made when shock waves reach the ground from a jet airplane or other craft exceeding the speed of sonic velocity.

soot — agglomerations of very finely divided, tar-impregnated carbon particles that form from the incomplete combustion of carbonaceous material. *See also* carbon particulate matter.

soot-blowing — periodic cleaning of the heat transfer equipment necessitated because during combustion the particulate matter sticks to the surface of the equipment and reduces its heat-exchanging ability; done by means of a jet of air or steam while the equipment is in use.

SOP — *See* standard operating procedure.

sorbent — an inert and insoluble material used to remove oil and hazardous substances from water.

sorbent barrier — a partition or an obstruction that prevents the attraction and retaining of substances by absorption or adsorption.

sorbitol — $C_6H_8(OH)_6$ — a crystalline, higher alcohol produced by the reduction of sugar with hydrogen. It is used in the manufacture of paper, tobacco, textiles, adhesives, pharmaceuticals, cosmetics, and in food.

sorption — a general term including absorption, adsorption, chemisorption, and persorption.

sound — auditory sensation produced by the oscillations, stress, pressure, particle displacement, and particle velocity in a medium with internal forces; pressure variation that the human ear can detect.

sound absorption — the process of removing sound energy by use of special materials.

sound absorption coefficient — the ratio of the sound absorbed by a surface exposed to a specific sound field to the sound energy on the surface.

sound analyzer — a device for measuring the band, pressure level, or pressure-spectrum level of a sound as a function of frequency.

sound energy — energy added to the medium in which sound travels; consisting of potential energy in the form of deviations from static pressure and kinetic energy in the form of particle velocity.

sound energy density — sound energy per unit volume, usually measured in ergs per cubic centimeter.

sound exposure level — the level in decibels calculated as 10 times the common logarithm of time integral of squared A-weighted sound pressure over a given time period or event divided by the square of the standard reference sound pressure of 20 micropascals and a reference duration of one second.

sound intensity — the rate at which sound energy flows through a unit area.

sound level — the weighted sound pressure level measured in decibels by using a metering characteristic; the weighing is in the scale A, B, or C.

sound level meter — a sensitive electronic voltmeter that measures the electric signal from a microphone which is ordinarily supplied with and attached to the instrument; the alternating current electric signal from the microphone is amplified sufficiently so that after conversion to direct current by means of a rectifier, the signal can deflect a needle on an indicating meter; an attenuator controls the overall amplification of the instrument; the instrument contains A-, B-, and C-scale networks; the high-frequency noise passed by the A-weighing network correlates well with annoyance effects and hearing damage effects on people; most of the studies that are done are read on the A-scale; on the A-scale, the meter reading is known as the dBA value.

sound navigation and ranging — *See* sonar.

sound power — the rate at which acoustic energy is radiated.

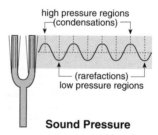

Sound Pressure

sound pressure — the difference between the actual pressure at any point in the field of a sound wave at any instant, and the average pressure at that point.

sound pressure level (SPL) — the value in decibels equal to 20 times the logarithm to the base 10 of the ratio of the pressure of the sound to a reference pressure, which shall be explicitly stated.

sound transmission class (STC) — single figure rating system designed to yield a preliminary estimate of the sound insulation properties of a partition or preliminary rank ordering of a series of partitions.

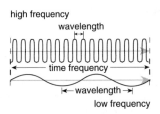

Sound Waves

sound waves — longitudinal waves causing particles of a gas, liquid, or solid elastic medium to vibrate along the direction of motion.

sour — (*oil*) pertaining to a gasoline or refined oil that contains hydrogen sulphide or mercaptans and has an offensive odor.

source — (*wells*) the site from which water is extracted for the purpose of supplying water to a private water system.

source reduction — (*hazardous waste*) any activity that reduces or eliminates the generation of hazardous waste within a process.

source separation — the separation of waste materials at the source by the public, business, or industry.

source specific — a source of pollution which can be identified in a community.

sour cream — a cream with an acidity of more than 0.20%, expressed as lactic acid.

SO$_x$ — *See* sulfur oxides.

spa — a hydrotherapy unit of irregular or geometric shell design.

space bomb — a container having a pesticide chemical plus a chemical aerosol under pressure which forces the pesticide out as a spray or a mist into a room or building.

space heater — a self-contained heating device of either the convection or radiant type intended primarily to heat only one room, two adjoining rooms, or some other limited space or area.

space spray — a pesticide chemical forced out of an aerosol container or sprayer as tiny droplets which form as a mist.

space telescope — an instrument launched into space from a spacecraft where the high resolution and faint object spectographics provide a wide range of spectral resolutions while the photometer measures the brightness of stars and the galactic background.

spasm — tightening or contraction of any set of muscles.

spatial resolution — the ability to resolve the spatial distribution of air pollution concentrations in air by modeling or monitoring; affected by distance to the source, number of sources, emission height, and emission strength.

special operation — the use of an area or building for a special purpose or certain circumstances

such as schools, hospitals, nursing homes, and shopping centers by people.

special waste — waste that requires extra treatment before entering into the normal plumbing system.

species — one of the smaller taxonomic subgroupings in a classification system of plants or animals consisting of interbreeding organisms of a single kind showing continuous morphological variations within the group but distinct from other such groups.

species diversity — a number which relates the density of organisms of each type present in a habitat.

specific activity (SA) — the activity or decay rate of a radioisotope per unit mass of a sample.

specific chemical identity — the chemical name, the Chemical Abstract Service registry number, or any precise chemical designation of a substance.

specific energy — the actual energy per unit mass deposited per unit volume in a given event.

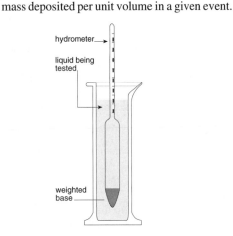

Instrument for finding
specific gravity of a liquid

specific gravity (sp gr) — (*weight*) the number expressing the ratio between the weight of certain volumes of a substance and the weight of an

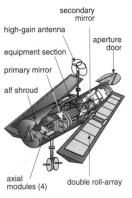

Space Telescope

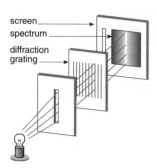

Grating Spectroscope

equal volume of some standard material, such as water or air, at standard temperature and pressure. Also known as relative density.

specific heat — the ratio between the amount of heat required to raise a given weight of a substance 1 degree in temperature and the amount of heat required to raise the same amount of a standard material, such as water, 1 degree in temperature.

specific humidity — the weight of water vapor per unit weight of dry air.

specific ionization — the number of ion pairs per unit length of path of ionizing radiation in a medium per centimeter of air or per micrometer of tissue.

specificity — the ability of a screening test to identify persons who do not have a disease, that is the ratio of persons that test negative to the total number of persons testing negative and testing positive.

specific rates — a ratio that refers to some particulate sub-group of events founded in the numerator and/or to some particular sub-group of a population counted in the denominator.

specific response — response of an air pollution analyzer to a single pollutant species without interference from other pollutants.

specific volume — the volume occupied by one pound of a substance under any specified condition of temperature and pressure.

specific weight — the weight per unit volume of a substance. *See* density.

spectrograph — a type of spectroscope which makes a photographic record of the spectrum of an object viewed through a telescope.

spectrophotometry — a direct-reading procedure used to measure the wavelength range of energy absorbed by a sample under analysis; similar to photometry, except that a prism made of glass is used in the visible range, quartz is used in the ultraviolet range, and sodium chloride or potassium chloride is used in the infrared range.

spectroscope — an optical instrument that separates a beam of light into its various component colors using a prism or grating, and a telescope, camera, or counter for observing the dispersed radiation.

spectrum — (*physics*) the band of colors formed when a beam of white light is broken up by passing through a prism or by other means; description of a function of time of its resolution into components each having a different frequency, amplitude, and phase.

speech interference level (SIL) — average of the sound-pressure levels of a sound expressed in decibels in three octave bands which have a center frequency of 500, 1,000, and 2,000 hertz.

speech perception test — a test designed to measure hearing acuity by administering a carefully controlled list of words.

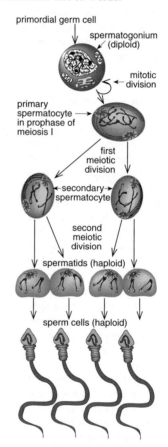

Spermatogenesis

spermatogenesis — the process of male gamete formation, including formation of a primary spermatocyte from a spermatogonium, meiotic division of the spermatocyte, and transformation of the four resulting spermatids into spermatozoa.

sp gr — *See* specific gravity.

Sphalerite

sphalerite — (Zn,Fe)S — an ore of zinc sulfide; brownish, yellowish, or black mineral. Also known as blende; false galena; lead marcasite; mock lead; mock ore.

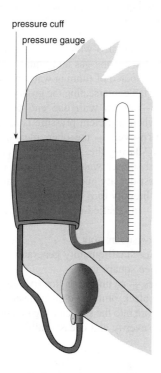

Sphygmomanometer

sphygmomanometer — an apparatus for measuring blood pressure.

spinal cord — the main dorsal nerve of the central nervous system in vertebrates extending down the back from the medulla.

spindle — (*genetics*) the microtubular apparatus that appears in many eukaryotic cells at the beginning of nuclear division and is responsible for the ordered separation of the chromosomes.

spirillary fever — (*disease*) a sporadic rat-bite fever caused by *spirillum minor, S. minus, Spirochaeta morsus muris*; a common form of rat-bite fever in Japan and Asia with fever rash of red or purple plaques. Incubation time is 1–3 weeks with healed wound reactivating when symptoms appear. The reservoir of infection is rats. It is transmitted by rat bite. It is not communicable from person to person. It is controlled by destroying rats and by treatment of the patient.

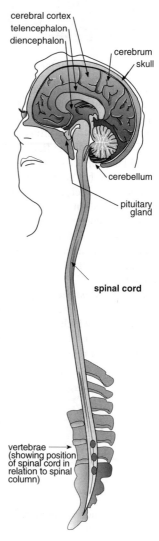

Spinal Cord

S
T

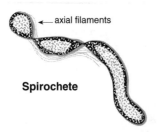

Spirochete

spirochete — any of an order of slender, spiraling, undulating bacteria of the order *Spirochaetales*.

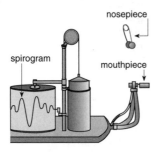

Spirometer

spirometer — an instrument that measures the flow of air in and out of lungs during respiration.

SPL — *See* sound pressure level.

splash-proof goggles — eye protection made of a noncorrosive material that fits snugly against the face and has indirect ventilation ports.

spoil — soil or rock that has been removed from its original location such as in mining.

spontaneous combustion — the process in which slow oxidation produces enough heat to raise the temperature of a substance to its kindling temperature without direct application of flame, as with the autoignition of rags soaked in flammable liquids.

spontaneously combustible — pertaining to a material that ignites as a result of retained heat from processing, or which will oxidize to generate heat and ignite, or which absorbs moisture to generate heat and ignite.

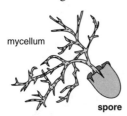

Spore

spore — a tiny mono- or multi-celled asexual reproductive structure produced by fungi or bacteria; often able to survive drought, cold, and other unfavorable environmental conditions.

sporophyte — the stage that produces spores in an organism having alterations of generations.

Sporozoea — a class of parasitic protozoa with both sexual and asexual phases.

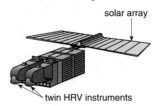

SPOT (French Satellite)

SPOT — an earth resource satellite with high resolution sensors launched by France in February, 1986.

spot check — supplemental test performed on a random basis.

spray — a mixture of water or other liquid plus a pesticide chemical applied in small droplets.

spray-drift — the tiny droplets of a pesticide chemical that are carried by wind or air currents away from the place of application.

spray tower — a vertical column for the spraying of the liquid through the gases in a variety of chambers of various designs; each liquid droplet contacts a volume of gas equal to the droplet cross-sectional area which creates a maximum contact between the gas and liquid.

spring — a source of running water produced by the water table intersecting the ground's surface.

springer — a can with one end permanently bulged when sufficient pressure is applied.

spring turn-over — the phenomenon in lakes occurring during the spring when surface ice melts and surface water temperature warms to its greatest density and then sinks, causing the water to literally turn over.

spurious — a falsified or erroneously attributed origin.

sq ft — *See* square foot.

SQG — *See* small quantity generator.

sq in — *See* square inch.

squall lines — an unbroken line (typically 12–30 miles) of black, ominous clouds towering 40,000 feet or more into the sky, causing thunderstorms of great severity and occasionally tornadoes.

Squamous Epithelial Tissue

squamous — covered or composed of scales.

square foot (sq ft) — a unit of length equal to 144 square inches.

square inch (sq in) — an area one inch wide by one inch long.

square mile — a unit of length equal to 640 acres.

square wave — a time-dependent function consisting of a series of discontinuous changes.

square yard — a unit of length equal to 9 square feet.

SRT — *See* solids retention time.

SS — *See* suspended solids; stainless steel.

SSA — *See* Social Security Administration.

St — *See* stoke.

stability — (*meteorology*) the atmospheric condition existing when the temperature in the air rises rather than falls with altitude, allowing for little or no vertical air movement; (*matter*) the ability of a material to remain unchanged.

stabilization — the treatment of active organic matter in sludge to produce inert, harmless material.

stabilization lagoon — *See* stabilization pond.

stabilization pond — a shallow oxidation pond with a continuous current used for oxidation of organic material present in the sewage. Also known as stabilization lagoon.

stabilizer-conditioner — a chemical compound, normally cyanuric acid, used to fortify and prolong the chlorine residual in pool water by reducing chlorine consumption lost by evaporation, dissipation, and oxidation.

stable — referring to a substance that does not change without the application of an external force.

stable air — an air mass that remains stationary rather than moving in its normal horizontal and vertical directions, not allowing for the dispersion of pollutants. Also known as stagnation.

stable isotope — a nonradioactive isotope of an element.

stack — a vertical passage above the roof through which products of combustion are conducted into the atmosphere; any chimney, flue, vent, roof monitor, conduit, or duct arranged to vent emissions to the ambient air; (*plumbing*) a general term for any vertical line of soil, waste, vent, or inside conductor piping.

stack effect — pressure-driven air flow produced by convection as heated air rises creating a positive pressure area at the top of the building and a negative pressure area at the bottom; the stack effect can overpower the mechanical system and disrupt ventilation and circulation in the building.

stack emissions — the particulate matter captured and released to the atmosphere through a stack, chimney, or flue.

stack gas — the air emitted to the atmosphere through a chimney after a production process or combustion takes place. Also called flue gas.

stack sampling — the collecting of representative samples of gases and particulate matter that flows through a duct or stack.

stack testing — a monitoring system which determines the emission characteristics and quantities at one particular point in time.

stack transmissometer — a monitoring device used to determine the opacity of gases leaving a smoke stack.

stagnant water — polluted water that is motionless or unable to drain off into the streams or into the ground.

stagnation — *See* stable air.

stain — a dye used to color microorganisms.

stainless steel (SS) — corrosion-resisting steel containing 15% or more chromium; used for cutlery, furnace parts, chemical plant equipment, valves, stills, turbine blades, and ball bearings.

standard — a desired or prescribed level of achievement or reference with respect to a criterion representing acceptable performance.

standard air — the dry air at 29.92 inches of mercury, absolute pressure and 70°F, weighing 0.075 pounds per cubic foot.

Standard Ambient Air Quality Standards (SAAQS) — standards established by the Environmental Protection Agency which describe the concentration limits that a given pollutant may reach annually and on a 24-hour basis.

standard candle — a standard measure of light brightness where one standard candle radiates about 12-1/2 lumens; replaced by the candela.

standard conditions — *See* standard temperature and pressure.

standard deviation (SD) — a measure of a dispersion or variability in a frequency distribution equal to the square root of the arithmetic mean of the squares of the deviations from the arithmetic mean of the distribution.

standard error — an estimate of the standard deviation of the means of many samples calculated as the standard deviations divided by the square root of the number of individuals in a sample.

standard gravity — a gravitational force which will produce an acceleration equal to 32.17 feet per second.

Standard Industrial Classification Code (SIC) — a classification system used to identify places of employment by major types of activity.

standardized incidence ratio (SIR) — a ratio of the crude rate of an exposed population to a weighted average of the category specific rate in the unexposed population, weighted by the distribution of the exposed population.

**S
T**

standardized mortality ratio (SMR) — the ratio of the number of deaths in a specific population to the number of deaths that would have been expected when the population was compared with the same mortality experience in a standard population.

standard metropolitan statistical area (SMSA) — an urbanized region with at least 100,000 inhabitants and strong economic and social ties to a central city of at least 50,000 people.

standard notation — *See* scientific notation.

standard operating procedure (SOP) — a document which describes in detail an operation, analysis, or action commonly accepted as the preferred method for performing certain routine or repetitive tasks; a routine procedure which has been designed by a manager or physician to be used when a specific type of problem occurs.

standard pressure — pressure equivalent to 760 millimeters of mercury at sea level at a temperature of 0°C.

standard rate — a rate calculated to permit a more accurate comparison of the state of health of two or more populations that differ substantially in some important respects, such as distribution by age or sex. Also called adjusted rate.

standard sample — the aliquot of finished drinking water that is examined for the presence of coliform bacteria.

standard temperature — the measurement arbitrarily set at 0°C at a pressure of 760 millimeters of mercury.

standard temperature and pressure (STP) — a temperature of 0°C at one atmosphere of pressure. Also known as standard conditions; normal temperature and pressure.

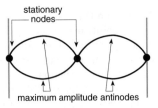

Standing Waves

standing wave — a periodic wave that has a fixed distribution in space. Also known as a stationary wave.

staphyloccal food poisoning — (*disease*) an intoxication of abrupt and sometimes violent onset with severe nausea, cramps, vomiting, and prostration; accompanied frequently by diarrhea for 1–2 days. Incubation time is usually 2–4 hours, but may vary up to 7 hours. It is caused by *Staphylococcus aureus* enterotoxins; found worldwide. The reservoir of infection is people. It is transmitted by ingestion of food containing the enterotoxin. It is not communicable from person to person; general susceptibility. It is controlled by proper heating and cooling of food, educating food handlers, proper hygiene, and the exclusion of people with boils, abscesses, runny noses, or other drainage conditions.

staphylococcal infection disease — (*disease*) any of a variety of diseases caused by bacterium that produces clinical symptoms ranging from a small boil to septicemia and death. Skin lesions such as impetigo, boils, carbuncles, and abscesses may occur; cellulitis, pneumonia, osteomyelitis, and endocarditis may also occur. Incubation time varies, but it is commonly 4–10 days. It is caused by coagulase-positive strains of *Staphylococcus aureus*, which may have different phage types, antibiotic resistance, or serological agglutination; found worldwide. The reservoir of infection is people. It is transmitted from the anterior nares, skin, and sites of infection in people; and is communicable as long as the lesions on the skin are present, where there is drainage from the nose, or while the carrier state persists. Susceptibility is highest among newborns and the chronically ill, elderly and debilitated people. It is controlled by good personal hygiene, especially hand washing, disinfection, proper placement and removal of all dressing and discharges in the hospital, treatment of the patient, and isolation in the hospital if it is a hospital-borne infection.

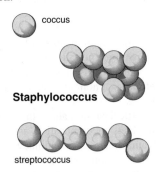

coccus

Staphylococcus

streptococcus

staphylococcus — a genus of gram-positive bacteria made up of spherical microorganisms tending to occur in grape-like clusters; constantly present on the skin and in the upper respiratory tract, and the most common cause of localized suppurating infections.

starch — $(C_6H_{10}O_5)_x$ — any of a mixture of polymers made up of the sugar glucose found in all assimilating (green) plants.

starch iodide method (iodometric) — a titration test in which chlorine displaces iodine from potassium iodide in an acid solution and forms a blue color with starch; it is decolorized by the addition of standard sodium thiosulfate; the test is generally used to measure high residuals of chlorine.

statcoulomb — *See* electrostatic unit of charge.

State Emergency Response Commission (SERC) — a committee of various state agencies, public and private groups, and associations appointed by the governor to assist in emergency preparedness and response capabilities through better coordination and planning.

static electric charge — the positive or negative stationary electric charge resulting from rubbing unlike bodies together.

static electricity — an accumulation of positive or negative charges on an object.

static pressure — (*acoustics*) pressure of the medium on which the alternating sound pressure is superimposed; (*fluid mechanics*) the pressure exerted by a fluid in a direction normal to the direction of flow; for a fluid at rest it is exerted uniformly in all directions; (*ventilation*) usually the difference between the absolute pressure in an exhaust system and atmospheric pressure such that static pressure less than atmospheric pressure is termed negative static pressure.

stationary front — a front which moves little or not at all, bringing mild conditions with some rain.

stationary grate — *See* fixed grate.

stationary source — any building, structure, facility, or installation which emits or may emit any air pollutant subject to regulation under the Clean Air Act.

stationary wave — *See* standing wave.

statistical significance — the probability of observing a certain result or class of results by chance alone is less than some predetermined value.

statistical technique — the use of information gained from an appropriate sample of a larger group to make inferences concerning the group in a specific matter.

statute of limitation — the period during which any criminal prosecution contemplated must be brought.

statutory law — a rule passed by a state or national legislature.

STC — *See* short-term concentration; sound transmission class.

STD — acronym for sexually transmitted disease.

steam — hot water vapor often used as a physical sanitizing agent for treating food equipment.

steam cycle — *See* Rankine cycle.

steam injection and vacuum extraction (SIVE) — a process used to remove volatile organic compounds and semivolatile organic compounds from contaminated soils.

steam-injection flares — flares that mix steam with exhaust gases at the top of the stack, separating the hydrocarbon molecules and reducing unwanted polymerization.

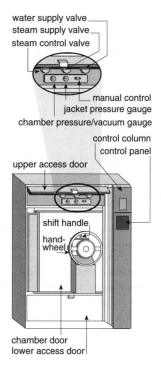

Steam Sterilizer

steam sterilization — a process conducted in an enclosed vessel at 121–123°C for 30 minutes to kill all bacteria and spores.

steel — an iron base alloy containing about 2% carbon that is malleable under certain conditions.

STEL — *See* short-term exposure limit.

STEM — acronym for scanning transmission electron microscope.

stenosis — narrowing of a body passage or opening, as of a blood vessel.

sterile — free of living microorganisms and spores.

sterilization — a process that kills all living organisms, including spores, on and in an object.

stibine — SbH_3 — a colorless gas with a disagreeable odor similar to hydrogen sulfide; MW: 124.8, BP: –1°F, Sol: slight. It is used as a chemical intermediate and in chemical synthesis; liberated from purification of antimony. It is hazardous to the blood, liver, kidneys, and lungs; and is toxic by inhalation. Symptoms of exposure include headache, weakness, nausea, abdominal pain, lumbar pain, hemoglobinuria, hemolytic anemia, jaundice, lung irritation. OSHA exposure limit (TWA): 0.1 ppm [air] or 0.5 mg/m^3.

still — a closed chamber in which heat is applied to vaporize a liquid, and the vapor is then condensed.

still air — air in a building with a velocity of 25 feet per minute or less.

St. Louis encephalitis — (*disease*) an acute inflammatory disease of short duration involving parts of the brain, spinal cord, and meninges with symptoms ranging from mild cases to high fever, disorientation, coma, convulsions, and possible death. Incubation time is usually 5–15 days. It is caused by a flavivirus, found in the United States, Canada, and parts of South America. Possible reservoirs of infection include birds, rodents, bats, reptiles, amphibians, mosquito adults or eggs. It is transmitted by the bite of a mosquito but not from person to person; usually affects young children, but can affect others and be more severe. It is controlled by destroying mosquito larvae, eggs, breeding places, adult mosquitos, and using mechanical control techniques.

stochastic — pertaining to the assumption that the actions of a chemical substance result from probabilistic events.

Stoddard solvent — a colorless liquid with a kerosene-like odor; MW: varies, BP: 428–572°F, Sol: insoluble, Fl.P: 110°F, sp gr: 0.78. It is used as a solvent in the dry cleaning industry; in the paint and varnish industries; as a solvent for printing inks and textile printing industries; in the manufacture of aerosol sprays; in the manufacture of sprays for pesticides, herbicides, household cleaners, and silicone compounds; as a solvent and thinner in protective coating materials; in metal cleaning and degreasing; in leather degreasing; as a general solvent in fabric water-proofing, processing of synthetic yarns, in rubber cements. It is hazardous to the skin, eyes, respiratory system, and central nervous system; and is toxic by inhalation, ingestion, and contact. Symptoms of exposure include dizziness, dermatitis, irritation of the eyes, nose, and throat. OSHA exposure limit (TWA): 100 ppm [air] or 525 mg/m^3.

stoichiometric air — *See* theoretical air.

stoichiometry — the mass relationship of elements and compounds in a chemical reaction.

stoke (St) — the unit of kinematic viscosity.

stoker — a mechanical device to feed solid fuel or solid waste to a furnace.

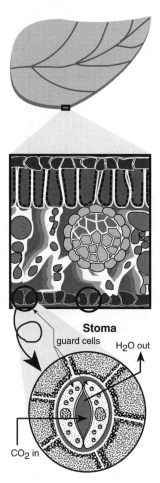

Stoma
guard cells　　H$_2$O out
CO$_2$ in

stoma — a tiny pore-like opening on the underside of a leaf through which a plant exchanges gases.

stone coal — *See* anthracite.

stool — fecal matter.

stopping power — the average rate of energy loss of a charged particle per unit thickness of a material or per unit mass of material transversed; a measure of the effect of a substance upon the kinetic energy of a charged particle passing through it.

storage — (*hazardous materials*) any method of keeping raw materials, finished goods, or products while awaiting use, shipment, or consumption; (*computer science*) the parts of a computer system used for storing data and programs.

Storage and Retrieval System (STORET) — an Environmental Protection Agency data bank of a variety of environmental information.

storage pit — a hole below ground in which solid waste is held before processing.

storage tube — a visual display device used for displaying maps and graphic information that retains the image continuously.

stored energy — *See* potential energy.

STORET — *See* Storage and Retrieval System.

storm sewer — a sewer used for conveying rain water, surface water, condensate, cooling water, or liquid waste off of a surface.

storm water — any water resulting from precipitation mixed with the accumulation of dirt, soil, and other precipitation which has fallen or which flows.

STP — *See* standard temperature and pressure.

STP flow rate — the rate of flow of fluid by volume corrected to standard temperature and pressure.

STP volume — the volume that a quantity of gas or air would occupy at standard temperature and pressure.

straight-chain hydrocarbon — a hydrocarbon in which the carbon atoms are linked together in a long, straight chain.

strategy — the science and art of employing political and psychological courses of organizations to afford maximum support to adopt a policy.

stratification — (*hydrology*) the phenomenon occurring when a body of water becomes divided into distinguishable layers; (*geology*) the deposition of sedimentary materials in layers; (*air*) the condition occurring when the air at different levels in the atmosphere is at a different temperature and little mixing occurs between the layers.

stratocumulus — irregular masses of clouds from near the earth's surface to 6,500 feet up, spread out in a rolling or puffy layer, gray with darker shading; does not produce rain but sometimes changes into rain clouds.

stratosphere — the region extending from tropopause to a height of approximately 33 miles from the earth where the temperature reaches a maximum; where clouds of water never form.

stratum — a single bed or layer, regardless of thickness, that consists generally of the same kind of rock or other material.

stratus — low clouds quite uniform, like fog, with an obvious base above the ground; when dull-gray, the sky looks heavy and ready to rain; fine drizzle can fall from these clouds; formed when a layer of air is cooled below the saturation point without vertical movement; these are sheet-like or layered.

street refuse — material picked up when streets and sidewalks are swept manually and mechanically.

streptobacillary fever — (*disease*) a bacterial disease caused by the bite of a rat with an onset of within 10 days after the bite has healed; characterized by chills, fever, headache, muscle pain, and rash; joints become swollen, red, and painful. Incubation time is 3–10 days. It is caused by *Streptobacillus moniliformis*, found worldwide but not common in North and South America. The reservoir of infection is the infected rat; transmission through secretion of the mouth, nose, or conjunctival sac of an infected animal, usually introduced by biting. It is not communicable from person to person; susceptibility is not known. It is controlled through specific pharmaceutical treatment. Also known as Haverhill fever.

streptococcal infection — (*disease*) disease or infection caused by *Streptococcus pyogenes*, group A streptococci of approximately 75 serological types; produces strep sore throat, scarlet fever, impetigo, erysipelas, puerperal fever, tonsillitis, pharyngitis. The incubation period is 1–3 days; common in temperate climates; seen in semitropical climates and less common in tropical climates. The reservoir of infection is people. It is transmitted by direct or close contact with a patient or carrier's nose, skin, hands, clothing, dust, or contaminated food. It is communicable for 10–21 days untreated; 24–48 hours if treated with penicillin; general susceptibility. It is

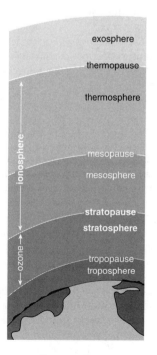

Stratosphere

controlled by proper health habits which incorporate an understanding of transmission, pasteurizing milk, and protecting and refrigerating foods.

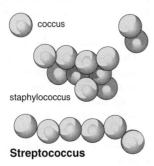

coccus

staphylococcus

Streptococcus

streptococcus — a genus of gram-positive facultatively aerobic cocci occurring in pairs or chains.

stress — the nonspecific response of the body to any demand made upon it; may result in a fight-or-flight response to potentially dangerous situations; the body reacts through elevated heart rate and blood pressure, and a redistribution of blood flow to the brain and major muscle groups and away from the distal body parts.

stressor — any agent causing a condition of stress.

strip — to remove friable asbestos materials from any part of a facility.

strip mining — *See* surface mining.

stripping — the process of removing one substance from another by heating it until it is loosened and then subjecting it to a gas stream that carries it away.

stroke — *See* cerebrovascular accident.

stroke counter — a flow-rate meter which measures air flow by counting the number of strokes of the piston and equating it to air flow.

strong acid — an acid capable of a high degree of ionization in water solution.

strong base — a base capable of a high degree of ionization in water solution.

strongyloidiasis — (*disease*) a helminthic infection of the duodenum and upper jejunum which is asymptomatic in most cases; dermatitis may occur when the larvae penetrates the skin; also cough, râles, and pneumonitis may occur as the larvae passes through the lungs; abdominal symptoms may occur when the adult female is in a mucosa of the intestine; the chronic infection may be either mild or severe with symptoms of diarrhea, weight loss, vomiting, weakness, or constipation. Incubation time is approximately 2 weeks from the penetration of the skin until the larva are found in the feces; the symptoms vary. It is caused by *Strongyloides stercoralis*, a nematode found in tropical and temperate areas. Reservoirs of infection include people and possibly dogs. It is transmitted from feces or moist soil contaminated with feces and penetration of the skin, carried to the lungs, and then to the stomach; communicable up to 35 years; general susceptibility. It is controlled by proper disposal of human feces, good health habits, good public education.

structural formula — a schematic representation of the arrangement of molecules of organic compounds.

structural geology — the study of the earth's structures and their relationship to the forces that produce these structures.

structural isomers — molecules that contain the same number and types of atoms but have a different arrangement of bonds.

Strongyloides stercoralis
(parasitic roundworm causing strongyloidiasis)
Strongyloidiasis

structure-activity relationship (SAR) — the association between a chemical structure and carcinogenicity.

structure analysis and design technique (SADT) — a tool for graphic representation of a method of analysis for the simplification of problem solving of complex systems.

strychnine — $C_{21}H_{22}N_2O_2$ — a colorless to white, odorless, crystalline solid; MW: 334.4, BP: decomposes, Sol: 0.02%, Fl.P: not known, sp gr: 1.36. It is used in the formulation of medicinals and pesticides. It is hazardous to the central nervous system; and is toxic by inhalation, ingestion, and contact. Symptoms of exposure include stiff neck and facial muscles, restlessness, apprehension, increased acuity of perception, increased reflex excitability, cyanosis, tetanic convulsions with opisthotonos. OSHA exposure limit (TWA): 0.15 mg/m³ [air].

stupor — partial unconsciousness or nearly complete unconsciousness.

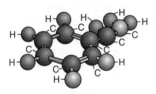

Styrene

styrene — $C_6H_5CH{=}CH_2$ — a colorless to yellow, oily liquid with a sweet, floral odor; MW: 104.2, BP: 293°F, Sol: slight, Fl.P: 88°F, sp gr: 0.91. It is used in the manufacture of tires and other rubber goods; in the manufacture of glass fiber; in the manufacture of surface coatings; as an intermediate material. It is hazardous to the central nervous system, respiratory system, eyes, and skin; and is toxic by inhalation, ingestion, and contact. Symptoms of exposure include drowsiness, weakness, unsteady gait, narcosis, defatting dermatitis, irritation of the eyes and nose. OSHA exposure limit (TWA): 50 ppm [air] or 215 mg/m³.

subacute — referring to stimulus not severe enough to bring about rapid response.

subacute toxicity — the property of a substance or mixture of substances to cause adverse effects in an organism upon repeated or continuous exposure within less than one-half of the lifetime of that organism.

subangular blocky — (*soils*) nut-like soil structure where the aggregates have sides from obtuse angles; corners are rounded, usually more permeable than blocky soil, nearly block-like with six or more sides, all three dimensions

about the same; usually found in subsoil or B horizon.

subchronic — of intermediate duration, usually used to describe studies on levels of exposure between 5–90 days.

subclinical — without clinical manifestations; the early stages or a very mild form of a disease.

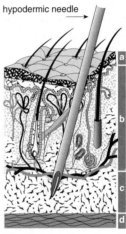

hypodermic needle

(**a**) epidermis   adipose layer (**c**)
(**b**) dermis   muscle layer (**d**)

Subcutaneous Injection

subcutaneous — beneath the layers of the skin.

subdivision ordinance — an ordinance adopted by a local entity for the appropriate establishment of rules relating to tracts of land.

subduction — (*geology*) the action or process of one crustal plate descending below the edge of another.

sublethal — referring to a stimulus below the level that causes death.

sublimation — the process by which matter passes from a solid to a gas without passing through a liquid phase.

sublime — to vaporize directly from the solid to the gaseous state.

S
T

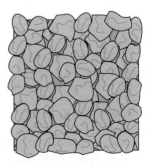

Subangular Blocky Soil

sublittoral — the region in a lake between the deepest-growing rooted vegetation and the part of the lake below the thermocline.

submerged arc welding — a process that produces coalescence of metals by heating them with an arc between a bare metal electrode and the work.

submersible pump — a centrifugal pump driven by a close coupled electrical motor constructed for submerged operation as a single unit.

subpoena — a court order requiring a witness to appear and give testimony.

subpoena duces tecum — a court order requiring a witness to attend a proceeding and bring certain documents.

subsidence — a settling or sinking of the earth's surface caused by natural means, such as chemical weathering in limestone areas, or by artificial means, such as mining.

subsoil — a layer of soil beneath the top soil with a lower organic content and higher concentration of fine mineral particles.

substitution — a technique used to reduce pollutants by changing or improving the process or altering the fuels used in the process.

substrate — (*biology*) a layer of material on which an organism can grow and multiply; (*chemistry*) any surface on which a layer of a different material can be deposited.

subsurface soil injection — the injection of hazardous waste under the surface of the ground and away from water supplies.

suction filter — *See* vacuum filter.

sudden infant death syndrome (SIDS) — the sudden and unexpected death of an apparently healthy infant typically occurring between the ages of three weeks and five months; not explained by postmortem studies.

sugar — the common term for a water-soluble, crystalline mono- or oligosaccharide; examples are sucrose ($C_{12}H_{22}O_{11}$) or cane sugar. *See also* glucose.

sulfate — a compound containing the $-SO_4$ group in which the hydrogen of sulfuric acid is replaced by either a metal or organic radical to become a sulfate salt or ester.

sulfhemoglobin — the reaction product of oxyhemoglobin and hydrogen sulfide.

sulfur dioxide — SO_2 — a colorless gas with a characteristic, irritating, pungent odor; a liquid below 14°F; shipped as a liquefied compressed gas; MW: 64.1, BP: 14°F, Sol: 10%. It is used as a bleaching agent and preservative agent in wood and pulp, sugar and food preparation industries; as a solvent in oil refineries; during synthesis in chemical, petroleum, textile, pharmaceutical, tanning, photography, metal, and rubber industries; during extraction, enrichment, and recovery processes in mining and metallurgy; in miscellaneous processes during leather tanning, special glass manufacture, water treatment, textile processing, and chrome waste treatment. It is hazardous to the respiratory system, skin, and eyes; and is toxic by inhalation and contact. Symptoms of exposure include rhinorrhea, choking, cough, reflex bronchoconstriction, eye and skin burns, irritation of the eyes, nose, and throat. OSHA exposure limit (TWA): 2 ppm [air] or 215 mg/m³.

sulfuric acid — H_2SO_4 — a colorless to dark-brown, oily, odorless liquid; pure compound is a solid below 51°F; often used in an aqueous solution; MW: 98.1, BP: 554°F, Sol: miscible, sp gr: 1.84. It is used in the manufacture of phosphoric acid and fertilizers; in petroleum refining; during the manufacture of pigments and dyes, and dyestuff intermediates; in the manufacture of industrial and military explosives; in the production of alcohols, phenols, and inorganic sulfates; in ore leaching and processing; in metal cleaning and plating; in the manufacture of detergents; in coke-oven gas refining; in the plastics industry for the manufacture of rayon, cellophane, cellulose, acetate, and others; in food processing; for the preparation of insecticides; in the manufacture of natural and synthetic rubber; in the treatment of industrial water for pH control; in the manufacture of textiles and leather; as a laboratory reagent as a solvent for chemical analysis; in chemical synthesis. It is hazardous to the respiratory system, eyes, skin, and teeth; and is toxic by inhalation, ingestion, and contact. Symptoms of exposure include pulmonary edema, bronchitis, emphysema, conjunctivitis, stomatitis, dental erosion, tracheobronchitis, skin and eye burns, dermatitis, irritation of the eyes, nose, and throat. *See also* oleum. OSHA exposure limit (TWA): 1 mg/m³ [air].

sulfur monochloride — S_2Cl_2 — a light amber to yellow-red, oily liquid with a pungent, nauseating, irritating odor; MW: 135.0, BP: 280°F, Sol: decomposes, Fl.P: 245°F, sp gr: 1.68. It is used for the production of white vulcanized oils; as natural and synthetic rubber extenders; as a cross-linking catalyst; in chemical synthesis for the production of intermediates for dyes, pharmaceuticals, insecticides, and war gases; for treatment of drying oils; for cold vulcanizing; as a solvent for sulfur and sulfur compounds. It is hazardous to the respiratory system, skin, and eyes; and is toxic by inhalation, ingestion, and contact. Symptoms of exposure include lacrimation, cough, pulmonary edema, eye and skin burns. OSHA exposure limit (TWA): ceiling 1 ppm [air] or 6 mg/m^3.

sulfur oxides — SO$_x$ — oxides of sulfur; any of several pungent, colorless gases formed primarily from the combustion of fossil fuels; major air pollutants and cause of damage to the respiratory tract, as well as vegetation.

sulfur pentafluoride — S$_2$F$_{10}$ — a colorless liquid or gas (above 84°F) with an odor like sulfur dioxide; MW: 254.1, BP: 84°F, Sol: insoluble, sp gr (32°F): 2.08. It is used during synthesis of sulfur hexafluoride; not produced commercially. It is hazardous to the respiratory system and central nervous system; and is toxic by inhalation and contact. Symptoms of exposure in animals include pulmonary edema, hemorrhage. OSHA exposure limit (TWA): ceiling 0.01 ppm [air] or 0.1 mg/m^3.

sulfur recovery plant — any plant that recovers elemental sulfur from a gas stream.

sulfur trioxide (SO$_3$) — an oxide of sulfur produced by the action of oxygen on sulfur dioxide in the presence of a catalyst such as iron oxide.

sulfuryl fluoride — SO$_2$F$_2$ — a colorless, odorless gas; shipped as a liquefied compressed gas; MW: 102.1, BP: –68°F, Sol (32°F): 0.2%. It is used as an insecticidal fumigant for control of dry wood termites and other structural pests. It is hazardous to the respiratory system and central nervous system; and is toxic by inhalation and contact. Symptoms of exposure include conjunctivitis, rhinorrhea, pharyngeal, paresthesia; in animals: narcosis, tremor, convulsions, pulmonary edema, kidney injury. OSHA exposure limit (TWA): 5 ppm [air] or 20 mg/m^3.

summary judgment — a judgment by the court made without a trial, based on written interrogatories, depositions, and sworn affidavits; the motion for summary judgment is filed when prosecution can substantiate that there are not genuine issues in dispute.

summons — a court order requiring the person named in an action to appear before the court and answer the complaint.

sump — a tank or pit which receives sewage liquid or liquid waste located below the normal grade of the gravity system that must be emptied by mechanical means.

sun — a star which is the source of almost all energy on earth transmitted as waves, some of which are visible and others that are invisible on earth; a globe of glowing gas, 8.7×10^5 miles in diameter held together by its own gravity; located 93 million miles away from the earth.

superchlorination — chlorination where the doses are large enough to complete all chlorination

reactions and to produce a free chlorine residual, as in a swimming pool.

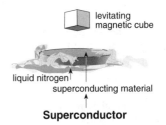

levitating magnetic cube

liquid nitrogen

superconducting material

Superconductor

superconductor — a conductor which has no resistance to electrical current; examples include lead, tin, and mercury.

Superfund Act — *See* Comprehensive Environmental Response Compensation and Liability Act.

Superfund Amendments and Reauthorization Act (SARA) — a 1986 law amending and reauthorizing the Comprehensive Environmental Response Compensation, and Liability Act of 1980; provides 8.5 billion dollars for cleanup of superfund sites, one-half-billion dollars for cleanup of underground tank leaks, and the establishment of alternative technologies for measuring, monitoring, and treatment of hazardous waste.

Superfund Innovative Technology Evaluation (SITE) — an Environmental Protection Agency program in which research is used to determine the state-of-the-art clean-up methods for hazardous waste sites around the nation and the techniques for removing hazardous substances from soils.

supernatant liquid — (*sewage*) the liquid remaining above a layer of settled solids after the solids collect at the bottom of a storage tank or container.

supersaturated solution — a solution that contains a greater quantity of solute than is normally possible at that temperature.

supersaturation — a condition in which air reaches more than 100% relative humidity.

supersonic — *See* ultrasonic.

supplied air respirator — an air lined respirator or self-contained breathing apparatus.

supply — the number of qualified personnel available to practice a given occupation, including those employed (or self-employed) and those seeking employment in the occupation.

suppuration — formation or discharge of pus.

SUR — acronym for safe use rule.

surface-active agent — any of a group of compounds added to a liquid to modify surface or interface tension; used in synthetic detergents to reduce the interfacial tension providing cleansing action. Also known as surfactant.

S
T

surface coating — coating used to cover paint, lacquer, varnish, and other chemical compositions used to protect and/or decorate surfaces.

surface condenser — a heat transfer device used to condense vapor by contact with a cooling surface behind which the cooling liquid is located.

surface impoundment (SI) — a facility or part of a facility which is a natural topographic depression, an excavation, or dike area formed primarily of earthen materials, which is designed to hold an accumulation of liquid waste or waste containing free liquids.

surface mining — the practice of removing coal by excavation of the surface without an underground mine. Also known as strip mining.

surface moisture — water that is not chemically bound to a metallic mineral or metallic mineral concentrate.

surface runoff — any material that is dissolved or suspended in free liquids, usually water, which can leach or flow across the surface of the land from the point of application to the point of discharge.

surface skimmer — a box in the side of the pool which contains a floating weir and a basket through which overflow water from the pool passes in order to clean the surface of the water.

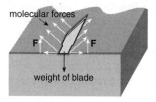

Surface Tension

surface tension — a property possessed by liquids in which the water surface in contact with the air acts like an elastic skin; the phenomenon is due to unbalanced molecular cohesive forces near the surface.

surface water — any water on the surface of the earth.

surface weather maps — maps used to plot wind direction, speed, pressure, temperature, dew point, visibility, current weather conditions, amount and types of clouds and their heights, pressure changes in the past 3 hours, weather in the past 6 hours, and the amount and kind of precipitation at a given weather station.

surfactant — See surface-active agent.

surgical asepsis — destruction of organisms before they enter the body, such as in the caring for open wounds and in surgical procedures.

surgical waste — the waste generated by surgical techniques in the treatment of disease, injury, or deformity.

surveillance and analysis (S & A) — the determination of facts related to environmental situations and the analysis of the data to determine a means of prevention and control.

surveillance system — a series of monitoring devices designed to determine environmental quality.

survey — a rapid means of gathering information concerning potential problems relating to the environment or of accurately determining position, contour, etc., of an area.

susceptance — the ratio of current-to-voltage in a nondissipative circuit; that part of the admittance which is related to the storage of electric energy.

susceptibility — the degree to which an organism is affected by a chemical or microorganism at a particular level of exposure.

suspended solids (SS) — solids physically suspended in water, sewage, or other liquids and which resist separation from the water by conventional means.

suspension — a mixture of finely divided solid material in a liquid from which the solid settles out on standing; the action of holding insoluble particles in a solution; (*pesticides*) the U.S. Environmental Protection Agency's prohibition of the use of a pesticide to prevent an imminent hazard from occurring from the continued use of the pesticide.

sv — See sievert.

SVOC — acronym for semivolatile organic compounds.

sweat — See perspiration.

sweetening — the process of improving the odor and color of petroleum products by the oxidation or removal of unsaturated, sulfur-containing compounds.

swill — semiliquid waste material consisting of food scraps and free liquids; usually fed to hogs.

swimmer load — the number of persons in the pool area at any given moment or during any related period of time.

swimmer's itch — See cutaneous larva migrans.

swimming pool — a body of water of such size in relation to the bathing load that the quality and quantity of the water confined must be mechanically controlled for the purpose of purification and contained in an impervious structure.

SWPF — See simulated workplace protection factor.

symbiosis — the relationship in which two organisms live together for the mutual advantage of each organism. Also known as consortism.

symmetrical — having a definite shape.

S
T

sympathetic nervous system — a division of the autonomic nervous system controlling the smooth muscles and glands of the body.

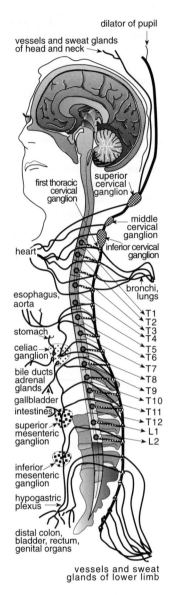

Sympathetic Nervous System

symptom — a warning that something is abnormal; an indication of a disease, infection, or poisoning.

synapse — the space between nerve endings where impulses can pass from one to another.

syncope — a temporary suspension of consciousness due to cerebral anemia. Also known as faint.

syndrome — a group of symptoms that together are characteristic of a specific condition or disease.

synecology — the branch of ecology dealing with communities rather than individual species.

synergetic — *See* synergism.

synergism — the cooperative action of separate substances so that the total effect is greater than the sum of the effects of the substances acting independently.

synergist — a chemical that produces greater effects when mixed with another chemical than when the added effects of each are used alone.

synergistic effect — a combined effect of two or more substances or organisms which is greater than the sum of the individual effect of each.

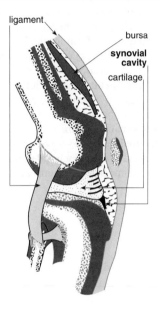

Synovial Fluid

synovial fluid — a secretion of cartilage that lubricates a joint.

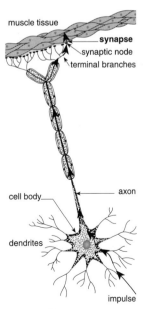

Synapse

S
T

syntax — (*computer science*) a set of rules governing the way statements can be used in a computer language.

synthesis — the reaction or series of reactions by which a complex compound is obtained from simpler compounds or elements.

synthesize — to form a material by combining parts of elements.

synthetic — referring to chemicals or other substances made artificially.

synthetic detergent — a liquid or solid chemical used as a cleaning agent which is soluble in water and does not form insoluble precipitates.

synthetic fiber — fiber derived from petrochemicals or coal chemicals, such as polyester or nylon.

synthetic pyrethrum — *See* allethrin.

system — a group of separate organs performing a similar function, such as the digestive system.

systematic error — (*statistics*) *See* bias.

systematic maintenance — preventative maintenance carried out on a schedule established according to a time period or number of units in use.

Système International d'Unités — *See* International System of Units.

systemic — a condition or process affecting the whole body.

systemic pesticide — a pesticide chemical that is carried to other parts of a plant or animal after it is injected or taken up from the soil or body surface.

systemic poison — a poison which spreads throughout the body affecting all body systems and organs.

systemic toxicity — adverse effects caused by a substance which affect the body in a general rather than local manner.

systems design — (*ergonomics*) a study of the principle of an individual as an integral part of a total person-machine system where there is an allocation between the functions carried out by machines and people.

system sensitivity — a measure of the lowest concentration of a chemical that can be detected in the collection side of permeation tests.

S
T

2,4,5-T — *See* 2,4,5-trichlorophenoxyacetic acid.

T — *See* tesla; tritium.

TA — *See* temporary abeyance.

tablespoon (tbsp) — a measurement of volume equal to 3 teaspoons.

tableware — all multi-use eating and drinking utensils.

tachypnea — very rapid, abnormal respirations.

taeniasis — (*disease*) a human intestinal infection characterized by nervousness, insomnia, weight loss, abdominal pain, and digestive disturbances. Incubation time is 8–14 weeks. It is caused by *Taenia sagnata* (beef tapeworm) or *Taenia solium* (pork tapeworm). Found worldwide, the reservoir of infection is humans with the immediate hosts being cattle or pigs. It is spread by the consumption of raw or inadequately cooked beef or pork, or the consumption of water containing the eggs of adult worms; *Taenia sagnata* is not transmitted from person to person; *Taenia solium* may be found in the feces for 30–40 years; general susceptibility. It is controlled by the proper cooking of beef and pork, proper sewage disposal, inspection of livestock, and proper water treatment. Also known as beef or pork tapeworm disease.

Taenia spiralis — a large tapeworm species found in the striated muscle of various animals; a common cause of infection in people as a result of ingestion of poorly cooked pork.

tailings — the remaining mining waste materials after a substance, such as uranium or iron, has been mechanically or chemically extracted from ore.

talc (non-asbestos form with less than 1% quartz) — $Mg_3H_2(SiO_3)_4$ — an odorless, white powder; MW: varies, BP: not known, Sol: insoluble, sp gr: 2.70–2.80. It is used in the cosmetic industry as face powder, body or dusting powder, soap, and toilet preparations; in the pharmaceutical industry in tablets, pills, salves, and lotions; as a filler and selective absorbent in the paper industry; in the ceramic industry; in cleaning up oil spills. It is hazardous to the lungs and cardiovascular system; and is toxic by inhalation and contact. Symptoms of exposure include fibrotic pneumoconiosis. OSHA exposure limit (TWA): 2 mg/m³ [resp].

tank — a stationary device designed to contain an accumulation of hazardous waste or other chemicals or fuels, which is constructed primarily of nonearthen materials, such as wood, concrete, steel, or plastic and provides structural support for the chemicals.

tantalum (metal and oxide dust) — Ta — a steel-blue to gray metal or a black powder; MW: 180.9, BP: 9797°F, Sol: insoluble, sp gr: 16.65 (metal) 14.40 (powder). It is used in the manufacture of electronic equipment; in the fabrication of metals; in fabrication of refractory non-ferrous alloys for nuclear and aerospace applications; in the manufacture of surgical metals, mesh, and clips; in the manufacture of special optical glass. No known hazards to humans; and is toxic by inhalation. Symptoms of exposure in animals include pulmonary irritation. OSHA exposure limit (TWA): 5 mg/m³ [air].

tape sampler — a device used in the measurement of fine particulates allowing air sampling to be done automatically at predetermined times by drawing air across the filter strip and determining what is on the strip.

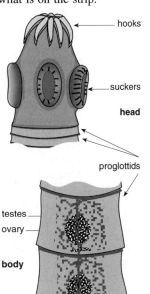

Tapeworm
details of head and body

tapeworm — a parasitic cestode worm of the class of *Cestoidea* (phylum *Platyhelminthes*) having a flattened band-like form which lodges in the intestines of animals and humans; they are transmitted to people in larval form imbedded in cysts of improperly cooked meat or fish; in humans they develop to maturity and attach themselves to the wall of the intestine where they grow and release eggs.

tar — a viscous, complex mixture of chemicals of top fractional distillation systems.

tar camphor — *See* naphthalene.

tare — a deduction from gross weight made to allow for the weight of an empty container.

target organ effect — the health effect on a specific organ, such as the liver, caused by a specific chemical or substance.

target organ toxicity — a broad range of adverse effects on target organs or physiological systems extending from those or rising through a single limited exposure to those assumed during the lifetime of exposure to a chemical.

target organ toxin — a toxic substance that attacks a specific organ of the body.

target theory — the theory explaining some biological effects of radiation on the basis that ionization occurring in a discrete (target) volume within a cell causes a lesion which subsequently results in a physiological response to the damage at that location. Also known as hit space theory.

tar sand — a sand coated with bitumen.

tar sand deposit — a geological formation that can be used to convert bitumen into crude oil.

taxonomy — the science of classification.

tbsp — *See* tablespoon.

TC — notation for carbon present.

TCBD — notation for tretrachlorodibenzo-t-dioxin.

TCDD — notation for tretrachlorodibenzo-p-dioxin.

TCE — *See* trichloroethylene.

TCL — *See* toxic concentration low.

TD — notation for toxic dose.

TD$_{50}$ — the dose that produces 50% of the toxic effect of a chemical.

TDI — notation for toluene diisocyanate.

TDL — *See* toxic dose low.

teaspoon (tsp) — measurement equal to 50 drops.

technical material — the pesticide chemical as it is manufactured by the chemical company.

technician — an environmental health or safety practitioner qualified by education, training, and practice to operate sampling, monitoring, and other data-gathering equipment in the conduct of environmental health or safety control and prevention activities. The minimum education level is specialized training.

technology assessment — the systematic analysis of the anticipated impact of a particular technology in regard to its safety and efficacy as well as its social, political, economic, and ethical consequences.

teepee burner — *See* conical burner.

TEL — *See* tetraethyllead.

**Television and Infrared
Observation Satellite**

television and infrared observation satellite (TIROS) — a satellite sent into orbit on April 1, 1960 to determine weather patterns.

tellurium and compounds — Te — an odorless, dark-gray to brown, amorphous powder or grayish-white, brittle metal; MW: 127.6, BP: 1814°F, Sol: insoluble, Fl.P: not known, sp gr: 6.24. It is used as an insecticide, germicide, fungicide, and photographic print toner. It is hazardous to the skin and central nervous system; and is toxic by inhalation, absorption, ingestion, and contact. Symptoms of exposure include garlic odor on breath, sweating, dry mouth, metal taste, somnolence, anorexia, dermatitis. OSHA exposure limit (TWA): 0.1 mg/m^3 [air].

tellurium hexafluoride — TeF$_6$ — a colorless gas with a repulsive odor; MW: 241.6, BP: sublimes, Sol: decomposes. It is used in scientific studies on physical and chemical properties. It is hazardous to the respiratory system; and is toxic by inhalation. Symptoms of exposure include headache, dyspnea, garlic odor on breath; in animals: pulmonary edema. OSHA exposure limit (TWA): 0.02 ppm [air] or 0.2 mg/m^3.

telomer — (*chemistry*) a reduced polymer formed by the reaction between a substance capable of being polymerized and an agent that arrests the growth of the chain of atoms.

telophase — the phase of nuclear division by mitosis or meiosis which begins as chromosomes reach the centrioles.

TEM — *See* transmission electron microscope.

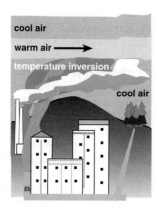

Temperature Inversion

temperature — a measure of the tendency of objects to transfer heat to or absorb heat from other objects. Its unit is the degree, and it can be read directly on a thermometer.

temperature gradient — the change of temperature along a given line in space.

temperature inversion — *See* inversion.

tempered air — the supply of air which has been heated sufficiently to prevent cold drafts. Also known as conditioned air.

tempering — the preheating and rapid cooling of a metal to increase its hardness.

temporal resolution — the resolution of temporal distribution of air pollution concentration in a study area by modeling or monitoring. It is affected by the sources of air pollution which seldom emit material into the air at a constant and continuous rate; changing wind direction which has a profound effect on the distribution of air pollutants; changing wind speed which has a major effect on the distribution of pollutants; changing atmospheric stability which can produce fluctuations in the ambient air pollution levels; and topography which can affect the distribution of air pollutants by affecting air movements that might not otherwise occur.

temporary abeyance (TA) — the holding of sample or samples in abeyance for possible future regulatory action.

temporary food establishment — a food establishment that operates in one location for not more than 14 consecutive days in conjunction with a single event.

temporary housing — any tent, trailer, mobile home, or any other structure used for human shelter as designed to be transportable and not attached to the ground, another structure, or to any utility system on the same premises for more than 30 consecutive days.

temporary partial disability — a condition under Worker's Compensation Insurance which results in partial loss of earning power but from which recovery can be expected.

temporary restraining order (TRO) — the first step of most injunctions in which the government or plaintiff requests the imposition of immediate temporary restraint of the defendant lasting for 10 days. *See also* preliminary injunction.

temporary total disability — a condition under Worker's Compensation Insurance which disables the employee from working but from which a complete or partial recovery can be expected.

1080 — *See* sodium fluoroacetate.

1081 — *See* fluoroacetamide.

tensile strength — the capacity of a metal to withstand a stretching force.

tensiometer — a device for measuring a negative hydraulic pressure or tension of water in soil involving a porous, permeable ceramic cup connected through a tube to a manometer or vacuum gauge.

TEPP (ethyl pyrophosphate) — $(C_2H_5)_4P_2O_7$ — a colorless to amber liquid with a faint, fruity odor; a solid below 32°F; MW: 290.2, BP: decomposes, Sol: miscible, sp gr: 1.19. It is used as a contact insecticide on agricultural and ornamental crops. It is hazardous to the central nervous system, respiratory system, cardiovascular system, and gastrointestinal tract; and is toxic by inhalation, absorption, ingestion, and contact. Symptoms of exposure include eye pain, blurred vision, lacrimation, rhinorrhea, headache, tight chest, cyanosis, anorexia, nausea, vomiting, diarrhea, weakness, twitch, paralysis, Cheyne-Stokes respiration, convulsions, low blood pressure, cardiac irregularities, local sweating. OSHA exposure limit (TWA): 0.05 mg/m³ [skin].

teratogen — a chemical that causes structural defects that affect the development of a fetus.

teratogenesis — the study of malformation induced during development of an animal or human from the time of conception.

terminal — (*computer science*) a device, usually including a cathode-ray tube and a keyboard, for communicating with the computer.

terminated — ended or adjudicated.

terphenyls — $C_6H_5C_6H_4C_6H_5$ — a colorless or light-yellow solid; consists of three different isomers; MW: 230.3, BP: 630–761°F, Sol: insoluble, Fl.P (open cup): 325 to 405°F, sp gr: 1.10–1.23. It is used in the formulation of waxes, polishes, and resin body paints; in the manufacture of

S
T

plastic scintillators. It is hazardous to the skin, respiratory system, and eyes; and is toxic by inhalation, ingestion, and contact. Symptoms of exposure include thermal skin burns, headache, sore throat, irritation of eyes and skin. OSHA exposure limit (TWA): ceiling 0.5 ppm [air] or 5 mg/m^3.

terracing — dikes built along the contour of agricultural land to hold runoff and sediment, reducing water pollution.

tertiary butyl valone — *See* pival.

tertiary treatment — *See* advanced wastewater treatment.

tertiary wastewater treatment — *See* advanced wastewater treatment.

tesla (T) — the International System unit of magnetic field; a unit of magnetic flux density equal to one weber per square centimeter.

test hole — any excavation, regardless of design or method of construction, done for the purpose of determining the most suitable site for excavation, such as for groundwater from an aquifer for use in a private water system.

tetanic — pertaining to tetanus.

tetanus — (*disease*) an acute disease caused by the bacterium *Clostridium tetani*; the spore of the organism is introduced into the body in an anaerobic state and then produces the exotoxin, which causes painful muscle contraction of the neck and trunk, abdominal rigidity occurs, as well as generalized spasms. Incubation time is 3–21 days; found worldwide. The reservoirs of infection include intestines of animals, including people, or soil contaminated with animal feces. It is transmitted through a puncture wound; not transmitted from person to person; general susceptibility. It is controlled through active immunization with tetanus toxoid given at least once every ten years. Also known as lockjaw.

tetanus antitoxin — an equine antitoxin used against the toxins of *Clostridium tetani*.

2,3,7,8-tetrachlorodibenzo-p-dioxin (PCDD) — C$_{12}$H$_4$Cl$_4$O$_2$ — a colorless solid with no distinguishable odor; MW 321.97, BP 412.2°C, Sol: 7.91 mg/L at 20–22°C in water, Fl.P: not known, density: 1.827 g/ml estimated. It is used in Germany for flame-proofing polyesters; as a control against insects and wood-destroying fungi. It is hazardous to the skin, liver, digestive system, immune system, and fetus; and is toxic by inhalation, ingestion, and contact. Symptoms of exposure include chloracne, hepatoxicity and immunotoxicity, hyperpigmentation, hyperkeratosis, hirsutism of the skin, hypertriglyceride hypercholesterolemia, insomnia, aching muscles, weight loss, loss of appetite, digestive disorders, headaches, neuropathy, sensory changes, loss of libido, developmental disorders; probable carcinogen. Environmental Protection Agency drinking water advisory: 3.5 × 10^{-8} milligrams per liter.

1,1,1,2-tetrachloro-2,2-difluoroethane — CCl$_3$CClF$_2$ — a colorless solid with a slight, ether-like odor; a liquid above 105°F; MW: 203.8, BP: 197°F, Sol: 0.01%. It is used in dry cleaning in combination with other solvents; in the polymer and plastics industries; as a dye solvent; as a solvent extractant for the separation and purification of biological material; as a corrosion inhibitor in brake fluid and surface coatings. It is hazardous to the respiratory system and skin; and is toxic by inhalation, ingestion, and contact. Symptoms of exposure include central nervous system depression, pulmonary edema, drowsiness, dyspnea, irritation of the skin and eyes. OSHA exposure limit (TWA): 500 ppm [air] or 4,170 mg/m^3.

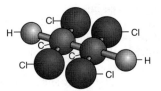

1,1,2,2–Tetrachloroethane

1,1,2,2-tetrachloroethane — CHCl$_2$CHCl$_2$ — a colorless to pale yellow liquid with a pungent, chloroform-like odor; MW: 167.9, BP: 296°F, Sol: 0.3%, sp gr (77°F): 1.59. It is used as a chemical intermediate; in cleaning and extraction processes; as a fumigant in greenhouses; in the manufacture of lacquers and varnishes, paint and varnish removers; in cleaning and degreasing metals; as a solvent in the preparation of adhesives; in refining waxes and resins. It is hazardous to the liver, kidneys, and central nervous system; and is toxic by inhalation, absorption, ingestion, and contact. Symptoms of exposure include nausea, vomiting, abdominal pain, tremor of the fingers, jaundice, tenderness and enlargement of the liver, dermatitis, monocytosis, kidney damage. OSHA exposure limit (TWA): 1 ppm [skin] or 7 mg/m^3.

tetrachloroethylene — Cl$_2$C=CCl$_2$ — a colorless liquid with a mild, chloroform-like odor; MW: 165.8, BP: 250°F, Sol (77°F): 0.02%, sp gr: 1.62. It is used as a dry cleaning solvent; as a degreasing and metal cleaning agent; as a chemical intermediate; as a scouring, sizing, desizing, solvent, and grease remover in the processing and finishing of textiles; as a general industrial solvent; as an extraction agent; as a drying medium in the metal and wood industries.

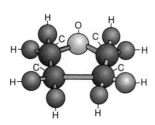

Tetrahydrofuran

It is hazardous to the liver, kidneys, eyes, upper respiratory system, and central nervous system; and is toxic by inhalation, ingestion, and contact. Symptoms of exposure include nausea, flushed face and neck, vertigo, dizziness, incoordination, headache, somnolence, skin erythema, liver damage, irritation of the eyes, nose, and throat. OSHA exposure limit (TWA): 25 ppm [air] or 170 mg/m^3.

tetrachloronaphthalene — $C_{10}H_4Cl_4$ — a colorless to pale yellow solid with an aromatic odor; MW: 265.9, BP: 593–680°F, Sol: insoluble, Fl.P (open cup): 410°F, sp gr: 1.59–1.65. It is used during the manufacture of electric equipment; as an inert compound of resins or polymers for coating or impregnating textiles, wood, and paper; as an additive for cutting oil; as an additive to special lubricants. It is hazardous to the liver and skin; and is toxic by inhalation, absorption, ingestion, and contact. Symptoms of exposure include acne-form dermatitis, headache, fatigue, anorexia, vertigo, jaundice, liver injury. OSHA exposure limit (TWA): 2 mg/m^3 [skin].

tetraethyllead (TEL) — $Pb(C_2H_5)_4$ — a colorless liquid (unless dyed red, orange, or blue) with a pleasant, sweet odor; MW: 323.5, BP: 228°F (decomposes), Sol: insoluble, Fl.P: 200°F, sp gr: 1.65. It is used in the preparation of antiknock agents for fuels. It is hazardous to the central nervous system, cardiovascular system, kidneys, and eyes; and is toxic by inhalation, absorption, ingestion, and contact. Symptoms of exposure include insomnia, lassitude, anxiety, tremors, hyper-reflexia, spastic, bradycardia, hypotension, hypothermia, pallor, nausea, anorexia, weight loss, disorientation, hallucination, psychosis, mania,

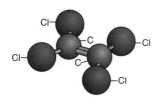

Tetrachloroethylene
(Tetrachlorethane)

convulsions, coma, eye irritation. OSHA exposure limit (TWA): 0.075 mg/m^3 [skin].

tetrahydrofuran — C_4H_8O — a colorless liquid with an ether-like odor; MW: 72.1, BP: 151°F, Sol: miscible, Fl.P: 6°F, sp gr: 0.89. It is used as a solvent in the preparation of printing inks, adhesives, lacquers, and other coatings; as a Grignard reagent in synthesis of motor fuels, vitamins, hormones, pharmaceuticals, synthetic perfumes, organometallic compounds, and insecticides; as an intermediate in the preparation of various chemicals. It is hazardous to the eyes, skin, respiratory system, and central nervous system; and is toxic by inhalation, ingestion, and contact. Symptoms of exposure include nausea, dizziness, headache, irritation of the eyes and upper respiratory system. OSHA exposure limit (TWA): 200 ppm [air] or 590 mg/m^3.

tetramethyllead — $Pb(CH_3)_4$ — a colorless liquid (unless dyed red, orange, or blue) with a fruity odor; MW: 267.3, BP: 212°F (decomposes), Sol: insoluble, Fl.P: 100°F, sp gr: 2.00. It is used as an antiknock additive for gasoline. It is hazardous to the central nervous system, cardiovascular system, and kidneys; and is toxic by inhalation, absorption, ingestion, and contact. Symptoms of exposure include insomnia, bad dreams, restlessness, anxiousness, hypotension, nausea, anorexia, delirium, mania, convulsions, coma. OSHA exposure limit (TWA): 0.075 mg/m^3 [skin].

tetramethyl succinonitrile — $(CH_3)_2C(CN)C(CN)(CH_3)_2$ — a colorless, odorless solid; MW: 136.2, BP: sublimes, Sol: insoluble, Fl.P: not known, sp gr: 1.07. It is used in the manufacture of polymers by and in the processing of products expanded with azobisisobutyronitrile. It is hazardous to the central nervous system; and is toxic by inhalation, absorption, ingestion, and contact. Symptoms of exposure include headache, nausea, convulsions, coma. OSHA exposure limit (TWA): 0.5 ppm [skin] or 3 mg/m^3.

tetranitromethane — $C(NO_2)_4$ — a colorless to pale yellow liquid or solid (below 57°F) with a pungent odor; MW: 196.0, BP: 259°F, Sol: insoluble, Fl.P: not known, sp gr: 1.62. It is used as an ingredient in the manufacture of liquid explosives; as an octane number improver in diesel fuels; as a laboratory analytical reagent; in research in rocket propellants. It is hazardous to the respiratory system, eyes, skin, blood, and central nervous system; and is toxic by inhalation, ingestion, and contact. Symptoms of exposure include dizziness, headache, chest pain, dyspnea, methemoglobinuria, cyanosis, skin burns, irritation of the eyes, nose, and throat.

S
T

OSHA exposure limit (TWA): 1 ppm [air] or 8 mg/m^3.

1,1,2,2-tetrochloro-1,2-difluoroethane — CCl$_2$FCCl$_2$F — a colorless solid or liquid (above 77°F) with a slight, ether-like odor; MW: 203.8, BP: 199°F, Sol (77°F): 0.01%, sp gr: 1.65. It is used in the dry cleaning industry in combination with other solvents; in the polymer and plastics industries; as a dye solvent; as a solvent extractant for the separation and purification of biological material; as corrosion inhibitors in brake fluid and surface coatings. It is hazardous to the lungs and skin; and is toxic by inhalation, ingestion, and contact. Symptoms of exposure include conjunctivitis, pulmonary edema, skin irritation. OSHA exposure limit (TWA): 500 ppm [air] or 4170 mg/m^3.

tetryl — (NO$_2$)$_3$C$_6$H$_2$N(NO$_2$)CH$_3$ — a colorless to yellow, odorless, crystalline solid; MW: 287.2, BP: 356–374°F (explodes), Sol: insoluble, Fl.P: explodes, sp gr: 1.57. It is used in the manufacture of explosives; as a pH indicator. It is hazardous to the respiratory system, eyes, central nervous system, skin; in animals: liver and kidneys; and is toxic by inhalation, absorption, ingestion, and contact. Symptoms of exposure include sensitization dermatitis, itching, erythema, keratitisis, sneezing, anemia, fatigue, cough, coryza, irritability, malaise, headache, lassitude, insomnia, nausea, vomiting, edema on nasal folds, cheeks, and neck. OSHA exposure limit (TWA): 1.5 mg/m^3 [skin].

Tfx — *See* toxic effects.

TGA — *See* nitrilotriacetic acid.

Th — *See* thorium.

thalidomide — C$_{13}$H$_{10}$N$_2$O$_4$ — a chemical substance used extensively as a sedative in sleeping pills in Europe and the United States in the early 1960s; its use was discontinued when it was determined that the drug caused phocomelia, or failure of limbs to develop in fetuses exposed to the drug at a certain time during pregnancy.

thallium — Tl — (soluble compounds) — appearance and odor vary depending upon the specific soluble solution; properties vary depending upon the specific soluble solution. It is used in the manufacture of special lenses, plates and prisms, medicinals, pyrotechnic products, and fuel additives; as a rodenticidal agent. It is hazardous to the eyes, central nervous system, lungs, liver, kidneys, gastrointestinal tract, body hair; and is toxic by inhalation, absorption, ingestion, and contact. Symptoms of exposure include nausea, diarrhea, abdominal pain, vomiting, ptosis, strabismus, peripheral neuritis, tremors, chest pain, pulmonary edema, seizures, chorea, psychosis, liver and kidney damage, alopecia, paresthesia of the legs. OSHA exposure limit (TWA): 0.1 mg/m^3 [skin].

thallium sulfate — Tl$_2$SO$_4$ — toxic, tasteless, and odorless crystals used as a rat poison acting similarly to arsenic.

thallus — a plant body that lacks differentiation into distinct members, such as stem, leaves, and roots, and does not grow from an apical point.

thematic map — a map showing special information such as soil, land-use, population, density, hazardous waste sites, and sewerage systems.

theoretical air — the amount of air circulated from the chemical composition of a waste as required to completely burn the waste. Also known as stoichiometric air; theoretical combustion air.

theoretical combustion air — *See* theoretical air.

therapeutic index — the ratio of the dose required to produce toxic or lethal effects to the dose required to produce nonadverse or therapeutic response.

therapy — the treatment of disease.

therm — a quantity of heat energy equivalent to 100,000 British thermal units.

thermal capacity — *See* heat capacity.

thermal conductivity — a specific rate of heat flow per unit time through refractories or other substances, expressed in British thermal units per square foot of area for a temperature difference of 1°F and for a thickness of one inch.

thermal efficiency — the ratio of actual heat used to total heat generated.

thermal pollution — a discharge of heat into bodies of water to the point that increased warmth activates organic material, depleting necessary oxygen and eventually destroying some fish and other aquatic organisms.

thermal processing — *See* thermal treatment.

thermal radiation — the transmission of energy by means of electromagnetic waves longer than visible light.

thermal regenerative system — an incineration system that relies on the heat radiation capability of an inert ceramic.

thermal shock resistance — the ability of a material to withstand sudden heating, cooling, or both without cracking.

thermal staple — a material that is resistant to change induced by heat.

thermal treatment — the treatment of a hazardous waste in a device which uses elevated temperatures as the primary means to change the chemical, physical, or biological character or composition of the hazardous waste. Also known as thermal processing.

thermal turbulence — turbulence caused by rising air heated from the earth's surface.

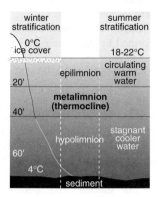

Thermocline

thermocline — a layer in a thermally stratified body of water of which the temperature changes rapidly relative to other layers of the water.

thermoconductivity — the specific heat of conduction of a gas or vapor in a carrier gas, such as air, is a measure of its concentration; the concentration of the gas is then determined by measuring the specific heat.

thermocouple — a device composed of dissimilar metals joined to form a circuit in which current flows when heat is applied at a junction of the metals; it develops an electromotive force when one junction is at a different temperature than the other.

thermoduric — bacteria that can survive the pasteurization process, but does not reproduce or grow at high temperatures.

thermograph — a self-recording thermometer.

thermoluminescence detector (TLD) — a personal dosimeter made of small chips of lithium fluoride; the absorbed ionizing radiation energy displaces electrons from their balance state; the electrons are then trapped in a metastable state but can be returned to the balance state by heating; when this occurs, light is emitted relating to the absorbed radiation dose.

thermonuclear reaction — a fusion reaction in which two light nuclei combine to form a heavier atom, releasing a large amount of energy.

thermophile — an organism that thrives at temperatures exceeding 55°C; causes spoilage of pasteurized milk.

thermoplastic — (*material*) a plastic that can be softened by heat and hardened when cooled and remolded; examples include vinyls, acrylics, and polyethylenes.

thermosetting material — a plastic that is heat-set to its final processing in a permanently hard state and cannot be remolded; examples are phenolics, ureas, and melamines.

thermostable — resistant to changes by heat.

Thévenin equivalent circuit — the representation of a circuit as a single voltage source and series resistance.

thinner — a liquid used to increase the fluidity of paints, varnishes, shellacs, and other materials.

thiram — $(CH_3)_2NC(S)SSC-(S)N(CH_3)_2$ — a colorless to yellow, crystalline solid; commercial pesticide products may be dyed blue; MW: 240.4, BP: decomposes, Sol: insoluble, Fl.P: not known, sp gr: 1.29. It is used in the rubber industry as an accelerator, peptizing agent, and vulcanizing agent; as a bacteriostat in commercial and surgical soap, antiseptics, sunburn oils, and fats; as an animal repellent on plants and trees. It is hazardous to the respiratory system and skin; and is toxic by inhalation, ingestion, and contact. Symptoms of exposure include dermatitis, irritation of the mucous membrane; with ethanol: consumption, flush, erythema, pruritus, urticaria, headache, nausea, vomiting, diarrhea, weakness, dizziness, dyspnea. OSHA exposure limit (TWA): 5 mg/m^3 [air].

thirty-day average — the arithmetic mean of pollutant parameter values of samples collected in a period of 30 consecutive days.

thixotropy — a property of materials which appear and act as a solid when undisturbed but will change to a semi-liquid when agitated.

THM — *See* trihalomethane.

thoracic — pertaining to the chest cavity.

thoracic cavity — the space lying above the diaphragm and below the neck, enclosed within the walls of the thorax containing the heart and lungs.

thoracic particulate mass threshold limit values (TPM-TLVs) — those materials which are hazardous when deposited anywhere within the lung airways and the gas-exchange region.

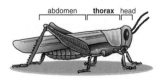

Grasshopper Thorax

thorax — (*zoology*) the middle region of the body of some insects between the head and the abdomen.

thorium (Th) — a naturally-occurring, radioactive metal; MW: 232.04, BP: 4,500°C, Sol:

S
T

insoluble, density: 9.7 g/cm³ to 11.7 g/cm³. It is used in the generation of nuclear energy; in refractory applications, lamp mantles, aerospace alloys, welding electrodes, nuclear weapon production; in ceramics, and in sun lamps. It is hazardous to the respiratory system, blood, pancreas, bone marrow; and is toxic by inhalation. Symptoms of exposure (in animals) include cirrhosis of the lungs, hematological effects, hepatic effects, pancreatic cancer. Nuclear Regulatory Commission maximum permissible concentrations in air per 40-hour week: 9 × 10⁻¹² microcuries per milliliter.

threadworm — any of a group of nematode worms.

threshold — the lowest dose of a chemical at which a specific measurable effect is observed; (*physiology*) the point at which a stimulus is of sufficient intensity to begin to produce an effect.

threshold dose — the minimum absorbed dose that will produce a detectable degree of any given effect.

threshold limit value (TLV) — the average referring to airborne concentration of a toxic substance to which nearly all workers may be repeatedly exposed 8 hours a day without adverse effects; also can be stated as any concentration of a toxic substance irrespective of how it makes its way into a living organism.

threshold limit value-ceiling (TLV-C) — the average concentration of a toxic substance which should not be exceeded during any part of working exposure.

threshold limit value-short-term exposure limit (TLV-STEL) — a concentration to which workers can be exposed continuously for a short period of time without suffering from irritation, chronic or irreversible tissue damage, narcosis, accidental injury, impair self-rescue, or materially reduce work efficiency, provided that the daily threshold limit value-time weighted average is not exceeded.

threshold limit value-time-waited average (TLV-TWA) — a concentration based on a time-weighted average for a normal eight-hour work day and forty-hour work week which nearly all workers may be repeatedly exposed without adverse effects.

threshold of audibility — the minimum effective sound pressure level of a signal capable of generating an auditory sensation in a specified condition; it is assigned a value of 0 decibels, 1,000 hertz.

threshold of pain — the upper intensity level for audible sounds which cause intense discomfort in the average human listener, usually between 130 and 140 decibels.

thrombin — a substance formed in blood clotting as a result of the reaction of prothrombin, thromboplastin, and calcium.

thrombocyte — *See* platelet.

thromboplastin — a substance essential in blood clotting formed by the disintegration of blood platelets.

thrombus — a plug or clot in a blood vessel remaining at the point of its formation and causing damage to the area.

thrush — (*disease*) *See* candidiasis.

thumb separation — a technique which employs the tendency of certain surface active solutes, such as alkyl benzene sulfonate, to collect at a gas-liquid interface.

thunder — the sudden tremendous heat from lightning causing the audible compression of shock waves in the air.

thunderheads — *See* cumulonimbus clouds.

thunderstorm — a violent vertical movement of air with clouds sometimes reaching in excess of 75,000 feet with periodic lightning, thunder, rain, and sometimes hail.

thyroid gland — an endocrine gland that produces, secretes, and stores thyroxin; its basic function is to control the rate of metabolism in the body; insufficient secretions cause sluggish, apathetic, and emotionally depressed behavior; excessive secretions cause a thyroid crisis where all of the bodily processes are accelerated to dangerous levels which may cause death. *See also* thyrotoxicosis.

thyrotoxicosis — the condition resulting from presentation to the tissues of excessive quantities of the thyroid hormones which may cause the pulse to increase to 200 beats per minute, a sharp rise in respiration, and a sharp rise in temperature because of a loss of control by the temperature control center of the brain.

tick — a blood-sucking arachnid parasite of either the hard or soft type; important vectors of significant arthropod-borne disease.

tidal marsh — low, flat, marshland traversed by interlaced channels and tidal sloughs and subject to tidal inundation.

tight soil — a compact, relatively impervious and tenacious soil or subsoil which may or may not be plastic.

tile — (*geographic information system*) a part of the database in a geographic information system representing a discrete part of the earth's surface.

timbre — characteristic of hearing in which the listener is capable of distinguishing between two sounds, even though the two sounds are of equal pressure and pitch.

time-weighted average — the average value of a parameter; a concentration of a chemical in air that varies over time.

time-weighted average concentration — a concentration of contaminant which has been weighted for the time duration of the sample.

time-weighted average exposure (TWA) — the average exposure over a given working period used in determining the contaminants within a given sample during that period.

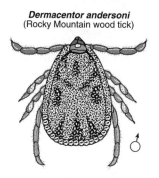

Dermacentor andersoni
(Rocky Mountain wood tick)

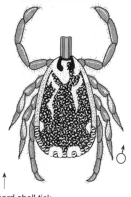

Amblyomma americanum
(Lone Star tick)

hard shell tick

soft shell tick

Ornithodoros sp.

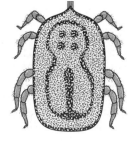

Ticks

tin — (*organic compounds*) — Sn — appearance and odor vary depending upon the specific organic compound; properties vary depending upon the specific organic compound. It is used as a stabilizing agent; as a catalyst in organic and inorganic synthesis; in the manufacture of insecticides, bactericides, fungicides, and molluscides; on agricultural crops; in the manufacture of rodent repellent for wire cable; as a fuel additive and lubricant; as veterinary medicine as poultry anthelmintic. It is hazardous to the central nervous system, eyes, liver, urinary tract, skin, and blood; and is toxic by inhalation, absorption, ingestion, and contact. Symptoms of exposure include headache, vertigo, eye irritation, psycho-neurologic disturbances, sore throat, cough, abdominal pain, vomiting, urine retention, paresis, focal anesthesia, skin burns, pruritus; in animals: hemolysis, hepatic necrosis. OSHA exposure limit (TWA): 0.1 mg/m^3 [skin]; (*inorganic compounds, except oxides*) a gray to almost silver-white, ductile, malleable, lustrous metal; MW: 118.7, BP: 4545°F, Sol: insoluble, sp gr: 7.28. It is used in the manufacture of blueprint paper; as a perfume stabilizer in soap; in the preparation of lubricating oil additives; in treating glass surfaces; as a catalyst in the manufacture of pharmaceuticals; as a bleaching agent for sugar; as a food additive; in dyeing silk; in metallurgy. It is hazardous to the eyes, skin, and respiratory system; and is toxic by inhalation and contact. Symptoms of exposure include skin and eye irritation. OSHA exposure limit (TWA): 2 mg/m^3 [air].

tinnitus — a ringing sound in one or both ears.

tipping — the unloading of refuse from a collection truck.

TIROS — *See* television and infrared observation satellite.

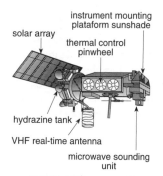

solar array

instrument mounting
plataform sunshade

thermal control
pinwheel

hydrazine tank

VHF real-time antenna

microwave sounding
unit

TIROS-N Spacecraft

TIROS-N — a group of TIROS satellites designed to carry high resolution radiometers to map clouds, sea-surface temperatures and the earth's surface by day and night; these satellites contain sensors to obtain vertical temperature soundings, moisture, and carbon dioxide measurements.

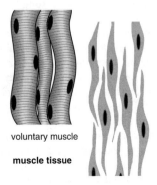

voluntary muscle

muscle tissue

involuntary muscle

Tissue

simple squamous

epithelial tissue

simple columnar

tissue — a group of cells having the same or related functions and structure.

tissue dose — the absorbed dose received by tissue in the region of interest expressed in rads.

tissue fluid — the fluid which bathes the cells of the body and is called lymph when contained in vessels.

titanium dioxide — TiO_2 — a white, odorless powder; MW:79.9, BP: 4532 to 5432°F, Sol: insoluble, sp gr: 4.26. It is used as a pigment in the manufacture of paints, varnishes, and lacquers to impart whiteness, opacity, and brightness; in the manufacture of paper, photographic paper, and cellophane coatings; in the manufacture of elastomers; in the manufacture of floor coverings; in the manufacture of ceramics and glass for capacitors; in the manufacture of printing inks; in the manufacture of coated fabrics and textiles; in the manufacture of building materials in roofing granules, ceiling tiles, cement-curing aids, and titanium carbide cutting tools; in the manufacture of cosmetics, food color additives, and synthetic diamonds. It is hazardous to the lungs; and is toxic by inhalation. Symptoms of exposure include slight lung fibrosis; carcinogenic. OSHA exposure limit (TWA): 10 mg/m³ [air].

titrant — a standard solution of known concentration and composition used during titration.

titration — the process of adding one reactant to another a drop at a time until the two are present in exactly equivalent amounts or a given reaction occurs.

titrimetric procedure — procedure in which a sample is reacted with a titrant solution containing a known concentration of a reactive chemical.

TL_{50} — *See* medial tolerance limit.

TLC — *See* total lung capacity.

TLD — *See* thermoluminescence detector.

TLV — *See* threshold limit value.

TLV-C — *See* threshold limit value-ceiling.

TLV-STEL — *See* threshold limit value-short-term exposure limit.

TLV-TWA — *See* threshold limit value-time-waited average.

TOC — *See* total organic carbon.

TOD — *See* total oxygen demand.

toe — the bottom of the working face of a sanitary landfill where deposited solid waste is in contact with virgin ground.

togavirus — an arbovirus spread by mosquitos which may cause fever, encephalitis, and rashes.

tolerance — the allowable concentration of a chemical on or in any food, plant, or animal meant for human or animal consumption considered to be safe by the Food and Drug Administration; (*physiology*) the ability of an organism to withstand environmental stress.

tolerance level — *See* permissible dose.

toluene — $C_6H_5CH_3$ — a colorless liquid with a sweet, pungent, benzene-like odor; MW: 92.1, BP: 484°F, Sol (61°F): 0.05%, Fl.P: 40°F, sp gr: 0.87. It is used as a solvent in pharmaceutical, chemical, rubber, and plastics industries; as a thinner for paints, lacquers, coatings, and dyes; as a paint remover; in insecticides; as a starting material and intermediate in organic chemical and chemical synthesis industries; in the manufacture of artificial leather; in photogravure ink production; as a constituent in automotive and aviation fuels. It is hazardous to the central nervous system, liver, kidneys, and skin; and is toxic by inhalation, absorption, ingestion, and contact. Symptoms of exposure include fatigue, weakness, confusion, euphoria, dizziness, headache, dilated pupils, lacrimation, nervousness, muscle fatigue, insomnia, paresthesia, dermatitis. OSHA exposure limit (TWA): 100 ppm [air] or 375 mg/m^3.

Toluene-2,4-Diisocyanate

toluene-2,4-diisocyanate — $CH_3C_6H_3(NCO)_2$ — a colorless to pale yellow solid or liquid (above 71°F) with a sharp, pungent odor; MW: 174.2, BP: 484°F, Sol: insoluble, Fl.P: 260°F, sp gr: 1.22. It is used in the manufacture of surface coatings and finishes, polyurethane paints, and electrical and thermal insulation; in the manufacture of flexible polyurethane foams and elastoplastics, adhesives, and sealants. It is hazardous to the respiratory system and skin; and is toxic by inhalation, ingestion, and contact. Symptoms of exposure include nose and throat irritation, choking, paroxysmal cough, chest pain, nausea, vomiting, abdominal pain, bronchial spasms, pulmonary edema, dyspnea, asthma, conjunctivitis, lacrimation, dermatitis, skin sensitization; carcinogenic. OSHA exposure limit (TWA): 0.005 ppm [air] or 0.04 mg/m^3.

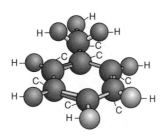

Toluene

o-toluidine — $CH_3C_6H_4NH_2$ — a colorless to pale yellow liquid with an aromatic, aniline-like odor; MW: 107.2, BP: 392°F, Sol: 2%, Fl.P: 185°F, sp gr: 1.01. It is used in the manufacture of dyes and dye intermediates; in the manufacture of organic chemicals and chemical intermediates; as a reagent chemical in blood sugar determinations. It is hazardous to the blood, kidneys, liver, cardiovascular system, skin, and eyes; and is toxic by inhalation, absorption, ingestion, and contact. Symptoms of exposure include anoxia, headache, cyanosis, weakness, dizziness, drowsiness, microhematuria, eye burns, dermatitis; carcinogenic. OSHA exposure limit (TWA): 5 ppm [skin] or 22 mg/m^3.

ton — a unit or weight measurement equal to 2,000 pounds. Also known as short ton.

tone — sound sensation having pitch and capable of causing an auditory sensation.

tonic spasm — a slight, continuous contraction of a muscle.

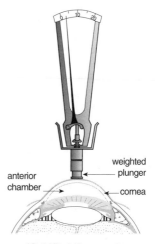

(Schiötz) Tonometer
(not proportionate to eye)

tonometer — an instrument used to measure pressure in the eyeball; used in glaucoma testing.

topical — meant for local application on a body surface.

topographic map — (*mapping*) a map showing topography in great accuracy and detail relative to the scale of the map.

topography — the detailed mapping or charting of the surface features of a relatively small area, district, or locality to show as much accurate detail as possible.

topology — the way in which geographical elements are linked together.

S
T

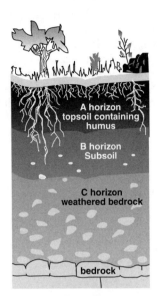

Topsoil
in the A horizon

topsoil — (*soils*) the uppermost few inches of soil consisting of organic matter and plant nutrients; usually the A horizon.

torch brazing — a brazing process in which the heat required is furnished by a fuel gas flame.

torr — a unit of pressure equal to 1/760 atmosphere.

tort — a civil wrong, other than a breach of contract, where the court is asked to provide a remedy in form of damages for actions taken against the individual bringing the tort.

total alkalinity — a measure of the sum total of all alkaline materials dissolved in water.

total body burden — the concentration of pollutants remaining in the body from previous exposures.

total dissolved solids — (*sewage*) the total dissolved (filterable) solids in wastewater.

total exchange — *See* cation-exchange capacity.

total ionization — the total positive or negative charge on the ions produced by radiation in the process of losing its kinetic energy.

total lung capacity (TLC) — the volume of air left in the lungs after maximum inspiration.

total organic carbon (TOC) — a measure of the kind, type, and amount of organic pollution occurring in a body of water.

total oxygen demand (TOD) — a combination of the biochemical oxygen demand and chemical oxygen demand of a body of water, or a quantity of sewage or organic waste.

total pressure — the algebraic sum of static pressure and velocity pressure.

total solids — (*sewage*) all solid constituents of sewage; the total of organic and inorganic solids suspended and dissolved.

total suspended nonfilterable solids (TSS) — (*wastewater*) total suspended solids as measured by a standard technique using glass fiber disks.

total suspended particulates (TSP) — all suspended particles of varying sizes.

toxaphene — $C_{10}H_{10}Cl_8$ (approximate) — an amber, waxy solid having an odor similar to chlorine or camphor; MW: 414, BP: dechlorinates at 155°C, Sol: 0.0003 g/100cc in water, density: 1.65 gm/cm³. It was one of the most widely used insecticides in the Unites States until it was banned by the Environmental Protection Agency in 1982. It is hazardous to the respiratory system, liver, adrenal gland, and central nervous system; and is toxic by inhalation and ingestion. Symptoms of exposure include acute pulmonary insufficiency with extensive miliary shadows, high serum globulin, liver damage, damage to the adrenal gland, salivation, restlessness, hyperexcitability, muscle tremors, convulsions, loss of consciousness. OSHA exposure limits (PEL TWA): 0.5 mg/m³ [air].

toxemia — a condition of poisoning by toxins or bacterial products in the blood.

toxic — relating to poisonous or deadly effects on the body by inhalation, ingestion, absorption, or contact with a toxin.

toxicant — any chemical that can injure or kill humans, animals, or plants; a poison.

toxic concentration low (TCL) — the lowest concentration of a gas or vapor capable of producing a defined toxic effect in a specified test species over a specified time.

toxic dose (TD_{50}) — the calculated dose of a chemical introduced by a route other than inhalation which is expected to cause a specific toxic effect in 50% of a defined experimental animal population.

toxic dose low (TDL) — the lowest administered dose of a material capable of producing a defined toxic effect in a specified test species.

toxic effects (Tfx) — the health effects that occur due to exposure to a toxic substance; any poisonous effect on the body, either reversible or irreversible, resulting in an undesirable disturbance of physiological function.

toxicity — the degree to which a substance is injurious or poisonous.

toxicity assessment — the process of characterizing the inherent toxicity of a chemical.

toxicological endpoint — the biological effect judged to be the most meaningful and specific in relation to the toxic chemical being studied.

toxicology — the study of how natural or synthetic poisons cause undesirable effects in living organisms.

Toxicology Information On-Line (TOXLINE) — the National Library of Medicine's collection of computerized references on human and animal toxicologic studies, toxicology, effects of environmental chemicals and pollutants, adverse drug reactions, and analytical methods.

toxic shock syndrome (TSS) — (*disease*) a severe illness caused by *Staphylococcus aureus* characterized by a sudden onset of high fever, vomiting, profuse watery diarrhea, and hypotension. Several body organs are typically involved; a rash appears during the acute phase and then the scaly skin on the palms and soles of the feet peels off. It can be prevented by not using tampons during the menstrual cycle, or the risk can be reduced by using tampons for short periods of time only.

toxic substance — any substance which can cause acute or chronic injury to the human body or which is suspected of being able to cause disease or injury under some conditions; substances such as convulsant poisons, central nervous system depressants, peripherally acting nerve poisons, muscle poisons, protoplasmic poisons, poisons of the blood, and blood forming poisons.

Toxic Substances Control Act (TSCA) — a 1976 law amended through 1986, it allows for the regulation of chemicals in commerce as well as before entering commerce; establishes a policy in which chemical manufacturers and processors are responsible for developing data about the health and environmental effect of their chemicals; it allows for government regulation of chemical substances that pose an unreasonable risk of injury to health or the environment.

toxic symptom — any feeling or sign indicating the presence of a poison in the system.

toxin — any poisonous substance of microbial, vegetable, animal, or synthetic chemical origin that reacts with specific cellular components to kill cells, alter growth or development, or kill the organism.

TOXLINE — *See* Toxicology Information On-Line.

toxoid — a modified, microbial, vegetable, or animal toxin used for immunization.

Toxoplasma — a genus of sporozoan parasites in mammals and some birds.

toxoplasmosis — (*disease*) a systemic protozoal disease which may be asymptomatic or produce fever with an increase in the number of lymphocytes in the blood. Incubation time may vary from 5–23 days. It is caused by *Toxoplasma gondii*; found worldwide in mammals and birds. Reservoirs of infection include rodents, pigs, cattle, sheep, goats, and birds. It is transmitted from eating raw or undercooked infected meat containing the cysts or ingestion of the oocysts, in food, water, or dust which is contaminated with cat feces; not directly transmitted from person to person; general susceptibility. It is controlled through proper cooking of meats, proper care of cats in feeding, disposal of cat feces and litter, and thorough hand washing.

TPM-TLVs — *See* thoracic particulate mass threshold limit values.

TPQ — acronym for threshold planning quantity.

trace analysis — the determination of a substance present in a sample at a relatively low concentration.

trace element — *See* micronutrient.

tracer — a minute quantity of radioactive isotope used in medicine or biology to study the chemical changes within living tissues.

tracer gases — compounds, such as sulfur hexafluoride, which are used to identify suspected pollutant pathways and to quantify ventilation rates.

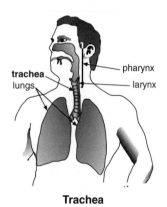

trachea
lungs
pharynx
larynx

Trachea

trachea — the windpipe for air breathing in vertebrates; an air tube in insects and spiders.

trade name — the trademark name or commercial name for a material or product.

transducer — any system or device which is actuated by waves from one or more transmission systems or which converts an input signal into an output signal of a different form.

transfer — the loading and unloading of chemicals between transport vehicles and storage vessels and the sending of chemicals by way of pipes between storage vessels and process reactors.

transfer facility — any transportation related facility including loading docks, parking areas,

S
T

storage areas, and other similar areas in which shipments of hazardous waste or solid waste are held during the normal course of transportation.

transfer station — a site in which solid waste is concentrated and then taken to a processing facility or a sanitary landfill.

transform — (*mapping*) the process of changing the scale, projection, or orientation of a mapped image.

transformation — the chemical alteration of a compound by a process such as biodegradation, hydrolysis, photodegradation, or reaction with other compounds.

transient equilibrium — the state reached if the half-life of the parent is short enough so the quantity present decreases appreciably during the period under consideration, but is still longer than that of successive members of the series; a stage of equilibrium will be reached after all members of the series decrease in activity exponentially with the period of the parent.

transient heat fatigue — a temporary condition of discomfort and mental or psychological strain resulting from prolonged heat exposure.

transistor — an electronic device used in place of a vacuum tube to control electric current.

transition — a change in the cross-sectional shape or area of a duct or hood.

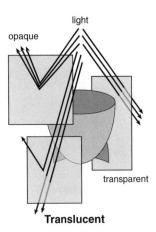

Translucent

translucent — the ability to transmit light so that objects beyond are entirely visible.

transmission electron microscope (TEM) — an electron microscope that routinely can resolve fibers of 0.0025 micrometer diameter; 100 times more sensitive than phase contrast microscopy.

transmission loss — the ratio expressed in decibels of the sound energy on a structure to the sound energy transmitted.

transmissometer — an instrument used to measure visibility; consists of a fixed light projector and a receiver; the light projected at a fixed intensity either reaches the receiver as projected, or decreases in intensity in proportion to the visibility restricting particles in the air between the projector and the receiver; the receiver translates the light it receives in visibility readings in hundreds of feet.

transmittance — the fraction of incident light that is transmitted through an optical medium.

transmutation — a process in which the nuclei of a radioactive element decays at a fixed rate by emitting nuclear particles; the element is changed to another element, such as when radioactive radium becomes radon. *See also* nuclear transformation.

transovarial — pertaining to the passing of an infection from one generation to another through the egg.

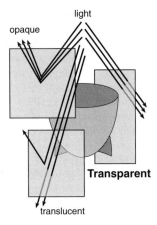

Transparent

transparent — the ability to transmit light so that objects can be seen clearly through it.

transpiration — the process in which green plants lose water by evaporation through their stomata; (*biology*) the passage of a gas through a membrane.

transplacental — pertaining to an agent that causes physical defects in a developing embryo.

transponder — a unit which receives pulses from a radar set or interrogator and automatically responds to the received pulse by transmitting a pulse or sequence of pulses which can be recognized by the interrogating station.

transport — the hydrologic, atmospheric, or other physical processes that convey the mass of a pollutant through and across media from source to receptor; (*engineering*) to carry or convey goods from one place to another using ships, trucks, trains, pipelines, or planes; (*solid waste*) the movement of solid waste subsequent to collection.

Transportation Research Information Service
(TRIS) — a database of transportation research information from 1968 on air, highway, rail, transport, mass transit, and other transportation modes; information included on environmental and safety concerns.

transport mode — any method of transportation, such as highway, rail, water, pipelines, or air.

transport velocity — the velocity required to prevent the settling of a contaminant from an air stream, usually related to the flow of air in a duct.

transstadial — pertaining to the transmission of organisms from infected adult ticks through the eggs to the larval, nymphal, and adult stages.

Transpiration

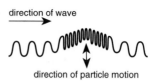

Transverse Wave

transverse wave — a wave in which the particles of a medium meet at right angles to the direction of the motion of the wave.

trauma — an injury caused by accident or violence.

treatment — any method, technique, or process designed to change the physical, chemical, or biological character or composition of any hazardous waste so as to neutralize or render it non-hazardous or less hazardous, or to recover it, make it safer to transport, store, dispose of, amenable for recovery, storage, or volume reduction; means include neutralization, precipitation, evaporation, incineration, filtration, and ion exchange; a technique used to improve a medical problem.

tremor — involuntary shaking, trembling, or quivering.

trench fever — (*disease*) a nonfatal febrile louse-borne bacterial disease characterized by headache, malaise, pain, and tenderness ranging in severity. Incubation time is 7–30 days. It is caused by *Rickettsia quintana*; found during World War I and World War II among troops and prisoners of war and in other occasional conditions of crowded living. The reservoir of infection is people, with the body louse *Pediculus humanus* the intermediate host; not directly transmitted from person to person but through the feces of the louse in a break in the skin. It is communicable as long as the organisms circulate in the blood, which may be weeks to years; general susceptibility. It is controlled by using delousing procedures, dusting clothing and bodies.

trench foot — a painful foot disorder resembling frostbite and resulting from exposure to cold and wet.

trench sanitary landfill method — a landfill method in which the waste is spread and compacted in a trench; the excavated material is spread and compacted over the waste to form the basic cell structure.

TRI — acronym for toxic release inventory.

triage — the sorting out or screening of patients seeking care to determine which service is initially required and with what priority.

Tribolium castaneum — *See* Red Flower Beetle.

Tribolium confusum — *See* Confused Flower Beetle.

S
T

tributyl phosphate — $(CH_3[CH_2]_3O)_3PO$ — a colorless to pale yellow, odorless liquid; MW: 266.3, BP: 552°F (decomposes), Sol: 0.6%, Fl.P (open cup): 295°F, sp gr: 0.98. It is used as an antifoaming agent or plasticizer in the manufacture of surface coatings and adhesives; as a solvent in the extraction of metals; as a heat-exchange medium. It is hazardous to the respiratory system, skin, and eyes; and is toxic by inhalation, ingestion, and contact. Symptoms of exposure include headache, nausea, irritation of the eyes, skin, and respiratory system. OSHA exposure limit (TWA): 0.2 ppm [air] or 2.5 mg/m³.

Trichinella — a genus of nematode parasites.

Trichinosis

trichinosis — (*disease*) an infection caused by *Trichinella spiralis*; an intestinal nematode whose larvae migrate and become encapsulated in muscles; asymptomatic to symptoms of thirst, profuse sweating, chills, weakness, and prostration. It may become a fulminating fatal disease. Incubation time is usually 10–14 days with a range of 5–45 days. It is found worldwide in pork or wild meat. Reservoirs of infection include pigs, dogs, cats, rats, and wild animals; transmitted by eating raw or improperly cooked meat containing larvae; not transmitted from person to person; general susceptibility. It is controlled by cooking meat thoroughly.

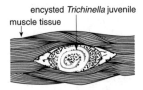

1,2,4 Trichlorobenzene

1-2-4-trichlorobenzene — $C_6H_3–C_{13}$ — a colorless liquid; MW: 181.46, BP: 415.4°F, Sol: not available, Fl.P: 222°F, sp gr: 1.463. Other information not available. ACGIH ceiling exposure limit (TLV-C): 5 ppm (40 mg/m³).

1,1,2-trichloroethane — $CHCl_2CH_2Cl$ — a colorless liquid with a sweet, chloroform-like odor; MW: 133.4, BP: 237°F, Sol: 0.4%, sp gr: 1.44. It is used in organic synthesis in the production of vinylidene chloride. It is hazardous to the central nervous system, eyes, nose, liver, and kidneys; and is toxic by inhalation, absorption, ingestion, and contact. Symptoms of exposure include central nervous system depression, liver and kidney damage, irritation of the nose and eyes; carcinogenic. *See also* methyl chloroform. OSHA exposure limit (TWA): 10 ppm [skin] or 45 mg/m³.

Trichloroethylene
(Trichloroethene)

trichloroethylene — $ClCH=CCl_2$ — a colorless liquid (unless dyed blue) with a chloroform-like odor; MW: 131.4, BP: 189°F, Sol (77°F): 0.1%, Fl.P: 90°F, sp gr: 1.46. It is used as a cleaning solvent in cold cleaning and vapor degreasing operations; as a scouring and cleaning agent in textile processing; in the extraction and purification of animal and vegetable oils in the food and pharmaceutical industry; in the manufacture of adhesives, anesthetics and analgesics; as a fumigant and disinfectant for seeds and grains. It is hazardous to the respiratory system, heart, liver, kidneys, central nervous system, and skin; and is toxic by inhalation, ingestion, and contact. Symptoms of exposure include headache, vertigo, visual disturbances, tremors, somnolence, nausea, vomiting, eye irritation, dermatitis, cardiac arrhythmia, paresthesia; carcinogenic. OSHA exposure limit (TWA): 50 ppm [air] or 270 mg/m³.

trichloromethane — *See* chloroform.

trichloronaphthalene — $C_{10}H_5Cl_3$ — a colorless to pale yellow solid with an aromatic odor; MW: 231.5, BP: 579–669°F, Sol: insoluble, Fl.P (open cup): 392°F, sp gr: 1.58. It is used as an insulating material; as an inert component for resins or polymers as a flame-resistant, water-proofing and insecticidal/fungicidal agent in coatings for wood, textiles, and paper; in catalytic chlorination of naphthalene; in the manufacture of special lubricants; as an additive for cutting oil. It

1,1,2–Trichloroethane

is hazardous to the skin and liver; and is toxic by inhalation, absorption, ingestion, and contact. Symptoms of exposure include acne-form dermatitis, anorexia, nausea, vertigo, jaundice, liver injury. OSHA exposure limit (TWA): 5 mg/m^3 [skin].

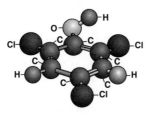

2,4,6–Trichlorophenol

2,4,6-trichlorophenol — $C_6H_3Cl_3O$ — a synthetic yellow solid with a strong, sweet smell; MW: 197.46, BP: 246°C, Sol: 800 mg/L in water at 25°C, density: 1.4901 g/cm^3. It is used in the preservation of wood and glue; as a disinfectant, sanitizer, bactericide, germicide; in pesticides; as an anti-mildew treatment for textiles; as a fungicide, herbicide, defoliant. It is hazardous to the respiratory system; and is toxic by inhalation. Symptoms of exposure include cough, chronic bronchitis, chest wheezing, reduced expiratory flow; may be carcinogenic for soft tissues. Environmental Protection Agency guidelines: 2×10^{-2} mg/Kg/d [air].

2,4,5-trichlorophenoxyacetic acid (2,4,5-T) — $C_6H_2Cl_3OCH_2COOH$ — a colorless to tan, crystalline solid; MW: 255.5, BP: decomposes, Sol: 0.03%, Fl.P: not known, sp gr: 1.80. It is used in the formulation of herbicides and plant hormones. It is hazardous to the skin, liver, and gastrointestinal tract; and is toxic by inhalation, ingestion, and contact. Symptoms of exposure in animals include ataxia, skin irritation, acne-like rash. OSHA exposure limit (TWA): 10 mg/m^3 [air].

1,2,3 Trichloropropane

1,2,3-trichloropropane — $CH_2ClCHClCH_2Cl$ — a colorless liquid with a chloroform-like odor; MW: 147.4, BP: 314°F, Sol: 0.1%, Fl.P (open cup): 180°F, sp gr: 1.39. It is used as a solvent and extractant for resins, oils, fats, waxes, and chlorinated rubber; as a commercial solvent for degreasing metal parts; in organic synthesis for the manufacture of other chemicals; as a co-polymer, telomere, or cross-linking agent for sealing compounds. It is hazardous to the eyes, skin, respiratory system, central nervous system, and liver; and is toxic by inhalation, absorption, ingestion, and contact. Symptoms of exposure include central nervous system depression, liver injury, skin irritation, irritation of the eyes and throat; carcinogenic. OSHA exposure limit (TWA): 10 ppm [air] or 60 mg/m^3.

1,1,2-trichloro-1,2,2-trifluoroethane — CCl_2FCClF_2 — a colorless to water-white liquid with an odor like carbon tetrachloride at high concentrations; a gas above 118°F; MW: 187.4, BP: 118°F, Sol (77°F): 0.02%, Fl.P: not known, sp gr: 1.56. It is used as a selective solvent in degreasing electrical equipment, photographic films, magnetic tapes, precision instruments, plastics, glass, elastomers, or metal components; as a dry cleaning solvent; as a refrigerant in commercial/industrial air conditioning and industrial process cooling; as a chemical intermediate for dechloronization of chemicals in the manufacture of polymers and copolymers; as a solvent in the textile industry. It is hazardous to the skin and heart; and is toxic by inhalation, ingestion, and contact. Symptoms of exposure include drowsiness, dermatitis, throat irritation; in animals: cardiac arrhythmia. OSHA exposure limit (TWA): 1,000 ppm [air] or 7,600 mg/m^3.

Trichomonas — a genus of parasitic protozoan occurring in the digestive and reproductive system of humans and other animals.

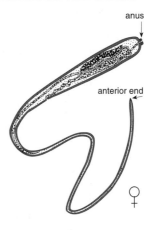

Trichuriasis
(*Trichuris trichuria*–whipworm which causes trichuriasis)

trichuriasis — (*disease*) a nematode infection of the large intestine, usually without symptoms; can have bloody, mucous stools and diarrhea. Incubation time is indefinite. It is caused by

Trichuris trichiuria; found worldwide. The reservoir of infection is people. It is transmitted through the eggs of the whipworm found in feces and hatches into larvae in contaminated soil; communicable for several years; general susceptibility. It is controlled by proper disposal of feces and good health habits, especially hand washing. Also known as whipworm.

trickling filter system — a system of secondary sewage treatment which is similar to self-purification action of streams; it is more accurately a biological oxidizing bed; the effluent is placed on the stones in the bed and microorganisms present consume the solids as a food supply.

triethylamine — $(C_2H_5)_3N$ — a colorless liquid with a strong, ammonia-like odor; MW: 101.2, BP: 193°F, Sol: 2%, Fl.P (open cup): 16°F, sp gr: 0.73. It is used in polymer technology as a catalyst; as a stabilizing agent and chain-transfer agent; in the preparation of film-forming resins; as a solvent pH stabilizer and to improve water solubility in polymers; in extraction and purification of antibiotics; in herbicides and pesticides; in organic synthesis; as a corrosion inhibitor; as a dye assist; as an ingredient of photographic development; in carpet cleaners. It is hazardous to the respiratory system, eyes, and skin; and is toxic by inhalation, absorption, ingestion, and contact. Symptoms of exposure include irritation of the respiratory system, skin, and eyes. OSHA exposure limit (TWA): 10 ppm [air] or 40 mg/m³.

trifluoromonobromomethane — $CBrF_3$ — a colorless, odorless gas; shipped as a liquefied compressed gas; MW: 148.9, BP: –72°F, Sol: 0.03%. It is used as a commercial and military fire extinguishant; as a refrigerant for food processing and storage; in organic synthesis; during the manufacture of hydraulic fluids; for special purposes in bubble chambers for ionization studies, in quark detection, and in radiation counters. It is hazardous to the heart and central nervous system; and is toxic by inhalation and contact. Symptoms of exposure include lightheadedness, cardiac arrhythmias. OSHA exposure limit (TWA): 1,000 ppm [air] or 6,100 mg/m³.

trihalomethane (THM) — any member of the family of organic compounds named as derivative of methane, wherein three of the four hydrogen atoms are each substituted by a halogen atom in the molecular structure.

trimester — a period of time divided into three sections.

2,4,6 Trinitrotoluene

2,4,6-trinitrotoluene — $CH_3C_6H^2(NO_2)_3$ — a colorless to pale yellow, odorless solid or crushed flakes; MW: 227.1, BP: 464°F (explodes), Sol (77°F): 0.01%, Fl.P: not known (explodes), sp gr: 1.65. It is used in the manufacture of shells, bombs, grenades, and mines; in commercial explosives; in propellant compositions; in the production of intermediates for synthesis of dyestuffs and photographic chemicals. It is hazardous to the blood, liver, kidneys, eyes, cardiovascular system, central nervous system, and skin; and is toxic by inhalation, absorption, ingestion, and contact. Symptoms of exposure include liver damage, jaundice, cyanosis, sneezing, cough, sore throat, peripheral neuropathy, muscular pain, kidney damage, cataract, sensitization dermatitis, leukocytosis, anemia, cardiac irregularities. OSHA exposure limit (TWA): 0.5 mg/m³ [skin].

triorthocresyl phosphate — $(CH_3C_6H_4O)_3PO$ — a colorless to pale yellow, odorless liquid or solid (below 52°F); MW: 368.4, BP: 770°F (decomposes), Sol: slight, Fl.P: 437°F, sp gr: 1.20. It is used as a flame retarder and plasticizer; in hot extrusion molding or bulk forming of plasticized polyvinyl chloride; in coatings and adhesives; as a gasoline additive to control preignition; as a hydraulic fuel and a heat exchange medium; as a synthetic lubricant; as a waterproofing agent; as a solvent mixture for nitrocellulose and other natural resins; as an extraction solvent in recovery of phenol from gas-plant effluents and coke-oven wastewaters; in grinding media for pigments; as an intermediate in the synthesis of pharmaceuticals. It is hazardous to the central and peripheral nervous systems; and is toxic by inhalation, absorption, ingestion, and contact. Symptoms of exposure include gastrointestinal distress, peripheral neuropathy, cramps in calves, paresthesia in feet or hands, weak feet, wrist drop, paralysis. OSHA exposure limit (TWA): 0.1 mg/m³ [skin].

triphenyl phosphate — $(C_6H_5O)_3PO$ — a colorless crystalline powder with a phenol-like odor;

MW: 326.3, BP: 776°F, Sol: insoluble, Fl.P: 428°F, sp gr: 1.29. It is used as a flame-retardant and plasticizer in compounding chlorinated rubber; on coatings and adhesives; in hot extrusion moldings; as a fireproofing agent in the manufacture of roofing paper. It is hazardous to the blood; and is toxic by inhalation and ingestion. Symptoms of exposure include minor changes in blood enzymes; in animals: muscular weakness, paralysis. OSHA exposure limit (TWA): 3 mg/m^3 [air].

TRIS — *See* Transportation Research Information Service.

tritium (T) — (^{3}H) — the heavy hydrogen isotope with one proton and two neutrons in the nucleus.

trophic accumulation — the passing of a substance through a food chain such that each organism retains all or a portion of its food and eventually acquires a higher concentration in its flesh than in the food.

trophic level — a system of categorizing organisms by the way they obtain food; examples are producers or organic detritus.

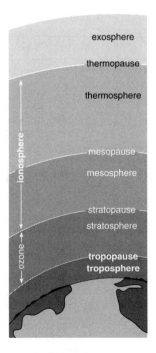

**Tropopause
Troposphere**

tropopause — the boundary between the troposphere and stratosphere, usually characterized by an abrupt change of lapse rate.

troposphere — the layer of the atmosphere from the earth's surface extending outward about 5 miles at the poles and 10 miles at the equator; most weather activities occur in this layer.

true color — the color of water resulting from substances which are totally in solution.

true negative — a negative test result for an individual who does not have the disease tested for.

true positive — a positive test result for an individual who does have the disease tested for.

true respiration — the use of oxygen by the cells with the release of carbon dioxide.

truncating — *See* round off.

TSCA — *See* Toxic Substances Control Act.

TSD — notation for treatment storage for disposal facility.

tsp — *See* teaspoon.

TSP — *See* total suspended particulates.

TSS — *See* total suspended nonfilterable solids; toxic shock syndrome.

tsunami — an unusually large sea wave produced by a sea quake or undersea volcanic eruption and not of tidal origin.

tsutsugamushi disease — (*disease*) *See* scrub typhus.

tuberculosis — (*disease*) a mycobacterial disease with lesions, involvement of lymph nodes, pleura, pericardium, kidneys, bones, joints, and larynx. Incubation time is 4–12 weeks until a skin test may be positive; disease symptoms may appear within a year or two after infection. It is caused by *Mycobacterium tuberculosis*; found worldwide. Reservoirs of infection include people and diseased cattle; transmitted by airborne droplet nuclei from sputum of people with infectious tuberculosis or from raw unpasteurized milk or dairy products from contaminated cattle; communicable as long as the tubercle bacilli are being discharged in the sputum; general susceptibility. It is controlled by limiting overcrowding, the use of diagnostic medical tests and X-rays, and proper control of cattle.

tularemia — (*disease*) an infectious zoonotic disease with a variety of symptoms including swelling of the regional lymph nodes, intestinal pain, diarrhea, and vomiting. Incubation time is 2–10 days, usually 3 days. It is caused by *Francisella tularensis*; found throughout North America, Europe, and parts of Asia. Reservoirs of infection include wild animals (including rabbits and hard ticks); transmitted by piercing the skin with blood or tissue which has been infected, or by the bite of infected ticks or deer flies; not communicable from person to person;

S
T

general susceptibility. It is controlled through proper use of gloves when skinning or handling animals, especially rabbits; and avoidance of bites of deer flies, mosquitos, and ticks. Also known as rabbit fever.

tumor — an abnormal swelling or growth. *See also* neoplasm.

tumorigenic — able to produce a tumor.

tundra — an area, including the arctic and antarctic zones, which can support vegetation such as lichens, grasses, and dwarf woody plants; characterized by long, severe winters and short, warm summers.

tuple — *See* database.

turbidimeter — a device that measures the amount of suspended solids in a liquid.

turbidity — an expression of the optical property of a sample which causes light to be scattered and absorbed rather than transmitted in straight lines through the sample.

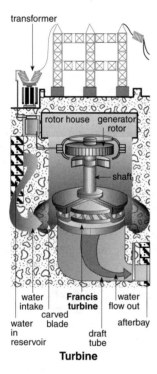

Turbine

turbine — an engine that forces a stream of liquid through jets at high pressure against the curved blades of a wheel forcing the blades to turn.

turbulent flow — a state of fluid flow in which the instantaneous velocities exhibit irregular and apparent random fluxuations so that, in practice, only statistical properties can be recognized and subjected to analysis.

turbulent loss — the pressure or energy lost from a ventilation system through air turbulence.

turnover rate — the period of time, usually in hours, required to circulate a volume of water equal to the pool capacity.

turpentine — $C_{10}H_{16}$ (approx) — a colorless liquid with a characteristic odor; MW: 136 (approx), BP: 309–338°F, Sol: insoluble, Fl.P: 95°F, sp gr: 0.86. It is used in the manufacture of synthetic pine oil; in insecticides, flavors, and perfumes; in the preparation of polishes; in the manufacture of synthetic camphor; in paints. It is hazardous to the skin, eyes, kidneys, and respiratory system; and is toxic by inhalation, absorption, ingestion, and contact. Symptoms of exposure include headache, vertigo, albuminuria, skin irritation and sensitization, irritation of the eyes, nose, and throat. OSHA exposure limit (TWA): 100 ppm [air] or 560 mg/m³.

tuyere — an opening or port in a grate through which air can be directed to improve combustion.

TVOC — notation for total volatile organic compound.

TWA — *See* time-weighted average exposure.

typhoid fever — (*disease*) a highly infectious, systemic, bacterial disease with sustained fever, headache, malaise, anorexia, bradycardia, nonproductive cough, constipation, occasionally diarrhea, and involvement of the lymphoid tissues. Incubation time is 1–3 weeks. It is caused by 106 types of *Salmonella typhi*; found worldwide. The reservoir of infection is people. It is transmitted by food or water contaminated by feces or urine from an infected person or carrier, shellfish, sewage contaminated waters, raw fruits, vegetables, and milk products. It is spread by flies; communicable for 3 months or more; general susceptibility. It is controlled through the proper disposal of human feces, fly control, protection of private and public water supplies, appropriate food handling, and removal of any infected persons or carriers from food handling.

typhus fever — (*disease*) any of three epidemic louse-borne human diseases transmitted by lice to people with symptoms of headaches, chills, prostration, fever, and general pains. Incubation time is 1–2 weeks, usually 12 days. It is caused by *Rickettsia prowazekii*; found in the United States, Mexico, and Central and South America. The reservoir of infection is people; transmitted by the body louse *Pediculus humanus*; general susceptibility. It is controlled by proper use of insecticides in clothes and people to kill lice, and immunization of special groups.

S
T

U

U-blade — a bulldozer blade with an extension on each side to increase moving capacity.

UC — notation for uniformity coefficient.

UEL — *See* upper explosive limit.

UF — *See* uncertainty factor.

UFFI — acronym for urea-formaldehyde foam insulation.

UFL — *See* upper flammable limit.

UHF — *See* ultrahigh-frequency.

UHV — *See* ultrahigh-voltage.

ulcer — the destruction of skin or mucous membrane with or without infection or pain.

ulceration — the formation of an open sore or ulcer.

ultimate analysis — the chemical analysis of a solid, liquid, or gaseous fuel to determine the amounts of carbon, hydrogen, sulfur, nitrogen, oxygen, and ash.

ultrahigh-frequency (UHF) — the band of frequency between 300–3,000 megahertz in the radio spectrum.

ultrahigh-temperature pasteurization — pasteurization accomplished by steam injection or by use of vacuum to heat milk to any of the following combinations of temperature and time: 191°F for 1 second, 194°F for 0.5 seconds, 201°F for 0.1 seconds, 204°F for 0.05 seconds, 212°F for 0.01 seconds.

ultra low volume (ULV) — (*pesticides*) a technique used in mosquito larvaciding where adulticiding equipment is set at a low level, typically 2 pints per acre, and the product is allowed to drift down into otherwise inaccessible areas such as wetlands; appears to be a successful technique for control of the Asian tiger mosquito.

ultrasonic — those sounds whose frequencies are above the normal range of human hearing, about 20,000 hertz. Also known as supersonic.

ultrasonics — the science of sound with frequencies above the audible range.

ultrasound — mechanical radiant energy of a frequency greater than 20,000 cycles per second; the upper limit of human hearing.

ultraviolet — radiations above the visible light spectrum with wavelengths shorter than those of violet light.

ultraviolet absorption spectrophotometry — the study of the spectra produced by the absorption of ultraviolet radiant energy during the transformation of an electron.

ultraviolet light — *See* ultraviolet radiation.

ultraviolet radiation (UV) — the invisible part of the light spectrum whose rays have wavelengths shorter than those of the bowed end of the visible spectrum and longer than those of the X-rays; the ultraviolet spectrum is usually considered to extend from about 50–380 nanometers; the maximum reddening of the skin occurs at 260 nanometers; bacteria and viruses may be killed at ultraviolet rays shorter than 320 nanometers. Also known as ultraviolet light.

U.L.V. — *See* ultra low volume.

uncertainty factor (UF) — a factor used in operationally deriving the reference dose from experimental data; intended to account for variation in sensitivity among members of the human population, uncertainty in extrapolating animal data to humans, uncertainty in extrapolating from data obtained in a study that is of less than lifetime exposure and uncertainty in using lowest observed adverse effect level data rather than no observed adverse effect level.

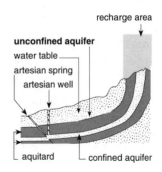

Unconfined Aquifer

unconfined aquifer — a geological formation that can be readily contaminated since there is no protective stratum or aquitard above it.

uncoupling — the process of prohibiting a phosphate from joining ADP to form ATP.

underdrain — the distribution system at the bottom of the sand filter to collect filtered water during a filter run and to distribute the backwash water.

underfire air — any forced or induced air under control as the quantity and direction is supplied beneath a grate and passes through a fuel bed.

underground storage tank (UST) — a tank placed underground as a fire-preventive measure and used to store gasoline, other oil products, or hazardous substances.

underwater light — a lighting fixture designed to illuminate a pool from beneath the water surface; may be "wet-niched," or located in the pool water, or "dry niched," located in the pool side wall beyond or behind a waterproof window and serviced from outside the pool.

underwrite — to insure; to scrutinize the risk and decide on its eligibility for insurance.

Underwriters Laboratories, Inc. — an independent, nonprofit organization for public safety testing of devices, systems, and materials.

unified classification — a system of classification of soils according to particle size, distribution, plasticity, liquid limit, and organic matter.

uniform corrosion — deterioration over the entire metal surface at the same rate.

United States Coast Guard (USCG) — a branch of the Armed Forces of the United States at all times and is a service within the Department of Transportation, except during wartime or when the President directs it to become part of the Navy.

United States Code (USC) — the codification of all laws of the United States.

United States Geological Survey (USGS) — a major program of the Department of Interior whose function is to classify public lands and examine geological structure, mineral resources, and products of the national domain.

United States Pharmacopeia (USP) — the official book of standard drugs in the United States.

United States Public Health Service (USPHS) — originating in 1798 in an Act of Congress authorizing marine hospitals for the care of American merchant seamen; reorganized in 1944 and expanded subsequently to support and protect the nation's physical and mental health.

unit risk — (*quantitative characterization of risk*) the excess lifetime risk due to continuous constant lifetime exposure to one unit of carcinogen concentration.

unleaded gasoline — gasoline containing not more than 0.05 grams of lead per gallon and not more than 0.005 grams of phosphorus per gram.

unsaturated flow — the movement of water in a soil which is not filled to capacity with water.

unsaturated zone — *See* zone of aeration.

unstable — tending toward decomposition or other unwanted chemical change during normal handling or storage.

unstable reactive — a chemical that in the pure state or as produced or transported vigorously will polymerize, decompose, condense, or become self-reactive under conditions of shock, pressure, or temperature.

upper atmosphere — the general term applied for the area from 60 kilometers in the air upward.

upper explosive limit (UEL) — the highest concentration of a vapor or gas in air that will produce a flash of fire when an ignition source, heat arc, or flame is present. Also known as upper flammable limit.

upper flammable limit (UFL) — *See* upper explosive limit.

uracil — $C_4H_4N_2O_2$ — a pyrimidine base; one of the four bases coding genetic information in the polynucleotide chain of ribonucleic acid.

uranium (insoluble compounds) — U — a silver-white, malleable, ductile, lustrous metal; weakly radioactive; MW: 238.0, BP: 6895°F, Sol: insoluble, Fl.P: not known, sp gr (77°F): 19.05, atomic number: 92. It is used as a chemical intermediate in the preparation of uranium compounds; for nuclear technology; in nuclear reactors as fuel to pack nuclear fuel rods; in the ceramics industry for pigments, coloring porcelain, painting on porcelain, and enameling; as a catalyst for many reactions; in the production of fluorescent glass. It is hazardous to the skin, bone marrow, lymphatics; and is toxic by inhalation, ingestion, and contact. Symptoms of exposure include dermatitis; carcinogenic; in animals: dermatitis, lung, and lymph node damage. OSHA exposure limit (TWA): 0.2 mg/m³ [air].

urban runoff — storm water from city streets, usually carrying litter and organic wastes.

urea — $CO(HN_2)_2$ — a nitrogenous waste substance found chiefly in the urine of mammals, formed in the liver from metabolized proteins; synthesized as white crystals or powder.

uremia — the accumulation in the blood of excessive amounts of waste products normally eliminated in the urine; a condition resulting from kidney failure.

urethra — the passageway through which urine leaves the bladder in most mammals.

uricosuria — excretion of uric acid in the urine.

urine — the liquid waste filtered from blood in the kidneys and excreted by the bladder.

urinogenital — pertaining to the urinary tract.

urticaria — elevated itchy patches of skin. Also known as hives.

USC — *See* United States Code.

USCG — *See* United States Coast Guard.

USDA — *See* Department of Agriculture.

use-dilution — a dilution as specified on labeling, which produces the concentration of the pesticide for a particular effect.

USGS — *See* United States Geological Survey.

USP — *See* United States Pharmacopeia.

USPHS — *See* United States Public Health Service.

utensil — any food-contact implement used in the storage, preparation, transportation, or dispensing of food.

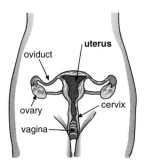

Uterus

uterus — a muscular organ which holds and nourishes the growing fetus during gestation in mammals.

UV — *See* ultraviolet radiation.

U
V

V — *See* volt.

vaccination — *See* immunization.

vaccine — a preparation made from a killed or attenuated pathogen to protect the individual from the disease by producing active immunity.

vaccinia — a contagious disease of cattle believed to be a form of small pox; transmissible to humans. Also known as cow pox.

vacuum — a space which contains no gas, liquid, or solid; devoid of matter; any pressure less than that exerted by the atmosphere.

vacuum cleaning — (*swimming pools*) a system of lines leading from the main drain and skimmers throttled or shut off until the vacuum connection is opened, permitting the pump to pull dirt from the pool through the lines.

vacuum filter — a filter which operates under a vacuum from the suction of a pump. Also known as a suction filter.

vadose — the unsaturated soil zone.

vadose zone — the unsaturated soil zone between the water table and the surface of the ground. Also known as the zone of aeration.

valence — the chemical-combining power of an atom; indicates the number of electrons that can be lost, gained, or shared by an atom in a compound. Also known as bond; chemical bond.

validate — to test the accuracy of mathematical models used to predict pollution concentrations by comparing predictive concentrations to measured concentrations under equivalent conditions.

validity — a measure of the extent to which an observed situation reflects the actual situation.

valve — a device which controls the direction of fluid flow, rate, and pressure.

valve stop — a guide which permits turning the valve plug to but not beyond the fully closed position.

vanadium pentoxide (respirable dust) — V_2O_5 — a yellow-orange powder or dark gray, odorless flakes dispersed in air; MW: 181.9, BP: 3182°F (decomposes), Sol: 0.8%, sp gr: 3.36. It is used as a catalyst in the preparation of organic and inorganic compounds; in chemical synthesis; in the manufacture of ultraviolet filter glass to prevent radiation injury and fading of fabrics; in the manufacture of afterburners for automobiles; in the textile industry as a catalyst to yield

intensive black dyes; in the printing industry; in the manufacture of ceramic pigments; as a component of special steels in electric furnace steels, welding rods, and permanent magnets. It is hazardous to the respiratory system, skin, and eyes; and is toxic by inhalation, ingestion, and contact. Symptoms of exposure include eye irritation, green tongue, metallic taste, eczema, cough, fine râles, wheezing, bronchitis, dyspnea, throat irritation. OSHA exposure limit (TWA): 0.05 mg/m^3 [air].

vanadium pentoxide (respirable fume) — V_2O_5 — a finely divided particulate dispersed in air; MW: 181.9, BP: 3182°F (decomposes), Sol:

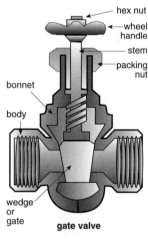

gate valve

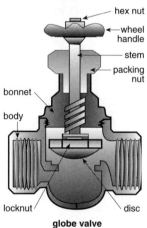

globe valve

Valves

390

0.1%, sp gr: 3.36. It is liberated from the production of pellets from electric furnaces and from the manufacture of semiconductors fused with sodium oxide. It is hazardous to the respiratory system, skin, and eyes; and is toxic by inhalation and contact. Symptoms of exposure include eye irritation, green tongue, metallic taste, throat irritation, cough, fine râles, wheezing, bronchitis, dyspnea, eczema. OSHA exposure limit (TWA): 0.05 mg/m³ [air].

Van Allen belt — the zone of intense, ionizing radiation circling the earth and containing charged particles trapped in the earth's magnetic field.

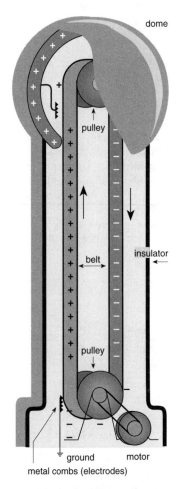

Van de Graaff Accelerator

Van de Graaff accelerator — an electrostatic generator.

van der Waals force — the intermolecular and interatomic forces that are electrostatic in origin. Also known as dispersion force.

vapor — the gaseous form of a substance that is a liquid at ordinary temperatures and pressure.

vapor density — the weight of a vapor or gas compared to the weight of any equal volume of air.

vapor dispersion — the movement of vapor clouds in air due to wind, gravity, spreading, and mixing.

vapor/hazard ratio — the relationship between the toxicity of a substance in its vapor phase and in its vapor pressure.

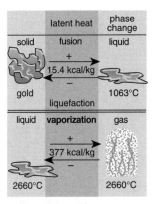

Vaporization

vaporization — the change of a substance from a liquid to a gaseous state.

vapor pressure — the pressure at any given temperature of a vapor in equilibrium with its liquid or solid form; (*soils*) the pressure exerted by a vapor in a confined space as a function of the temperature; (*meteorology*) the partial pressure of water vapor in the atmosphere; (*matter*) the partial pressure of any liquid. Also known as saturated vapor pressure.

vapor recovery system — a closed system of pipes, valves, and compressors in which vapors that might otherwise escape into the atmosphere are compressed into liquid form and returned to their source.

variable air volume system (VAV) — an air handling system that conditions the air to a constant temperature and varies the outside air flow to ensure thermal comfort.

variance — a difference between that which is required or specified and that which is permitted.

varicose — pertaining to a blood vessel that is unnaturally swollen.

vascular — relating to a channel for the conveyance of a body fluid, such as blood, or to a system of such channels.

vasoconstriction — a decrease in the cross-sectional area of blood vessels.

vasopressin — *See* antidiuretic hormone.

vat pasteurization — pasteurization of milk at 145°F for 30 minutes.

U
V

vats — large containers or tanks.

VAV — *See* variable air volume system.

VC — *See* vital capacity.

VDT — *See* visual display terminal.

VDU — *See* visual display unit.

vector — (*mathematics*) a quantity which has both direction and magnitude; (*medicine*) an agent, such as an insect, capable of transmitting an infection or disease from one organism to another.

vector control specialist — an environmental health practitioner responsible for surveillance and control of rodents, insects, and other vectors of public health and economic significance.

vector graphics structure — (*geographic information system*) a means of coding line and area information in the form of units of data expressing magnitude, direction, and connectivity.

vehicle — water, food, milk, or any other substance or particle serving as an intermediate means by which the pathogenic agent is transported from a reservoir and introduced into a susceptible host through ingestion, inhalation, inoculation, skin contact, or mucous membrane.

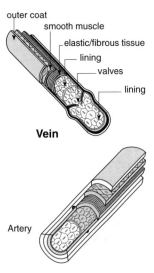

Vein

Artery

vein — a vessel that conveys blood to the heart in vertebrates.

velocity — a vector specifying the time-rate-of-change of displacement with respect to some reference plane. *See also* potential difference.

velocity pressure — the algebraic difference between total pressure and static pressure; the kinetic pressure equal to that required to bring a fluid at rest to flow at a given velocity; the velocity pressure is always positive and in the direction of the flow.

velometer — a device for measuring air velocity.

vending machine — any self-service device, which, upon insertion of a coin, paper currency, token, card, or key, dispenses unit servings of food either in bulk or in packages without the necessity of replenishing the device between each vending operation.

venom — a toxic or poisonous agent of animal origin.

ventilate — to change the quality of air by using a fan or opening windows or vents.

ventilated air sampler — an instrument constantly in contact with the air sample.

ventilated cell composting — a composting method in which compost is mixed and aerated by being dropped through a vertical series of ventilated cells.

ventilation air — the total air or combination of air brought into the system from the outdoors and the air that is recirculated within the building.

ventilation rate — the rate at which outside air enters and leaves a building, expressed as the number of changes of outdoor air per hour, or the number of cubic feet per minute of air entering the building.

ventricular fibrillation — a cardiac arrhythmia marked by fibrillar contractions of the ventricular muscle due to rapid, repetitive excitation of myocardial fibers without coordinated ventricular contraction. Also known as fibrillation. *See also* atrial fibrillation.

venturi flares — flares mixing air with the gas stream before it is ignited; usually used at ground level.

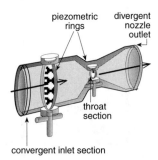

Venturi Meter

venturi meter — a fluid-flow measuring device consisting of a short tube with a tapering constriction in the middle that causes an increase in the velocity of the fluid flow and a corresponding decrease in fluid pressure. Also known as venturi tube.

venturi scrubber — a scrubber system in which dust-laden gas passes through a duct incorporating a venturi section at the throat causing an increase in velocity of the gas flow and a decrease in gas pressure resulting in removal of dust from the gas.

venturi tube — *See* venturi meter.

venturi wet scrubber — a collection device where an aqueous stream or slurry is used to remove particulate matter and/or gaseous pollutants.

venue — place of trial.

verdict — the ruling of the judge or jury in a criminal case.

verification — methods, procedures, and tests used to determine that a proposed system is in compliance with specific rules and regulations.

vermiculite — $(Mg,Fe,Al)_3(Al,Sl)_4O_{10}(OH)_2 \cdot 4H_2O$ — an expanded mica used as a sorbent for spill control and clean-up.

vernon globe — a thin-walled, flattened, black globe made up of a copper sphere 15 centimeters in diameter, with a temperature sensing device at its center and used to measure the radiant energy transferred between the surrounding surfaces and the convective heat exchange of the surrounding air. *See also* globe thermometer.

verruca — a wart-like elevation on the endocardium of a plant or animal.

vertical and horizontal spread (VHS) — the transport of landfill wastes to nearby receptors.

vertigo — a sensation of dizziness.

very high frequency (VHF) — the band of frequencies from 30–300 megahertz in the radio spectrum.

veryllium (compounds) — Ve — a hard, grayish element that does not occur naturally but as a chemical component of certain soils, volcanic dust and rocks, such as bertrandite and beryl; MW: 9.012, BP: 2,970°C, Sol: insoluble, density $1.8469/cm^3$. It is used in aircraft disc brakes, X-ray transmission, windows, space vehicles and instruments; in satellite structures, missile parts, nuclear reactors, and weapons; in fuel containers, precision instruments, rocket propellants, navigational systems, heat shields, and mirrors. Veryllium oxide is used in high-technology ceramics, electronic heat, sinks, electrical insulators, microwave oven components, gyroscopes, military vehicle armor, rocket nozzles, crucibles, thermalcouple tubing, and laser structural components; veryllium alloys are used as electrical connectors and relays, springs, precision instruments, aircraft engine parts, nonsparking tools, submarine cable housings and pivots, wheels and pinons, automotive electronics and molds for injection-molded plastics, automotive, industrial, and consumer applications. It is hazardous to the respiratory system and the endocrine system; and is toxic by inhalation. Symptoms of exposure include nonneoplastic respiratory disease, acute pneumonitis, weight loss, and effects on the endocrine system. Veryllium workers have a death rate higher than the national average for cardiovascular and pulmonary diseases, and increased incidence of lung cancer. OSHA exposure limits (TWA): 0.002 mg/m³ [air].

very poorly drained soil — (*soils*) soil in which the water is removed so slowly that the water table often remains at or on the surface.

very severe limitations — (*soils*) limitations that preclude development of a tract of land without the use of sanitary sewer systems.

vesicant — any substance or material that produces blisters on the skin.

vesicle — a small blister on the skin; a bladder-like cavity.

vesiculation — formation of vesicles.

Vessel Sanitation Program (VSP) — a cooperative activity between the cruise ship industry and the United States Public Health Service that results in an evaluation of food, water, sewage disposal, housing, and helps prevent disease and injury.

VHAP — *See* volatile hazardous air pollutant.

VHF — *See* very high frequency.

VHS — *See* vertical and horizontal spread; refers to the transport of landfill wastes to nearby receptors.

viable — living; capable of growing or evolving.

vibrating screen — an inclined sizing screen that is vibrated mechanically.

vibration — a periodic motion of the particles of an elastic body or medium in alternately opposite

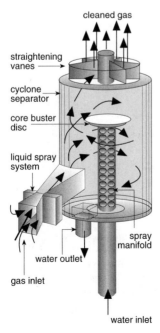

cleaned gas

straightening vanes

cyclone separator

core buster disc

liquid spray system

spray manifold

water outlet

gas inlet

water inlet

Venturi Scrubber

U V

directions from the position of equilibrium when that equilibrium has been disturbed; any movement of the earth, ground, other surface or equipment by temporal and special oscillation of displacement, velocity, or acceleration in any mechanical device or equipment located upon, attached to, affixed to, or in conjunction with that surface.

Vibrio — any of a genus of short, rigid, motile bacteria typically shaped like a comma or an S.

vibrio parahaemolyticus **food poisoning** — (*disease*) an intestinal disorder with watery diarrhea, abdominal cramps, and sometimes nausea, vomiting, fever, and headache. Incubation time is 12–24 hours but can range from 4–96 hours. It is caused by *Vibrio parahaemolyticus*, a halophilic vibrio; approximately 72 antigen types have been identified. Outbreaks have occurred in Japan, Southeast Asia, and the United States primarily in warm months. Reservoirs of infection include marine coastal environments where the organisms congregate in marine silt in the cold season and are free swimming in coastal waters, and in fish and shellfish in the warm season. It is transmitted by ingestion of raw or improperly cooked seafood or through the handling of raw seafood; not communicable from person to person; general susceptibility. It is controlled by proper cooking of all seafood and handling of the cooked seafood.

Vibrio vulnificus — (*disease*) a bacterium that causes gastroenteritis, "primary septicemia," or wound infections; generally with symptoms of diarrhea and bullous skin lesions; consumption of the microorganisms in raw seafood by people with chronic diseases (especially liver disease) may result in septic shock rapidly followed by death. Incubation time is 16 hours for gastroenteritis; the disease is found sporadically in the United States. The reservoir of infection is humans and other primates. It is transmitted through water, sediment, plankton, and consumption of raw shellfish; people with open wounds or lacerations can become contaminated in seawater. It is not communicable from person to person; general susceptibility. It is controlled by avoiding contact of open cuts with seawater, the proper disposal of feces, and cooking all shellfish thoroughly. Also known as primary septicemia.

villi — microscopic finger-like projections of the intestinal mucosa.

villus — a small, finger-like outgrowth from the surface of a membrane.

vinyl — a general term applied to a class of resins such as polyvinyl chloride, acetate, and butyryl.

Vinyl Acetate

vinyl acetate — $CH_3COOCH:CH_2$ — a clear, colorless liquid with a sweet, pleasant, fruity smell; MW: 86.09, BP 72–73°C, Sol: in water at 20°C 2g/100 mL, density: at 20°C 0.932, Fl.P –8°C (closed cup) and –1.1°C (open cup). It is used in the production of polyvinyl acetate and polyvinyl alcohol; with polyvinyl chloride to produce copolymers for adhesives; in architectural painting, nonwoven textile fibers, textile sizing and finishing; in adhesives, as a raw material for polyvinyl butyryl; as a barrier in packaging paint and in coating applications, in plastic floor coverings; phonograph records; flexible film, sheeting, and molding; and coextrusion compounds. It is hazardous to the respiratory system; and is toxic by inhalation. Symptoms of exposure include irritation of the nose, throat, and mucous membrane; respiratory damage in animals causing death due to lung congestion, hemorrhage froth in the trachea, and excess pleural fluid. OSHA exposure limits (PEL TWA): 10 ppm [air].

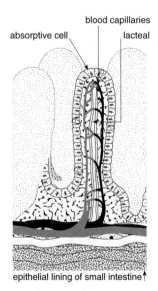

Villi

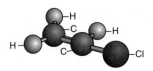

Vinyl Chloride

vinyl chloride — $CH_2{:}CHCl$ — a colorless gas or vapor with a mild, sweet odor; MW: 62.5, BP: –13.4°C, Sol: in water at 25°C, 2,763 mg/L, density: at 20°C, 0.9106 g/cm^3. It is used in organic synthesis and in adhesives; automotive parts, furniture, packaging materials, pipes, wall coverings, wire coatings, and in copolymer products, such as films and resins. It is hazardous to the pulmonary system, hepatic system, liver, kidneys, reproductive system, gastrointestinal system, central nervous system, and in fetal development; and is toxic by inhalation. Symptoms of exposure include dizziness, disorientation, a burning sensation in the soles of the feet, headaches, gastritis, ulcers, upper gastrointestinal bleeding, lack of blood clotting, hepatic abnormalities, burns on the conjunctiva and cornea, angiosarcoma of the liver, muscle pain, scleroderm-like skin changes, ataxia, visual and/or hearing disturbance, nausea, developmental toxicity, birth defects, decreased sexual function. OSHA exposure limits (PEL TWA): 1 ppm [air].

vinyl toluene — $CH_2{=}CHC_6H_4CH_3$ — a colorless liquid with a strong, disagreeable odor; MW: 118.2, BP: 339°F, Sol: 0.009%, Fl.P: 120°F, sp gr: 0.89. It is used in spray applications of vinyl toluene polyester surface coatings; during the preparation of unsaturated polyester resins and alkyd coatings; in the manufacture of thermoplastic moldings via extrusion, injection, or other processes. It is hazardous to the eyes, skin, and respiratory system; and is toxic by inhalation, ingestion, and contact. Symptoms of exposure include drowsiness, irritation of the eyes, skin, and upper respiratory system. OSHA exposure limit (TWA): 100 ppm [air] or 480 mg/m^3.

violation — (*pollution*) any incident of excess emissions, regardless of the circumstances of the occurrence; (*law*) an infringement of rules or laws.

viral gastroenteritis — a generic term for diseases caused by astrovirus, calicivirus, enteric adenovirus, parovirus, acute nonbacterial infectious gastroenteritis.

viral hepatitis A — (*disease*) a type of hepatitis with sudden onset; fever, malaise, anorexia, nausea and abdominal discomfort is followed in a few days by jaundice; varies from a mild condition of 1–2 weeks to severely disabling lasting many months. Incubation time is 15–50 days with an average of 28–30 days. It is caused by hepatitis A virus, a 27 nanometer picornavirus with the characteristics of an enterovirus; found worldwide, especially in areas where there is poor environmental control and in institutions, day care centers, low cost housing, rural areas, and in the military. It is transmitted from person to person by oral-fecal route and through food and water; communicable until a few days after the onset of jaundice; general susceptibility. It is controlled through public health education, good personal hygiene, proper management of day care centers and other institutions, proper sterilization of syringes and needles, and vaccines. Also known as infectious hepatitis.

viral hepatitis B — (*disease*) a type of hepatitis with a gradual and cumulative effect; anorexia, nausea, and vomiting leading to jaundice. Incubation time is 45–180 days with an average of 60–90 days. It is caused by hepatitis B virus, a 42 nanometer double-stranded deoxyribonucleic acid virus composed of a 27 nanometer nucleocapsid core surrounded by an outer lipoprotein coat containing the surface antigen; found world wide commonly in high risk groups such as parenteral drug abusers, homosexual men, institutional staff and patients and hemodialysis individuals. The reservoir of infection is people; transmitted by all body secretions and excretions; blood, saliva, semen, and vaginal fluids are the only fluids to have been shown to be infectious; communicable for many weeks and a chronic carrier state may exist for years; general susceptibility. It is controlled by vaccines, good personal hygiene, nonuse of reusable needles, and protection against infected bodily fluids. Also known as serum hepatitis.

viremia — presence of a virus in the blood.

virgin material — a raw material.

virulence — a relative measure of the frequency with which a microorganism will produce disease once it has established itself in the host; infectiousness.

virulent — capable of overwhelming bodily defense mechanisms.

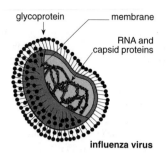

glycoprotein   membrane

RNA and
capsid proteins

influenza virus

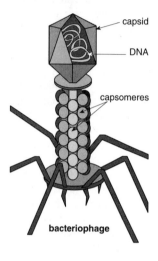

capsid

DNA

capsomeres

bacteriophage

RNA

**tobacco mosaic
virus**

capsid

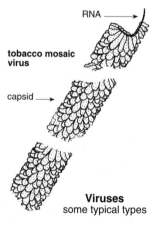

Viruses
some typical types

virus — the smallest organism known ranging in sizes of about 0.025–0.25 micrometers characterized by lack of independent metabolism; a disease-producing organism that needs living cells to grow.

viscera — the internal organs of the body.

viscosity — the property of a fluid or gas that offers resistance to flow; the resistance to flow due to the friction between molecules.

viscous — having a heavy or gluey consistency.

visible infrared spin scan radiometer (VISSR) — an instrument used on the synchronous me-

teorological satellite to provide both day and night images of the earth and cloud cover.

visible light — the light that can be detected by the human eye ranging from extreme red, with a frequency of 4.3×10^{14} sec^{-1}, through orange, yellow, green, blue, indigo to violet, with a frequency of 7.5×10^{14} sec^{-1}. *See also* spectrum.

visible radiation — the wavelengths of the electromagnetic spectrum between 10^{-4} centimeter and 10^{-5} centimeter.

VISSR — *See* visible infrared spin scan radiometer.

visual display terminal (VDT) — *See* cathode-ray tube.

visual display unit (VDU) — a terminal with a cathode-ray tube.

vital capacity (VC) — the largest volume of air measured on complete expiration after taking the deepest inspiration without a forced or rapid effort.

vital records — those records relating to birth, death, marriage, health, and disease.

vital statistics — data collected from contemporaneous registration of vital events, such as births, deaths, fetal deaths, marriages, divorces, etc.

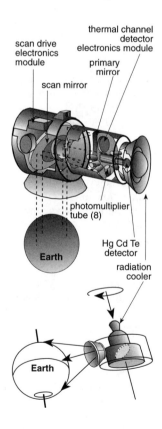

scan drive
electronics
module

thermal channel
detector
electronics module

primary
mirror

scan mirror

photomultiplier
tube (8)

Earth

Hg Cd Te
detector

radiation
cooler

Earth

**Visible Infrared Spin Scan
Radiometer (VISSR)**

vitamin — an organic substance found in natural foodstuffs and essential for normal body activity.

vitamin D milk — milk with added vitamin D content which has been increased by the approved method to at least 400 United States Pharmacopeia units per quart.

vitiligo — a condition in which destruction of melanocytes occurs in small or large areas resulting in patches of depigmentation.

vitrification — a process whereby high temperatures create permanent chemical and physical changes in a ceramic body converting it into glass or a glassy substance.

vitrification technology — a system of converting contaminated soils, sediments, and sludges into oxide glasses rendering them nontoxic and suitable for landfilling as a nonhazardous material.

V/m — *See* voltmeter.

VOC — *See* volatile organic compound.

voids — spaces in or between particles or fibers of a filtering medium; (*environmental engineering*) spaces found in an existing landfill.

volatile — referring to a substance that evaporates rapidly and explodes readily.

volatile hazardous air pollutant (VHAP) — volatile organic compounds released into the air, such as benzene, perchlorethylene, and carbon tetrachloride, and are potential health hazards.

volatile matter — the gaseous constituents of solid substances.

volatile organic compound (VOC) — organic chemicals that evaporate readily and exist as gases in the air.

volatile solid — the quantity of a solid in water, sewage, or other liquid lost on ignition of total solids.

volatility — the quality of a solid or liquid to evaporate quickly at ordinary temperatures when exposed to the air.

volcanic ash — a fine, white, porous powder similar to diatomite but lighter in weight; used as a filter media or filter aid; particle diameter is less than 4 millimeters.

volcanic glass — natural obsidian formed by the cooling of molten lava too rapidly to allow crystalization.

volcanism — the phenomenon and activities associated with the origin of volcanoes and the movement of both lava and other volcanic material.

volt (V) — the meter, kilogram, second, ampere unit of voltage potential equal to the potential difference across a conductor when 1 ampere of current in it dissipates 1 watt of power.

voltage — *See* potential difference.

voltage gain — the ratio of the output signal level with respect to the input level in decibels.

voltage-voltage difference — the electric potential; the work required to transport a unit of electric charge from one point to another.

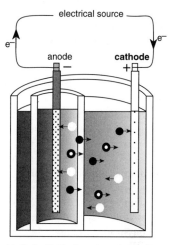

Voltaic Cell showing Cathode

voltaic cell — a device which changes chemical energy into electrical energy by the action of two dissimilar metals immersed in an electrolyte. *See also* primary voltaic cell.

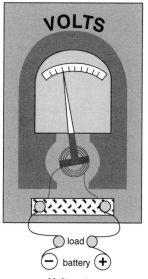

Voltmeter
showing galvanometer

voltmeter — an instrument used to measure in volts the difference of potential between two points in an electrical circuit.

volume flow rate — the quantity of a unit flowing per unit time.

volume reduction program — (*solid waste*) a technique developed by the generator that identifies waste minimization activities and goals; includes a series of steps from data gathering and analysis to identification of deficiencies, recommendations for a schedule, and its incorporation into the ongoing corporate management program.

volumetric analysis — (*chemistry*) a technique of quantitative analysis in which the volume of a liquid reagent of known concentration required to react completely with a substance whose concentration is being determined is measured by titration.

voluntary muscle — muscle tissue, including all the skeletal muscles, that appears microscopically to consist of striped myofibrils and are composed of bundles of parallel, striated fibers under voluntary control; the heart, which is a striated muscle, is under involuntary control.

vps — acronym for vibrations per second.

Vruchus pisorum — *See* Pea Weevil.

VSD — acronym for virtually safe dose.

VSP — *See* Vessel Sanitation Program.

VSS — acronym for volatile suspended solid.

vulcanism — all of the processes associated with the movement of molten rock material.

vulcanization — the process of combining natural, synthetic, or latex rubber with sulfur and various other additives, usually under heat and pressure to eliminate tackiness when warm and brittleness when cool, and to change permanently the material from a thermoplastic to a thermosetting composition; it improves the strength, elasticity, and abrasive resistance of the material.

Voluntary Muscle

involuntary muscle

W

W — *See* watt.

wading pool — an impervious structure less than three feet deep containing water and mechanically controlled for purification purposes; usually designed for children's use or wading.

waiting period — a period which must elapse to show that a person is disabled during which income benefits are not paid; refers only to compensation, not medical or hospital care, which must be provided immediately.

warewashing — (*food*) the cleaning and sanitizing of food-contact surfaces of equipment and utensils.

warfarin — $C_{19}H_{16}O_4$ — a colorless, odorless crystalline powder; MW: 308.3, BP: decomposes, Sol: 0.002%, Fl.P: not known. It is used as a rodenticide. It is hazardous to the blood, cardiovascular system; and is toxic by inhalation, absorption, ingestion, and contact. Symptoms of exposure include back pain, hematoma of the arms and legs, epistaxis, bleeding lips, mucous membrane hemorrhage, abdominal pain, vomiting, fecal blood, petechial rash, abnormal hematologic indices. OSHA exposure limit (TWA): 0.1 mg/m³ [air].

warm air mass — a large mass of stable air which produces steady winds, and creates poor surface visibility with stratus clouds and drizzle; it is warmer than surrounding air.

warm-blooded — *See* homoiothermal.

Warm Front

warm front — the advancing edge of a warm air mass.

warning device — a sound emitting device used to alert and warn people of the presence of a dangerous situation.

warrant — a court order directed to the marshall or other proper officer requiring the arrest and appearance before the court or United States Magistrate of the defendant named in the order.

waste disposal — the orderly process of discarding useless or unwanted material.

waste heat boiler — a boiler that uses the heat of exhaust or processed gas to generate steam or to heat water.

waste minimization audit — a survey conducted by in-plant personnel or consultants which identifies and evaluates opportunities to reduce hazardous waste generation.

Waste Pollution Abstract — an index of water pollution articles from 1971 to the present.

waste reduction — a policy mandated by the United States Congress in the 1984 Hazardous and Solid Wastes Amendments to the Resource Conservation and Recovery Act (RCRA); part of waste management which includes source reduction and recycling.

waste stabilization pond — a large, shallow basin, usually 2–4 feet in depth, where the biological oxygen demand of wastewater is reduced; the organics are converted to inorganics or stabilized because of the metabolic activity of bacteria, algae, and surface aeration.

waste stream — a general term for the total waste output of an area, location, or facility.

wastewater drain — a receptor and appurtenances for the disposal of liquid waste.

wastewater specialist — an environmental health practitioner who controls water pollution from municipal and industrial sources in order to assure water quality that fosters the health, welfare, and comfort of the community.

wastewater treatment plant (WWTP) — a facility designed and utilized to remove impurities from water through a series of aerobic processes and the production of an effluent released to a stream which will not be polluting in nature.

water activity (Aw) — (*food*) a measure of the free moisture in a food; the quotient of the water vapor pressure of the substance divided by the vapor pressure of pure water at the same temperature.

water column — a unit used in measuring pressure.

water cycle — (*geology*) the cycle by which water evaporates from oceans, lakes, and other bodies of water, forms clouds, and is returned to those bodies of water through rain and snow, the runoff from rain and snow, or through groundwater. *See also* hydrologic cycle.

Water Environment Federation (WEF) — formerly known as the Water Pollution Control Federation; a group of autonomous regional associations concerned with the treatment and disposal of domestic and industrial wastewater, groundwater, and toxic and hazardous waste.

water hardness — *See* hard water.

water ice — a food which is prepared by freezing water while stirring a mix composed of one or more optional fruit ingredients, optional sweetener ingredients, sugar, dextrose, or corn syrup.

water miscible — the quality of a substance to mix readily with water.

water of hydration — the water present as a structural part of certain crystals.

water pollution — the addition of harmful or objectionable material causing an alteration of water quality.

Water Pollution Control Act — a 1948 law funding state water pollution control agencies and providing technical assistance to states containing limited provisions for legal actions against polluters.

Water Quality Act — a 1970 law updated through 1987 mandating that states establish water quality standards and regulating discharges into river systems.

water quality criteria — acceptable levels of pollutants in bodies of water that are consistent with uses such as drinking, fishing, industry, and swimming.

water-reactive — a chemical that reacts with water to release a gas that is either flammable or presents a health hazard.

water recreational facility (WRF) — any artificial basin or other structure containing water used or intended to be used for recreation, bathing, relaxation, or swimming where body contact with the water occurs or is intended to occur; includes auxiliary building and appurtenances.

Water Resources Act — a 1984 law passed which declares the necessity of the existence of an adequate supply of quality water for production of materials and energy for the nation's needs and efficient use of the nation's energy and water resources which are essential to the economic stability, national growth, and citizens' well-being.

Water Resources Planning Act — a 1965 law updated through 1988 which declares the policy of Congress to encourage conservation, development, and utilization of water and related land resources of the United States on a comprehensive and coordinated basis by the federal government, states, counties, and private enterprises.

watershed — the region drained by a stream, lake, or other body of water.

water softening — the removal of minerals, primarily calcium and magnesium, to reduce the amount of soap needed, scale on cooking utensils or laundry basins, and materials on the interior pipes or water; increases heat transfer efficiency through walls of heating elements or exchange units of water tanks.

water-supply engineering — the study of the sources of water supply including water treatment, storage, transmission, and distribution facilities.

water supply specialist — an environmental health practitioner responsible for the supply of safe and acceptable water to the public to prevent disease.

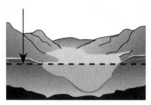

Water Table

water table — the upper boundary of the groundwater below which all spaces within rock are completely filled with water; the area between the zone of saturation and the zone of aeration.

watt (W) — an International System of Units unit of power equal to one joule per second.

watt/cm² — *See* watts per centimeter squared.

watts per centimeter squared (watt/cm^2) — a unit of power density used in measuring the amount of power per absorbing area.

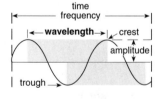

Wavelength

wavelength — the distance between any two corresponding points of two consecutive waves.

wave period — the time taken for two successive crests of a wave to pass a given point.

way bill — a record accompanying the shipment during transport.

WBGT — *See* wet bulb globe temperature index.

weak electrolyte — an electrolyte only slightly ionized in moderately concentrated solutions.

weather — description of the short-term physical conditions of the atmosphere including the moisture, temperature, pressure, and wind.

weathering — a physical disintegration of rocks that are slowly decomposed by exposure to either physical or chemical agents.

weather radar — a radar system designed to detect the physical characteristics of precipitation or clouds, their distribution, and movement.

weber — named after Wilhelm E. Weber, German physicist; the practical meter-kilogram-second unit of magnetic flux equal to that flux which in linking a circuit of one turn produces in it an electromotive force of one volt as the flux is reduced to zero at a uniform rate of one ampere per second.

weed — any plant growing where it is not wanted or growing to the detriment of some crop; vegetation which has attained a height of 12 inches or more and which constitutes a potential rodent harborage or any other health or safety hazard.

weed eradication — the destruction or removal of all weeds.

weevil — any of various beetles usually having a long snout whose larvae destroy crops.

WEF — *See* Water Environment Federation.

weight (wt) — the gravitational force with which a body is attracted toward the earth.

weighted moving average — an average value of an attribute computed for a point from the values at surrounding data points taking into account their distance or importance.

weighting network — (*sound*) the electrical networks associated with sound level meters.

Weil's disease — (*disease*) *See* leptospirosis.

Weir — a device in a waterway or swimming pool surface skimmer used to control the water level or divert the flow.

well — any excavation regardless of design or method of construction done to remove groundwater from an aquifer or to determine the quality, quantity, or level of groundwater.

well drained soil — (*soils*) soil in which the water is removed readily but not rapidly.

western equine encephalitis — (*disease*) an acute inflammatory encephalitis of short duration involving parts of the brain, spinal cord, and meninges. It is caused by an alpha virus; found chiefly in western and central United States, Canada, and Argentina. Possible reservoirs of infection include birds, rodents, bats, reptiles, amphibians, surviving mosquito eggs or adults; transmitted by the bite of infected mosquitos, not from person to person; usually a disease of children, but may affect others. It is controlled by destroying mosquito larvae, eggs, breeding places, adult mosquitos, and using mechanical mosquito control techniques.

wet bulb globe temperature index (WBGT) — a heat index recognized by the National Institute of Occupational Safety and Health to establish threshold limit values for heat stress, utilizing wet bulb temperature, dry bulb air temperature, and globe temperature.

wet-bulb temperature — the temperature at which liquid evaporating into the air can bring the air into saturation.

wet-bulb thermometer — a thermometer in which the bulb is covered with a water-saturated cloth or other material.

wet chemistry — the chemical analysis of pollutants dissolved in a chemical solution.

wet collector — one of a variety of devices used for cleaning, cooling, and deodorizing gas particulates and vapor emissions to the atmosphere. *See also* scrubber.

wet digestion — a solid waste stabilization process in which mixed solid organic wastes are placed in an open digestion pond to decompose anaerobically.

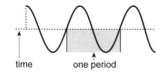

Wave Period

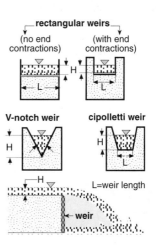

Weir

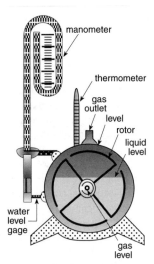

Wet–Gas Meter

wet-gas meter — a meter that measures the total gas volume by entrapping the gas in inverted cups or veins under a liquid; the buoyancy of a gas causes a rotation of the cups or veins that is proportional to the volume indicated on a precalibrated meter.

wetlands — a piece of land or area, such as a tidal flat or swamp, which contains much soil moisture because of its location at the land/water interface; provides ecological, biological, economic, scenic, and recreational resources; among the most productive ecosystems on the earth, especially estuaries and mangrove swamps. At least one half of the biological production of the oceans occur in the coastal wetlands; 60–80% of the world's commercially important marine fish either spend time in the estuary or feed upon the nutrients produced there. Coastal wetlands help protect inland areas from erosion, flood, waves, and hurricanes; provides cash crops such as timber, marsh, hay, wild rice, cranberries, blueberries, fish, and shellfish.

wet milling — mechanical size reduction of solid waste that has been wetted to soften the paper and cardboard constituents.

wet soil — (*soils*) nonsticky to very sticky; nonplastic to very plastic soil.

wettable powder — a chemical in dust form that can be mixed with water to form a suspension.

wetting — a means of reducing surface tension of soil and allowing detergent solutions to penetrate and spread out.

wetting agent — a material that decreases the surface tension of a liquid causing it to spread out easily across a surface.

wheal — a localized area of edema on the body surface; often associated with severe itching.

Wheatstone bridge — (*electricity*) a bridge used for measuring an unknown resistance by comparing it with a known resistance.

whey — a liquid or semi-liquid material remaining after the removal of fat and casein from milk or cream in the process of making cheese.

whipped cream — cream containing sugar and flavoring to which a harmless gas has also been added causing whipping of a product.

whipping cream — cream which contains not less than 30% milk-fat.

whipworm — *See* trichuriasis.

white blood cell — *See* white corpuscle; leukocyte.

white corpuscle — a colorless cell of the blood. Also known as a leukocyte; white blood cell.

WHO — *See* World Health Organization.

wholesale food processing establishment — any food establishment or portion thereof undertaking any of the activities of an establishment for purpose of sale or delivery to another person for the purpose of resale; includes warehouses, food manufacturing plants, beverage manufacturing plants, bakeries, etc.

wholesome — food which is nutritious, clean, free from adulteration, and otherwise suitable for human consumption.

Wilson's disease — a rare, progressive disease inherited in an autosomal recessive trait and due to a defect in the metabolism of copper with accumulation of copper in the liver, brain, kidney, cornea, and other tissues. Also known as hepatolenticular degeneration.

wind — the natural horizontal movement of air over the earth's surface; speed may be given in meters per second, miles per hour, or knots; movement of the atmosphere brought about not only by the rotation of the earth, but by unequal temperatures on the earth.

window — (*computer science*) an area on the screen that can be used to view specific data.

windrow — a low-technology operation using long rows of shredded refuse, normally 10 feet wide at the base and 4–6 feet high, which allows sufficient moisture and oxygen to support aerobic life.

windrow composting — an open-air composting method in which compostable material is placed in windrows, piles, or ventilated bins or pits, and is occasionally turned or mixed; the process may be anaerobic or aerobic.

wind vane — an instrument used to measure the direction of the wind.

witness — the party having personal knowledge of the facts who appears to present that knowledge as testimony under oath in any proceeding in a lawsuit. An expert witness is one who is skilled in some arts, science, trade, profession, or other human activity and possesses particular knowledge concerning it.

WL — *See* working level.

WLM — *See* working level month.

WMO — *See* World Meteorological Organization.

wood alcohol — *See* methyl alcohol.

wood pulp waste — wood or paper fiber residue resulting from a manufacturing process.

word — (*computer science*) a set of bits, usually 16 or 32, that occupies a single storage location and is treated by the computer as a unit of information.

work — a force acting against resistance to produce motion in a body.

work area — a service area or areas where major activities are conducted; a room or defined space in a workplace where hazardous chemicals are produced or used while employees are present.

worker's compensation — coverage required by state law for compensation to workers who have been injured on the job regardless of the employer's negligence.

Worker's Compensation Law — rules of conduct or action prescribed and enforced by a controlling authority governing employer and employee relations and handling occupational disabilities.

working face — the portion of a sanitary landfill where waste is discharged by collection trucks and is compacted primary to placement of cover material.

working level (WL) — any combination of short-lived radon daughters in one liter of air that will result in the ultimate emission of 1.3×10^5 megaelectronvolts of potential alpha energy.

working level month (WLM) — the air concentration in working levels multiplied by exposure duration in multiples of the 170-hour occupational month with one working level equal to 200 picocuries per liter.

workplace — an establishment at one geographical location that contains one or more work areas.

workplace protection factor (WPF) — a measure of a protection provided in the workplace by a properly functioning respirator when correctly worn and used.

worksheet — a document on which an analyst records analytical work and findings.

work situation design — (*ergonomics*) the assignment of appropriate work hours, rest periods, training, supervision, and management to provide a good working environment and reduce the potential for disease and injury.

workspace design — (*ergonomics*) the design of the physical environment to fit the needs of the worker to perform without excessive effort within a range of healthy postures while standing, sitting, or exerting force.

workstation — (*computer science*) the desk, keyboard, digitizing tablet, and cathode-ray tubes connected together and working as a single unit.

World Health Organization (WHO) — an international organization of countries whose objective is the attainment by all people of the highest possible level of health; physical, mental, and social; chartered by the United Nations.

World Meteorological Organization (WMO) — successor to the International Meteorological Committee and International Meteorological Organization; helps explain long-term global acid precipitation trends.

WPCF — *See* Water Environment Federation.

WPF — *See* workplace protection factor.

WRF — *See* water recreational facility.

writ — a court order commanding the performance of some act by the person to whom it is directed.

writ of mandamus — literally "we order"; a court order to a public official or board to do their duty by virtue of office or position, such as compelling the issuance of a license, the payment of salaries, or expenses improperly withheld.

wt — *See* weight.

WWTP — *See* wastewater treatment plant.

WWW — acronym for world weather watch.

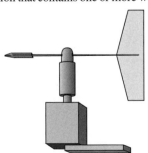

Wind Vane

xenobiotic — pertaining to a chemical compound, such as a drug, pesticide, or carcinogen, that is formed in a living organism.

x-radiation — *See* X-ray.

X-ray — a penetrating, electromagnetic radiation whose wavelength is shorter than that of visible light; usually produced by bombarding a metallic target with fast electrons in a high vacuum. Also known as x-radiation; roentgen ray.

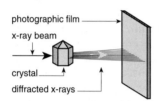

X-Ray Diffraction
in crystallography

X-ray diffraction (XRD) — the scattering of X-rays by matter causing varied intensity due to interference.

X-ray fluorescence analyzer (XRF) — an instrument containing a small, radioisotopic source which when placed against a surface stimulates lead atoms if present in paint to re-emit characteristic X-rays.

XRD — *See* X-ray diffraction.

XRF — *See* X-ray fluorescence analyzer.

xylene — $C_6H_4(CH_3)_2$ — a colorless liquid with an aromatic odor; any of three isomeres, O-, M-, P-; MW: 106.2; BP: 292/269/281°F; Sol: insoluble; Fl.P: 63/84/81°F; sp gr: 0.88/0.86/0.86. It is present in coal and wood, tar, naphtha, etc. It is used in the manufacture of dyes and as a raw material for polyester fibers. It is hazardous to the eyes, skin, respiratory system, and digestive system. Symptoms of exposure include dizziness; excitement; incoherence; staggering gait; irritation of the eyes, nose, and/or throat; vomiting; abdominal pain; and nausea. OSHA exposure limit (TWA): 100 ppm [air].

xylidine — $(CH_3)_2C_6H_3NH_2$ — a pale-yellow to brown liquid with a weak, aromatic, amine-like odor; MW: 121.2, BP: 415–439°F, Sol: slight, Fl.P: 206°F, sp gr: 0.98. It is used in the manufacture of dyes; in the manufacture of curing agents, antioxidants, and antiozonants; in polymer production; as a gasoline additive; in the manufacture of pharmaceuticals; in organic synthesis in the preparation of wood preservatives, wetting agents in textiles. It is hazardous to the blood, lungs, liver, kidneys, and cardiovascular system; and is toxic by inhalation, absorption, ingestion, and contact. Symptoms of exposure include anoxia, cyanosis, damage to the lungs, liver, and kidneys. OSHA exposure limit (TWA): 2 ppm [skin] or 10 mg/m^3.

Y

yard (yd) — a unit of length measurement equal to 3 feet or 0.9144 meters.

yard waste — plant clippings, prunings, and other discarded materials from yards and gardens.

yd — *See* yard.

yellow fever — (*disease*) an acute, viral infectious disease of short duration and different severities with jaundice, sudden onset of fever, headache, backache, prostration, nausea, and vomiting. Incubation time is 3–6 days; caused by the flavivirus of yellow fever; found in Africa, South America, Central America, and formerly found in the United States. It is transmitted by the bite of the infected *Aedes aegypti* mosquito. Reservoirs of infection include people and *Aedes aegypti* mosquitos. It is communicable during the initial stages of the disease when a mosquito bites an infected person and then bites another person; this is usually about 9–12 days; some immunity exists in infants, and lifelong immunity exists in individuals who have had yellow fever. It is controlled by immunization and/or control of *Aedes aegypti* mosquitos.

yersiniosis — (*disease*) an acute, enteric disease with acute watery diarrhea, enterocolitis, false appendicitis, fever, headache, pharyngitis, anorexia, and vomiting. Incubation time is 3–7 days, usually under 10 days; caused by *Yersinia pseudotuberculosis* and *Y. enterocolitica*, with numerous serotypes and biotypes; found worldwide, especially in infants and children. It is associated with household pets, especially sick puppies and kittens. Reservoirs of infection include domestic animals, birds, and mammals. It is transmitted by oral-fecal route from contact with infected persons or animals, or eating or drinking food or water contaminated with feces; communicable as long as symptoms last, which may be 2–3 months or a chronic carrier state; general susceptibility. It is controlled by proper disposal of human, cat, and dog feces; protection of water supplies; proper hand washing; proper preparation of food; pasteurization of milk.

yttrium compounds — Y — a dark gray to black metal; MW: 88.9, BP: 5301°F, Sol: not known, sp gr: 4.47. It is used in the manufacture of red phosphorus used in color picture tubes, fluorescent and mercury vapor lamps; as simulated diamonds; in the manufacture of ceramics for use in high temperature furnaces and high intensity incandescent lamps; in the manufacture of refractories for stable, high temperature, high strength compositions; in the manufacture of lasers; in the manufacture of electronic components in telephones, radar and space communications networks. It is hazardous to the eyes and lungs; and is toxic by inhalation, ingestion, and contact. Symptoms of exposure include eye irritation; in animals: pulmonary irritation, eye injury, possible liver damage. OSHA exposure limit (TWA): 1 mg/m^3 [air].

Y
Z

Z

Z — a symbol indicating the number of protons.

Z+ — *See* impedance.

ZBA — *See* Zoning Board of Appeals.

zeolite — any one of a group of hydrated aluminum, calcium, or sodium silicates used to soften water; frequently made synthetically.

zero — (*mapping*) the origin of all coordinates defined in an absolute system.

zero gas — a gas containing less than 1 ppm sulfur dioxide.

zero population growth (ZPG) — the maintenance of a given population at a constant level so that there is no growth.

zinc chloride fume — $ZnCl_2$ — a white particulate dispersed in air; MW: 136.3, BP: 1350°F, Sol (77°F): 432%, sp gr (77°F): 2.91. It is liberated from arc welding of galvanized iron and steel pipes; from fluxing iron/steel prior to galvanizing; from vulcanizing and reclaiming processes for rubber; from solutions in glass and metal etching; from the manufacture of dry cell batteries; from petroleum refining operations. It is hazardous to the respiratory system, skin, and eyes; and is toxic by inhalation and contact. Symptoms of exposure include conjunctivitis, nose and throat irritation, cough, copious sputum, dyspnea, chest pain, pulmonary edema, bronchopneumonia, pulmonary fibrosis, fever, cyanosis, tachypnea, burning skin, eye and skin irritation. OSHA exposure limit (TWA): 1 mg/m³ [air].

zinc oxide fume — ZnO — a fine, white, odorless particulate dispersed in air; MW: 81.4, BP: not known, Sol: insoluble, sp gr: 5.61. It is liberated during brazing, welding, burning, and cutting of zinc and galvanized metals; during use as ceramic flux; from the manufacture of glass to increase brilliance and luster of glass; from use as an intermediate in the manufacture of other zinc compounds; in the manufacture of electronic devices. It is hazardous to the respiratory system; and is toxic by inhalation. Symptoms of exposure include sweet metallic taste, dry throat, cough, chills, fever, tight chest, dyspnea, râles, reduced pulmonary function, headache, blurred vision, muscle cramps, low back pain, nausea, vomiting, fatigue, lassitude, malaise. OSHA exposure limit (TWA): 5 mg/m³ [air].

zirconium compounds — Zr — a soft, malleable, ductile, metal or a gray to gold amorphous powder; MW: 91.2, BP: 6471°F, Sol: insoluble, sp gr: 6.51. It is used in the manufacture of ceramics, glass, and porcelains; in the synthesis of pigments, dyes, and water repellents; in tanning operations; as abrasives and polishing materials; as an igniter in the manufacture of munitions and other items as detonators, photoflash bulbs, and lighter flints; in the manufacture of skin ointments and antiperspirants; as a gas getter in the manufacture of high vacuum tubes and radio valves; in chemical synthesis. It is hazardous to the respiratory system and skin; and is toxic by inhalation and contact. Symptoms of exposure include skin granulomas; there is evidence of retention in lungs, and irritation of the skin and mucous membranes in animals. OSHA exposure limit (TWA): 5 mg/m³ [air].

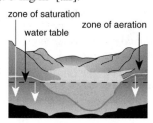

Zone of Aeration
Zone of Saturation

zone of aeration — the area above the water table and below the ground surface where the interstices of rocks and soil are filled with water and air. Also known as vadose zone; unsaturated zone.

zone of capillarity — the area above a water table where some or all of the pores are filled with water that is held by capillary action.

zone of decomposition — a zone in a stream where the dissolved oxygen is exhausted, allowing for anaerobic decomposition putrefaction.

zone of degradation — part of the self-purification process in a stream where visible evidence of pollution including floating solids, pieces of garbage, sticks, papers, and fecal materials are seen.

zone of leaching — the layer of soil just beneath the top soil where water percolates, removing soluble nutrients that accumulate in the subsoil.

zone of recovery — a zone in a stream where the dissolved oxygen appears to increase and the organic solids decrease causing the microorganisms, especially anaerobes, to die.

zone of saturation — a region beneath the surface of the soil in which all spaces between rocks and soil are filled with water. Also known as saturated zone.

zoning — a means of planning to ensure that community land will be used in the best possible way for the health and general safety of the community.

Zoning Board of Appeals (ZBA) — a board established by a local government to hear appeals concerning a change of zoning for a parcel of land or structure to allow different usage than originally approved by law.

zoning ordinance — an ordinance adopted by a local entity to prohibit or approve certain uses of land.

zoom — (*computer science*) the ability to proportionately enlarge or reduce the scale of a figure or map on a visual display terminal or other optical instrument.

zoonoses — a disease adapted to and normally found in animals which can also affect humans.

zooplankton — microscopic aquatic animals that fish passively feed on.

ZPG — *See* zero population growth.

zygomycosis — any infection caused by fungi of the class *zygomycetes*.

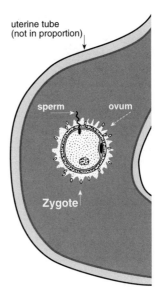

Zygote

zygote — a cell produced by the joining of two gametes which are either sex or germ cells.

zymosis — *See* fermentation.

Y
Z